New Materials and their Applications 1990

The Materials and Testing Group of the Institute of Physics with the following co-sponsoring institutions:

- Applied Solid State Chemistry Group
- British Board of Agrement
- British Composites Society
- British Cryogenic Council
- British Institute of NDT
- British National Committee for NDT
- Brunel University
- Macro Group
- Institute of Ceramics
- Institute of Metals
- Institute of Physics –
  - Industry Committee
  - Instrument Science and Technology Group
  - Polymer Physics Group
  - Thin Films and Surfaces Group
  - Tribology Group
- Institution of Electrical Engineers
- Plastics and Rubber Institute
- Society of Glass Technology

Organising Committee

- Dr S G Burnay (Chairman)
- Mr D J Ando
- Dr D Dew-Hughes
- Dr M J Folkes
- Dr M G Gee
- Dr K T V Grattan
- Dr P C Hewlett
- Dr D Holland
- Dr R G Jones
- Dr D C Phillips
- Dr M J Richardson
- Dr D P Thompson
- Dr G J Turvey
- Dr J W Wood
- Dr K R Whittington

New Materials and their Applications 1990

Proceedings of the 2nd International Symposium on New Materials and their Applications held at the University of Warwick, 10–12 April 1990

Edited by D Holland

Institute of Physics Conference Series Number 111
Institute of Physics, Bristol, Philadelphia and New York

CODEN IPHSAC 111 1–534 (1990)

British Library Cataloguing in Publication Data

International Symposium on New Materials and their
Applications (*2nd : 1990 : University of Warwick*)
New materials and their applications, 1990
1. New materials
I. Title II. Holland, D. III. Institute of Physics IV. Series
620.11

ISBN 0-85498-049-0

Library of Congress Cataloging-in-Publication Data are available

Published under The Institute of Physics imprint by IOP Publishing Ltd
Techno House, Redcliffe Way, Bristol BS1 6NX, England
335 East 45th Street, New York, NY 10017-3483, USA
US Editorial Office: 1411 Walnut Street, Philadelphia, PA 19102, USA

Printed in Great Britain by Galliard (Printers) Ltd, Great Yarmouth, Norfolk

Preface

The second conference in the series New Materials and their Applications took place at Warwick University from the 9th to the 12th of April 1990. On this occasion, the organisation took the form of one day symposia, running in parallel over the 3 days, with each symposium being organised by one or more of the co-sponsoring bodies as follows:—

Physics of Materials Forming Processes	—organised by M J Foulkes (Institute of Physics, Materials and Testing Group)
Tribology of New Materials	—organised by M G Gee (Institute of Physics, Tribology Group) and chaired by B J Briscoe (Imperial College)
Processing and Properties of Ceramic Matrix Composites	—organised by D C Phillips (British Composites Society) and chaired by B Harris (Bath University)
Thin Film Superconductors	—organised by D Dew-Hughes (Institute of Metals) and R G Jones (Institute of Physics, Thin Films and Surfaces Group)
Application of New Materials in Sensor Technology	—organised by K T V Gratton (Institute of Physics, ISAT Group)
Glasses and Ceramics	—organised by D Holland (Society of Glass Technology) and D P Thompson (Institute of Ceramics)

This organisation gave the opportunity to assemble experts from a wide range of technologies who could then give an up to the minute appraisal of the New Materials to an audience drawn from industry and academe.

In addition to oral presentations, there were a number of posters and the majority of all these contributions are presented in this text.

All the co-sponsors, and the members of the organising committee, must be thanked for their contribution to the success of this conference and special thanks are due to the Chairman of the committee, Dr S G Burnay.

D Holland

Contents

Inst. Phys. Conf. Ser. No 111
Paper presented at Int. Conf. on New Materials and their Applications, Univ. of Warwick, 1990

The quiet revolution in materials: The challenges

T F Page

Materials Division; Department of Mechanical, Materials and Manufacturing Engineering; The University of Newcastle upon Tyne; Newcastle upon Tyne, NE1 7RU, UK

ABSTRACT: The enormous proliferation in both the spectrum and sophistication of structural materials poses considerable challenges if the enabling opportunities presented by new and improved engineering materials are not to be wasted. Essentially, new approaches are needed for the ways in which we incorporate materials selection, specification, processing and usage into design. Other key issues involve the ways in which quality and reliability may be sought and the means by which new materials are introduced into service. These and other key issues of strategy are presented and discussed.

1. OPPORTUNITY AND CHALLENGE

The development, selection and use of materials have always been critical areas unpinning, and often enabling, virtually all progress in engineering and technology. Everything is made of something! However, over the last decade or more, the apparently ceaseless quest for new and improved materials has led to an ever-increasing spectrum of solids available for structural engineering. Largely, this expansion in materials has been driven by the perceived need either to improve specific properties, to achieve better extremes of properties or to improve microstructural control of performance and reliability. The resultant range of new metallic alloys, polymers, ceramics, surface-engineered materials, nano-scale materials and composites (perhaps the ultimate *designer materials*) is almost bewildering in its breadth and complexity.

Consequently, users, manufacturers and designers now have a wider range of increasingly sophisticated materials at their disposal with many materials having the potential to be *tailor made*, through innovative specification design and processing, for particular applications: indeed there is a **revolution in materials** in process. The fruits of this revolution are increasingly apparent (and, perhaps, increasingly taken for granted) in all aspects of our lives - fibres for optical communications; compact discs; cheaper, smaller and more powerful computers; better prostheses; lighter, more fuel-effective and crash-tolerant cars; high performance fibre-reinforced racing cars and everyday sports equipment; composites and non-metallic materials in aircraft structures; wear-resistant coatings in engines, scratch-resistant coatings on spectacles; hollow fan blades and single crystal turbine blades in aircraft engines. The list is virtually inexhaustible and society's expectations are ever increasing.

However, if we are to obtain the maximum benefits from this revolution, it is clear that a number of important **strategies** are needed - not all of them new, but certainly made more urgent both by the pace of change and by the need to improve and keep our technological edge in a number of areas. Besides the scientific developments involved, it is these strategies which present many of the **challenges** to grasping the opportunities the materials revolution has to offer. In the extreme, failure to respond to these needs may well lead to our not even being *intelligent customers* - let alone suppliers - for some areas of tomorrow's materials.

At the outset, the materials revolution can be viewed in both positive and negative ways. For example, the last few years have seen enormous strides made in our approaches to the improvement, manufacture, selection and usage of established engineering materials (e.g. steels, light alloys, polymers etc). Thus, considerable advances have been made in improving quality and performance by better processing, in improving selection criteria, in computer-aided modelling, selection and performance prediction, in better understanding of both microstructure-property relations and the behaviour and demands of materials in complex engineering structures. While, on one hand, the rapid proliferation of materials can be seen to go far beyond our current capabilities in many of these areas, many of the conceptual needs can be seen as extensions of changes in attitude which are already taking place.

Also, at the outset, two central issues are clear, namely:

1. the use of many new materials (e.g. brittle ceramics and anistropic composites) will require our moving away from the *established wisdom*, phenomenological experience and areas of practise relevant to more conventional, more homogeneous and more tolerant engineering materials;

2. in many cases, ownership of conventional raw materials will be far less strategically important than access to the best precursor materials (e.g. fibres, prepregs, whiskers and powders) and *development of the most suitable and advanced process technologies;*

2. BASIC UNDERSTANDING VS APPLICATIONS

The wider spectrum of available materials - and the increased need to tailor materials to specific applications - make demands both upon those who seek to use materials and those whose role is to provide the fundamental understanding relevant to enabling that use. We will need improved levels of dialogue - often of an interdisciplinary nature - with *users* more able to specify the fundamental problems inhibiting particular applications and *researchers* more responsive to providing that knowledge within the context of appreciating the limitations imposed by manufacturing capability, component design and the particular demands of the application itself.

However, a further implication is the need to drive the quest for *precompetitive* fundamental understanding well in advance of the *competitive* research necessary for exploitation. In recent years, too much of our research funding for industrial-academic interactions has been targeted at projects too near eventual exploitation. We would do well to learn the lessons of how the balance of competitive and non-competitive research is organised in, for example, Japan.

Also, the constant quest for improvement must be targeted. All too often, end-users are reluctant to provide a quantitative specification as a target against which a material can be judged. If only we were prepared to ask *when is good, good enough?* we might learn that several current new materials match many perceived applications.

3. APPLICATIONS: REVOLUTION OR PROGRESSIVE CHANGE?

In the past we have tended not to change too much too quickly. Materials developments have taken place within existing process and design capabilities, new process routes have been applied to established materials or new designs have used existing materials and manufacturing routes. We may need radical changes to this attitude.

Essentially, there are two different strategies for introducing new materials into engineering applications. One route involves *progressive replacement* of existing components by similar components in new materials, (e.g. replacement of metallic valve seats in engines by identical ceramic components). This has the advantage of not moving away from experience and established practices too quickly but may fail to fully (or even correctly) exploit the performance possibilities (and changes in design) which a new material should offer. Even, *surface engineering* comes into this category since it seeks to gain the advantages of new materials without either compromising existing designs or moving away from the known behaviour of the substrate.

The other route involves fully exploiting the possibilities of new materials by a major move away from established practices, (e.g. the complete ceramic heat engine). Such an extensive move may involve elaborate demonstrator projects to prove the feasibility of new processes, the performance of new materials, the design of a new component or assembly, or all of these.

Obviously the two routes differ in timescale, costs and commitment and either one or the other may be the best for different components. However, an early and clear strategy decision is necessary.

4. THE QUEST FOR QUALITY

There is increasing concern regarding the quality, reliability and reproducibility of components. Not only are these topics relevant to performance and life-time prediction, but they also underlie standards of acceptance, insurance and litigation. Besides, affecting end usage, they also now pervade almost all stages of the manufacturing process. For example the introduction of Just-In-Time (JIT) manufacturing methods can only minimise inventories, stock levels and rejection rates if every component and sub-assembly is of sufficient and reproducible quality for incorporation into the final assembly.

However, ensuring the mechanical quality of many high strength new materials involves inspection for impossibly small defects. Thus, a necessary strategy might be to move away from quality control and inspection towards quality assurance through well-founded mechanistic understandings of defect behaviour, good process modelling and adequate processing control such that quality is *built-in* rather than measured.

5. THE NEED FOR INTEGRATED DESIGN

One theme running through this paper is that, increasingly, materials specification, fitness-for-purpose, and process route are inextricably interrelated parts of the overall design concept for a particular application. As more emphasis is placed on tailoring materials to their applications - and even individually designing a suitable material and its necessary process path for a given component - the design process will move away from conventional approaches (where materials selection often comes well down the activity list) towards a full integration of conceptual design, performance requirements, materials optimisation and the selection of the most suitable process route. This is the philosophy of **integrated design** and obviously demands a much more open-minded and interdisciplinary approach than in the past.

However, this particular challenge comes at an opportune time when our abilities to use computer models to optimise process routes, to provide a preliminary assessment of materials, to analyse operational conditions and to predict performance are rapidly increasing. Thus, much preliminary screening of designs, specification of production routes, materials tailoring and in-depth performance assessment will be carried out as computer exercises rather than relying on prototypes and experiment.

Despite this, the challenges remain of creating new standards and codes of practice for brittle, anistropic, unforgiving materials when we yet have little experience on which to build.

6. RESOURCES

The way forward certainly demands that we make the very best use of our existing material resources, particularly **people**, and that we create resources such as national and international data bases and expert systems (for example on thermochemical data). *However, the urgent need is to educate people in the skills and attitudes necessary for success.*

A mix of long-term basic research underpinning later product-oriented development is also necessary. A new material may take 10-15 years to finally become accepted!

Sadly, this all comes at a time when our education system is under pressure and when the industrial (and even academic) research and development base is increasingly of a short term nature. These must be areas for urgent consideration.

7. THE WAY FORWARD

The challenges to our successfully exploiting new materials for engineering applications are many. However, my belief is that the way forward involves the following issues:

1. carefully defining and targeting those areas of materials development and of application where the UK has the skills and resources to compete successfully in the new materials markets of tomorrow;

2. changing the way in which we educate people to think about materials. Increasingly new approaches - and interdisciplinary approaches - are necessary;

3. improving our resources of educated people, data bases and precompetitive research programmes;

4. integrating materials selection and process route into design philosophies: indeed *design the optimum material and its process route* rather than select it;

5. improving the dialogue (and levels of appreciation of each others problems) between materials researchers, developers and end-users;

6. increasing the use of computer-aided modelling and analysis in optimising materials processing, usage and performance;

7. urgently establishing design strategies for materials which increasingly diverge from established experience;

8. not only overcoming innate conservatism in the way in which we use materials, but also being prepared to develop materials against quantitative specifications.

These are some of the challenges imposed by the materials revolution, only time and determination will tell whether we can successfully respond to them.

Novel forming methods (for composites)

A Kelly

Vice-Chancellor, University of Surrey, Guildford, Surrey GU2 5XH

ABSTRACT: The recognition of the composite principle as a means of providing very strong, stiff materials of light weight and high temperature resistance coupled with high toughness has led to a number of quite new forming processes, not previously envisaged for composites, plastics or ceramics.

These are considered under the headings of:

- reconstructive processing, where the desired material is re-assembled, eg inorganic foams;
- in situ processing, where a fibre is grown or aligned within a matrix, eg directionally solidified metals and some forms of co-polymerised polymers;
- reactive processing, in which one material is totally transformed into another of the required shape, eg assisted oxidation.

Many high performance metals and metal composites are being made by spray deposition. The speeds with which these new processes can be carried out will be considered as well as novel fibre geometries which are necessary.

1. INTRODUCTION

The importance of new forming processes and the study of the most rapid and economical and assessible (quality control) methods of making a given artefact are assuming the greatest importance for the materials sciences and engineering because a given artefact can nowadays be made, from widely differing generic types of starting material eg one can make a motor car completely from plastic, instead of a metal or an electric motor from ceramics. This is because the differences between the physical properties attainable by the different classes of materials are disappearing, eg thin film super-conductors operating at $77^{O}K$ need not be made of metal (eg Koch and Chaudhari, 1988). The best thermal conductors are not metals but ceramics (eg Berman, 1976). Glassy metals are possible.

Stiffness, density, strength, toughness, wear resistance and fatigue resistance are very important properties for most mechanical and mundane uses of materials. Here the difference between the properties attainable by members of the different classes is practically eliminated.

Concomitant with these changes has been the discovery of the composite principle which not only allows hitherto unattainable property combinations. It also enables continuous variation of a physical property without relying on discrete values shown by specific materials. Of course new methods of processing are necessary to make the composites. At the same time recognition of the principle of compositing itself suggests new forming methods.

2. RECONSTRUCTIVE PROCESSING

Reconstructive Processing - a term introduced by Ray (1978) describes processes whereby a naturally occurring substance, a mineral, a rock or a tree is dissected by chemical means - perhaps also with some mechanical force and then reassembled into the required shape. Asbestos has been processed in this way for many years and paper is made in this way with the introduction of fillers such as chalk or clay into the mat of cellulose fibres produced from trees after the elimination of the lignin. A similar type of process is used to make reproducible sheets of graphite crystals consisting of thin packets of graphite. Graphite crystals are decomposed into thin lamellae by treatment with a strong acid so as to introduce water between the layer planes and then heating to boil the water and expand the graphite which is subsequently sheared into sheets and reassembled. The result is a superb sealing gasket. Similar very clever methods have been used to expand vermiculite by judicious ion exchange to separate the layer planes by chemical swelling. The chemically swollen material is then mechanically delaminated in a high shear mixer in water. Subsequent removal of the water leads to an extremely adherent film - a foam can also be made at this stage by introducing air into a suspension with maximum solids content.

In reconstructive processing the dispersing fluid or gas is introduced in order to disperse the material and put it in the required form and is then removed. Closely related to this of course are processes whereby the dispersing fluid which has the property of penetrating between particles of the desired solid and of causing them to flow adequately is itself used as the matrix of the composite. Synthetic granite which consists of 80-90% vol% of granite in epoxy resin forms a castable composite which may be used for machine tool beds. Similarly very highly filled polymers consisting of 55 Vol% of sand in methylmethacrylate fluid (72% by mass) is very easily cast if attention is paid to the range of particle sizes of the sand and the rheology of the polymer coupling agent. The material called Asterite is a fine substitute for kitchen and sanitary ware ceramics.

The important physical questions here are the rheology of the fluid and the surface tension relations between the fluid and solid. The use of a particular range of particle sizes in specific volume fractions is also of enormous utility in assisting rapid flow and some of the "lore" has recently been subject to empirical analysis (Cheng et al 1990). There are very intriguing hints in the patent literature to easily shearable mixtures of solids of high packing fraction ($>$80%) provided specific volume fractions of definite size ratios are used, eg Ritter (1971).

The use of a polymer to fashion material to shape and its subsequent removal is the basis of much processing of ceramics. A recent good example due to the organisers of this conference is found in a recent paper by Wright et al (1990) who have made thin springs from ceramics, eg Al_2O_3 and $YBa_2Cu_3O_x$. The suspension consists of some 86 to 88% (by wt) cf ceramic with a mixture of various phthalates with stearic acid and sometimes other additives. This is subject to high shear mixing. The softening point may be controlled by the composition as well as the volatility of the diluent so as to effect removal of the binder. The physics needed concerns not only the particle sizes of the powder of ceramic but the influence of additives on softening point and volatility. Attention to the range of particle sizes is necessary so as to avoid shrinkage (maximum volume fraction) and to maintain the ability to extrude (minimum volume fraction).

3. REACTIVE PROCESSING

It is the difficulty of plastically deforming or even of inducing change of shape of the hard high melting point ceramic materials which otherwise have many desirable properties that leads, I believe, to what I call reactive processing in which one material is totally transformed into another of the required properties and sometimes shaped at the same time. A process which has attracted a good deal of attention is illustrated in Figures 1a and 1b which is a novel technique for fabricating ceramics or composites with a ceramic matrix by the directed oxidation of a molten metal either to produce the ceramic alone, or through a mass of filler material if a ceramic matrix composite is required. The figure illustrates the outward growth of a ceramic-metal reaction product from a molten metal exposed to a vapour phase oxidant. Additives will be made to the metal to assist oxidation , eg Si, Ge, Sn and P plus a little Mg in aluminium. A wide range of composites comprising both particulate and

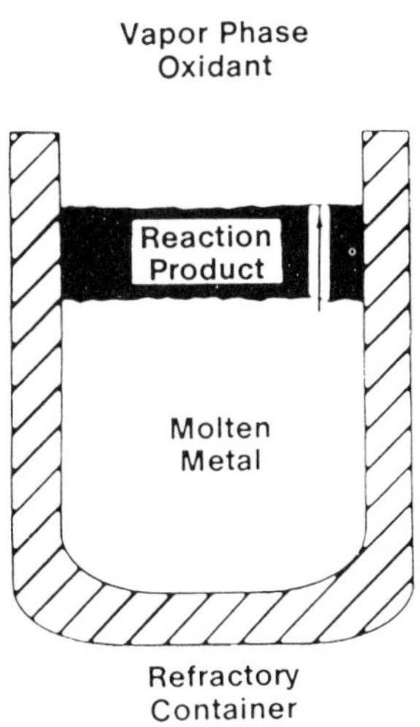

Figure 1a Schematic of the matrix formation reaction illustrating the outward growth of a ceramic/metal reaction product from a molten metal exposed to a vapor phase oxidant.

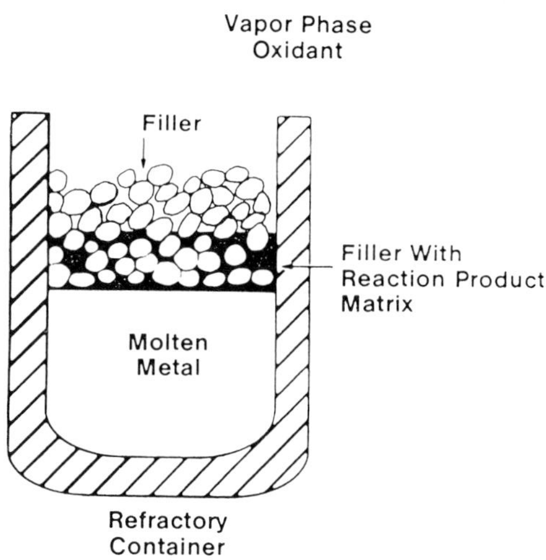

Figure 1b Schematic of the formation of a ceramic composite by the outward growth of a ceramic/metal reaction product matrix through a mass of filler material.

fibrous filler materials have been demonstrated, eg Al_2O_3/Al, AlN/Al, TiN/Ti and ZrN/Zr. The putative attractive features of the technology, in comparison with conventional ceramic technology, include the ease of making a ceramic matrix composite directly at low temperature, the elimination of shrinkage, the potential to make large ceramic bodies and projected low cost. An obvious problem is the final elimination of the metal (Newkirk et al 1988).

Of course the reaction need not be with a material extraneous to the composite (in the above case oxygen). Tani and Wada (1990a) describe a process which they call internal synthesis for preparing (particulate) ceramic composites such as SiC reinforced with internally synthesized TiB_2. The process gets over the difficulty that TiB_2 is not usually available in small enough particle sizes and furthermore reacts with water. In one method SiC powder mixed with TiN and amorphous Boron was hot pressed at above 2000^oC in an argon atmosphere. The molar content of B is more than twice that of TiN. During hot processing the reaction

$$TiN + 2B \longrightarrow TiB_2 + \tfrac{1}{2}N_2$$

occurred at between 1100^oC and 1700^oC. The excess boron aids the densification (sintering) of the SiC so that the process provides hot pressed SiC with small TiB_2 particles dispersed within it. The necessity to eliminate a gas is, of course, a disadvantage, if hot pressing has to be used and a proposed process shows promise in which the TiB_2 is synthesised by a pressureless sintering reaction involving SiC, TiO_2, B_4C and carbon. Again excess B_4C and C are needed as sintering aids for the SiC (Tani and Wada 1990b).

Of course reaction injection moulding to make reinforced plastics is a very well-known process.

4. IN-SITU PROCESSING

Most types of strengthening and stiffening methods used in metal and ceramic processing consist of the introduction of an extraneous phase, eg to block dislocation motion. The same is true of polymers par excellence where a filler stiffens and, of course, in a composite, reinforces as well. In situ processing to produce the reinforcement directly within the matrix has been accomplished in the metal and ceramic case by directional solidification of eutectics where the temperature gradient at the solid liquid interface and the rate of growth are both closely controlled so as to produce a rod or lamellar eutectic of the desired orientation. The process is well described by McLean (1983). The disadvantages of this elegant conception is that the volume fraction of reinforcing phase is controlled by the equilibrium phase diagram and although the addition of other elements may vary this a little, a large change is impossible. Other problems needing to be attacked are that the processing is very sensitive to the presence of trace elements whereas a viable commercial method should not require super purity of the ingredients. The other problem - one where a better understanding of the physical limitation of the process would be useful - is that the current rates of production possible with eutectic alloys is very slow relative to directionally solidified and single crystal superalloys.

In situ processing also naturally suggests itself as a means of introducing reinforcement directly into a polymer. Takanyagi et al (1980) describe the production of a thermoplastic (nylon 6 or 66) containing fibres of polyparaphenylene terephthalamide (PPTA) directly, essentially by co-polymerising and then extruding the sulphuric acid solution of the reaction product into a large amount of water and ethanol. The molecular mass and the aspect ratio of the rigid rod component (similar to some form of Kevlar) has of course a marked effect on the mechanical properties. As with metallic eutectics processing speeds would need to be greatly increased if the method were to show real commercial possibilities. Prevorsek and Mills (1989) discuss other possibilities and point out why melt processing is not possible in this case.

A diagram of the type of microstructure produced with PPTA in nylon is shown in Figure 2.

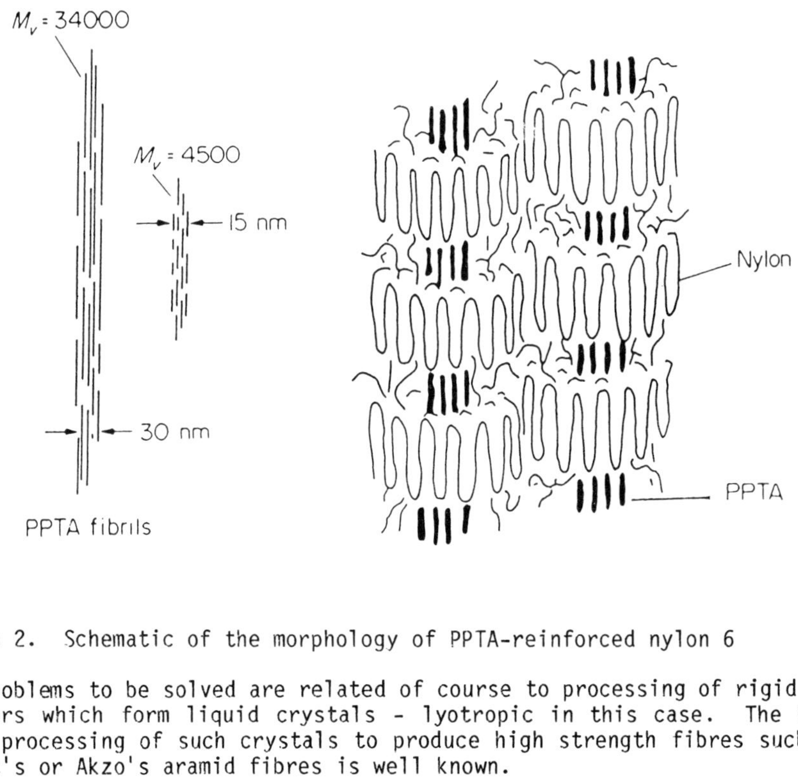

Figure 2. Schematic of the morphology of PPTA-reinforced nylon 6

The problems to be solved are related of course to processing of rigid rod polymers which form liquid crystals - lyotropic in this case. The high speed processing of such crystals to produce high strength fibres such as Dupont's or Akzo's aramid fibres is well known.

The question also naturally arises of the use of liquid crystals to assist the processing of more conventional polymers. These are the thermotropic liquid crystals which are used to produce moulded parts with a nice surface finish. Tupperware made of Amoco's Xydar is a good example and others are Vectra LCP of Celanese.

So-called structural liquid crystal polymers are thermoplastic polyesters or polyamides of aromatic acids, amines, or hydroxyl compounds. They retain some order in the molten state. On solidification they tend to be more highly ordered than other polymers and so have superior properties.

Sometimes these can be deliberately increased by incorporation of rigid rod aromatic molecules. The advantage is that they behave like ordinary thermoplastics, though with rather high processing temperatures: 340-370°C. Of course such polymers can be filled with up to 50% of glass or talc or other filler. LCPs compete with the higher temperature resins for use in electrical components where they must remain stable at soldering temperature.

There are many interesting problems in the rheology of such materials - see for example Doraiswamy and Metzner (1986).

5. METAL MATRIX COMPOSITES AND SPRAY METHODS

It is the ease of casting processes that seems to be the key to the commercial viability of metal matrix composites. The advantages in specific strength and stiffness of the material incorporating ceramic particles or fibres, over the unreinforced metal are well known. The overriding considerations to achieve commercial success are safety, speed (hence low cost) and subsequent processability of the composite by conventional means. Continuous fibres are too expensive, whiskers are toxic and so the reinforcement preferred is a flake form. Consideration must then be given to the size and shape and the method of introduction. Particles about 10-50 μm diameter and with aspect ratio 1-10 are considered. The possible methods of introduction are powder blending, mechanical alloying, melt stirring, compocasting and spray codeposition. The third and fifth of these are the ones chosen by the major aluminium producers. In the first the particles (up to 20% volume fraction) are stirred into the molten material under vacuum by a high shear mixer (at ~600°C!) the problems are clearly avoidance of oxidation and the presence of water. Cospray involves clever design of the atomiser of the aluminium alloys into drops - mainly between solidus and liquidus when the substrate is struck and 15 μm diameter particles of SiC (say) are introduced and the aluminium droplets are maintained at 100 μm size for safety reasons. A nice set of physical problems arise concerning the process of extracting heat from the aluminium - can one alter ΔH_m? - and again on the geometry of the relative particle shapes and sizes to obtain the various packings. The material builds up on the collector almost without voids although a subsequent extrusion is necessary in order to eliminate all porosity.

Without the introduction of a ceramic filler or reinforcer, the method has great advantage for conventional powder processing since the powder (spray) is produced and deposited with very little porosity in one operation. Most processes are developments of the so-called Osprey method invented about 1977. The atomisation produces particles with a density in the spray of that of a heavy gas - a few grams per litre moving very rapidly at 50-100 ms^{-1}. The process physics is fascinating involving measurement of the particle, size, speeds, number per unit volume, state, ie solid, liquid, mushy and the proportions of these and problems of course of the extraction of heat and the effects of impact.

6. FIBRE GEOMETRIES

As fibre composites acquire application in other shapes than thin sheets stressed in their plane, the need to consider three-dimensional arrays of fibres arises and a renewed interest in the physics of and computer control of weaving and braiding processes. Some surprisingly simple geometrical arrays do not appear to have been fully explored in the literature open to composite specialists.

Questions arise as to possible maximum density of packing of fibre bundles of various cross sections. If straight fibres are arranged along three orthogonal directions, the maximum packing fraction attainable is 0.75 (or ¾) for fibres of square cross section or $3\pi/16 = 0.589$ for fibres of circular cross section. Such a structure yields easily in shear parallel to any of the (orthogonal) axes and in any direction on these planes. It is well known to structural engineers that the rigid frame most economical of material to resist a given set of forces is tetrahedral (Michell 1904). An array of uniform fibres aligned along the direction normal to the faces of a regular tetrahedron has a surprisingly high packing fraction of 0.5 (or ½) provided fibres of triangular section are provided. Such an array resists all tensile stresses and most shears except for those in highly specific directions on specific planes. The directions turn out to be the edges of the tetrahedron defining the directions of the fibres. Fibres arranged solely along these directions, form a dodecahedral arrangement with the fibres running normal to the faces of a rhombic dodecahedron.

This dodecahedral arrangement has fibres running in 6 directions in space and among these six directions no more than 3 lie in any one plane. An array in these directions is such that any general strain can be resisted by the need to extend at least one set of fibres and there is no easy direction of shear or of extension. This would be called in weaving parlance a 6-D arrangement. Questions then arise such as:

(i) what is the densest array of fibres possible in this 6-D arrangement? The answer is 1/3;
(ii) what is the cross section of fibre bundle necessary to achieve this packing fraction? - answer rhombic;
(iii) what is the simplest technique of weaving or braiding that can attain this structure?;
(iv) can this technique be followed by a machine?

> Studies of this type are important because it may not be, for instance, necessary to resist a general strain by elongation of one set of fibres. A lack of resistance to shear is a very small number of clearly defined directions may not be of great import. Then it would follow that perhaps the tetrahedral arrangement of fibres where a packing fraction of ½ in a 4-D arrangement can be attained would be suitable in a particular application.

For all of these arrangements the tensile and shear moduli must be estimated without the complication of fibre-fibre interaction.

In all of these considerations there is a developing need to be able to display in two dimensions not only the directions in which the fibres are to run but also the cross-sectional form of the fibre. Of course, a stereogram will represent in 2 dimensions directions arranged in three dimensions in space and this representation is what is called angle true. Parkhouse and I find that the stereogram is of course fine for directions but is not good for finding the gaps between the arrays of fibres running in the various directions.

REFERENCES

Berman R 1976 Thermal Conductivity in Solids (Oxford University Press)
Cheng D C-H, Kruszewski A P, Senior J R and Roberts T A J. Mater. Sci. **25** 353
Doraiswamy D and Metzner A B 1986 Rheol. Acta **25** 580
Koch R H and Chaudhari P 1988 Nature **332** 682
McLean M 1983 Directionally Solidified Materials for High Temperature Service (London: The Metals Society)
Michell A G M 1904 Phil. Mag. Ser. **6 8** 589
Newkirk M S, Lesher H D, White D R, Kennedy C R, Urquhart A W and Claar T D 1987 Ceram. Eng. and Sci. Proc. 8 [7-8] 879
Prevorsek D C and Mills P J 1989 Polymer-Polymer Composites in Concise Encyclopedia of Composite Materials ed A Kelly (London: Pergamon Press)
Ray N H 1978 Inorganic Polymers (London: Academic Press)
Ritter J 1971 App. Polym. Symp. **15** 239 (New York: Wiley)
Takanyagi M, Takayuki O, Masanori M and Takashi K 1980 J. Macromolec. Sci. Phys. **B** 17(4) 591
Tani T and Wada S 1990a J. Mater. Sci. **25** 157
Tani T and Wada S 1990b J. Mater. Sci. Letters **9** 22
Wright J K, Thomson R M and Evans J R G 1990 J. Mater. Sci. **25** 149

Inst. Phys. Conf. Ser. No 111
Paper presented at Int. Conf. on New Materials and their Applications, Univ. of Warwick, 1990

Multiple live-feed processing as a route for fibre management in composite materials

P S Allan and M J Bevis

Wolfson Centre for Materials Processing, Brunel University, Uxbridge, Middx. UB8 3PH

ABSTRACT: In multiple live-feed moulding (MLFM) a macroscopic shear is applied to a solidifying melt, through the independent control of pressure at a multiplicity of gating points. MLFM has been applied successfully to a wide range of thermoplastic and thermosetting plastics matrix composites. The microstructure control and associated physical property enhancement that results from the process is reviewed.

1. INTRODUCTION

The basic principles of Multiple live-feed Moulding (Allan and Bevis 1987, 1984) are:-

i) To provide a multiplicity of live feeds to a mould cavity.

ii) To be able to displace sufficient molten material within the mould to create a macroscopic shearing of the melt.

iii)To be able to control the solidification process to ensure the optimum cycle times.

The application of a macroscopic shear, specified in magnitude and direction, at the melt-solid interface provides for:

a) Enhanced fibre alignment by design in fibre reinforced materials.

b) Elimination of mechanical discontinuities which result from the initial mould filling process, including internal weld lines.

c) Reduction in the detrimental effects of a change in moulded section thickness on the mechanical performance of fibre reinforced materials.

d) Elimination or reduction in defects resulting from the moulding of thick-sectioned components.

A two-live-feed device (B) is illustrated in Figure 1 below, located between the injection moulder screw/barrel (A) and mould cavity (C). The piston (2) is retracted allowing melt to be injected into the mould cavity. On completion of mould filling the pistons 1 and 2 may be moved back and forth to operate on the solidifying melt as follows:

Mode A: the pistons 1 and 2 can be pumped back and forth at the same frequency but with a phase difference of 180°.
Mode B: the pistons 1 and 2 can be pumped back and forth at the same frequency and in phase.
Mode C: the pistons 1 and 2 can be held down under a static pressure.

The frequency and displacement of the pistons and the sequence of operation of modes A,B and C are selected to optimise fibre orientation and moulding cycle time.

Fig.1

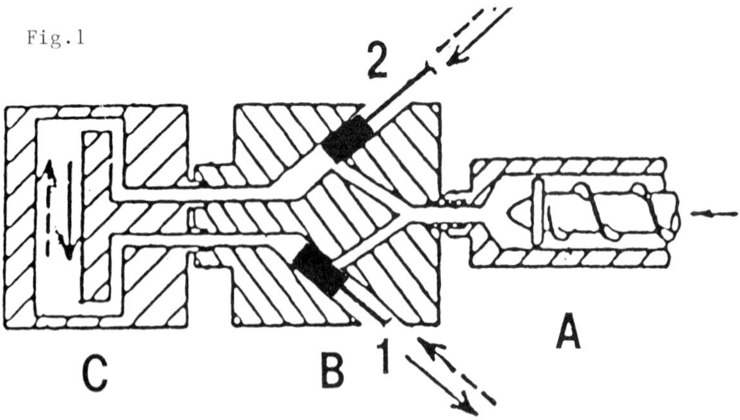

Figure 2a is an X-ray microradiograph (Allan and Bevis, 1987) illustrating the high level of uniaxial fibre alignment that can be produced in thick-section polymer matrix composite injection mouldings. In this case the glass fibre reinforced polypropylene moulding with a 20x20mm cross section was produced using the MLFM process, and may be compared with the equivalent in figure 2b, produced by conventional injection moulding.

Fig.2a

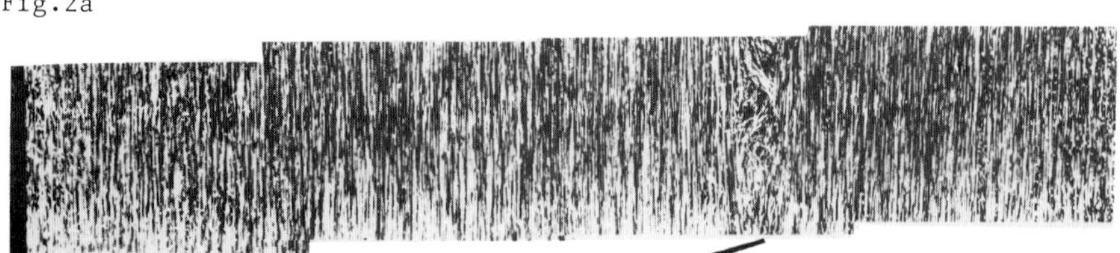

Edge Centre Fig.2b

The mechanical property enhancement that results from the application of the MLFM process is illustrated in Figure 3 (Allan and Bevis, 1987), which shows the change in tensile modulus through the thickness of mouldings produced by conventional injection moulding (broken line) and by multiple live-feed moulding (continuous line).

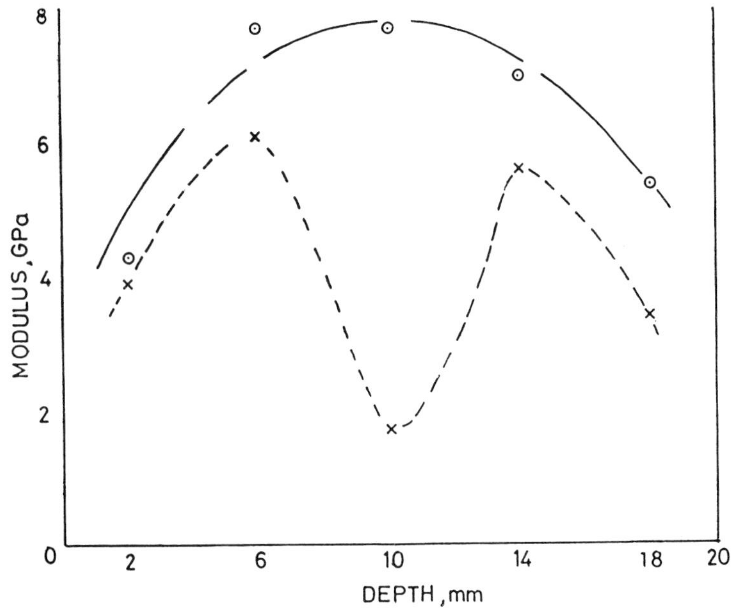

Fig.3

The significant enhancement of internal weld line strength resulting from application of MLFM has been reported (Allan and Bevis, 1987,1989) and illustrated here by application to a thermotropic liquid crystal polymer. Figure 4 shows two copolyester (Thapar and Bevis, 1989) liquid crystal polymer mouldings produced using a relatively low mould temperature.

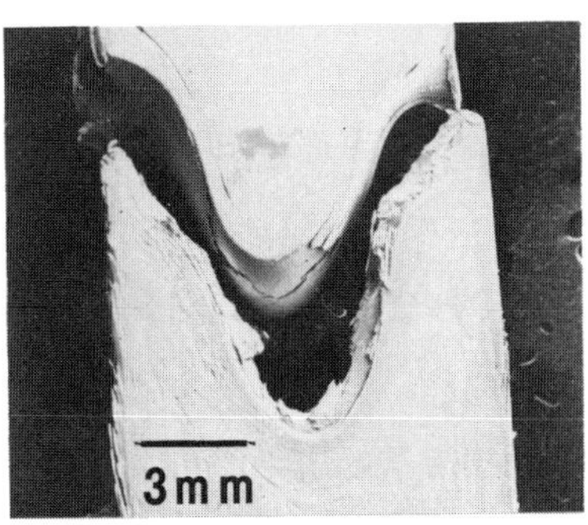

Fig 4a (after H Thapar)

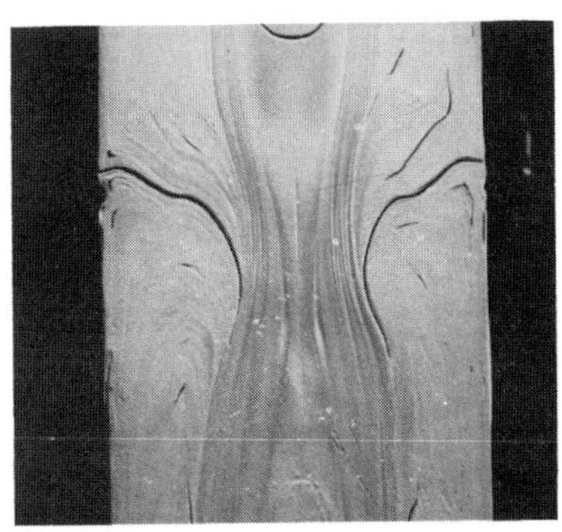

Fig.4b (after H Thapar)

The mouldings shown in figures 4a and 4b were produced respectively without and with the application of MLFM, and then subjected to tensile testing. These photographs, together with the stress-strain curves shown in Figure 5, illustrate the advantageous effect of MLFM on the tensile deformation of moulded LCP's.

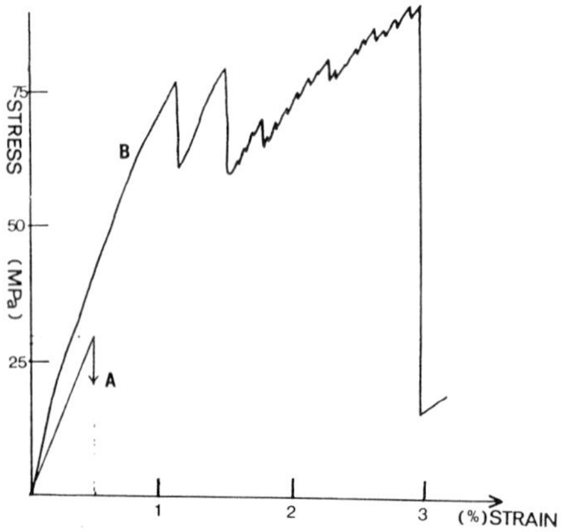

Fig.5 (after H Thapar)

Curves A and B relate to the mouldings illustrated in Figures 4a and 4b respectively.

2. EXAMPLES OF MLFM CONFIGURATIONS

The production of specific microstructures results from the strategic placing of live feeds into the mould cavity. The use of four live feeds has been described (Allan and Bevis,1989) in connection with the arrangement illustrated in Figure 6. In this case a two live-feed device (D) is attached to each injection unit on a two colour injection moulding machine. Alternating successively during solification between the out of phase operation of pistons 1 and 4 and then 2 and 3 produces a laminated structure of preferred fibre orientation, the number of laminates and the preferred orientation of fibres within each laminate being determined by the sequence of operation of the live feeds 1 to 4.

Figure 6 is an example of one embodiment of MLFM. Alternative arrangements are illustrated in Figures 7,8 and 9.

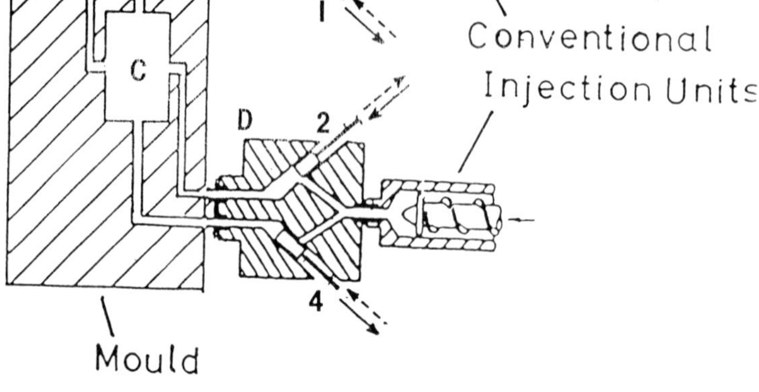

Fig.6

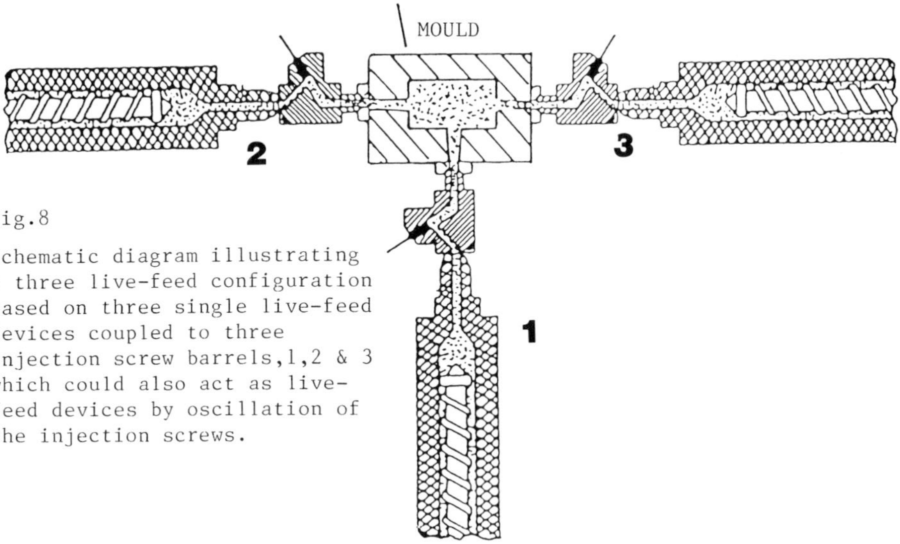

Fig.8

Schematic diagram illustrating a three live-feed configuration based on three single live-feed devices coupled to three injection screw barrels,1,2 & 3 which could also act as live-feed devices by oscillation of the injection screws.

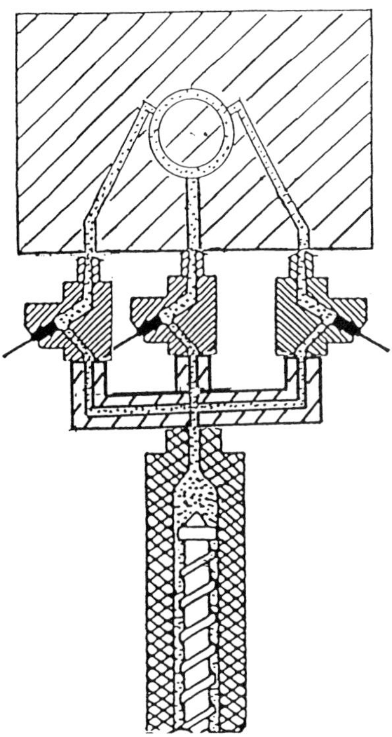

Fig.7
Schematic diagram illustrating a three live-feed configuration based on a manifold fed from a single injection screw/barrel.

Fig.9
Schematic diagram of a two live-feed configuration, built into the mould tool and located between the fixed and moving platens.

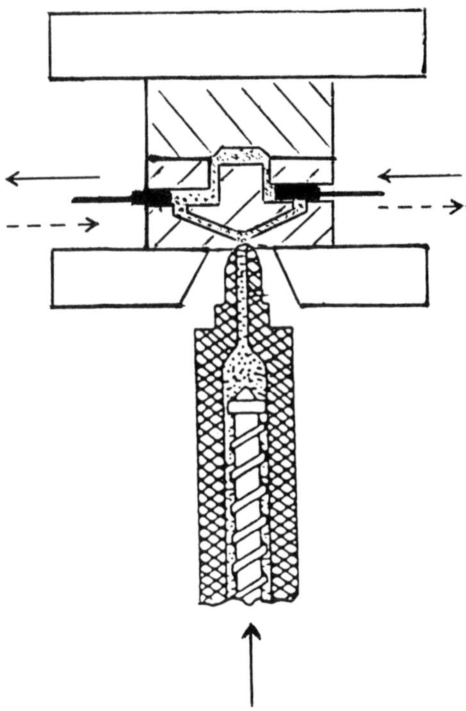

3. SOME RECENT APPLICATIONS OF MLFM

The application of MLFM leads to better control of the microstructure in mouldings, and in addition to enhancement of mechanical properties, also providing a route for control of other physical properties. These include coefficients of expansion (Allan, Bevis, Ess, Chuah, Thapar and Wang,1990) and dimension control. Figure 10 illustrates the cavity used to produce a 150mm diameter ring with a 10x15mm cross section. In the example reported here 30 per cent by weight glass fibre reinforced copolyester LCP was injected into the cavity through gate 1 to produce an internal weld line in the vicinity of gate 3. Mouldings were produced using a conventional moulding cycle and also MLFM, the latter based on the alternate operation of pairs of live feeds during solidification (Allan and Bevis, 1989).

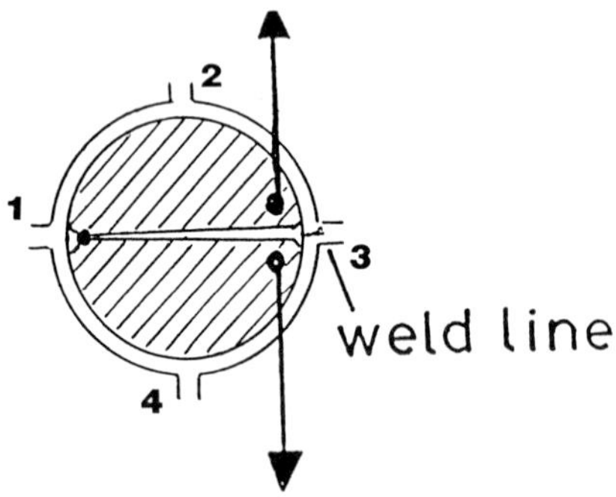

Fig.10

Schematic diagram of ring cavity with gating and mechanical testing geometries

The mechanical properties of the moulded rings where the quoted results represent the mean values gained from six mouldings were:

	Relative Stiffness Value	Load at Failure (KN)
Conventional	0.76	2.96
MLFM	1.08	8.71

The application of MLFM leads to a substantial enhancement in stiffness and internal weld line strength. The resultant uniformity of microstructure throughout the circumference of the ring also results in a substantial dimensional control. This is illustrated in Figure 11.

The use of polymer blends and compounding injection moulding, coupled with MLFM, provides a route for the in-situ formation and management of fibres in moulded parts. Figure 12 illustrates the formation of LCP fibres and their alignment within a polypropylene matrix, in essence following co-rotating twin-screw extrusion compounding and two live-feed moulding, as illustrated schematically in Figure 13.

Similar results are obtained when fibres and metal powders are compounded with a polymer binder, and then moulded using MLFM. Figure 14 illustrates the high level of fibre alignment attainable in a metal matrix following application of MLFM; glass fibres in an aluminium matrix and polyolefin binder in this illustrative example.

Fig.12a

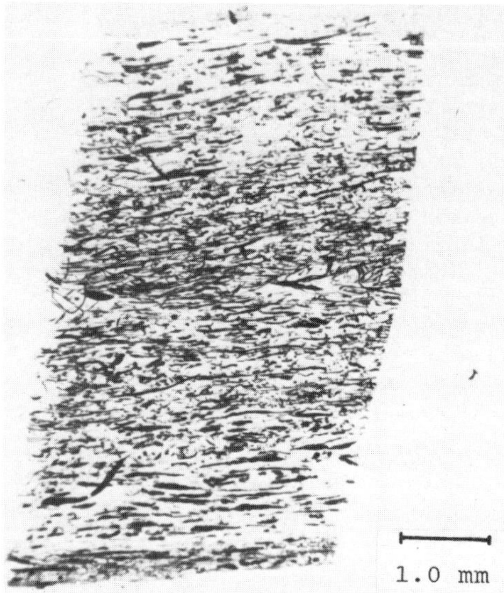

Fig.12b

Fig.12c

Figure 12 (after P Chuah)
Vectra B950 (Hoechst Celanese) 20 per cent by vol. in a Propathene GSM10 (ICI) polymer matrix. Light micrographs taken through the thickness microtomed sections:- a) conventional moulding, and
b) MLFM
c) Scanning electron micrograph illustrating LCP fibres exposed on a low temperature fracture surface, relating to the fracture of a moulding with a microstructure illustrated in Figure 12(b).

Fig.11 (after J Ess)

Fig.11a
Results of fifty point measurements of the internal diameter of a '150'mm diameter ring produced by conventional single gate '1' injection moulding.
The enlargement of the measured deviation(s) is indicated by the scale marker.

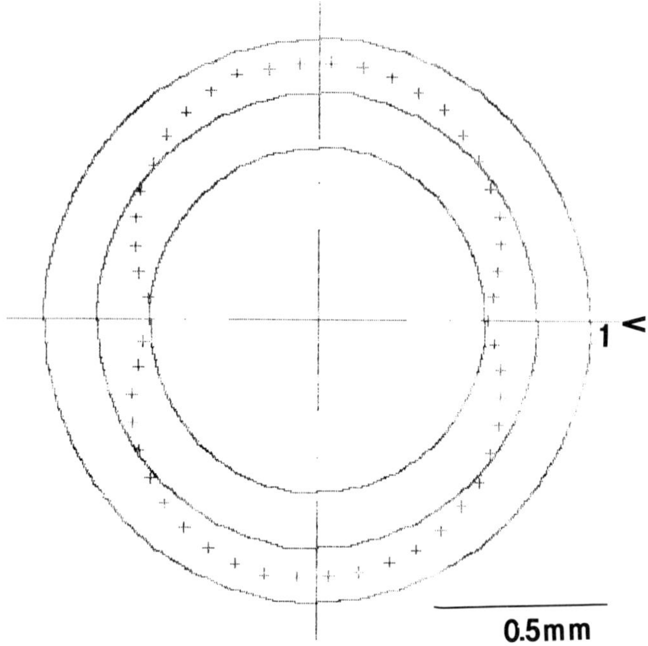

Fig.11b
Measurements of the internal diameter of a ring produced by multiple live-feed moulding, indicating a substantial improvement over the control of dimensions. This improvement is associated with the uniformity of processing history and microstructure throughout the volume of the ring.

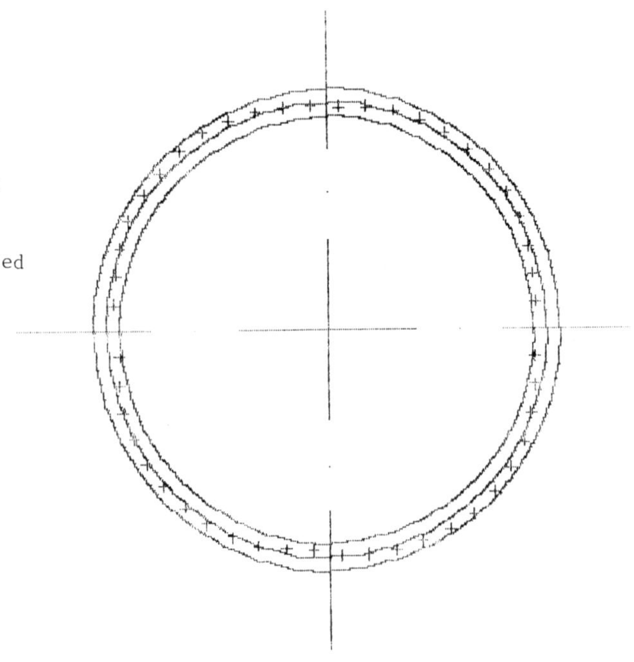

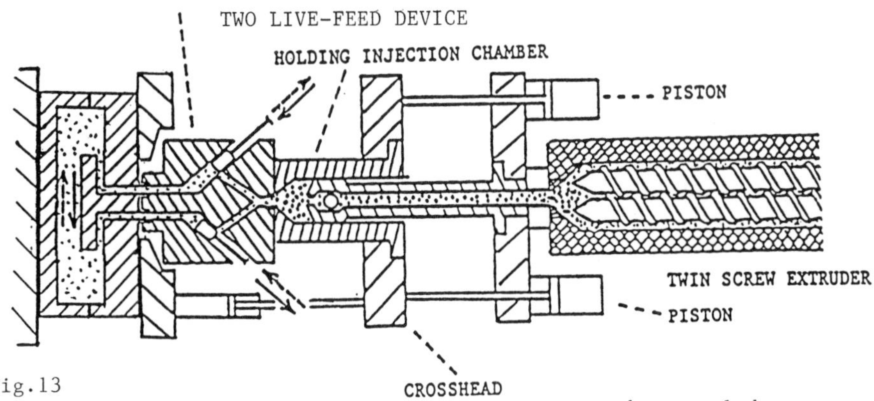

Fig.13
Schematic diagram of a co-rotating twin-screw extruder coupled to a two live-feed moulding device, to provide for the in-situ formation and management of LCP fibres with a polymer matrix.

4. APPLICATION TO CONTINUOUS EXTRUSION PROCESSES

The application of a macroscopic shear to solidifying melts also relates to the continuous formation, by extrusion, of a product in which constituent elements, normally randomly oriented, are capable of alignment or other ordered orientation to enhance the strength, formability or other useful quality of the extruded product (Allan and Bevis, 1989b). While the extruder is still molten and moving through the die in the direction of the extrusion axis, it is subjected to pulsating displacements in a direction transverse relative to that axis. These displacements give rise to a shearing action within the material, resulting in a degree of transverse orientation of the constituents. This orientation may be "locked in" to the extruded product when the melt solidifies, so improving the transverse strength of that product, and/or may correct an unwanted orientation that existed in the melt upstream of the point where the pulsating displacements were applied. The displacements may be applied by the synchronized operation of pistons, operating in chambers which open into the die cavity, and used for the management of microstructure in extruded products of rectangular, tubular and other sections.

ACKNOWLEDGMENTS

The authors are indebted to Drs J Ess and H Thapar,and F Ahmad and P Chuah for their contributions, and to British Technology Group, Ministry of Defence and the Wolfson Foundation for financial support.

Fig.14 (after F Ahmad)
a) Scanning electron micrograph of a polished section through the thickness of the core of an injection moulded polyolefin binder containing 55 per cent by volume of Al powder and 10 per cent volume of glass fibre. The Al exhibits light contrast.

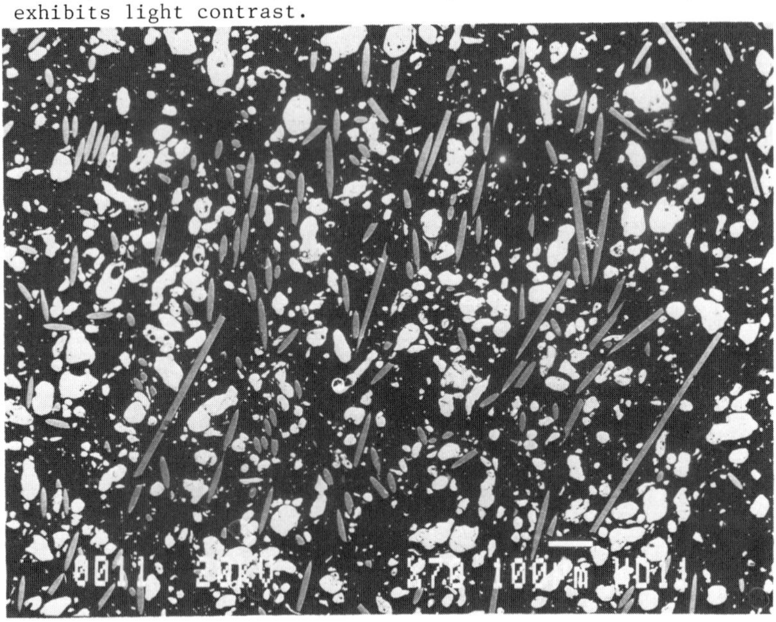

b) The equivalent micrograph from a moulding produced by MLFM, illustrating a high level of fibre alignment within the core of the moulding.

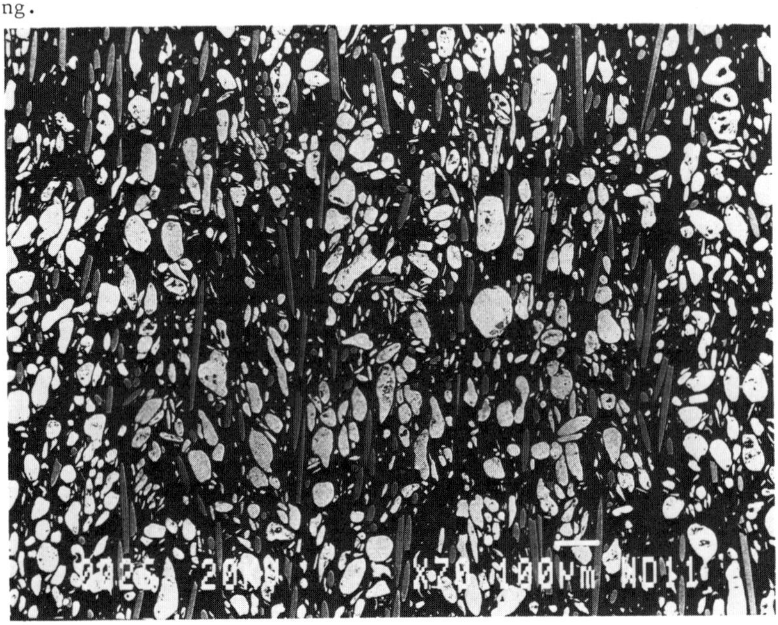

References

Allan P S and Bevis M J 1984. European Patent Applications 85 309 311. Priority 21.12.84.

Allan P S and Bevis M J 1987. Plast.Rubb.Proc.Appln.7 3-10.

Allan P S and Bevis M J 1989. Proc. 3rd Eur.Conf.on Composite Materials, ed. A R Bunsell, P Lamicq and A Massiah (London and New York: Elsevier Applied Science) pp 375-380.

Allan P S and Bevis M J 1989b. Patent Application GB 8900434.

Allan P S, Bevis M J, Ess J, Chuah P, Thapar H and Wang L 1990 (in preparation).

Thapar H and Bevis M J 1989. Plast. Rubb. Proc. App. **12** 39-52.

Inst. Phys. Conf. Ser. No 111
Paper presented at Int. Conf. on New Materials and their Applications, Univ. of Warwick, 1990

Plastic forming operations for ceramic suspensions

J.R.G. Evans
Department of Materials Technology, Brunel University, Uxbridge, Middlesex, UB8 3PH

ABSTRACT: Techniques drawn from the polymer industry can be applied, mutatis mutandis, to the shaping of ceramic powder assemblies before firing. Fine powders are now available with particle diameters in the 20-100nm range. Such particles are of a size comparable to the chain end to end distance of polymer molecules. This paper discusses the influences of molecular weight of the organic vehicle on plastic forming operations.

1. INTRODUCTION

By embracing meritorious developments in polymer and composite processing technology, the ceramics community stands to gain a rich portfolio of shaping operations, the full benefit of which has yet to be realized. These techniques have in common, the incorporation of ceramic powder in an organic vehicle. The resulting suspension may then be shaped or joined as though it was a polymeric material before removing the organic phase by pyrolysis and firing the ceramic. There are three reasons why the vehicle should be as crowded as possible: a) to reduce shrinkage during removal of organic matter, b) to reduce slumping during reheating, c) to reduce sintering shrinkage.

2. PLASTIC FORMING OPERATIONS

Injection moulding is the procedure which offers the greatest versatility of form to the ceramist (Edirisinghe and Evans 1986a, 1986b) and progress continues to be made in understanding the influences of material and machine parameters on the integrity of mouldings (Edirisinghe and Evans 1989, Zhang et al 1989a). Sections greater than 10-15mm suffer from defects which are characteristic of the process and result from non-uniform solidification in the cavity and from the evolution of volatile matter during pyrolysis. Figure 1 shows an injection moulded rotor which accommodates 60 vol% alumina in a polypropylene vehicle. Figure 2a shows cracks in the hub in the as moulded state which can be overcome by prolonging access to the solidifying core to allow compensation for shrinkage (Figure 2b) (Zhang et al 1986b). The exact compositions of organic vehicle are given in the relevant citation.

Extrusion has been used for the fabrication of ceramic components for many years but the correct choice of organic vehicle allows ceramic windings (Figure 3) to be prepared from a range of ceramic powders (Wright et al 1990). In addition to the absence of metallurgical

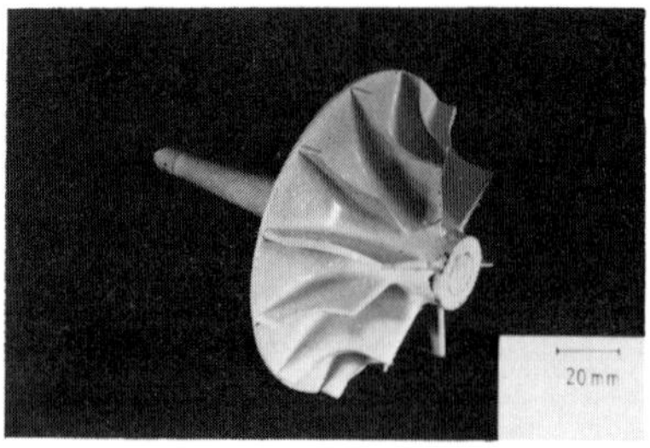

Fig.1 Moulded alumina rotor containing 40 vol% polypropylene based binder.

deterioration at high temperatures, ceramic springs offer resistance to corrosive environments in chemical plant. Provided adequate purity can be achieved the method may be of value in the fabrication of ceramic superconductors for power applications. In this process the high molecular weight of the organic vehicle helps to confer melt strength.

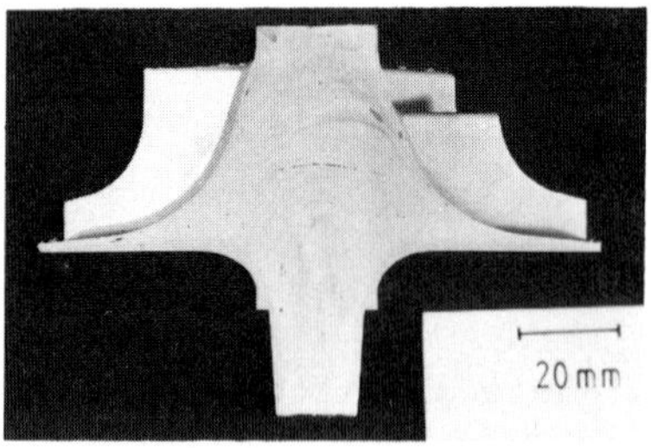

Fig.2a Cracks in the centre of moulded alumina rotor using a polypropylene binder.

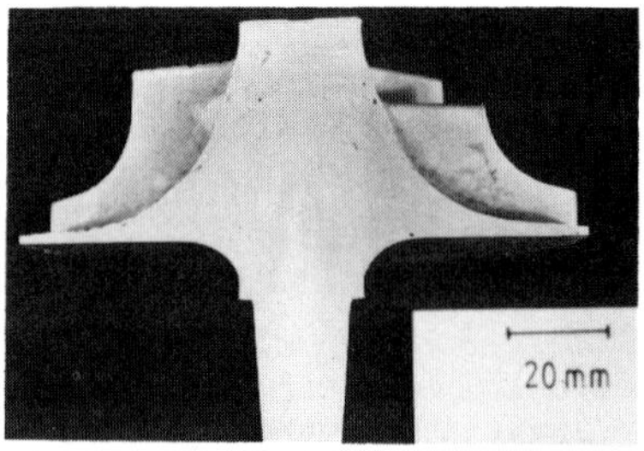

Fig.2b The avoidance of cracks by prolonging sprue solidification to allow fluid flow to compensate for shrinkage.

The vacuum forming of thin wall ceramic bodies is also possible (Haunton et al 1990). Figure 4 shows a fired alumina coupon in which detail has been incorporated by vacuum forming. Considerable melt extension is also possible; figure 5 shows a vacuum formed vessel immediately after pyrolysis of the organic vehicle. After forming, the sheet is heated for several minutes to allow molecular relaxation which would otherwise cause the form to collapse on reheating.

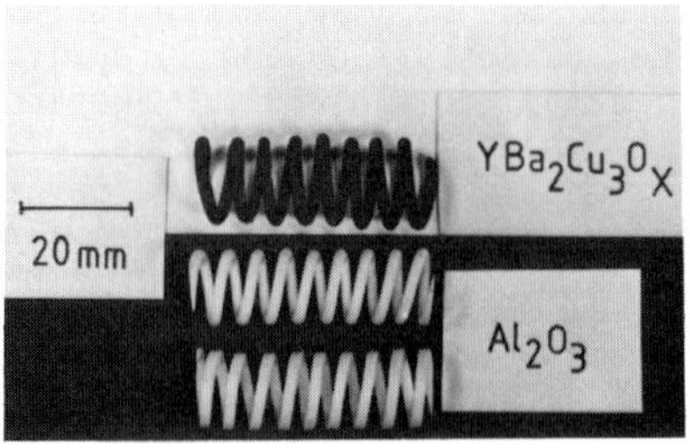

Fig.3 Ceramic windings prepared using a polystyrene binder diluted with phthalate esters.

The blow moulding of ceramics using water soluble polymers (Kobayashi et al 1981) and thermoplastic polymers (Renlund et al 1985) has also been reported and is of interest in the manufacture of alumina lamp envelopes. Figure 6 shows a blow moulded bulbous tube before binder removal (Hammond and Evans, to be published).

If ceramics can be shaped by polymer shaping methods can they not also be joined before firing by joining methods used for polymers? Initial work suggests that they can (Rashid et al) and this opens up scope for a higher order of complexity of form which could be achieved by no other route. Indeed multilayer ceramic technology uses this method in electrical application (Schwartz 1984).

Fig.4 Sintered alumina plate containing detail produced by vacuum forming.

Melt spinning is a technique which revolutionised synthetic fibres. It is also of interest in the manufacture of refractories and superconductors and would allow the study of grain boundary behaviour in fine polycrystalline ceramic wire by heat treating to produce grain boundary intersection with the surface.

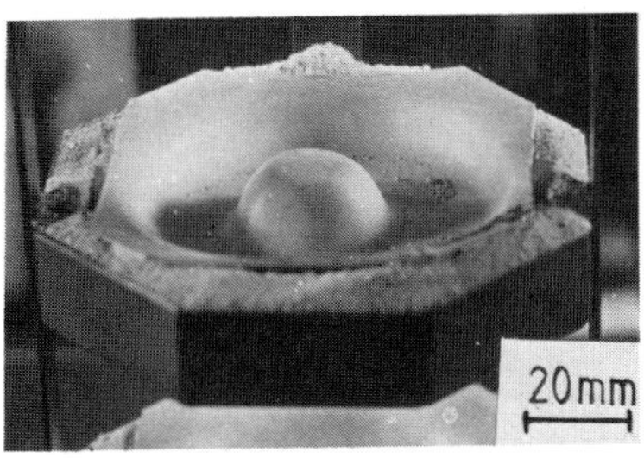

Fig.5 Vacuum formed alumina vessel immediately after removal of the polystyrene vehicle.

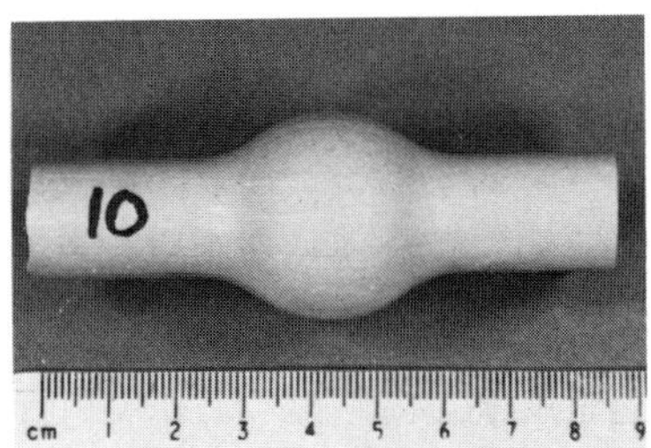

Fig.6 Bulbous tube formed by blow moulding before removal of the polystyrene vehicle.

The ceramic powder has a considerable influence on the maximum ceramic volume fraction that can be incorporated in suspensions (Farris 1968) and also on the removal of organic vehicle (Evans and Edirisinghe) but it is the organic phase which commands the broad range of properties which allow these operations to succeed.

Early work on ceramic and metal injection moulding favoured the use of waxes (Edirisinghe and Evans 1986a). More recent developments have selected high polymers (Johnsson et al 1983). However, during this period of development, the influence of metal-organic and gas phase reaction chemistry has presented the ceramist with ultrafine powders in the nanometre region. Such powders have a high surface excess energy to drive the sintering process and are uniform in size to minimize grain growth. Very small critical defects are possible from such powders. provided they are correctly and cleanly assembled and hence high strengths in 1-1.5 GPa region are possible. The situation now exists where ceramic particles in the 20-100nm region have comparable sizes to the dimensions of the random coil of polymer molecules. This means that it is timely to review the influence of molecular weight of organic vehicle on plastic forming operations to see how the arguments in favour of low or high molecular weight species balance out in the light of polymer adsorption studies.

3. INFLUENCE OF MOLECULAR WEIGHT ON FLOW BEHAVIOUR

The flow properties of ceramic suspensions are controlled by a) the particle characteristics (particle size distribution, particle shape, specific surface area) which determine the maximum packing fraction V_{max}, b) the viscosity and flow behaviour index of the organic vehicle and c) the properties of the particle-fluid interfacial region.

A particular powder can be characterised by measuring the viscosity of the unfilled organic vehicle η_o and of a minimum of two suspensions of different volume loadings V under identical conditions. Expressed as relative viscosity η_r, the viscosity of suspensions η, can be predicted by a large number of semi-empirical equations (Edirisinghe and Evans 1986a) the most versatile of which is (Zhang and Evans 1990):

$$\eta_r = \frac{\eta}{\eta_o} = \left(\frac{V_{max} - CV}{V_{max} - V}\right)^2$$

This is a modification of Chong's equation (1971) in which C = 0.25. C is an empirical constant while V_{max} is the volume fraction at which viscosity becomes infinite. At very low shear stresses V_{max} also shows a slight shear stress dependence (Wildemuth and Williams 1984). Unfortunately V_{max} obtained in this way, although primarily a function of particle characteristics, conceals the influence of the volume fraction of adsorbate. Since adsorbed layer thickness is roughly proportional to molecular weight (Priel and Silberberg 1978), high polymers can be expected to reduce the maximum volume fraction that can be incorporated in suspensions.

A direct procedure for measuring this volume fraction has been described by Doroszkowski (1968). The true relative viscosity-volume fraction relationship for a given powder is first obtained in a fluid of known adsorbed layer dimensions (oleic acid was used). The procedure is then repeated with the organic phase of interest. The displacement of the η_r-V curve gives the adsorbed layer volume fraction and hence thickness.

In low viscosity organic liquids suspensions can be stabilized by electrostatic repulsion achieved by controlling pH in aqueous fluids or by steric stabilization by incorporating amphipathic molecules. There is little evidence to suggest that steric stabilization operates in high viscosity fluids such as polymers. Dispersion is likely to be stabilized by the overlap of adsorbed polymer layers. This compression effect decreases the entropy in the interfacial region and induces a repulsion between particles (Gregory 1978). Thus high molecular weight species can be expected to increase effective particle size but to offer interparticle repulsion.

However the use of low molecular weight fluids, such as waxes, reduces η_o and hence the vicosity of suspensions. Sadly, however, waxes tend to show near-Newtonian flow behaviour and at high shear rates the risk of a shear thickening (dilatent) transition presents itself. At best this means that viscosity in certain parts of the processing machinery increases substantially and at worst that particle contact occurs resulting in a three dimensional structure through which the wax is filtered. Unfilled wax then emerges from the extruder or injection moulding machine barrel and a hard particulate deposit resides in the machine. An excellent review of shear thickening in suspensions based on Newtonian fluids is to be found in the work of Barnes (1989). Shear rate and particle characteristics are the parameters exerting greatest influence on the transition. High polymers offer, in addition to pseudoplastic (shear thinning) flow behaviour, the adsorbed layer repulsion that help to prevent particle contact.

3. SOLIDIFICATION

Intrinsic to the injection moulding process is the non-uniform cooling which occurs in the cavity. This results in residual stress which can cause cracks in the as-moulded state (figure 2a) or defects which only appear when the organic vehicle is removed but whose configuration proves that they originate at the solidification stage (Thomas and Evans 1988). A series of studies by Hunt et al (1990) presents numerical models to predict the incidence of defects during solidification in such a way that sensitivity analysis of material and machine parameters can highlight influential properties. Residual stress distribution is particularly sensitive to solidification temperature and to coefficient of thermal expansion of the suspension. In semicrystalline polymers the volumetric shrinkage associated with phase changes also influences defect formation (Edirisinghe and Evans 1987).

Crystalline melting temperature T_m of paraffins can be expressed as a function of number of carbon atoms X, as (Wunderlich 1980):

$$T_m = 414.3\ (X - 1.5)\ /\ (X + 5.0)$$

converging for polyethylene at 414.3K.

Glass transition temperature T_g of amorphous polymers is related to molecular weight by (R.J. Roe 1987):

$$T_g = T_g\ (\infty) - K/M_n$$

where T_g (∞) is the glass transition temperature at infinite molecular weight. Amorphous polymers are more versatile in that T_g can be controlled by diluent additions but they tend to suffer from a high temperature dependence of viscosity above the solidification temperature which semicrystalline polymers avoid. This reduces mould packing (Hunt et al 1990).

The other important effect of molecular weight is on mechanical strength in the solid state. Wax based compositions tend to be so fragile that thin sections break on ejection from the mould. Thomas (1988) used an ethylene vinyl acetate copolymer to increase the strength of wax based suspensions both by enhancing molecular weight and increasing polarity. Mutual solubility is necessary for such blending.

Similarly, the melt strength needed to vacuum form and blow mould the components shown if figures 3-6 was attained by a) using a polystyrene binder, the glass transition temperature of which was lowered by incorporating diluents and b) by ensuring uniform dispersion of the ceramic powder by twin screw extrusion to avoid low viscosity, powder-depleted regions.

4. REMOVAL OF THE ORGANIC PHASE

The weight loss mechanisms which allow organic material to be removed each depend on the molecular weight of the starting material.

a) Capillary flow into a fine surrounding powder or porous tile.

The flow of liquids in isotropic porous media is described by D'Arcy's law:

$$Q/A = -\frac{K}{\eta}\frac{\Delta P}{x}$$

where Q/A is the volume flow rate per unit area. Modifications to account for completing capillary pressure gradients $\Delta P/x$, and for differing permeabilities of the ceramic body and its surround (German 1987) can be used to model the flow rate of the departing fluid. Above a critical molecular weight melt viscosity at low shear rates is related to weight average molecular weight M_w by

$$\eta = K\, M_w^{3.5}$$

with the consequence that flow rates in porous media are considerably reduced when intermediate waxes or polymers are used. The use of elevated temperature to reduce viscosity is effective only if thermal degradation is prevented when thick sections are being treated. Once chain scission events have started, the problems associated with diffusion in the melt (discussed below) commence.

However, there is another potentially over-riding effect of molecular weight on capillary flow not accounted for by D'Arcy's law. The obstruction to fluid flow in fine glass capillaries has been used as an experimental method to measure adsorbed layer thickness for polymers adsorbed from solution (Priel and Silberberg 1978). An effective increase in viscosity can be attributed to a reduction in the hydrodynamic radius of the capillary from which the adsorbed layer thickness can be found. For polystyrene, the hydrodynamic layer was proportional to molecular weight and was in the range 5nm to 100nm for molecular weights in the range 20,800 to 1,700,000. If ceramic powders are treated as perfect spheres in close packed array the radius of the interstitial pore is 0.155r where r is the particle radius. Table 1 (Evans and Edirisinghe) shows that where a high polymer is combined with a fine powder interstitial pore space could be occupied entirely by immobile layers.

b) Evaporation without chain scission.

Figure 7a shows the thermogravimetric weight loss of a paraffin wax from a 60 vol% alumina composition for finely divided shavings and for small mouldings, measuring 7mm cube in both flowing air and nitrogen. The weight loss is independent of atmosphere and a zero order activation energy of 56 $kJmol^{-1}$ is recorded which corresponds to the activation energy for the evaporation of comparable paraffins (Wright et al 1989). Provided thermal or oxidative degradation are avoided, weight loss is controlled by a surface mass transfer coefficient dependent on gas flow rate and by Knudsen effusion through fine pores. It is a shrinking unreacted core process and therefore could prevent internal defects in large sections provided thermal degradation during prolonged heating can be prevented.

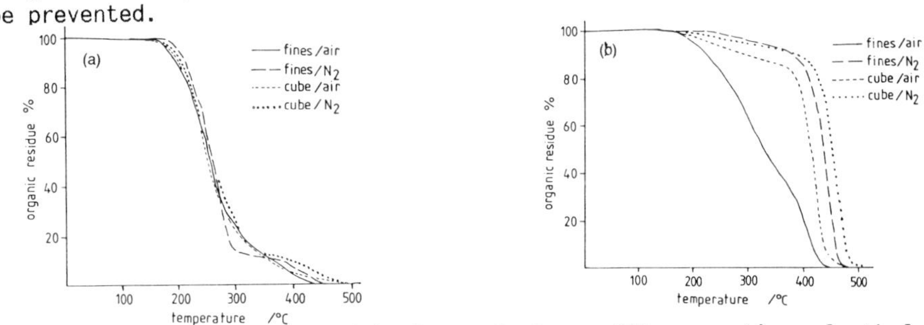

Fig.7 Thermogravimetric weight loss of a) paraffin wax b) polyethylene wax from an alumina suspension in nitrogen and air.

c) Oxidative degradation and evaporation of oxidation products.

When high molecular weight waxes and polymers are used, chain scission precedes evaporation and the reaction sequence determines the molecular weight and hence volatility of fragments. The reaction tends to be more pronounced at the surface region and shrinking unreacted core reaction kinetics can be applied (Wright and Evans to be published). Figure 7b shows the corresponding weight loss curves for a polyethylene wax. This time both sample size and atmosphere control weight loss kinetics.

Table 1. Approximate interstitial pore radius for a range of ceramic powders and approximate chain end to end distance for a range of organic binders, from (Evans and Edirisinghe, to be published).

Ceramic	Pore radius/nm	Organic vehicle	end to end distance/nm
Alumina MA2LS	460	microcrystalline wax	3
Alumina A16SG	30	atactic polypropylene	5
Zirconia HSY3	15	low density polyethylene	14
Zirconia DZF	8	polypropylene	18
		polystyrene	20

d) Thermal degradation and evaporation of products.

In a study of the molecular weight dependence of weight loss kinetics from mouldings, thermal degradation played a part for microcrystalline wax and higher molecular weight species (Wright et al 1989). It is not confined to the surface region and generates products of decomposition throughout large sections which must then diffuse to the surface and evaporate. Boiling at the centre occurs when the partial pressure of degradation products in solution reaches ambient pressure. Diffusion occurs in the melt within the pore architecture and is subject to impediments from adsorbed immobile layers as with capillary flow. The dimensions in table 1 can be used to predict the effective decrease in pore volume fraction caused by immobile layers and hence the reduction in effective diffusion coefficients in the composite (Evans and Edirisinghe to be published).

5. CONCLUSIONS

The interplay between particle size and polymer molecular weight in the plastic forming of ceramic suspensions has been examined at a level that paves the way for a strategic approach to materials selection. For fine ceramic powders, a two phase composite approach is inadequate and suspensions should be regarded as including immobile adsorbed layers which control both viscosity and mass transport during binder removal.

REFERENCES

Barnes H A 1989 J. Rheol. 33 329
Chong J S Christiansen E B and Baer A D 1971 J.Appl.Polym.Sci. 15 2007
Doroszkowski A and Lambourne R 1968 J. Coll. Interf. Sci. 26 214
Edirisinghe M J and Evans J R G 1986a Int. J. High Tech. Ceram. 2 1
Edirisinghe M J and Evans J R G 1986b Int. J. High Tech. Ceram. 2 249
Edirisinghe M J and Evans J R G 1987 J. Mater. Sci. 22 2267
Edirisinghe M J and Evans J R G 1989 Mat. Sci. Eng. 109 17
Evans J R G and Edirisinghe M J to be published
Farris R J 1968 Trans. Soc. Rheol. 12 281
German R M 1987 Int. J Powder Met. 23 237
Gregory J 1978 in The Scientific Basis of Flocculation K J Ives (ed.) (Alphen aan den Rijn, Netherlands; Sijthoff and Noordhoff) pp 101-130
Hammond P and Evans J R G to be published
Haunton K M Wright J K and Evans J R G 1990 Br. Ceram. Trans. J. in press
Hunt K N Evans J R G and Woodthorpe J 1990 J. Mater. Sci. in press
Johnsson A Carlstrom E Hermansson L and Carlsson R 1983 Proc. Brit Ceram. Soc. 33 139
Kobayashi K Furuta M and Maeno Y Euro Pat. 34056 19 Aug. 1981
Priel Z and Silberberg A 1978 J. Polym. Sci. Polym. Phys. 16 1917
Rashid H Lindsey K and Evans J R G to be published
Renlund G M and Johnson C A US Pat. 4551496 Nov 5 1985
Roe R J 1987 Encyclopaedia of Polymer Science and Engineering 7 Wiley pp 536-537
Schwartz B 1984 Amer. Ceram. Soc. Bull. 63 577
Thomas M S and Evans J R G 1988 Br. Ceram. Trans. J. 87 22
Wildemuth C R and Williams M C 1984 Rheol. Acta 23 627
Wright J K Evans J R G and Edirisinghe M J 1989 J.Amer.Ceram.Soc.72 1822
Wright J K Thomson R M and Evans J R G 1990 J. Mater. Sci. 25 149
Wright J K and Evans J R G to be published
Wunderlich B 1980 Macromolecular Physics 3 Academic Press pp 25-30
Zhang J G Edirisinghe M J and Evans J R G 1989a Industrial Ceramics 9 72
Zhang J G Edirisinghe M J and Evans J R G 1989b J. Mater. Sci. 24 840
Zhang T and Evans J R G 1989 J Euro. Ceram. Soc. 5 165

ACKNOWLEDGEMENTS

The author is grateful to the SERC for continued support for the ceramics fabrication programme.

Inst. Phys. Conf. Ser. No 111
Paper presented at Int. Conf. on New Materials and their Applications, Univ. of Warwick, 1990

Towards improved materials from powder processing

K Kendall

ICI PO Box 11, Runcorn, Cheshire WA7 4QE

Abstract

Powder processing of materials is a preferred route for moulding polymers, metals and ceramics which are difficult to melt or to forge. Unfortunately, the performance of powder processed products generally falls far short of that exhibited by melt processed articles made from the same material. This paper shows that such property deficiencies may be partly corrected by paying attention to the van der Waals forces which act between fine particles of powder. These forces cause agglomeration and consequent defective structure in the moulded compact. Ceramic products have been considerably improved both in strength and reliability through this approach.(Alford et al 1987, Clegg et al 1989, Kendall et al 1990).

1. INTRODUCTION

Many materials are most conveniently made from powders. Whereas the common metals, polymers and composites are normally formed by melt processes such as casting, extrusion or injection moulding, more intractable substances such as refractory metals and ceramics require moulding and sintering from the powder state. The purpose of this paper is to demonstrate that the products derived from such powder processes have always been deficient in properties compared to melt derived products because of the structural defects stemming from agglomerates in the powder, agglomerates formed by the same van der Waals forces which drive sintering. By controlling these interparticle forces during powder moulding operations, substantial improvement in strength and reliability of ceramics has been attained (Kendall et al 1990).

There are several incentives for powder processing of materials. A straightforward illustration is the manufacture of pharmaceutical tablets or catalyst pellets, where powder pressing gives a direct method for moulding special porous structures. Another similar case is the casting of silicate cement pastes or lead acid battery dispersions which harden at modest temperatures to form solid materials by reaction with the fluid dispersant medium. More generally, powder methods are preferred because melt methods are difficult or impossible.

For example, ultrahigh molecular weight polymers do not flow easily above the melting point and so are best densified by pressing fine powder. The same is true of PTFE which decomposes near its melt temperature. Refractory materials such as tungsten or zirconia exhibit very high melting points and consequently are best formed by sintering fine powders at around half the melt temperature. In this process, the grain size can also be controlled to preserve a micrometre scale structure ensuring high strength.

It is clear, then, that powder processing is the most desirable route for manufacturing many important materials. However, the products of powder processing have suffered from a poor reputation because of property deficiencies. This is exemplified in Figure 1 which contrasts the strength and reliability of vitreous silica made by melting and by powder sintering. The starting material was the same in each case, a fumed silica made by reaction of silicon tetrachloride, so the gross inferiority of the powder processed product can only be explained by some drawback of the compaction and sintering method. We shall see that the cause of this problem is the agglomeration of the fine powder grains as a result of van der Waals forces (Kendall et al 1989).

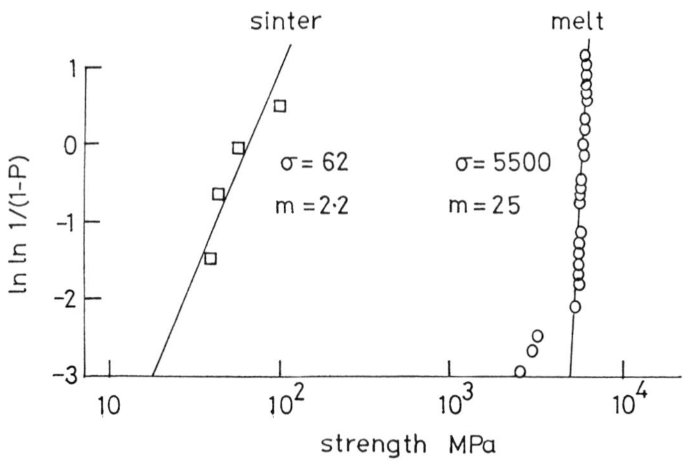

Fig. 1. Strength and reliability of vitreous silica made by sintering and by melting (Kruger and Mularie 1983)

2. THE PROBLEMS

There are several problems limiting the potential of powder processing for advanced materials. The first difficulty is that the compaction and sintering of fine powders is not a trivial matter. Powders finer than a certain critical size stick together and resist compaction and sintering. The second snag is that the strength of the densified material is always much lower and more variable than expected from the grain size of the product. This can be explained by the presence of large and variable agglomerates caused by strong adhesion between the powder grains. These agglomerates survive the powder compaction step and act as crack initiating defects in the sintered parts.

Both of these problems can be quantified in terms of the adhesion energy of the particles. In an ideal reversible experiment, in which two particles are brought into contact without energy dissipation, this energy is the thermodynamic work of adhesion W (Johnson et al 1971, Kendall 1988), the reversible energy required to make or break unit area of contact between the solid surfaces. It is twice the solid surface energy when defined in this way. However, in many examples of adhesive contact between particles, it is known that conditions are not ideal, and that the measured surface energies depend on the rate of testing and on other experimental conditions (Kendall 1974, Maugis and Barquins 1978). Then we have to define measured adhesion energies Γ_m required to make the contact, and Γ_b to break it. The equilibrium value W, the work of adhesion, lies between these two measured values, the differences being explained by kinetics and by adhesive hysteresis.

Consider how the adhesion between particles influences the processing and structure of powder-based products. In the first place, the adhesion drives sintering, the mechanism by which powder compacts densify according to Frenkel's (1945) equation (Kendall 1988b)

$$\dot{e} = 3\Gamma_m/8\eta D \tag{1}$$

which relates strain rate $\dot{e}$ to the adhesion energy Γ_m, the particle diameter D and the particle viscosity η. Mechanisms other than viscous sintering are described by slightly different equations (Herring 1950). Secondly, the adhesion influences the size of agglomerates which form in the powder during moulding, as predicted by the equation (Kendall et al 1989b)

$$\Delta = 0.064[2ED^{2.5}/\Gamma_m(1-\nu^2)]^{2/3} \tag{2}$$

The larger the agglomerate size, the more defective is the product and the lower its strength according to fracture mechanics (Griffith 1920). However, these agglomerates may be broken down during powder mixing, and this is predictable from the agglomerate strength σ^* which cannot exceed (Kendall 1988c)

$$\sigma^* = 15.6\phi^4\Gamma_b/D \tag{3}$$

where ϕ is the packing fraction of particles in the agglomerate, Γ_b is the adhesion energy for separation of the particles and D the particle diameter.

An example of the importance of adhesion to powder processing concerns the optimum particle size for making ceramics. From equation (1) it is evident that the maximum sintering rate is obtained for finer particles, as plotted in figure 2. However, equation (3) shows that finer particles lead to stronger agglomerates. Since the powder compaction machine can only apply a limited shear stress to break down the agglomerates, it is clear that sufficiently fine particles will not be disagglomerated during compaction and so will not sinter. From equation (3) the critical particle size D_c at which this occurs depends on the limiting shear stress σ_c^* exerted by the machine according to

$$D_c = 15.6\phi^4 \Gamma_b / \sigma_c^* \quad (4)$$

This is drawn in figure 2 to explain why sintering should occur most rapidly at a particular particle size, depending on the compaction machinery. Putting appropriate numbers for ϕ = 0.6, Γ_b=1 Jm^{-2} and σ_c^*=4 MPa gives a value of D_c=0.5μm which is near the observed value of particle size at which sintering is most rapid for alumina and titania, compacted by simple pressing equipment.

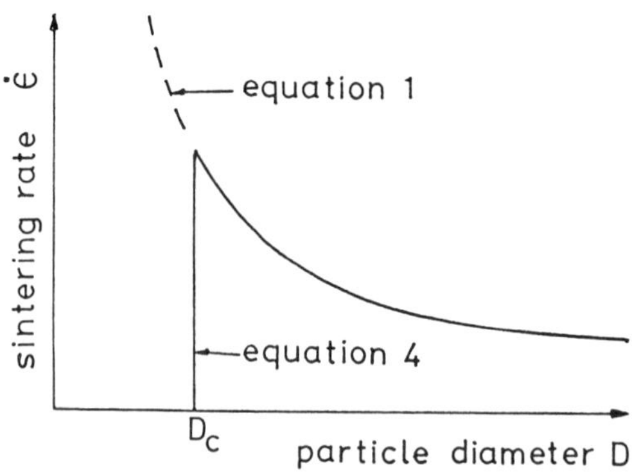

Fig. 2. Sintering strain rate as a function of particle size. Above D_c the sintering obeys Frenkel's equation. Below D_c sintering is prevented by agglomerates.

3. MEASURING ADHESION ENERGY

One might have thought that equation (1) could form the basis of a method for measuring the adhesion energy Γ_m through a determination of sintering rate, knowing the viscosity and diameter of the particles. Unfortunately, this is not successful because the viscosity is such a large and ill-defined quantity. In fact, there are no properly established methods for measuring the powder adhesion energy, despite its importance to sintering, flocculation and agglomerate strength. Moreover, the distinction between measuring Γ_m on making and Γ_b on breaking contact is rarely drawn.

However, in the past few years, new methods have been developed for determining Γ on the premise that the mechanical properties of powder compacts are a direct measure of the particle surface interactions (Kendall et al 1986, 1987). The model on which this idea is based is shown in figure 3. Two smooth elastic spheres which come into contact (Fig.3a) do not behave like ball-bearings, bouncing off each other according to Hertz elastic contact theory (Johnson 1987). Instead, the particles behave like smooth rubber spheres which adhere strongly, as demonstrated with 5 mm diameter synthetic polymer beads. Such adhesion, which explains why fine ceramic particles can be compacted to form cohesive tablets (Fig.3b), has been described in terms of the adhesion energy of the solid grains (Johnson et al 1971). The particles become flattened by the attractive surface forces at their contact points and the size of the contact spot is a measure of adhesion energy.

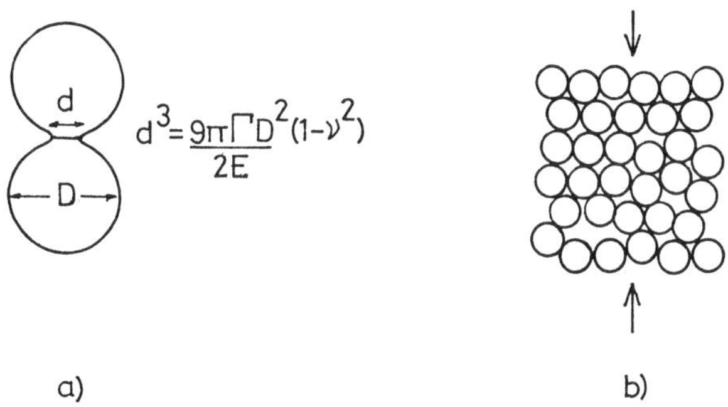

Fig. 3a. Model of adhesive elastic contact between two spheres.
3b. Model of cohesive tablet formed by multiple adhesive contacts between particles.

For very fine particles, the contact spots are usually too small to be defined directly. However, the contact spot size controls several properties of powder assemblies, including elastic modulus (Kendall et al 1987), electrical conductivity (Kendall 1990b) and thermal conduction (Batchelor and O'Brien 1977, Jagota et al 1990) and these properties may therefore be used as measures of contact energy. For example, the elastic modulus of a cubic array of uniform elastic spheres is

$$E^* = [9\pi\Gamma_m E^2/16D(1-\nu^2)^2]^{1/3} \quad (5)$$

allowing the contact energy Γ_m to be calculated knowing the diameter and the elastic properties of the spheres. A problem arises for random arrays of spheres and again for non-uniform, odd-shaped grains because there is a strong geometrical influence on the elastic modulus of a powder compact (Walton 1987). To a first approximation, the geometry can be defined in terms of the solids packing fraction ϕ. Then the elastic modulus of the compact is found to be

$$E^* = 17.1\, \phi^4 [E^2\Gamma_m/D]^{1/3} \quad (6)$$

Recently (Kendall 1990) a new method for measuring adhesion energy has been proposed which avoids the need to know the elastic properties of the grains. In this method, shown in figure 4, a block of compacted powder is addressed by a piezoelectric transducer to measure the speed of ultrasonic waves through the assembly. The powder compact is then stressed mechanically and the change in wave speed observed. For low stresses, the contact spots between particles are dictated mainly by the adhesion energy and so there is little change in wave speed. But at a critical stress , the speed begins to increase rapidly with load, as the external stress begins to dominate the surface energy and to push the grains into more extended contact. At this critical stress σ_c the contact energy is

$$\Gamma_m = \sigma_c D/40\pi\phi^4 \qquad (7)$$

Similar considerations apply to the electrical conductivity of carbon black particles in a compact (Kendall 1990b).

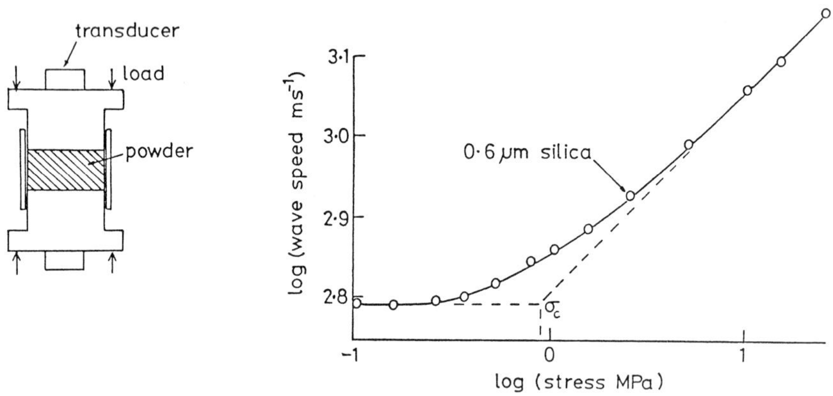

Fig. 4. Experiment to investigate the ultrasonic wave speed through a powder compact as the compressive stress is raised. The results show the critical stress at which wave speed begins to increase.

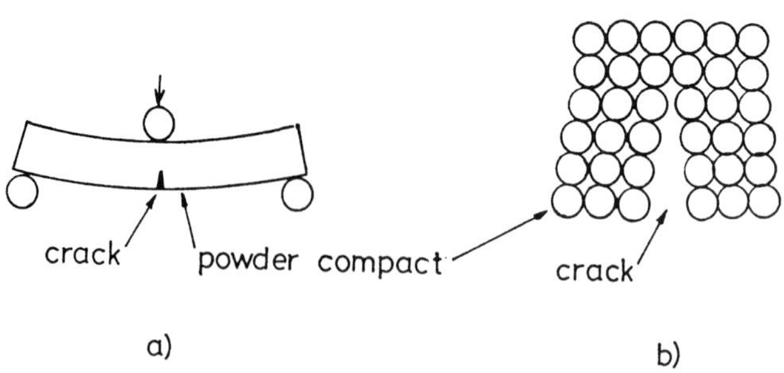

Fig. 5a. Bending test on a notched bar of compacted powder.
5b. Model of crack in a powder assembly.

The energy required to break interparticle contacts, Γ_b the contact breaking energy, is substantially larger than Γ_m, the contact making energy, as a result of energy dissipation during fracture or flow of the particle structure. Γ_b is most readily measured in a fracture experiment (fig.5) where a block of powder compact containing a notch of known length is bent until cracking occurs. From this determination (Kendall 1988c), the energy required to break one interparticle contact can be calculated from a model of a crack in the particulate assembly, knowing the particle diameter, elasticity and packing (fig.5b).

An alternative idea is to measure the flow stress of the compact during a friction (Kendall 1986) or hardness test (Kendall 1990c) and to deduce Γ_b from a model of shearing in a particle assembly.

4. IMPROVED CERAMICS

The understanding of interactions between fine solid particles pays off in the improvement of materials made by powder processing.

An interesting example of such an advance is the increase in performance of hydraulic cements which stemmed from the reduction of van der Waals forces in a wet cement mix (Birchall et al 1981, Kendall et al 1983, Kendall 1987). Addition of water soluble polymer to the cement dispersion allowed high solids loading in the mix, upgraded the rheology to give uniform intense shearing, prevented flocculation of grains, and facilitated defect removal from the slurry, thereby increasing bend strength, toughness, reliability and impermeability to a remarkable degree. The rise in bend strength was more significant than the critical temperature increase observed recently for ceramic superconductors (Fig.6). These improved cements are now being exploited in Japan.

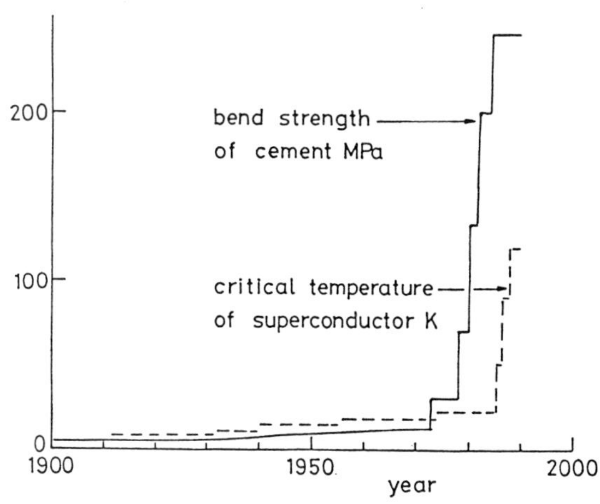

Fig. 6. Rise in bend strength of cement with time compared to the improvement in superconducting critical temperature.

Similar advances have been noted in the processing and properties of ceramics. It was first demonstrated in 1987 that certain ceramic powders could be disagglomerated by controlled shearing and that the resulting dispersed structure could be preserved through the drying, burn-out and firing sequence needed to produce dense ceramics (Alford et al 1987). The more perfect structure developed during the shearing stage led to several advantages:- improved moulding, better surface finish, stronger green bodies, lower temperature sintering, stronger products and higher reliability. These benefits were found in titania, alumina, silica, zirconia, yttrium barium copper oxide, and silicon carbide (Kendall et al 1990).

To illustrate the scale of improvement obtained, consider the fabrication by powder methods of an electrically conducting ceramic (trade-name Ebonex, Ebonex Technologies Inc) from its powder. There are two obvious ways of fabricating this material; by die pressing the powder in steel moulds, or by plastically mixing the powder with a polymer vehicle and extruding it. In the pressing method, the dry powder was poured into the die and compacted into a disc-shaped pellet at a pressure of 140 MPa. The solids packing fraction was 0.60 . By contrast, the plastic mixing method employed a twin-roll mill to blend the powder into a 40% solution of polyvinyl alcohol. This composition, which was designed to reduce the van der Waals interactions between particles, was de-aired and extruded into rods which were dried and pyrolised at 500°C to remove the polymer. An indication of the improved structure was the increased packing fraction of 0.71 in this case.

Sintering in argon showed a small but significant change in the structure of the compacts. Whereas the pressed tablet was slow to sinter and was only 85 % dense at 1300°C, the extruded material was 97 % dense at this temperature.

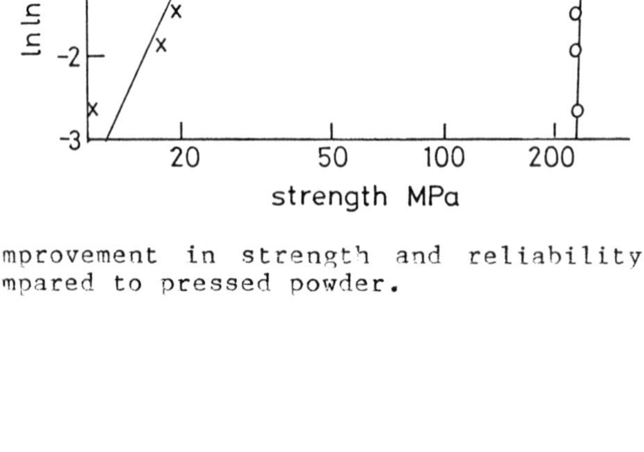

Fig. 7. Improvement in strength and reliability of extruded ceramic compared to pressed powder.

However, the most dramatic difference was in the microstructure and mechanical properties of the final sintered ceramic. The pressed products showed large defects in SEM pictures of polished sections whereas the extruded material showed much smaller and more uniform flaws. Bending strength and Weibull modulus determinations supported these observations (Fig.7). The strength of the plastically processed product was an order of magnitude higher than that of the pressed material, and the Weibull modulus was also substantially improved.

4. CONCLUSIONS

Powder processed materials are much influenced by the adhesive van der Waals forces which act between particles, causing imperfect structures in powder compacts, problems of sintering, and defects in densified parts. By identifying these adhesive forces in terms of the adhesion energy between grains, the main variables influencing this process have been isolated, and test methods devised for their quantification. These studies have led to better processes for fabricating cements and ceramics, leading to marked improvements in properties. Typically, strength and reliability have been raised by an order of magnitude through this approach.

REFERENCES

Alford N McN, Birchall J D and Kendall K 1987 Nature 330 51
Batchelor G K and O Brien R W 1977 Proc R Soc Lond A355 313
Birchall J D, Howard A J and Kendall K 1981 Nature 289 388
Clegg W J, Alford N McN, Birchall J D and Kendall K 1989 J Am Ceram Soc 72 432
Frenkel J 1945 J Phys USSR 9 385
Griffith A A 1920 Phil Trans R Soc Lond A221 193
Herring C 1950 J Appl Phys 21 301
Jagota A and Hui C Y 1990 J Appl Mech in press
Johnson K L 1987 Contact Mechanics (Cambridge University Press) pp 125-129.
Johnson K L, Kendall K and Roberts A D 1971 Proc R Soc Lond A324 301
Kendall K 1974 J Adhesion 7 55
Kendall K 1986 Nature 319 203
Kendall K 1987 European J Eng Education 12 21
Kendall K 1988 Sci Progr Oxford 72 155
Kendall K 1988b Materials Forum 11 61
Kendall K 1988c Powder Metallurgy 31 28
Kendall K 1990 Nature submitted
Kendall K 1990b J Phys D submitted
Kendall K 1990c to be submitted
Kendall K, Howard A J and Birchall J D 1983 Phil Trans R Soc Lond A310 139
Kendall K, Alford N McN and Birchall J D 1986 Special Ceramics 8 255
Kendall K, Alford N McN and Birchall J D 1987 Proc R Soc Lond A412 269.

Kendall K, Alford N McN, Clegg W J and Birchall J D 1989 I Chem E 5th Int Symp on Agglomeration pp355-359
Kendall K, Alford N McN, Clegg W J and Birchall J D 1989b Nature 339 130
Kendall K, Alford N McN, Clegg W J and Birchall J D 1990 Proc Inst Ceram in press
Kruger A A and Mularie W M 1983 Proc Soc Photo-Optical Instrumentation Engineers 404 2.
Maugis D and Barquins M 1978 J Phys D 11 1989
Walton K 1987 J Mech Phys Sol 35 213

Injection mouldable ceramic–polymer blends: Effect of some powder characteristics on the binder removal process

Ricardo E.F.Q. Nogueira and Mohan J. Edirisinghe
Department of Materials Technology
Brunel University
Uxbridge, Middlesex, UB8 3PH, U.K.

ABSTRACT: Ceramic suspensions containing four types of alumina powder at 60% vol. were prepared using a polypropylene binder system. 18 mm diameter cylinders moulded from the formulations were subjected to binder removal by thermal degradation and the results obtained are explained in terms of the characteristics of the powders used.

1. INTRODUCTION

Ceramics are an emerging engineering material. Therefore, in recent years there has been a great deal of interest in the automated mass production of ceramic components containing thick sections (>15 mm). In this respect injection moulding ceramic-polymer suspensions followed by removal of the polymeric binders before sintering the ceramic has been favoured (Mutsuddy 1983, Edirisinghe and Evans 1989). However, binder removal from thick section components still involves high scrap rates (Carlsson 1989, Edirisinghe 1989).

Methods of binder removal from moulded bodies have been reviewed (Edirisinghe and Evans 1986a). Pyrolysis is favoured and the mechanisms involved have been described (Woodthorpe et al 1989, Wright et al 1989). The effect of some material properties and process control parameters have also been discussed (Edirisinghe 1989). In the case of polyolefin binders removal involves degradation and outward diffusion of resulting products. This sets up a concentration profile for the degradation products and if the concentration exceeds a critical level defects can form in the moulded body. Recent investigations have revealed that binder removal from thick section components containing coarse ceramic powders have been more successful (Edirisinghe 1990) and therefore the role of powder characteristics such as particle size distribution, surface area and particle shape should be systematically investigated.

Shrinkage occurs during binder removal and particles pack causing an increase in the volume fraction of powder (Shaw et al 1990). Although the initial volume of ceramic (V) may not increase to the maximum packing volume, V_{max}, when particles touch each other, any increase in filler volume during binder removal results in the reduction of the pore channel network between ceramic particles and this hinders the outward diffusion of degradation products. This could cause defects.

Powder surfaces can contain adsorbed water (Bernett and Zisman 1968) which has to diffuse out during binder removal. Also, the polymeric binders adsorb onto the powder particles (Kwei 1965). The adsorbed layers could be immobile and therefore hinder the outward diffusion of both water and degradation products during binder removal causing defects. The presence of powder could also catalyse the thermal decomposition of some binder systems (Masia et al 1989).

In this investigation powder characteristics are varied in formulations which are used for binder removal experiments. The results are used to identify powder characteristics suitable for the binder removal stage in the ceramic injection moulding process.

2. EXPERIMENTAL DETAILS

The ceramic powders used were A152.SG*, CT2000.SG*, CT3000.SG* and MA250(2)**. Particle size distributions (psd) and scanning electron micrographs of the powders are shown in Figs. 1 and 2. The binder system consisted of isotactic polypropylene, microcrystalline wax and stearic acid in the weight ratio 6:2:1. Details of this binder system are given in Table 1.

Component	Grade	Source	Density kg m^{-3}
Isotactic Polypropylene	GY545M	I.C.I. Ltd. Welwyn Garden City, U.K.	905
Microcrystalline Wax	1865Q	Astor Chemicals Ltd. West Drayton, U.K.	910
Stearic Acid	30269 4Q	B.D.H. Chemicals Ltd. Essex, U.K.	941

Table 1. Details of the binder system used.

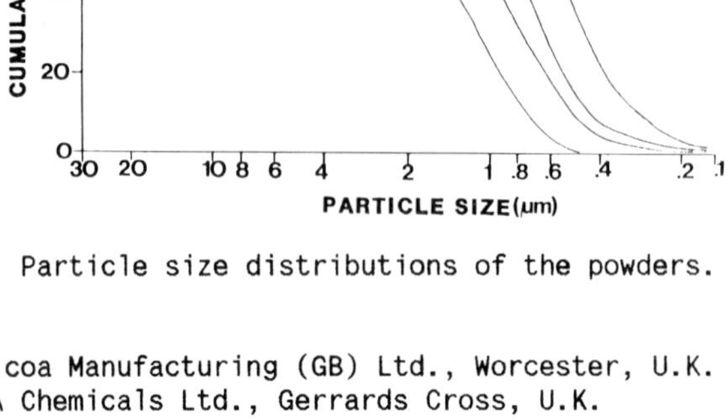

Figure 1. Particle size distributions of the powders.

* ex Alcoa Manufacturing (GB) Ltd., Worcester, U.K.
** ex BA Chemicals Ltd., Gerrards Cross, U.K.

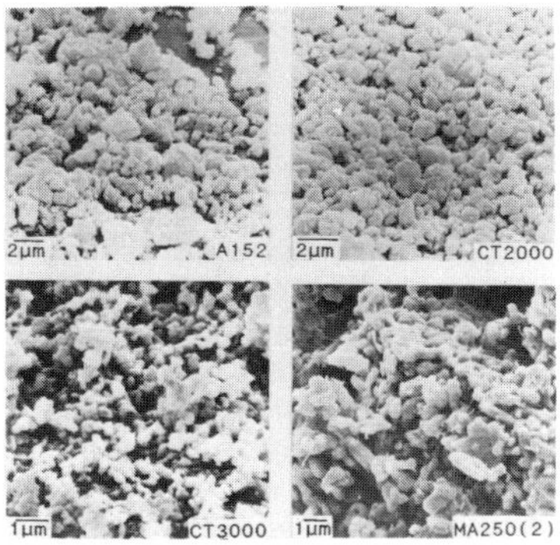

Figure 2. Scanning electron micrographs of the powders.

The powders and the binder system were blended in a high speed mixer and compounded using a co-rotating intermeshing twin screw extruder according to the procedure described before (Edirisinghe and Evans 1986b, 1987). The temperatures used in the extruder barrel were 220-225-235-225°C feed to exit. Three samples of each formulation were ashed after compounding and the true volume of ceramic was calculated. Exact compositions are given in Table 2. The molecular weight distribution of the major binder polypropylene in each formulation was determined using Gel Permeation Chromatography*.

Ceramic			Isotactic Polypropylene	Microcrystalline Wax	Stearic Acid
Type	%wt	%vol.		%wt	
A152	87.01	60.5	8.66	2.89	1.44
CT2000	86.72	59.9	8.85	2.95	1.48
CT3000	86.68	59.8	8.88	2.96	1.48
MA250(2)	86.93	60.3	8.71	2.90	1.46

Table 2. Exact compositions of the formulations made.

Fine shavings of each formulation were subjected to thermogravimetric studies in flowing oxygen free nitrogen at a heating rate 5°C min^{-1} using a Perkin-Elmer TGS-2 thermobalance. The ceramic powders were also heated in air in the thermobalance at 5°C min^{-1} to 600°C to find out the water desorption characteristics. Each powder sample was first heated to 600°C and left in ambient air for 48 hours before heating in the thermobalance in order to distinguish between the presence of adsorbed water and other losses on ignition.

Samples of each formulation were compression moulded in a heated press to obtain 25 mm diameter, 1 mm thick wafers which were used for shrinkage measurements. Four diameters were scribed on the wafers and

* performed by Rubber and Plastics Research Association, Shrewsbury,U.K.

identified. A travelling microscope was used to measure each diameter (d). The wafers were supported on a stainless steel mesh and heated in flowing oxygen free nitrogen using a slow ramp*. After pyrolysis the individual diameters were re-measured (d-Δd) to calculate the average linear shrinkage $\frac{\Delta d}{d}$.

18 mm cylinders from each formulation were compression moulded using an Apex Model M1/R press at 225°C under 12 MPa pressure. The cylinders were X-ray radiographed to ensure they were free of macro defects and then subjected to binder removal in flowing oxygen-free nitrogen using the same temperature ramp under which shrinkage measurements were obtained. After binder removal the cylinders were X-ray radiographed to reveal the internal structure.

3. RESULTS AND DISCUSSION

Powder	Median Particle Size (μm)	Specific Surface Area (m^2g^{-1})	%wt. Loss on Ignition	%wt. Adsorbed Water
A152	1.1	3.0	0.47	0.18
CT2000	0.9	4.3	0.20	0.57
CT3000	0.6	6.5	0.24	0.75
MA250(2)	1.7	3.5	0.32	0.27

Table 3. Some powder characteristics.

The information in Figs. 1,2 and Table 3 shows that A152, CT2000 and CT3000 contain equiaxed particles with monomodal size distributions. The important difference between these three powders, as indicated by the median particle size and surface area data, is that the particle sizes are different; in general, CT3000 and A152 contain the smallest and largest particles respectively and CT2000 has an intermediate size. Also compared to CT2000, CT3000 is loosely agglomerated as seen in the scanning electron micrograph (Fig. 1). The particle characteristics of MA250(2) are strikingly different to the other three powders used in that it has a wide multimodal size distribution of irregular particles. The psd (Fig. 1) shows that it contains some very coarse particles which according to the scanning electron micrograph (Fig. 1) are probably agglomerates.

The loss on ignition (Table 3) for all the powders was <0.5wt.% with A152 showing the highest value. The %wt. adsorbed water was dependent on the surface area. CT3000 which has approximately twice the surface area of A152 contained over four times more adsorbed water than A152 (Table 3). One of the reasons for the loose agglomeration observed in CT3000 (Fig. 2) could be the higher amount adsorbed water. Although the %wt. values of adsorbed water for the powders are small, it represents a large volume when converted to steam because of the low molecular weight of water. Also, the thermogravimetric traces (Fig. 3) show that

* 50°C/hour to 150°C, held for 24 hours at 150°C, then at 2°C/hour to 500°C.

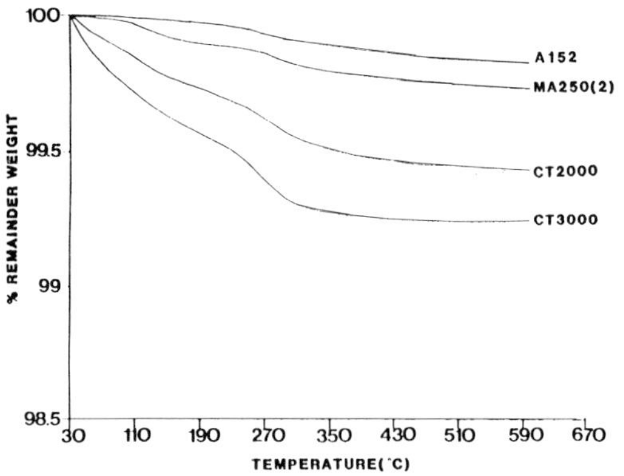

Figure 3. Thermograms of powders heated in air at 5°C min^{-1}.

removal of adsorbed water begins at a relatively low temperature during debinding when diffusion is difficult. Therefore the presence of adsorbed water in the ceramic could lead to defects during debinding. Some of the adsorbed water may be removed during the processing of the blends and if the polymeric binders coat the powder particles re-adsorption will be prevented. However, it is clear that adsorbed water is an unwelcome constituent in the powder and should be minimised as far as possible. Clearly, fine powders with a high surface area would pose problems in this respect.

Fig. 4 shows the polypropylene molecular weight distribution of each formulation after compounding. They are very similar except that the formulation containing MA250(2) shows a lower $\overline{M}_n$. This is due to the increase in mechanical degradation during compounding caused by its irregular particle shape and the presence of large particles. The increase in mechanical degradation is supported by the high level of Fe detected (Shaw et al 1990) in formulations containing MA250(2). A lower $\overline{M}_n$ implies an increase in low molecular weight species which could affect the thermal degradation behaviour at low temperatures.

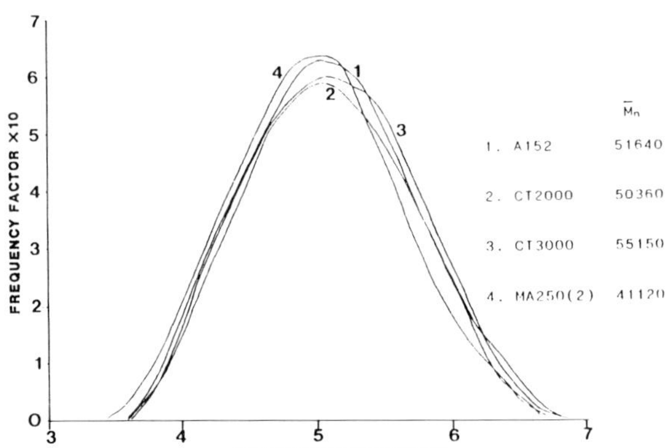

Figure 4. Polypropylene molecular weight distribution of each formulation after compounding.

The thermograms of the formulations shown in Fig. 5 indicates that

CT3000 showed a greater weight loss at a lower temperature. This could be attributed to catalyses (Masia et al 1989) due to the high surface area of this powder; approximately twice that of A152 which has the lowest surface area (Table 3). There is no appreciable difference in the thermal degradation characteristics of the formulations containing the other powders. This shows that the small difference in surface area between these powders was not significant enough to cause appreciable catalysis of thermal degradation. Also the level of increase in mechanical degradation during compounding (and the decrease in $\overline{M}_n$) was not sufficient to be effective in altering the low temperatures thermal degradation behaviour of the formulation containing MA250(2).

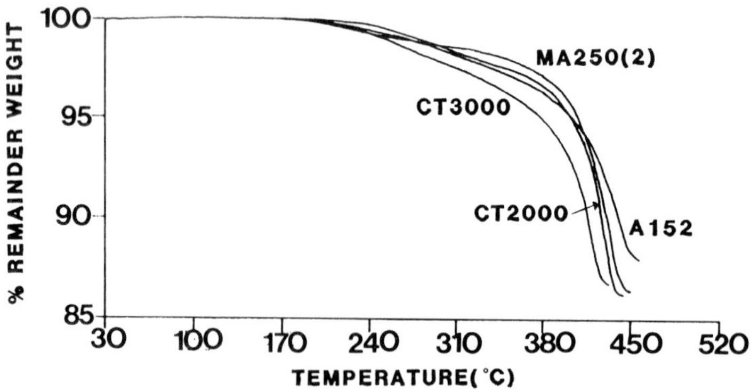

Figure 5. Thermograms of the formulations in nitrogen at a heating rate of 5°C min^{-1}.

The % linear shrinkage that occurs during binder removal and the resulting final volume fractions of the formulations (V*) are given in Table 4. V* can be calculated using standard definition of volumetric shrinkage:

$$1 - \frac{V}{V^*} = 3\left(\frac{\Delta d}{d}\right) - 3\left(\frac{\Delta d}{d}\right)^2 + \left(\frac{\Delta d}{d}\right)^3$$

where V represents the initial volume fraction of ceramic.

Formulation	% Linear Shrinkage	% V*
A152	0.35	61.1
CT2000	1.10	61.9
CT3000	3.20	65.9
MA250(2)	0.13	60.5

Table 4. % Linear shrinkage on binder removal and V* values of the formulations.

The shrinkage observed during binder removal is the result of competition between the forces tending to cause shrinkage and those that inhibit it. It is well known that a finer particle size increases the force tending to cause shrinkage. On the other hand, fine particles are associated with a high surface area which helps in the formation of adsorbed layers of water and polymer on the ceramic particles. Adsorbed layers could be immobile and restrict the movement, especially the rotation of moving particles. Shrinkage in formulations containing non-spherical particles depends on rotation of particles (McKenzie and Shuttleworth 1949) in addition to the approaching movement of particle centers. A wide, multimodal particle size distribution will help particle packing (Poslinski et al 1988) and this will help to increase shrinkage during binder removal but an irregular particle shape will have the opposite effect (Shukla 1989).

The shrinkage results shown in Table 4 can be discussed in terms of the particle characteristics given in Figs. 1 and 2 and Table 3. MA250(2), with a irregular particle shape, the highest mean particle size and a specific surface area of 3.5 m^2g^{-1} showed the lowest shrinkage, despite the fact that it had wide multimodal particle size distribution. Compared to MA250(2), A152 has a slightly lower surface area but the median particle size is 35% smaller. This, together with the fact that it has a more regular equiaxed particle shape caused the higher shrinkage during binder removal. CT2000 has a median particle size $<1\mu m$ and the specific surface area is 43% greater than A152. The linear shrinkage observed is approximately three times that of A152 indicating that the decrease in particle size was most effective in increasing the shrinkage during binder removal. CT3000 with the lowest median particle size and the highest surface area which showed the greatest shrinkage during binder removal also confirms this argument. The shrinkage values are also affected by the V_{max} values of each powder but research work in progress indicates that in the case of A152,CT2000 and CT3000 this value is between 0.7 to 0.8 (volume fraction) allowing comparison.

X-ray radiography of the 18 mm diameter cylinders after compression moulding indicated that they were free from cracks and voids. The internal structure of these cylinders after debinding is given in Fig. 6 below. The cylinder containing CT3000 was so badly bloated during binder removal that it was damaged during handling.

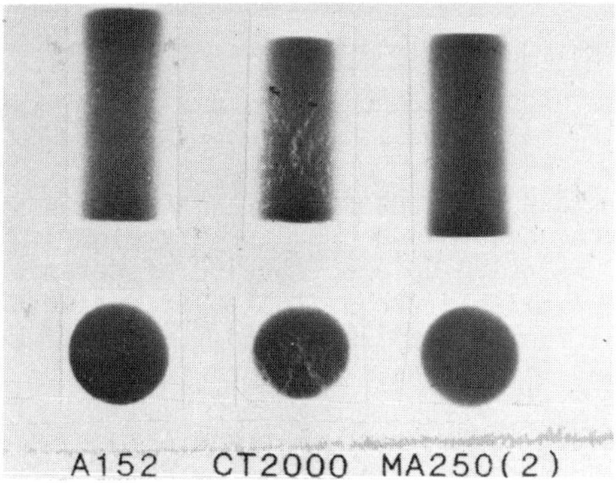

Figure 6. Cylinders of the formulations after binder removal: Contact print of the X-ray radiographs.

The radiograph indicates that the cylinders made using the formulation containing CT2000 showed bloating and extensive cracking. The cylinder made using the A152 formulation showed several fine surface flaws. The MA250(2) cylinder showed only a few surface flaws. Some of the defects could be avoided by slightly modifying the binder system. In fact, the use of atactic polypropylene, instead of microcrystalline wax (work in progress) has allowed the production of similar size MA250(2) cylinders without the presence of any macro defects. Also, the quality of the A152 cylinder has been improved. However, even with this modification the bloating and cracking in the 18 mm cylinders containing the CT2000 powder remained.

These observations indicate that binder removal from injection mouldable ceramic polymer blends is critically dependent on the characteristics of the ceramic particles. Firstly, the channels in the ceramic-polymer assembly available for outward diffusion of polymer degradation products depend on the arrangement of ceramic particles. This depends on ceramic particle size, size distribution and particle shape. A fine particle size, a wide size distribution and a regular particle shape reduces the size of the channels.

A152, CT2000 and CT3000 are powders with a near spherical regular particle shape and this allows some comparison of the channels available. Considering the median particle size given in Table 2 to represent the diameter (D) of spherical particles the pore diameter of the inscribed circle is 0.155D (Cumberland and Crawford 1987). This comparison given in Table 5 shows that the width of the pore channels in A152 could be 22% and 84% greater than CT2000 and CT3000 respectively.

Ceramic	Diameter of Pore Channel (nm)
A152	171
CT2000	140
CT3000	93

Table 5. Estimate of the channel size available for diffusion of degradation products.

In addition, it is clear from the results that blends containing fine, regular shaped powders undergo a higher shrinkage during binder removal thus increasing the effective volume fraction of ceramic. This further reduces the channels available for diffusion of degradation products and also decreases the diffusivity of the blend; the diffusivity decreases with increasing filler volume fraction (Barrer 1968).

Also two forms of binder could exist in these channels; immobile layers adsorbed onto the ceramic particles and "fluid" polymer which helps in the outward diffusion of degradation products. The proportion of binder adsorbed depends on the surface area of the ceramic particles. Therefore fine powders with a high surface area poses more difficulties during binder removal.

These factors explain why the cylinders containing the powders CT2000 and CT3000 with a finer particle size and a higher surface area showed bloating and cracking during binder removal. The unfavourable conditions resulting from the increase in particle packing caused by the factors described above impeded the outward diffusion of degradation products. On the other hand, the collective particle characteristics of A152 and MA250(2) described earlier were more favourable to binder removal. These observations are in conflict with the fact that fine powders are preferred (Barringer and Bowen 1982) for sintering which is the final phase of the ceramic injection moulding process. Therefore, it is essential to compromise on the particle characteristics to suit both the binder removal and the sintering stage and powders like A152 may provide the basis for the development of powders suitable for the ceramic injection moulding process.

4. CONCLUSIONS

The following factors which affect the binder removal stage should be taken into account when developing powders for the injection moulding process.

1. A higher amount of adsorbed water is present in finer powders which have a larger surface area. The removal of adsorbed water occurs at the beginning of the debinding schedule when temperatures are low and diffusion is difficult. This could lead to defects.
2. Although an irregular particle shape and a coarse particle size affected the mechanical degradation during compounding the ceramic-polymer blends, it did not influence the molecular weight distribution enough to alter the thermal degradation characteristics.
3. Only a large difference in specific surface area was effective in catalysing the thermal degradation of the binder system.
4. Shrinkage during binder removal was mainly dependent on particle size with fine powders giving rise to a higher value and therefore a larger increase in volume fraction of ceramic. An irregular particle shape helped in reducing shrinkage.
5. In addition to the fact that packing of powders with a fine particle size allows narrower channels for the outward diffusion of degradation products during binder removal, a larger surface area could help in the formation of adsorbed polymer layers. These layers could be immobile and could therefore hinder the diffusion of both binder degradation products and adsorbed water.
6. Although coarser powders like MA250(2) with a wide multimodal particle size distribution prove to be successful at the binder removal stage, they are not suitable for sintering which is the final stage of the ceramic injection moulding process. However, the encouraging results obtained from the binder removal experiments involving A152 which has a relatively narrow monomodal particle size distribution with a mean particle size of 1.1 µm, an equiaxed particle shape and a specific surface area of 3 m^2g^{-1} indicates that it could provide the basis for developing powders suitable for the fabrication of thick section ceramic components using injection moulding.

ACKNOWLEDGEMENTS

The help given by Mr J. Collier and Mr K. Salmon of BA Chemicals Ltd.

and Mr M. Chuck of Alcoa Manufacturing Ltd. is acknowledged. The authors also wish to thank Mrs Kathy Armstrong for typing the manuscript.

REFERENCES

Barrer R M Diffusion in Polymers Crank J and Park G S (Eds) (London: Academic Press) 1968 pp.165-217

Barringer E A and Bowen H K 1982 J. Amer. Ceram. Soc. 65 C199

Bernett and Zisman 1968 J. Coll. Interf. Sci. 28 243

Carlsson R 1989 Materials and Design 10 10

Cumberland D J and Crawford R J The Packing of Particles (Amsterdam: Elsevier) 1987 p.60

Edirisinghe M J 1990 Proc. Brit. Ceram. Soc. in press

Edirisinghe M J 1989 Euro-Ceramics de With G et al (Eds) Vol. 1 (Amsterdam: Elsevier) pp 285-291

Edirisinghe M J and Evans J R G 1989 Mater. Sci. and Eng. A109 17

Edirisinghe M J and Evans J R G 1987 Br. Ceram. Trans. J. 86 18

Edirisinghe M J and Evans J R G 1986a Int.J.High Tech.Ceram. 2 249

Edirisinghe M J and Evans J R G 1986b Proc. Brit. Ceram. Soc. 38 67

Kwei T K 1965 J. Polym.Sci. A3 3229

Masia S Calvert P D Rhine W E and Bowen H K 1989 J. Mater. Sci. 24 1907

McKenzie J K and Shuttleworth R 1949 Proc.Phys.Soc. (London) B62 833

Mutsuddy B C 1983 J. Ind. Res. Dev. 25 76

Poslinski A J Ryan M E Gupta R K Seshadri S G and Frechette F J 1988 J. of Rheol. 32 751

Shaw H Nogueira R E F Q and Edirisinghe M J 1990 Ceramic Powder Sci. III Messing G L (Ed) (USA: Amer. Ceram. Soc.) in press

Shukla V N 1989 J. Amer. Ceram. Soc. 72 1797

Woodthorpe J Edirisinghe M J and Evans J R G 1989 J. Mater. Sci. 24 1038

Wright J K Evans J R G and Edirisinghe M J 1989 J. Amer. Ceram. Soc. 72 1822

Paper presented at Int. Conf. on New Materials and their Applications, Univ. of Warwick, 1990

Application of x-ray microtomography to the study of SiC/Al metal matrix composite material

T M Breunig[1], J C Elliott[2], P Anderson[2], G Davis[2], S R Stock[1], A Guvenilir[1], and S D Dover[3]

[1] School of Materials Engineering and Mechanical Properties Research Laboratory, Georgia Institute of Technology, Atlanta, GA 30332-0245 USA;
[2] Department of Child Dental Health, The London Hospital Medical College, Turner Street, London E1 2AD, United Kingdom;
[3] Department of Biophysics, King's College, 26-29 Drury Lane, London WC2B 5RL, United Kingdom.

ABSTRACT

As the service requirements and performance demands on aircraft continue to increase, new materials will be required. One of the classes of new materials being developed is metal matrix composites. Before this class of materials can be safely utilized, a fundamental understanding of the damage evolution and failure mechanisms will be required. This study reports damage observations in a SiC/Al metal matrix composite using x-ray microtomography, a nondestructive technique. Interfacial shear damage and fibre fractures were observed with a spatial resolution of 25 μm.

1. INTRODUCTION

Utilization of metal matrix composite (MMC) materials in structural applications for high performance aircraft will require an understanding of the mechanical and thermal response over the anticipated lifetime of the structure. To evaluate the safe operational lifetime of a structure, an understanding of the overall material properties degradation must be established. The current study is a portion of a larger project to establish an understanding of the nucleation and growth of damage in MMCs. The project goals are to determine the damage initiation and accumulation mechanisms in a continuous fibre MMC, to correlate the damage with the associated structural response, and to determine the spatial resolution required to identify structural features of interest.

Unlike monolithic materials, damage in composites is frequently initiated internally. Failure modes which must be detected include matrix cracking, fibre-matrix debonding, and fibre fracture. Due to the large scale statistical variability of the macroscopic properties of composite components, it is very important to monitor the damage state throughout the lifetime of each specimen. In this manner, the structural response can be correlated with the internal damage state.

Current methods of damage state evaluation are based upon stiffness loss measurements and/or other nondestructive methods of measuring macroscopic properties (Johnson 1988). These techniques are relatively insensitive to many of the internal damage mechanisms which may be operative. Since the stiffness of a material is not a unique function of a given damage state, not all the components of the stress tensor may reflect the decreased stiffness. Since monitoring macroscopic properties response will not allow a quantification of damage present, an alternative method must be used.

This paper describes the use of the nondestructive testing technique, x-ray microtomography, to evaluate the damage accumulation in a SiC/Aluminum MMC. X-ray microtomography using laboratory radiation (Elliott *et al* 1982, Elliott *et al* 1984, Elliott *et al* 1990) and synchrotron radiation (Bowen *et al* 1986, Elliott *et al* 1987, Flannery *et al* 1987, Stock *et al* 1989, Kinney *et al* 1990) to obtain spatial resolutions of 10-20 μm has been under development for approximately 10 years. This nondestructive technique has been used successfully to image MMC structures using laboratory radiation (Elliott *et al* 90) and synchrotron radiation (Stock *et al* 89, Kinney *et al* 90). Due to recent positive results using x-ray microtomography to study composite materials, it is anticipated that this technique will be in wide use for nondestructive evaluation of high performance materials in the near future.

2. BACKGROUND

The two major components of a composite structure are the reinforcing and matrix materials. The reinforcing materials used take many different forms and shapes ranging from spheres and irregularly shaped particles to very long fibres and filaments. The function of the reinforcing material is to provide strength and stiffness to the structure. The matrix material provides dimensional stability, load transfer to the reinforcement and protection from external damage. Orientation of the reinforcing material allows the designer to provide strength and stiffness only in the loading direction(s), thereby reducing the overall structural weight.

The material system evaluated in this project is a unidirectional (UD) continuous fibre SiC/Al matrix material fabricated by AVCO Specialty Materials Division of Textron Corporation in 1986. The composite laminate contains 8 plies (*laminae*) of SCS-8 SiC fibres and a 6061-0 Aluminum matrix. Each ply was fabricated by a technique that uses a foil to support the fibre and a plasma sprayed matrix material to maintain fibre location and orientation. The SiC filaments were produced by a chemical vapour deposition technique (Chawla 1987, Nutt and Wawner 1985) in which a 33 μm diameter graphite fibre substrate was used. The final diameter of the SiC filaments was 142 μm.

The interface region between the fibre and matrix has been the focus of many investigations (Dhingra and Fishman 1986, Lee *et al* 1988). The secondary phases that may form at the interface can have a dramatic effect on the observed properties of a MMC. The strength of this region can dictate the dominant damage mode(s). In general, a strong interface does not allow debonding to occur. If a crack forms at, or approaches a strong interface, the notch formed will cause fibre fracture. If the interface is weak and the applied shear stress exceeds the interfacial shear strength, debonding will occur with no loss in fibre strength. The interface in SiC/Al is usually classified as weak, and the latter mechanism is expected to dominate.

The need to nondestructively evaluate damage and flaws present in a structure has increased, and will continue to do so, as higher performance demands are required. Conventional computed tomography (CT) and its high resolution variant, x-ray microtomography, are closely related to the medical computed axial tomography (CAT). These techniques can be very useful for evaluating damaged structures in a nondestructive manner.

Computed tomography records the projected image from a series of line projections of a cross-section of the object at different angular orientations of the object (figure 1). These projections are then combined via a reconstruction algorithm (Kak and Slaney 1987) to produce a map of the x-ray absorption in the cross-section. Figure 1 illustrates how the absorption of a feature can be imaged from various orientations to obtain information about the size, shape and location of the feature. In one projection, the lightly absorbing rectangle forms a narrow profile of very low absorption. In the other projection the absorption profile is wider, and the absorption is greater because the x-rays pass through a longer length of highly absorbing material.

The scheme used to collect the data in the present investigation is illustrated in Figure 2. This apparatus uses a micro-focus x-ray source with a pin hole collimator and a single detector. The projections are recorded by translation of the specimen through the beam and successive specimen rotations.

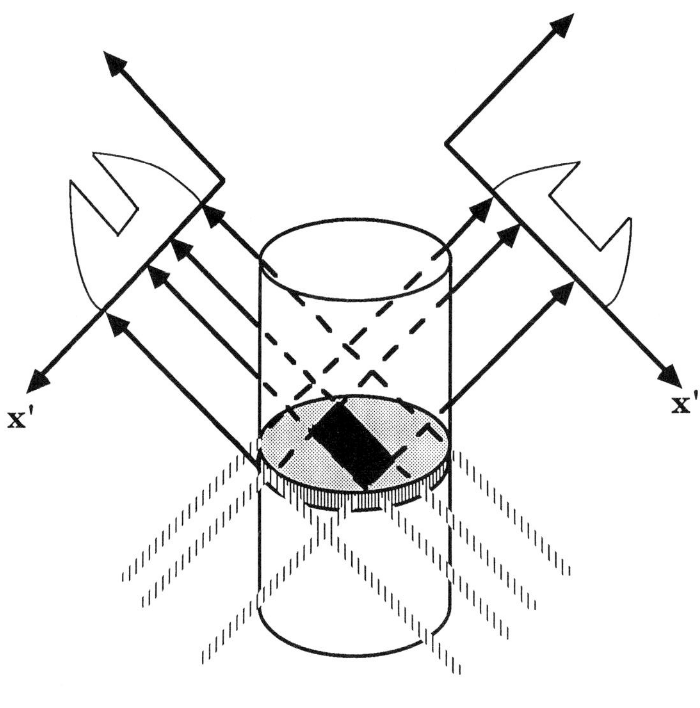

Figure 1 Illustration of the way in which information from projections of the cross-section of an object can give details about its internal structure.

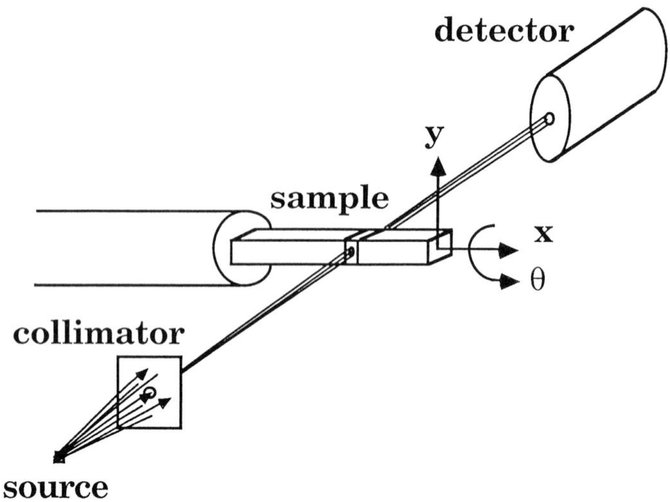

Figure 2 Method of data collection used in the present study.

3. EXPERIMENTAL PROCEDURE

The object of this study was to look for and resolve fibre breaks and interfacial shear in SiC/Al. The deformation mode to be evaluated is three-point bending of a notched sample. This deformation type was selected since the damage produced should be localized to the notched region of the sample. The three-point bend fixture has a 30 μm span with loading at the mid-point. The test specimen used was 1.5 x 1.5 x 40 mm with the fibre parallel to the longitudinal (40 mm) axis. A 0.6 mm deep notch was cut into the centre of the specimen with a diamond wafering blade. The notch orientation was perpendicular to the plies in a attempt to simulate a "through the thickness" flaw in a rib or stiffener under bending conditions. This flaw type is also similar to a fastener hole in a structure. A series of bend specimens were deformed to various levels. The sample deformation used for this paper was very large to highlight the fibre breakage, pull-out and other shear damage present.

The x-ray generator was operated at 2.0 mA and 40 kV with a silver target. A 60 μm thick palladium filter was positioned between the target and a 10 μm platinum aperture. The filter reduced the white and AgK_β (24.9 keV) radiations compared to the AgK_α (22.1 keV) radiation because its absorption edge is at 24.4 keV.

The specimen was placed immediately behind the 10 μm aperture on the kinematically designed rotation and translation stage (Elliott *et al* 1987). Each projection was comprised of 128 points separated by 25 μm. 251 projections at 1.434° angular intervals were measured and reconstructed onto a 256 x 256 grid by a filtered back projection algorithm that did not incorporate any smoothing. The time taken to record data for each section was about 15 hours. The data collection for figure 3 section 1 used 501 projections at an angular interval of 0.717°. A more detailed description of the above system can be found in reference (Elliott *et al* 1990).

4. RESULTS AND DISCUSSION

Eleven tomographic sections (figures 3 sections 1-11), each separated by 100 μm, were collected. In the figures, the ply orientation is horizontal with the tensile surface on the left side and the compressive (loading) surface on the right side of the image. The tomographic sections start 100 μm outside of the notch region and continue 800 μm into the notch damage region.

The features in the sections can be identified by their respective linear absorption coefficients. The calculated values for SiC and Al the matrix are 7.7 cm^{-1} and 7.1 cm^{-1} (McMaster *et al* 1969), respectively. The observed ranges for SiC and Al were 7.4 to 8.1 cm^{-1} and 6.8 to 7.6 cm^{-1}, respectively. The large range observed for the SiC was due to the radial composition variations reported for the fibres (Nutt and Wawner 1985). The Al variation was due to the composition range allowed for the matrix alloy 6061.

Section 1 (figure 3) shows some indications of interfacial shear damage. The cause of this damage appears to be the deformation process since this type of damage does not occur on other cut surfaces. The low density regions on the fibre surfaces in rows 1, 2, 3 and 5 on fibres 6, 7, 6 and 7 respectively indicate separation of the fibre and matrix material. (The starting point for numbering the rows and fibres is the upper right hand corner. The starting place on rows 2 and 7 are the half fibres.)

In section 2 (figure 3), the same fibres as in section 1 indicate signs of interfacial shear damage. Fibre 7 in row 7 and fibre 6 in row 8 also have low density interfacial shear regions present. This section can be identified as the beginning of the notch region with the section thickness overlaping the true cut edge because the lower density region on the left side of the section indicates that the voxels are only partially occupied by the composite material. In section 3, the location of the notch edge is well defined.

Most of the damage induced by the three-point bending deformation was confined to the next 700 μm of the specimen. Within this region, the damage features can be followed from section to section. The deformation induced in this sample was large and accentuates the fibre fractures and pull-out regions. An interesting feature to note is that the fibre-matrix shear zone extends approximately 200 μm to 300 μm beyond both ends of the fractured fibres. Concentrating on the fibres in row 8, a typical example of the fibre-matrix shear region can be seen starting in section 11. Moving into section 10, the sheared regions have progressed along the fibres, and have increased in size and number. In section 9, several of the fibres have been pulled out of the matrix. The average separation of fractured fibres was 400 μm to 500 μm. An example of this feature is fibre 7, row 2 in sections 4-9.

Matrix plastic deformation appears to occur after fibre fracture and pull-out. In section 9 (figure 3), fibres 5 and 6 in row 1 have pulled out and in that region the matrix has plastically deformed towards the gap left by the fibre. This type of matrix deformation can also be seen in sections 5-10 (figures 3).

The fibre fracture surface does not appear to be planar because the fibres 5 and 6 in row 1 are partially missing in section 11 (figure 3).

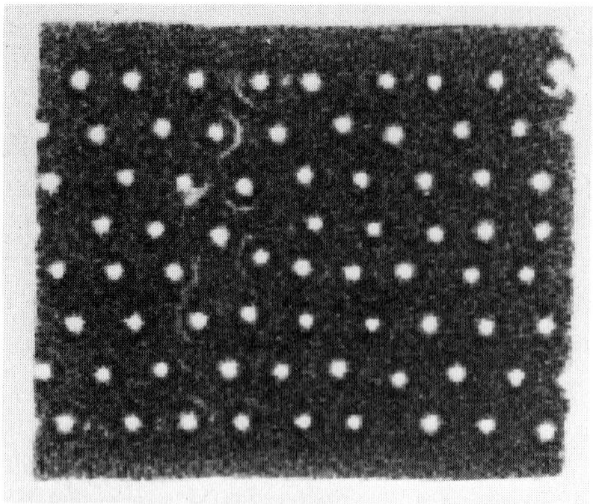

Section 1) 100 μm before notch.

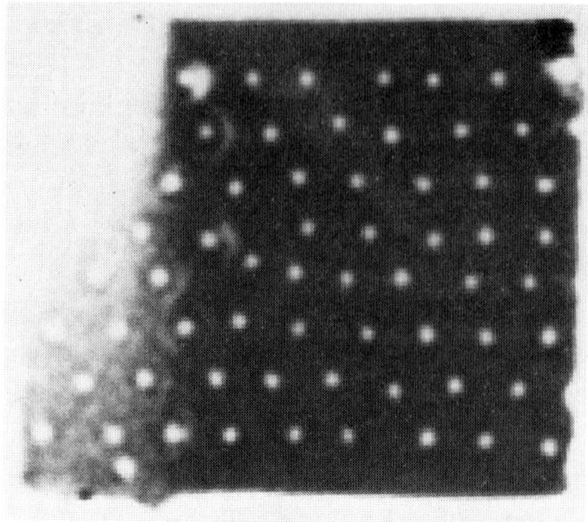

Section 2) At the edge of the notch.

Section 3) 100 μm

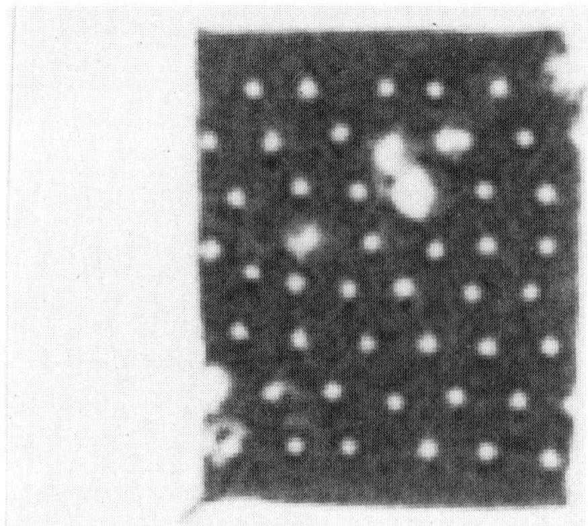

Section 4) 200 μm

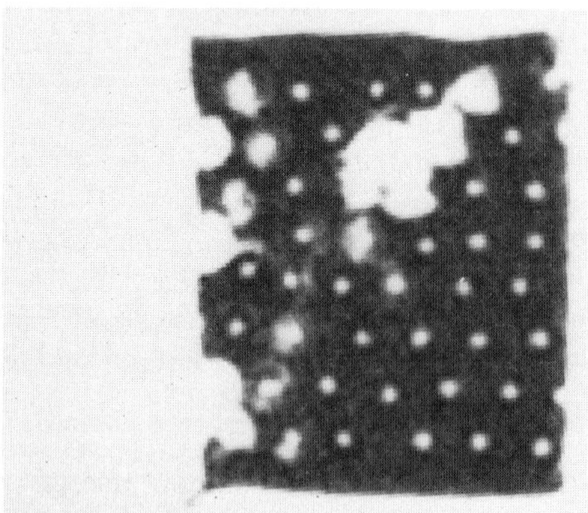

Section 5) 300 μm

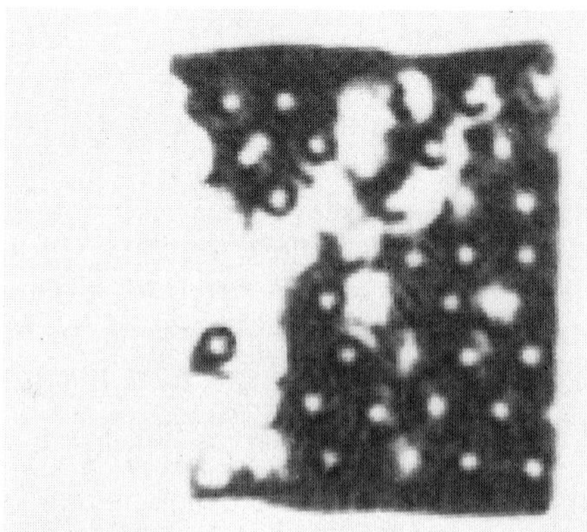

Section 6) 400 μm

Figure 3 (sections 1-6) Tomographic reconstruction of sections taken at 100 μm intervals starting 100 μm before the notch. The distance from the notch is given below each photograph. The vertical thickness is 1.5 mm in all reconstructed sections.

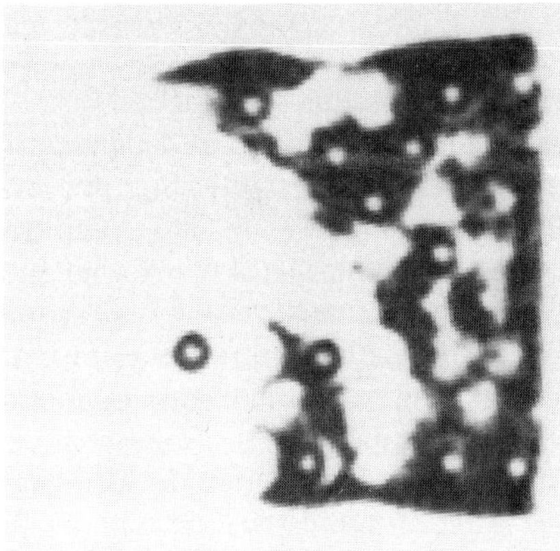

Section 7) 500 μm

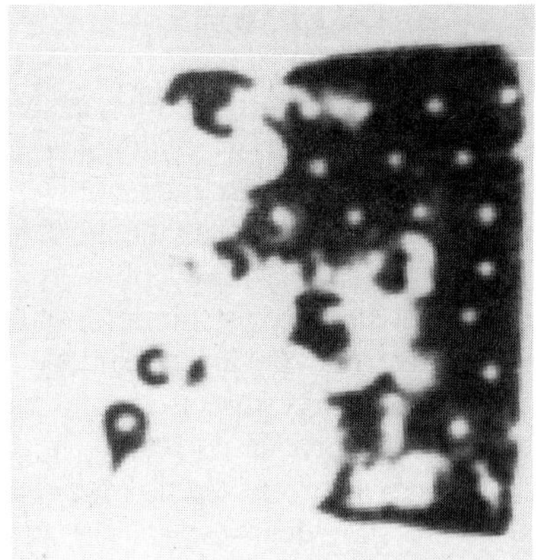

Section 8) 600 μm

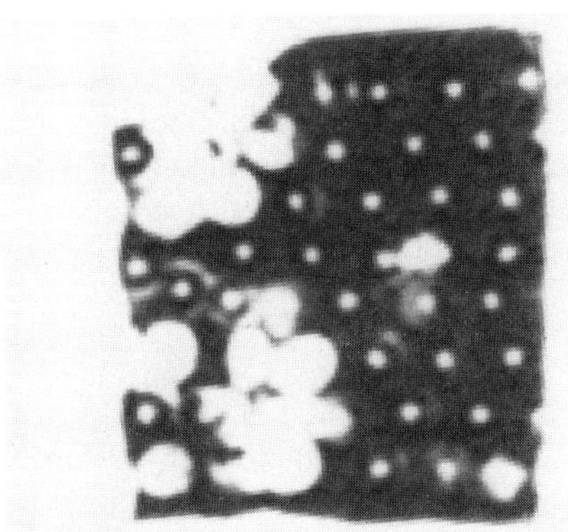

Section 9) 700 μm

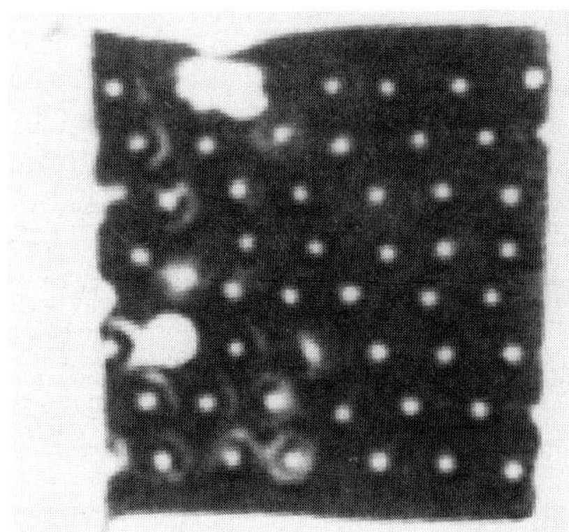

Section 10) 800 μm

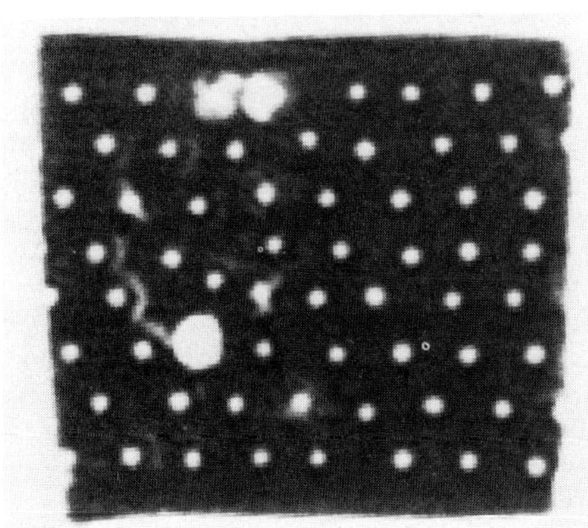

Section 11) 900 μm

Figure 3 *cont.* (sections 7-11) Tomographic reconstruction of sections taken at 100 μm intervals starting 100 μm before the notch. The distance from the notch is given below each photograph.

5. CONCLUSIONS

The 25 μm spatial resolution obtained with the laboratory microtomography apparatus used in this study was adequate for imaging the deformation induced damage in the SiC/Al MMC. The measured ranges of linear absorption coefficients are in good agreement with the calculated values for SiC and Aluminum alloy 6061. The fracture process was initiated by fibre fracture, propagated as fibre-matrix interfacial shearing and finally matrix plastic deformation. With the present spatial resolution it is possible to nondestructively image damage initiation and accumulation in a virtually any material. Other exciting prospects for the future include studying structural changes under *in situ* conditions, such as corrosion, fatigue and elevated temperatures. Another possibility is evaluation of electronic integrated circuit chips where the internal devices and track sizes are ever decreasing. The decreasing size scale of interest will be a major driving force behind development of equipment capable of 1 μm spatial resolution.

6. ACKNOWLEDGEMENTS

This work was supported by the Medical Research Council, Special Trustees of The London Hospital, National Science Foundation (through grants MSM-8614493 and INT-8513629), and Lockheed Aeronautical Systems Company. We would also like to thank S.D. Antolovich for valued discussions on deformation of composite materials.

REFERENCES

Bowen D K, Elliott J C, Stock S R and Dover S D 1986 SPIE X-ray Imaging II, **691** 94-8

Chawla K K 1987 *Composite Materials*, (New York: Springer-Verlag, Inc.) pp 41-4

Dhingra A K and Fishman S G eds 1986 *Interfaces In Metal Matrix Composites* (TMS AIME/ASM Metal Composites Committee)

Elliott J C and Dover S D 1982 J. Microsc. **126** 211-3

Elliott J C and Dover S D 1984 Metabolic Bone Disease Rel. Res. **5** 219-21

Elliott J C, Bowen D K, Dover S D and Davies S T 1987 Biological Trace Element Res. **13** 219-27

Elliott J C, Anderson P, Davis G, Dover S D, Stock S R, Breunig T M, Guvenilir A and Antolovich S D 1990 Submitted to J. X-ray Sci. Tech.

Flannery B P, Deckman H W, Roberge W G and D'Amico K L 1987 Science **237** 1439- 44

Johnson W S 1988 National Aeronautics and Space Administration Technical Memorandum 100628

Kak A C and Slaney M 1987 *Principles of Computerized Tomographic Imaging* (New York: IEEE Press)

Kinney J H, Stock S R, Nichols M C, Bonse U, Breunig T M, Saroyan R A, Nussardt R, Johnson Q C, Bush F and Antolovich S D 1990 J. Mater. Res. **5** (in press)

Lee D-J, Vaudin M D and Handwerker C A 1988 *High Temperature/High Performance Composites* Mater. Res. Soc. Symp. Proc. **120** 357-65

McMaster W H, Kerr Del Grande N, Mallett J H and Hubbell J H 1969 *Compilation of X-ray Cross Sections* (Lawrence Radiation Laboratory Report UCRL-50174 Rev. 1, University of California, Livermore, CA, USA)

Nutt S R, and Wawner F E 1985 J. Mater. Sci., **20** 1953-60

Stock S R, Kinney J H, Breunig T M, Bonse U, Antolovich S D, Johnson Q C and Nichols M C 1989 *Synchrotron Radiation in Materials Engineering* Mater. Res. Soc. Symp. Proc. **143** 273-8

Inst. Phys. Conf. Ser. No 111
Paper presented at Int. Conf. on New Materials and their Applications, Univ. of Warwick, 1990

Use of dynamic diffraction methods to study the synthesis of materials under various conditions

P.Barnes, D.Hausermann, & S.E.Tarling

Industrial Materials Group, Dept.Crystallography, Birkbeck College, (University of London), Malet Street, London WC1E 7HX, U.K.

ABSTRACT: Advances in detector technology and radiation sources have had an impact on the design of in-situ dynamic diffraction experiments. These advances have been exploited in dynamic studies on both novel and traditional materials being subjected to a range of environmental conditions. The environmental control has been geared towards mimicking conditions experienced in both solid state and hydrothermal synthesis processes. Examples are taken from the fields of ceramics and zeolites; the techniques used involve laboratory, synchrotron and neutron sources of radiation.

1. INTRODUCTION

The study of the synthesis of materials, using diffraction methods, raises a number of challenging problems that are associated with

a) providing the required environment for synthesis while sustaining diffraction from the materials under study;
b) ensuring that diffraction correctly probes the materials as they would be formed under full scale production;
c) capturing diffraction data at rates that are compatible with the kinetics of the synthesis processes being studied.

These three requirements are far from trivial. Firstly the conditions to be imposed during sythesis cover a demanding range of temperature, pressure and chemical environment that are not always compatible with the need to provide diffracting conditions. A manifestation of the second requirement is the disparity between the typical dimensions of scale (e.g. centimetres to metres) utilised in industrial operations and the limited penetration of radiation in diffraction experiments (e.g. 0.1mm to millimetres). Thus surface effects, that are not typical of bulk processes, must be avoided. The third problem is clearly most serious when rapid changes are involved.

Considerable progress has been made recently in all these three areas. This paper presents examples, covering a wide range of material and synthesis conditions, of how the problems can be minimised by correct choice and design of diffraction experiment. Three complementary dynamic diffraction systems have been developed to meet these needs:

2. A LABORATORY-BASED DYNAMIC POWDER DIFFRACTOMETER SYSTEM

A laboratory-based diffractometer equipped with a suitable position sensitive detector is capable of collecting powder diffraction patterns in 3-10 minutes (Mamott et al 1989; Anwar et al 1990). By surrounding the specimen region with a high temperature environmental chamber and with careful positioning of scattered radiation shields, dynamic diffraction patterns can be collected at high temperature with adequate peak:background ratios and angular resolution ($\Delta 2\theta \sim 0.1°$). An example is given in Figure 1a: the time-resolved patterns depict the calcination of zirconium hydroxide to tetragonal zirconia; the rates of transformation can be extracted from such data and studied as a function of hydroxide preparative conditions (Figure 1b).

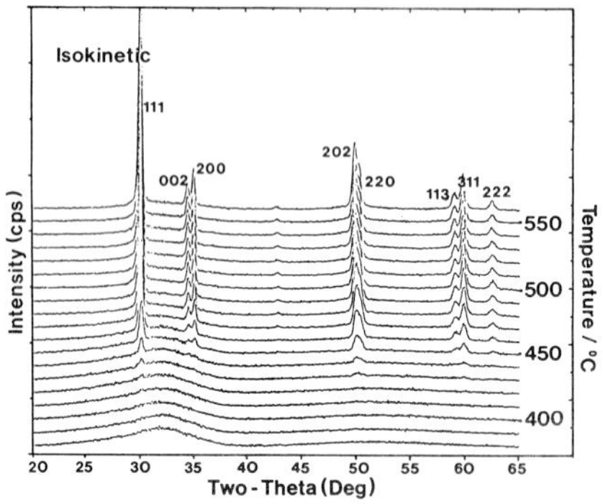

Fig.1a: Example of time-resolved diffraction patterns showing the calcination of amorphous $Zr(OH)_4$ to tetragonal zirconia: patterns were collected in 6 min. and are spaced 10 min. and 10°C apart in time and temperature.

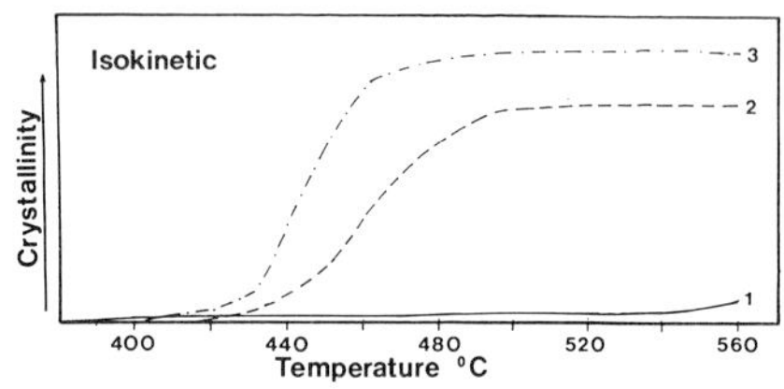

Fig.1b: Rates of crystallisation of tetragonal zirconia during calcination of $Zr(OH)_4$, for 3 different pH values used in the preparation. The data were obtained from time-resolved plots such as in Fig.1a (Mamott et al 1989 & 1990). A wide range of crystallisation behaviour is clearly evident.

3. SYNCHROTRON-BASED ENERGY-DISPERSIVE DIFFRACTION

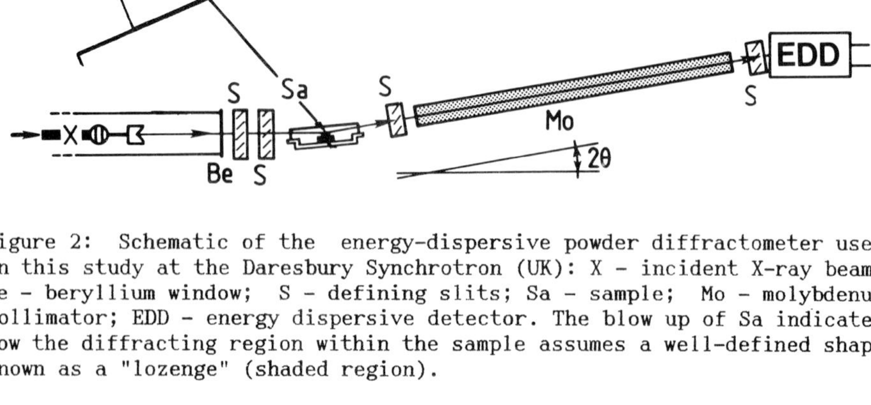

Figure 2: Schematic of the energy-dispersive powder diffractometer used in this study at the Daresbury Synchrotron (UK): X - incident X-ray beam; Be - beryllium window; S - defining slits; Sa - sample; Mo - molybdenum collimator; EDD - energy dispersive detector. The blow up of Sa indicates how the diffracting region within the sample assumes a well-defined shape known as a "lozenge" (shaded region).

The synchrotron provides a broad continuous spectrum of X-rays with energies up to 30keV in the case of 1.2 Tesla dipole radiation, or up to 70 keV (wavelength of ~0.18 Å) in the case of 5 Tesla Wiggler radiation. This continuous radiation can be exploited (Anwar et al 1990) by using the technique of energy-dispersive diffraction (EDD) as illustrated in Figure 1. With EDD the reflections are collected simultaneously at a fixed scattering angle 2θ which is optimised for each experiment with factors such as resolution, absorption etc. in mind. The diffraction patterns, collected using an energy dispersive detector, can be described by Bragg's law conveniently rewritten in energy-space as $hc/E = 2d.\sin\theta$. The main disadvantage of EDD is that it is a low resolution technique: in the case of synchrotron EDD, the resolution is detector limited to about 1% and the limit on the working count rate is from 10-40,000 c.p.s. depending on required accuracy. The over-riding advantages are those of rapid (sub-second) data capture rates and the relative ease, with a fixed geometry system, of building environmental chambers (Anwar et al 1990).

The field of zirconia-ceramics provides another appropriate example, this time of the tetragonal→monoclicic transformation during the cooling stage of the synthesis cycle: Figure 3a displays a series of energy-dispersive diffraction patterns which reveal both the calcination stage discussed earlier <u>and</u> the subsequent tetragonal to monoclinic transformation which occurs during the fast cool from 1,000°C. The superior time resolution of the energy dispersive method becomes crucial when kinetic data are required at high cooling rates: the degree of tetragonal to monoclinic conversion can be extracted from these diffraction data and plotted against temperature (Figure 3b) thus providing quantitative correlations on the dynamic aspects of synthesis (Mamott et al 1990).

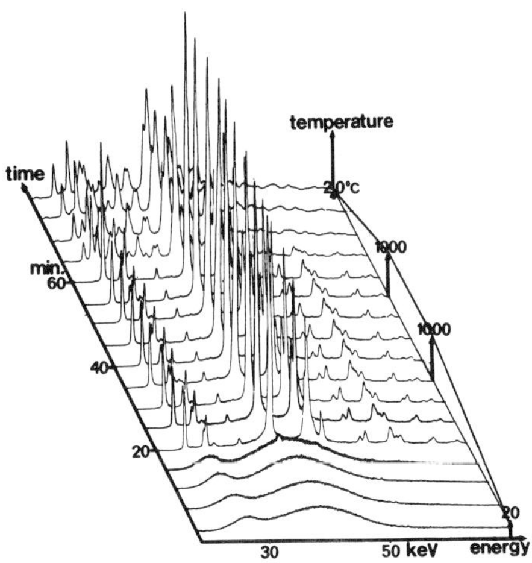

Figure 3a: Time-resolved EDD patterns showing the calcination of hydroxide up to 1,000°C; holding 1,000°C; then subsequent conversion on cooling to the monoclinic form.

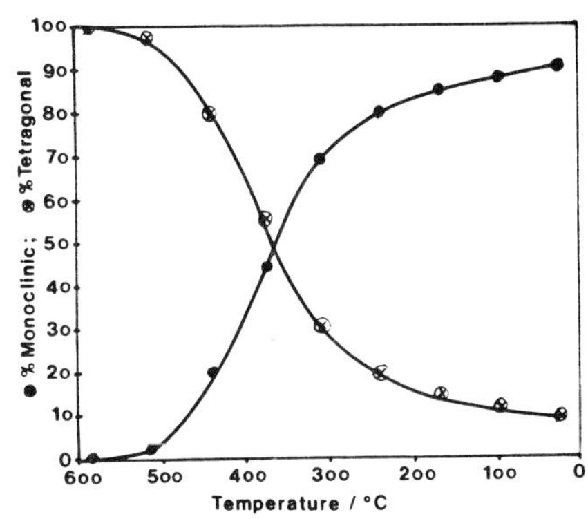

Figure 3b: Quantitative analysis obtained from the data (3a) on the cooling period showing percentage of tetragonal and monoclinic form as a function of degree of cooling.

The field of cement hydration provides an opportunity to demonstrate the versatility in specimen chamber design that is possible with a fixed geometry energy-dispersive diffractometer. The essential problems to be considered in this case are that:

a) the important reactions occur within the bulk: for example, surface effects need to be avoided;
b) the setting process ideally requires a sealed atmosphere;
c) good grain sampling statistics are required;
d) the setting process can induce movement in the specimen bulk.

The well defined parallel beam geometry of the ED-diffractometer permits a precise positioning of the diffracting "lozenge" (Figure 2) within the specimen (Hausermann and Barnes 1990). The size and position of this lozenge can be chosen according to needs, for example so that diffraction from the sample or chamber window can be avoided, or so that the sampling volume is maximised. In cases where significant specimen movement occurs, for example in first order transformations or reactions involving a volume change, there are two solutions involving either transmission or reflection geometries. For the transmission case, the lozenge is arranged to be long and narrow using a small scattering angle. On spinning the sample this lozenge explores a shallow and extensive discus-like region which, if positioned near the centre of the sample, ensures continuity of sample during movements of its extremeties. This method has been achieved using a heated spinning specimen cell shown in Figure 4a, and illustrated in Figure 4b for the case of a fast cement hydration sequence to the hydrate "ettringite" (Noor et al 1989).

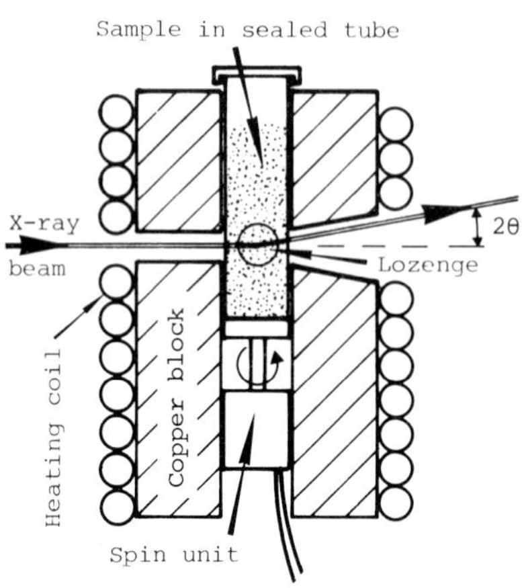

Figure 4a: Heated hydration cell (0 - 100°C) with sample spin for dynamic EDD. The position of the cell is optimised so that the lozenge is entirely inside the bulk of the sample.

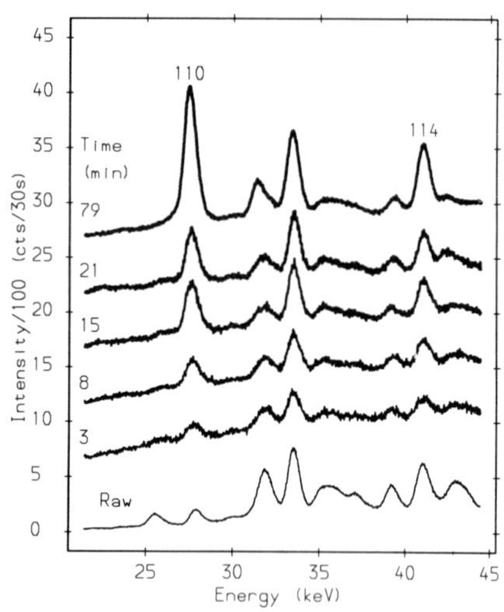

Figure 4b: Collection of complex energy - dispersive patterns for a hydration sequence involving high alumina cements: the labelled peaks growing with time are due to the ettringite phase.

The second method uses reflection geometry and is particularly useful if bulky or highly (X-ray) absorbing samples are involved. This method is based on the so called "Hart and Parrish" geometry (Parrish et al 1986) in which a stack of multiple slits is used as the diffracted beam collimation system to create a set of parallel lozenges (Fig.5a). The resulting multiple lozenge must be positioned near the specimen surface so that any surface movement, by shrinkage or expansion, merely causes different lozenges to become active and hence compensate for those which become inactive. An example illustrating this method is shown in Fig.5b.

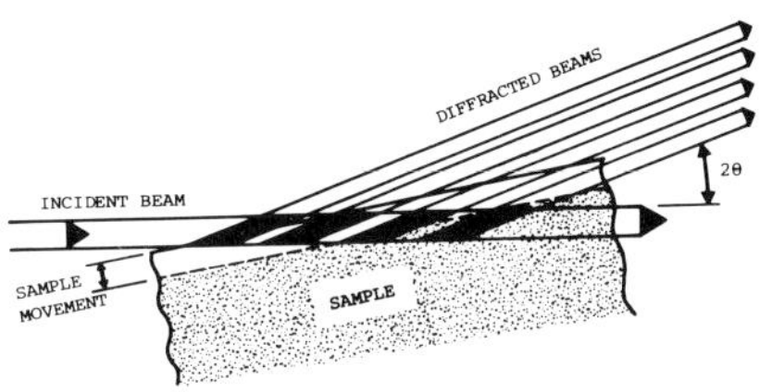

Figure 5a: Schematic diagram which illustrates the principle of the multiple slit system: the result is a stack of parallel lozenges (only 4 out of 25 are shown for clarity) which ensures that any sample movements are accomodated by this arrangement.

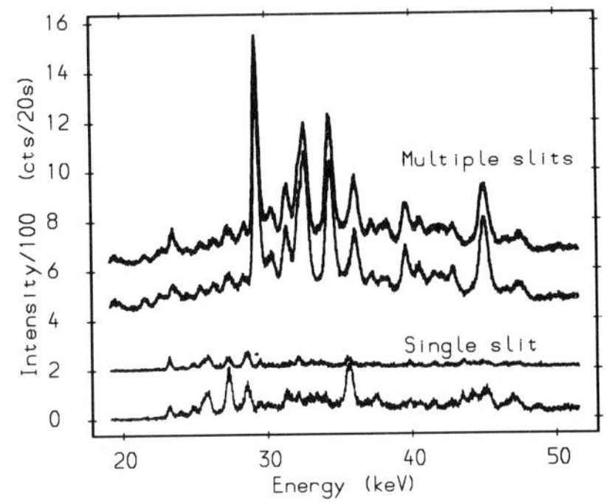

Figure 5b: The effect of sample movement. Lower 2 patterns - single slit system: shrinkage of sample in first 3 minutes of cement hydration leads to loss of pattern; upper 2 patterns - multiple slit system: a similar sequence shows that sample movement is accomodated.

4. TIME-RESOLVED NEUTRON DIFFRACTION

The scale of neutron diffractometer systems is well suited to the design of ambitious environmental systems such as the autoclave cell (Fig. 6a).

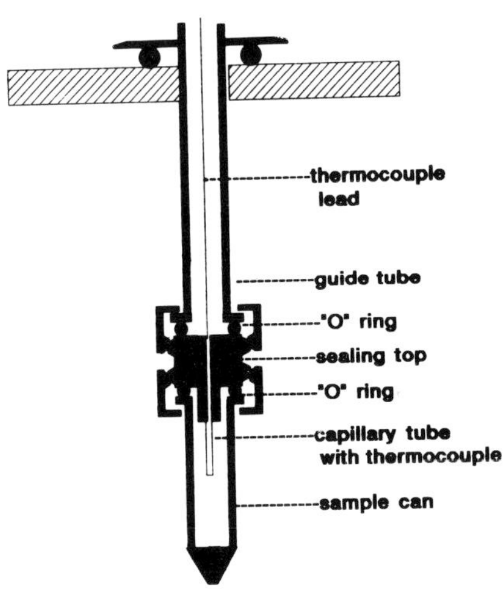

Figure 6a: Schematic of the special autoclave cell for use with neutron sources. Both 6a & 6b are reproduced with the permission of J.Appl.Cryst.

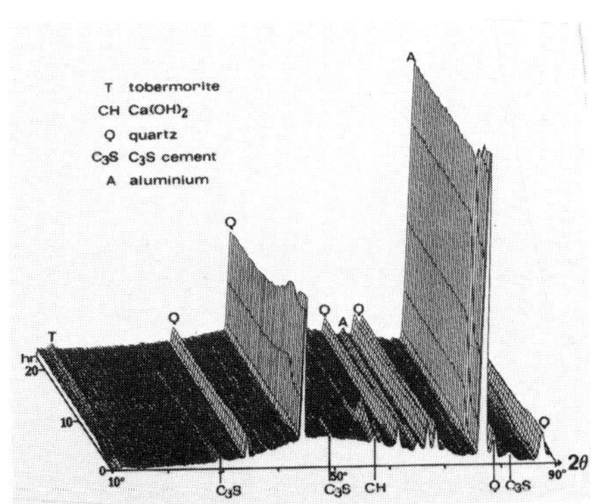

Figure 6b: Time-resolved patterns showing time/temperature sequence of events during the autoclave - curing of cement. Above ~185°C, the rate of $Ca(OH)_2$ and silica consumption increases followed by the formation of "tobermorite".

Also the high penetration of neutrons through machinable metals makes it somewhat easier to build cells that can operate under combined high temperature and pressure. A novel use for such cells has been found in the study of hydration of aerated cement under autoclave conditions: that is, with D_2O-steam pressurised to 10 bars at ~185°C (Polak et al 1990). The time-resolved patterns obtained (Figure 6b) clearly identify the rate-limiting steps in the synthesis: that is initial build up then decay of $Ca(OH)_2$; increase in dissolution of silica; then sigmoidal growth of the calcium silicate hydrate "tobermorite" above 178°C.

The autoclave cells have also been modified to study the kinetics of zeolite synthesis via hydrothermal routes (Polak et al 1990). However to study the solid state synthesis of the zeolitic pigment ultramarine, it was necessary to devise a quartz tube cell arrangement (Barnes et al 1988, 1990) that could withstand the temperature (800°C) and sulphurous environment involved: the resulting synthsis is shown in Figure 7.

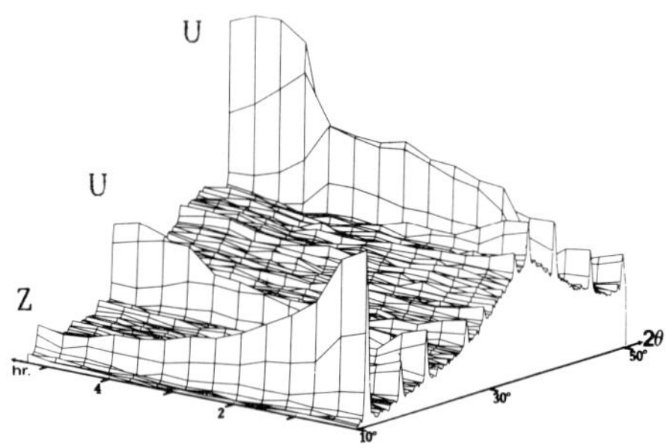

Figure 7: Time-resolved neutron diffraction sequence showing solid state synthesis of ultramarine from Na-sodalite-A between 20 and 800°C. The decrease in zeolite-A (Z) and rise in utramarine (U) is evident.

A comparison between the various synthesis pathways is particularly interesting in terms of Al,Si - order/disorder. It has now been demonstrated (Klinowski et al 1987, Tarling et al. 1988) that Al,Si-site ordering in these zeolites depends on the synthesis route chosen so that both extremes - from Loewenstein (1953) order to disorder - can be produced. Rietveld (neutron) refinement and magic angle spinning NMR have been the key techniques used in elucidating these important differences.

5. ACKNOWLEDGEMENTS

London University Environmental Powder Diffraction Service; SERC for synchrotron and neutron (ILL) beam time; ALCAN Chemicals Ltd., Aerated Concrete Ltd., and Reckitt's Colours Ltd.

6. REFERENCES

Anwar J, Barnes P, Clark S M, Dooryhee E, Hausermann D, Tarling S E 1990 J.Mat.Sci.Lett.25 436.

Barnes P, Tarling S E, Clark S M, and Polak E 1988 ILL Exp.Rep. & Theor. College Activities, Inst. von Laue-Langevin rep., Grenoble.

Barnes P 1990 Nondestr.Test.Eval.5, in press.

Hausermann D and Barnes P 1990 in preparation.

Klinowski J, Carr S W, Tarling S E and Barnes P 1987 Nature 330 56.

Loewenstein W 1953 Amer.Min.39 92.

Mamott G T, Barnes P, Tarling S E, Jones S L and Norman C J 1989 Powder Diffraction 3 234; 1990 subm. to J.Mat.Sci.

Noor M M and Fentiman C H, 1990 in preparation.

Parrish W, Hart M, Erickson C J, Masciocchi N and Huang T C, 1986 Adv. X-ray Anal.29, 243.

Polak E, Munn J, Barnes P, Tarling S E, and Ritter C, 1990 J.Appl.Cryst. in press.

Tarling S E, Barnes P and Klinowski J 1988 Acta Cryst.B44, 128.

Paper presented at Int. Conf. on New Materials and their Applications, Univ. of Warwick, 1990

Ultrasonic characterization of oriented roll-drawn polypropylene using shear wave birefringence

R Mloszewski, D A Hutchins and J Franceur*

Department of Physics, Queen's University, Kingston, Ontario, Canada K7L 3N6, and Department of Engineering, University of Warwick, Coventry CV4 7AL, England.
* Alcan International Ltd, Kingston, Ontario, Canada.

ABSTRACT: Ultrasonic shear waves have been used to estimate the degree of anisotropy present within roll-drawn polypropylene samples, using shear wave birefringence. These measurements have been correlated with the draw ratio used during manufacture. It is shown that the technique has promise as a monitoring tool for this material.

1. INTRODUCTION

Oriented polymers are being investigated increasingly as possible materials for structural applications. Of interest here is the increase in elastic modulus and tensile strength as a result of drawing. This process induces a degree of anisotropy into the structure of the previously isotropic material, giving it increased strength in the direction of drawing (Samuels 1974). In this project, ultrasonic techniques were studied for the characterisation of such materials in terms of the degree of anisotropy present and the effect on strength. There did not seem to be a technique available that could quickly and accurately give a nondestructive measurement of the degree of anisotropy present, which is often not known for a particular material if the production parameters are not strictly controlled. An ultrasonic method has thus been investigated, which uses shear wave propogation as a measure of anisotropy. Although other methods of characterization exist, they usually involve sectioning of the material and are not applicable directly to an industrsial environment. Ultrasonic methods are thus of advantage, and have been used in several previous studies (eg Aindow, Markham, Puttuck, Ruder and Rudman, 1986 and Markham 1970) to investigate the properties of fibre-reinforced polymer composites. In the present work, the technique has been applied to drawn polypropylene, to investigate changes in structure.

2. ROLL DRAWING OF POLYPROPYLENE

Roll drawing is a process whereby a billet of polymer is drawn through a set of rollers at different speeds, causing a simultaneous extension in the drawing process and a decrease in the material thickness. This process is illustrated in Fig. 1. It is common practice to estimate the amount of deformation that has been produced by using the Draw Ratio, which is defined in Fig. 2 as the ratio of the initial and final cross-sectional areas of the polymer sample. This draw ratio gives an indication of the amount of mechanical deformation, but only gives an approximate estimate of the degree of induced anisotropy.

The physical processes that occur when a sample of polypropylene is drawn are complex. The undeformed material is generally thought to consist of an assembly of spherulites. Each spherulite is composed of crystalline lamellae which grow radially outward in a

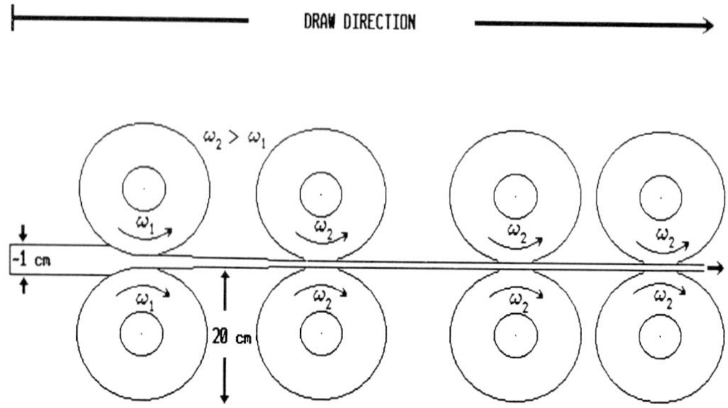

Fig. 1 Schematic diagram of the roll drawing process.

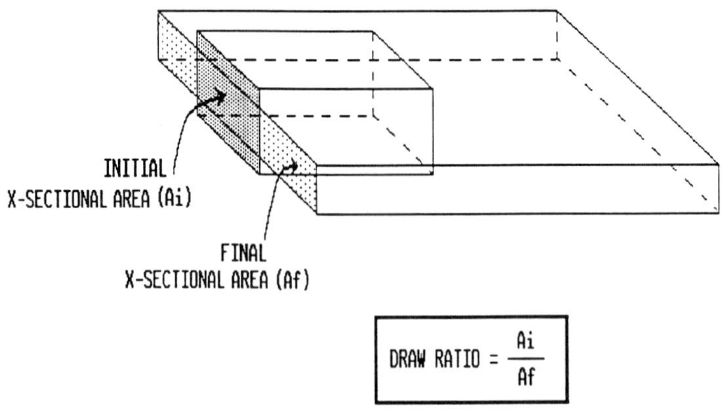

Fig. 2 Definition of draw ratio.

helical structure. As the polymer is drawn, the spherulites deform at constant volume, and become elongated. This is followed at higher draw ratios by disintegration of the spherulites into their composite lamellae, which then themselves tend to align in the direction of the applied force. Known as a microfibrillar structure, it exhibits anisotropy and an increase in strength in the drawing direction. This is the property of interest in structural applications.

3. ULTRASONIC SHEAR WAVE MEASUREMENTS

In a bulk isotropic solid material, there are in general two types of elastic wave that can exist: the longitudinal and shear (or transverse) waves. These waves travel at different speeds, with the longitudinal having the faster velocity. If anisotropy is present, then the shear wave splits into two orthogonally polarized components ("quasi-shear waves"), and the longitudinal component becomes the "quasi-longitudinal wave". The displacement vector polarization of the two shear modes allows anisotropy to be estimated from them, as their velocities are different. This phenomenon is known as SHEAR WAVE BIREFRINGENCE, and is analogous to the equivalent optical birefringence.

Experimentally, the two shear wave components can be generated and detected with a polarized shear wave transducer, as shown in Fig. 3. This is placed on the top surface of the polymer sheet, with its polarized direction at some angle θ to the drawing direction. The shear excitation is naturally resolved by the material into the two quasi-shear components as shown. In general, the directions of polarization are parallel and perpendicular to the anisotropy, with the former being the faster.

The excitation of the shear wave transducer with a short high voltage pulse causes the two shear modes to travel as separate transients through the thickness of the material. These can be resolved in time if the two transients are at a high enough frequency. In our case, we have used a 20 MHz Panametrics transducer, fitted with a quartz delay line, and excited using a Panametrics pulser/receiver. The waves travelled through the material, and reflected back to the transducer which then detected them in the "pulse-echo" mode. An example of a typical received signal in roll-drawn polypropylene is shown in Fig. 4. The "fast" and "slow" shear signals are visible, polarized parallel and perpendicular to the drawing direction, separated by some time difference Δt. It is this time difference that is used here to estimate the shear wave birefringence between the two shear modes, and hence the anisotopy in polypropylene caused by the drawing process.

4. RESULTS AND DISCUSSION

The shear birefringence has been measured for a set of roll-drawn polypropylene samples, manufactured under carefully-controlled conditions with known draw ratios. The aim was to establish a correlation between acoustic birefringence and both draw ratio and the strength of the resultant material.

The birefringence is plotted against draw ratio in Fig. 5. It is evident that up to a draw ratio of 10, the shear wave birefringence increases with draw ratio. Beyond this value, the birefringence data seemed to settle towards a constant value. This indicates that there is little change in anisotropy beyond this value of draw ratio. Using this data, a model is proposed of the change in structure of the polypropylene sample with increased draw ratio. With reference to Fig. 6, the data can be interpreted using a layered structure, where the roll-drawing causes the outer skin of the sample to be of a microfibrillar structure. The microfibrillar skin is separated from the sherulitic core by a transition zone. From detailed measurements of the birefringence data, it is postulated that the microfibrillar skin and the transition zone do not penetrate more than 0.8mm from either surface. The skin is thought to be formed to its final thickness at low draw ratios (<3). At higher draw ratios, more of the inner unoriented core is incorporated into the skin, with the sheet becoming thinner by depletion of the core.

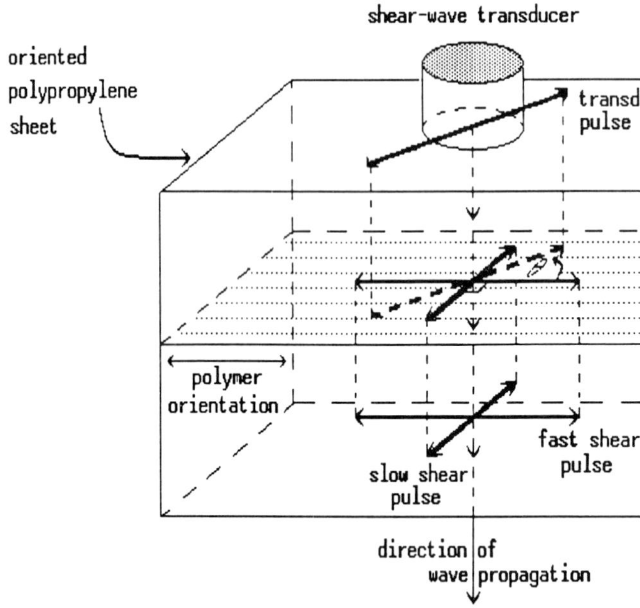

Fig. 3 Use of a polarized shear wave transducer for generating fast and slow shear waves.

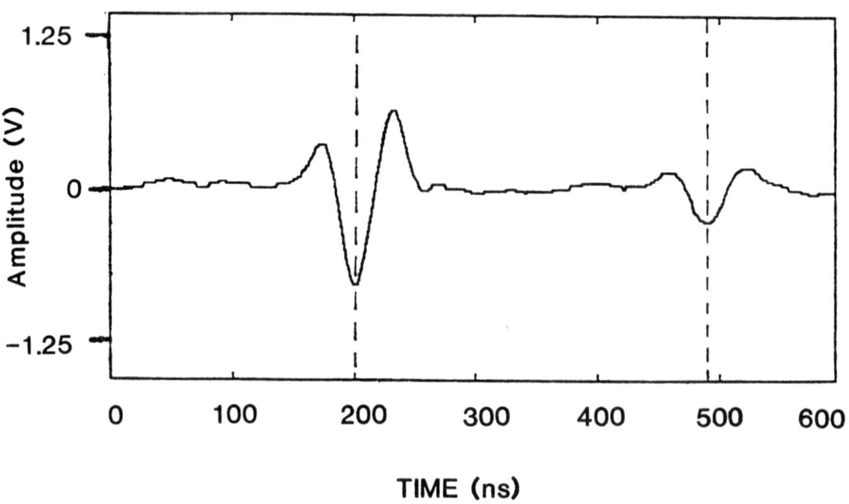

Fig. 4 Typical pulse-echo ultrasonic signal in roll-drawn polypropylene.

It is interesting to note that the angle θ between the shear transducer polarization direction and the anisotropy direction determines the amount of shear energy resolved into each component; this has been used to estimate the direction of anisotropy caused by drawing, a factor that is important in structural applications, where the strength of the material is critically dependent on direction within the material.

5. CONCLUSIONS

It has been demonstrated that ultrasonic shear waves can be used to monitor the changes of anisotropy in roll-drawn polypropylene. This uses the phenomenon of acoustic shear wave birefringence. Correlation of birefringence data with draw ratio and changes in the thickness has led to the conclusion that a microfibrillar structure exists to a finite limit in depth from each surface, and not throughout the total thickness of the sample. The shear wave data can be used as a nondestructive test method determination of both the DEGREE OF ANISOTROPY and the DIRECTION OF DRAWING during manufacture.

6. REFERENCES

Aindow J D, Markham M F, Puttuck K E, Ruder J G and Rudman M R 1986 *NDT International* **19** 24
Markham M F 1970 *Composites* **1** 145
Samuels R J 1974 *Structured Polymer Properties* (New York: John Wiley and Sons)

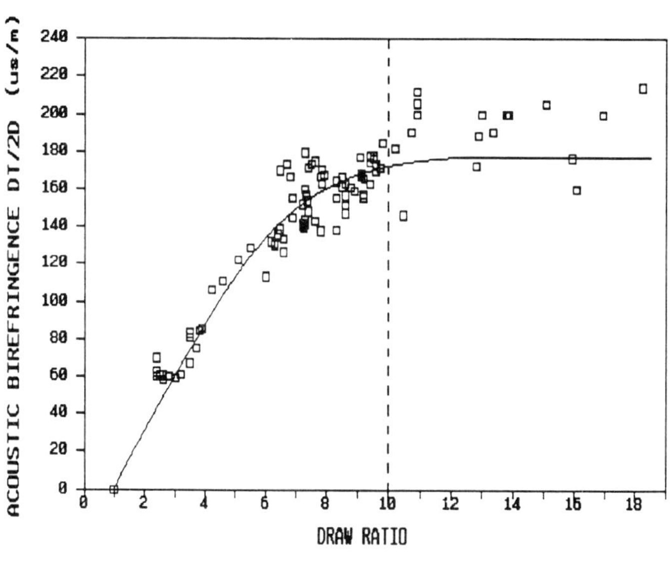

Fig. 5 Ultrasonic birefringence as a function of draw ratio in polypropylene.

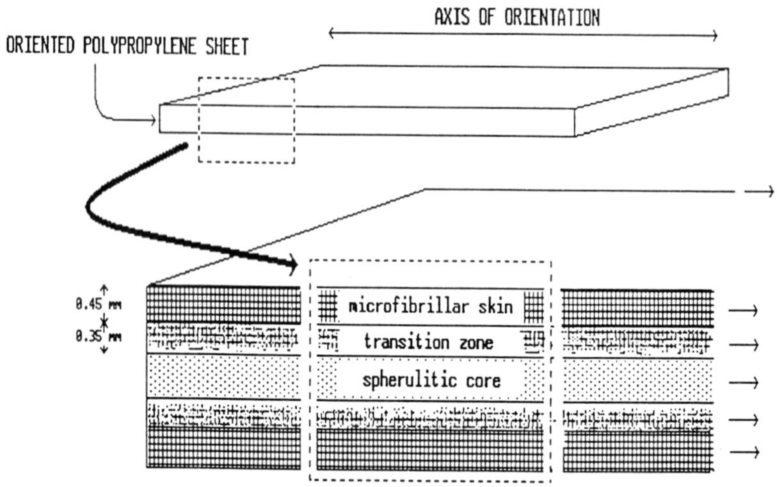

Fig. 6   Proposed layered structure for roll-drawn polypropylene.

Inst. Phys. Conf. Ser. No 111
Paper presented at Int. Conf. on New Materials and their Applications, Univ. of Warwick, 1990

Friction of expoxidized natural rubber

Alan D Roberts

Tun Abdul Razak Laboratory, Malaysian Rubber Producers' Research Association, Brickendonbury, Hertford SG13 8NL, UK

ABSTRACT: Natural rubber can be epoxidized to any desired level so that a range of properties is obtainable. We report studies of the friction of this modified rubber. An aim was to see how the variation in glass transition temperature afforded by different degrees of epoxidation bears upon frictional behaviour and whether data could be transformed by using the WLF equation.

Measurements of sliding friction were against tracks of wavy-surfaced glass. The results show that the tribological response of epoxidized grades broadly varies according to their glass transition temperature. The change in properties brought about by epoxidation opens up a range of new uses for natural rubber.

1. INTRODUCTION

Epoxidized natural rubber (ENR) is a new class of 'synthetic' rubber[1] obtained by the chemical treatment of natural latex (Figure 1). In principle, any desired epoxidation level can be made. For each mole percent of modification of natural rubber, the glass transition temperature T_g rises by approximately 1°C. Currently, two grades of ENR are available namely ENR-25 and ENR-50, containing 25 and 50 mole percent epoxidation levels respectively. Their physical properties differ from the parent rubber. They have higher damping than natural rubber and can behave quite differently; for example, ENR-50 is oil resistant. Epoxidation also improves wet-grip characteristics, and the ENR-25 grade offers an attractive balance of wet-grip and low rolling resistance in tyre treads[2].

Investigations of the frictional behaviour of ENR vulcanizates have so far referred to specific examples of use, such as in tyre treads or shoe soles. The object of laboratory experiments here presented is a broad study aimed at an understanding. The variation in glass transition afforded by different levels of epoxidation relates directly to adhesion and friction. It is of both scientific and technological interest to examine the details of the relation.

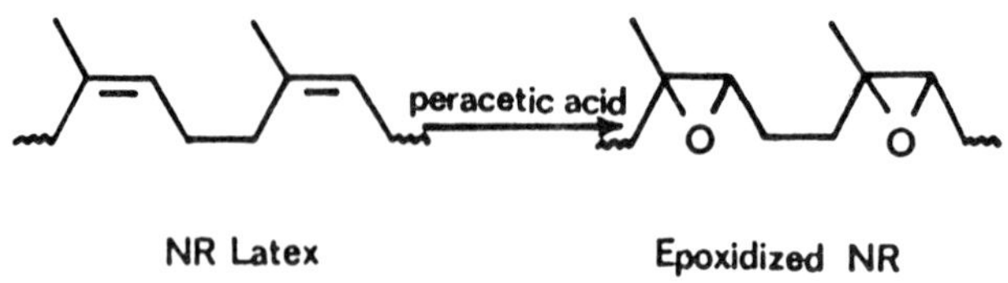

Fig.1 - Chemical modification of natural rubber.

2. Experimental

Rubber test pieces were prepared from designated grades of natural and epoxidized rubber, as indicated by the formulation details (Table 1), produced in a fully vulcanized state by hot compression moulding. They were made in a special steel former to give a vulcanizate thickness of 10mm with a sliding area of 25 x 25mm^2 (surface finish 0.2µm CLA).

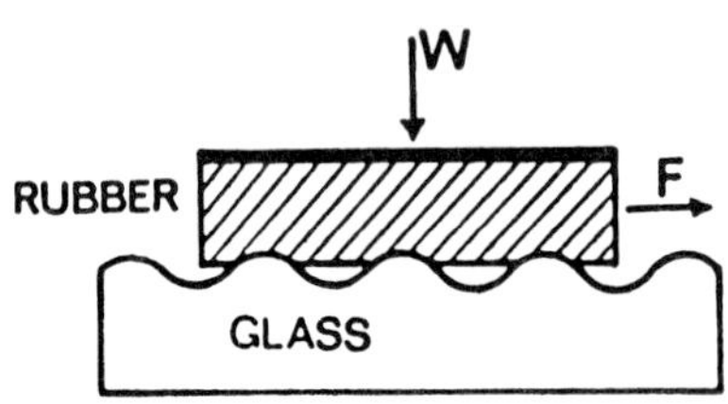

Fig.2 - Contact geometry for experimental studies of friction.

TABLE 1. Rubber compound formulation (parts by weight) and physical properties

Natural rubber, or Epoxidized NR	100
Stearic cid	1
Zinc oxide	5
Sulphur	1.45
N-cyclohexylbenzothiazole-2-sulphenamide	0.4
Poly-2,2,4-trimethyl-1,2-dihydroquinoline	1

Vulcanizates	NR	ENR-10	-25	-50	-60	-70
Cure 150°C time(min)t95	21	10	24	31	48	50
Hardness (IRHD)	34	31	34	36	32	47
Glass transition temperature (°C)	-68	-57	-42	-18	-12	-2
Young's modulus after 7s contact (MPa)	1.10	0.96	1.10	1.23	1.00	2.10

Friction measurements were made for the sliding of rubber on wavy-glass (Figure 2). The apparatus has been described in detail elsewhere([3]). It provides for the loading (W) together of test bodies and the measurement of the friction force (F) over a broad range of sliding speeds, from which the friction coefficient μ (= F/W) is calculated.

The friction data was transformed using the Williams-Landel-Ferry equation ([4]) for horizontal axis displacement

$$\log_{10} a_T = -8.86\ (T - T_g - 50)/(51.5 + T - T_g)$$

where T is the ambient temperature and T_g is the glass transition temperature of the rubber sample under test.

3. Sliding on track of wavy-glass.

The measured friction coefficient at different sliding speeds of smooth-surfaced natural and epoxidized rubber pads against wavy-glass are shown in figure 3, together with a comparison of surface roughened pads. To be more representative of practical circumstnces, some pads were deliberately abraded with sandpaper. The roughness produced was about 1 µm CLA, and this lowered the friction at low sliding speeds. All these tests were carried out at 20°C with rubber vulcanizates that ranged in glass transition temperature from -70°C to -2 °C. A transformed plot of the data was made using the WLF equation. The resulting plot (Figure 4) shows reasonably good transformation of the data into a single `mastercurve' with data points for vulcanizates of low glass transition temperature appearing to the left of the plot, and those of high to the right. The data show an increase in friction with rate (it doubles over 4 decades of $a_T v$) until the glassy region is reached, when it then declines. The maximum friction coefficient attained was at $\lg a_T v = 1$, which is similar to that obtained for natural rubber by varying the ambient temperature(5).

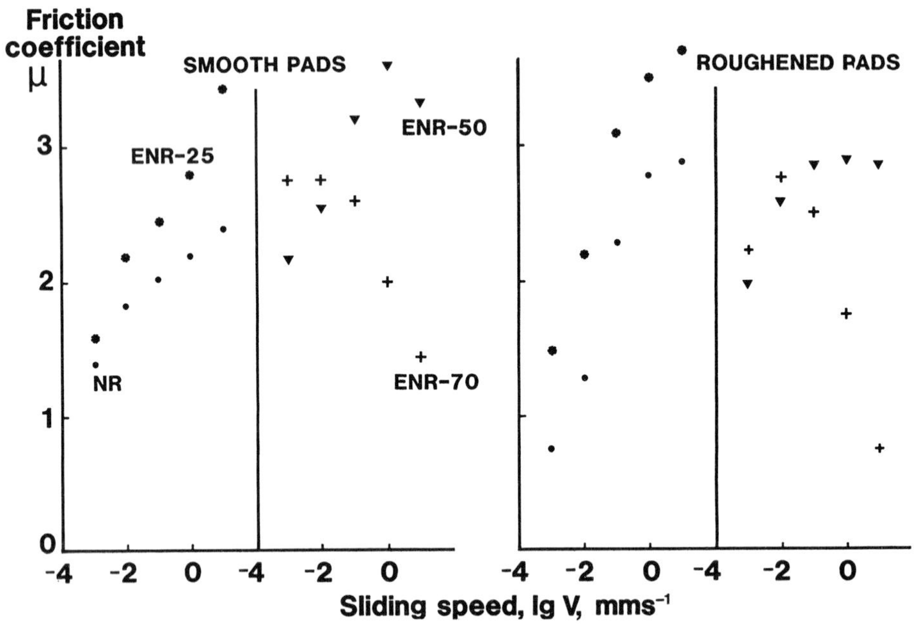

Figure 3. - Friction variation with speed for epoxidized NR sliding on wavy-glass at 20°C under a load of 10N.
● NR; * ENR-25; ▼ ENR-50; + ENR-70

If the rubber pad surface is deliberately roughened, then the friction is lowered for low $a_T v$ values (Figure 4 dotted line). Roughening appears to accentuate the peaked shape of the mastercurve.

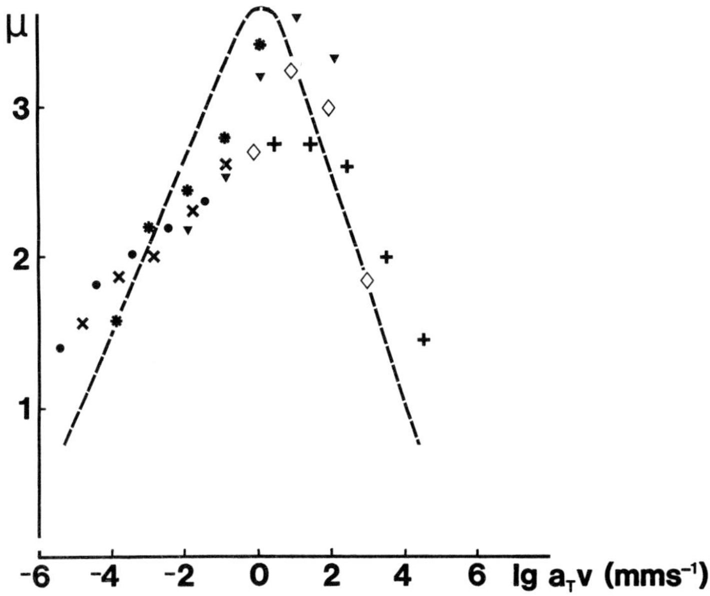

Figure 4 - Transformed friction results for sliding on wavy-glass at 20°C. Rubber surfaces were either smooth or rough. ● NR; X ENR-10; * ENR-25; ▼ ENR-50; ◇ ENR-60; + ENR-70; ------ roughened.

4. Chemical effects

The strength of adhesion of rubber to a substrate may depend on many factors. The energy required to peel rubber strips off a substrate has been shown[6] to be related to the thermodynamic work of adhesion and a hysteretic energy loss factor characteristic of the particular rubber compound under test. The reversible work of adhesion, which relates to free surface energies, arises from interactions that may be physical or chemical in nature. Reported here are some attempts to investigate the chemical aspects as they relate to the adhesion and friction of epoxidized rubber. Most polymers possess acidic and/or basic sites which can enhance wettability and adhesion. Natural rubber (cis-polyisoprene) is considered to be weakly basic because of the electon donating character of olefinic groups; the presence of electron donating epoxide groups makes it more strongly basic.

Sliding friction measurements with NR and ENR samples were carried out on the wavy-surfaced track made of soda glass. The object was to search for an acid-base treatment effect of the glass.

Soda glass is weakly basic, but when treated with hydrochloric acid the sodium in the surface region of the glass exchanges for hydrogen ions, thus temporarily creating an acidic surface. The aim was to see whether the frictional resistance of epoxidized rubber was greater on acidic glass.

Tests were carried out at 20°C for a range of sliding speeds at a constant normal load of 9.8N. Initial tests were made on untreated glass. Subsequently, the glass surface was rendered acidic by rubbing with cotton wool soaked with a 1:9 mixture of 2M HCI/IPA, cleaned with a brush soaked in the same solution and then dried by fanning with cool air. Later the glass surface was made basic using a solution of 2M NaOH applied in the same way, followed by a thorough water rinse and drying. It was found that of the range of rubber samples tested, ENR-70 gave the most marked response. Results shown in Figure 5 are for three sets of measurements in the acidic and in the basic states of glass surface treatment. Despite considerable scatter, there is a separation in the data for the two different states.

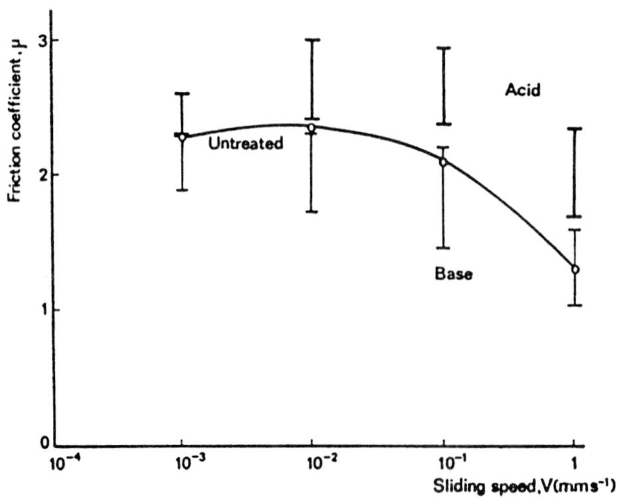

Figure 5. - Sliding friction of ENR-70 on wavy-glass showing the effect of acid and base treatments of the glass at room temperature (bars). Initial experiments for untreated glass (open symbols).

5. Possible applications

High hysteresis rubber is required for good wet grip of tyres on roads, and until recently it seemed that this must result in high rolling resistance [7]. The epoxidation of natural rubber gives a progressive increase in hysteresis, but, surprisingly, a rolling resistance as low as natural rubber. This is best shown on a Morton-Krol diagram[8]

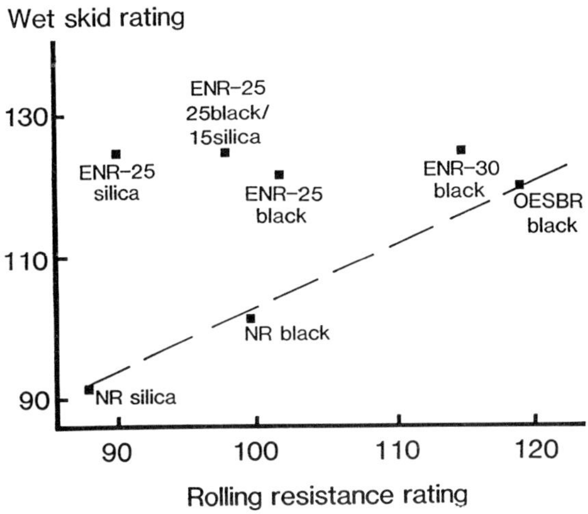

Figure 6. - Morton-Krol diagram of wet grip (asphalt) versus rolling resistance for steel belted radial tyres retreaded with various rubber compounds.

Another promising application for epoxidized natural rubber is in conveyor belt covering. The modified rubber provides for good abrasion resistance and good adhesion to a PVC core. Other possibilities are in hose, where epoxidized rubber can be used as an interlayer to improve bonding, and glazing strips that adhere well to glass. Another application could be for solid tyres where good grip and grease resistance is often sought, for example, for fork-lift truck tyres.

Besides potential in both pneumatic and solid tyres, the combination of properties exhibited by ENR suggest application in a wide range of general rubber goods. Its oil resistance characteristics are being evaluated in such applications as oil resistant cut thread and oil seals. Its high wet grip is being investigated for the production of non-slip flooring and sports shoe soling. ENR and PVC show high mutual adhesion which means that polymer blends of the two can be prepared, and these are also being developed for shoe soling. Potential in many other applications can be envisaged, including speciality adhesives.

The chemical route to the epoxidation of natural rubber was developed in the laboratories of MRPRA in the 1970's, followed in the early 1980's by physical property and technical capability evaluations. Scale-up chemical technology was carried out in the laboratories of the Rubber Research Institute of Malaysia in the middle 1980's. Commercial production of both 25 and 50 mole per cent epoxidized natural rubber commenced in Malaysia in 1989 (Kumpulan Guthrie Berhad). Initial production is limited to some 2000 tons p.a., but the plant has been designed for a rapid increase in production volume. The epoxide modification of natural rubber represents about the cheapest possible. This augurs well for its widespread adoption as a new material in a host of different applications, tribological and others.

6. Conclusion

The new polymer, Epoxidized Natural Rubber, with its different glass transitions as desired, can give different levels of adhesion, rolling resistance and sliding friction. The WLF equation offers a good guide as to likely levels of adhesion and friction for a given degree of epoxidation. Under certain circumstances, additional adhesion is manifest that accords with the `polar' nature of the rubber. As levels of adhesion and friction can be pre-arranged by the extent of epoxidation, the rubber should find wide application.

Acknowledgement

Thanks are due to Mrs C A Brackley for experimental assistance and to Dr I R Gelling for helpful discussion.

References

1. I.R. Gelling (1987) `Epoxidized Natural Rubber' Natural Rubber Technology 18, 21
2. C.S.L. Baker, I.R. Gelling and J. Palmer (1985) International Rubber Conference, Kuala Lumpur, Malaysia
3. A.D. Roberts and J.C. Richardson (1981) Wear, 67, 55
4. M.L. Williams, R.F. Landel and J.D. Ferry (1955) J.Amer.Chem.Soc. 77, 3701
5. K.A. Grosch (1963) Proc.Roy.Soc.London.A, 274, 21
6. A.N. Gent and J. Schultz (1972) J.Adhesion, 3, 281
7. C.S.L. Baker, I.R. Gelling and I.R. Wallace (1988) International Rubber Technology Conference, Penang, Malaysia
8. G.F. Morton and L.H. Krol (1982) ACS Rubber Division Meeting, Chicago, USA.

Paper presented at Int. Conf. on New Materials and their Applications, Univ. of Warwick, 1990

Aramid fibre friction: Asbestos replacement in high friction materials?

B.J. BRISCOE and P.J. TWEEDALE.

Dept. Chemical Engineering and Technology, Imperial College, London.

ABSTRACT: Historically asbestos has been used as the reinforcing fibrous component in composites used for high friction materials used in brakes and clutches. Health considerations on it's use have resulted in the search for replacements which are less hazardous. High temperature stable fibrous materials are popular,such as the aramids. The origins of their frictional performance are studied in a simple simulation of brake performance as high volume fraction fibre content composites, with other fibre reinforced composites and a brake material. An analysis of the experimental results using material characteristics and the implications for use are developed.

1. INTRODUCTION

The special composites used for frictional traction devices such as brakes and clutches are required to dissipate large amounts of frictional work as heat. This will induce large thermal stresses, mechanical and chemical changes and more subtle processes resulting from the thermal gradients in the contact. The nature of the frictional contribution of aramid fibres as reinforcing materials is investigated here. This is mainly in phenolic matrices, those most typical in use for these materials, at varying concentrations and in comparison to other fibre reinforcing systems. The emphasis is on isolating the aramid fibre as high volume fraction fibre reinforcement. The experiments reported here load a flat specimen of the composite against a rotating shaft and study the friction as function of a steadily increasing load ramp. The aim being to isolate the effect of aramid fibres friction and compare this to other fibre reinforcing systems as well as a commercially available brake material.

1.1 Friction Materials

The friction and wear of brake materials has been extensively investigated by Rhee (1970,1971,1974,1975,1976,1984,1985,1989a), and he has particularly emphasized recently the similarity with polymer composite materials (Rhee,1989b). Recent materials development have centred on producing improved semi-metallics and asbestos free formulations. The replacement of asbestos is no easy matter as no other single material combines the frictional performance, thermal stability, wear resistance, strength and cost in such an effective way.

In general brake materials are required to satisfy a wide range of conflicting requirements, for example: good wear resistance, high operating duty, ease of manufacture, low cost. This is further complicated by being required to offer a stable braking force and prolonged wear life over the wide ranges of working conditions which include speed, pressure, temperatures and environments. They are further required to be compatible with a range of counterfaces and free from side effects such as judder and squeal, particularly in this era of lightweight fabrication.

The material demands for disc brakes are more intense as the pad is smaller relative to the disc compared to drum brake linings to the drum. This results in higher pressures to develop the same retardation force e.g 1MPa compared to 0.2 MPa for a deceleration of 0.2g. A common way of assessing the performance of a material in a friction contact is the so called PV factor, the product of pressure and velocity in a contact. Bearing materials, especially polymers and their composites where the amount of frictional work to be dissipated is important, have been widely assessed in this way. Typical figure from the evaluation of brake materials are 2-24 MPa.m/s at contact speeds of 2-3.2 m/s and pressures of 1-6 MPa (Rhee,1985).

1.2 Polymer Composites

Polymer composite bearing materials have been used in many applications in preference to metallic systems because of their inherent self lubricating qualities. They are limited however in the amount of thermal work that they may dissipate. It is therefore common to run them against a smooth metallic counterface where a large amount of the problematic thermal work may be removed from the contact. Fillers and fibre reinforcement are added to improve the load carrying abilities of polymers, and may modify the friction and improve wear almost independently of that.

The frictional properties of polymer composites can be understood by careful application of the Adhesion Model of Friction (Briscoe,1987). The important point is the relative scale of the heterogenity in the composite compared to the adhesive friction dissipation zone. In brake materials this is the major dissipation process. A polymer composite will generate, at it's interface, a complex mixture or 'third body'. The friction arising from shearing this complex mixture is not a great function of the bulk as the 'weaker' phases are rapidly distributed over the counterface as a transferred layer and may be regarded as a 'lubricant'.

Particulate and fibre reinforced composites are similar behaviourally with the matrix determining tribology at low inclusion levels (~10%), whilst at higher levels (~50%) the reinforcement dominates. The nature of the addition is important, with carbon fibre being beneficial tribologically whilst glass is not (Lancaster, 1968,1970,1972a,b,c; Tanaka, 1977; Clerico, 1983; Jain, 1983; Chang, 1983). The type and extent of additions can bring synergistic effects (Fusaro,1986).

1.3 Friction Background

There are two generic models of friction which have been accepted since the time of

Bowden and Tabor (1959, 1964), and both are associated with the Two Term Model of Friction. These relate to zones of frictional dissipation, the bulk related 'ploughing component' and the interface 'adhesion component' over which deformational energy is dissipated. The first is larger and therefore involves lower rates, whilst the second is thought of as much narrower, say 100nm thickness, with correspondingly higher rates of strain. This is an idealised situation but helps rationalise an appreciable part of frictional data. Primarily these mechanisms are assumed to be non interacting and therefore the frictional contribution of each mechanism may be summed. This has been applied to polymers (Briscoe,1981) as well as polymer composites (Briscoe,1985).

The frictional data presented here will be interpreted using a plot of the 'load index' of friction against the imposed load. This is unusual as the load index will be used to evaluate the frictional temperature on the frictional behaviour rather than the load dependence of the friction as is normal. i.e. $F = k\ W^n$ where n is the load index and usually varies from 2/3 to 1 for differing modes of deformation. Elastic contacts may give results over this range, whilst plastic deformation is usually concentrated around 1 (Archard,1954;Lodge,1954). The variation of n from 2/3 to 1 has been interpreted as arising from the load dependence of the friction on the area of contact i.e. $A = k'\ W^{n'}$. It may therefore be shown that $\mu = \tau_0 / P + \alpha$ where τ and α are material constants that characterise the interfacial region. Further τ and P assume temperature of the following type $\tau = \tau_0 \exp(bT)$ and $P = P_0 \exp(cT)$.

2 EXPERIMENTAL DETAIL

2.1 Experimental Apparatus

This has been used in similar experiments in polymer tribology (Briscoe 1984, 1985). A flat specimen surface is loaded against a rotating mild steel shaft (25mm dia.) which is driven by a belt drive from an AC variable speed motor(120-1280rpm, shaft surface speed 0.16-1.7m/s) though the shaft speed was kept constant at 325rpm. The sample holder is mounted in strain gauged beams to a loading arm which is also strain gauged for the imposed load. The normal load on the specimen during the experiment is continuously ramped from 0 to a maximum of 200N by a small DC motor operating via a pulley system. Here a constant rate of 50N/min increase in the normal load was imposed. The signals from the normal load transducers and those for the frictional force were fed to an XY recorder to give a permanent record.

2.2 Polymer Composite Details

The unidirectionally oriented composite materials fabricated were:-
- Epoxy-Aramid (20/80 vol %)
- Phenolic-Aramid (50/50 & 65/35 wt %)
- Phenolic-Carbon (65/35 wt %)
- Phenolic-Glass (65/35 wt %)

From:- aramid fibre, Kevlar 49 roving type 968, supplied by DuPont; carbon fibre, type E/XAS high strength grade, supplied by Hysol Grafil; glass fibre, E type, from

Pilkington Glass; the epoxy resin was Poly/Bed 812, supplied by Polysciences; and the phenolic resin, a brake composition, Cellobond, supplied by BP Chemicals.

The epoxy /aramid specimens were made by pulling a large group of fibres through square cross section brass moulds and then allowing the resin to impregnate the fibres by a wicking process. These composites were then cured in a recirculating air oven at 110 C for 8 hrs.

All the phenolic composites were made by the same process, by pouring resin powder into 20mm x 6mm x 2mm stainless steel moulds containing unidirectionally oriented fibres. Several layers of fibres and powders were then laid on top of one another to fill the mould. These were then compressed at 30 MPa pressure, and the process repeated until the mould was filled. The specimens were then cured following the resin manufacturers recommended schedule. Initially a pressure of 30 MPa for 1.5 min, which was released for 15 seconds and then reapplied for 8.5 min, all at a mould temperature of 160 C. The samples were then post cured at 150 C for 10 hrs.

3. EXPERIMENTAL RESULTS

The results of the experiments performed here would normally be presented as a linear plot of the frictional force or it's normalised form the friction coefficient against imposed load. For this work they are more usefully presented as the double logarithmic plot of friction force against load, which allows the load indices of the materials to be determined during the experiments Figures 1-6. The high friction levels obtained generally indicate high bulk temperatures were generated.

3.1 Epoxy Systems

These were constructed to address as directly as possible the frictional properties of the aramid fibres, being of high fibre volume fraction. The neat epoxy resin underwent failure due to brittle fracture at low load, whilst only the fibre system with fibre orientation parallel to the shaft rotation was investigated, the other orientations were not fabricable. The load index results shown in Figure 1 show pronounced curvature at higher loads.

3.2 Phenolic/Aramid Systems

The various orientations at the two compositions of the aramid fibre/phenolic composites and the phenolic resin are shown in Figures 2 and 3. Comparison with the epoxy system reveals that the reduced fibre content and stronger tougher matrix have resulted in a more stable frictional response with load, with reduced curvature at high loads. The higher of the two fibre contents has little effect on the overall frictional level indicating the friction at large addition levels is independent of the reinforcement level. There are some subtleties of orientation evident too. The overall level of deformation for the experiments is within the elastic- plastic range from the experimental load indices. The transferred film formation demonstrates the link between brake and polymer composite technologies, being strongly adherent to the counterface, the 'repeat' experiment showing no reduction in frictional level.

3.3 Other Phenolic Systems

The simulation results for the carbon fibre and glass composites constructed for comparative purposes are shown in Figure 4 and 5. These do not show the low load rise in friction, as with the epoxy and aramid composites, which start at frictional levels characteristic of the matrix. The experiments were not carried out to the load level used for the aramid composites as they showed signs of failure, especially the glass/phenolic. This demonstrated a load index near 1 with the parallel and normal orientations over 1. The carbon system does not exhibit any fibre direction sensitivity and all the load indices are near to 0.8. The 'repeat' experiments show the frictional work to be greater in the second run.

3.4 Commercial Brake Material.

The commercially available brake material results are shown in Figure 6 and shows neither an initial increase in friction or decreasing dependence on the friction force at the higher loads. Failure in this contact was by cracking at a relatively light load, this due to the strains imposed on the specimen in this test. The strongly adherent transferred film observed on the shaft indicated that sufficiently high temperatures and loads had been generated to form a 'friction layer', which characterises effective brake materials. The load dependence of the friction here is very linear.

4 DISCUSSION

Calculations of the PV factor for the experiments showed these to be in the range of brake evaluations, for example the phenolic/aramid composite oriented parallel had a measured wear scar of 18 sq mm, the maximum load used was 11.1 MPa at the experimental velocity of 0.43 m/s, to give a PV of 4.7 MPa.m/s. This is toward the lower end of the range used for dynamometer testing and typical of service factors, with the speed being lower than normal and the pressure higher.

Examination of the load index of each experiment shows complex behaviour, this is shown particularly in the case of the epoxy/aramid composite. Averaged over the whole of the simulation the load index value is seen to be 0.67, but initially the value is nearer 1. The local load index (the gradient of the plot) during the experiment therefore varies from 1 (plastic deformation) through to 2/3 (elastic deformation) followed by further reduction to 0 and negative values; to yield an overall average of 0.67. The phenolic/aramid systems show great similarities to this with no subtleties depending on the fibre orientation, as do the phenolic/carbon materials over a more restricted load range. In direct contrast the phenolic/glass systems and the commercial brake material show a direct correlation of the frictional work with load, values of the load index close to 1, with surface damage and counterface damage.

The experimental data show the initial load index values are close to 1, see Table 1. The aramid and carbon systems have average values that are much less, and that the local values for these tends to decrease as the experiment proceeds. The latter effect is barely evident in the glass systems and completely absent in the brake material. It seems most unlikely that the apparent reduction in the load index can be due to a

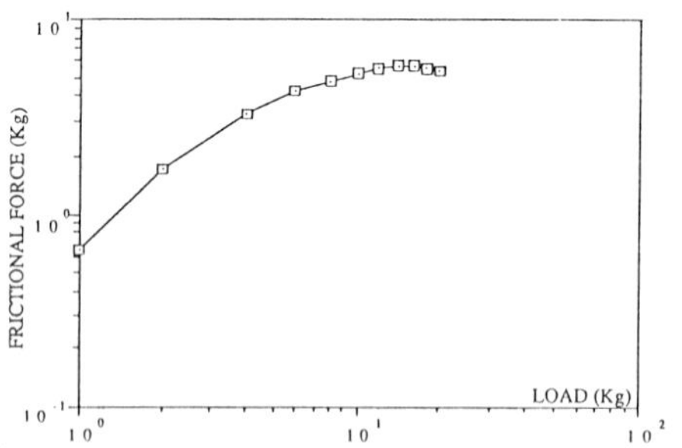

Figure 1. Epoxy/aramid (80%) composite experimental results.

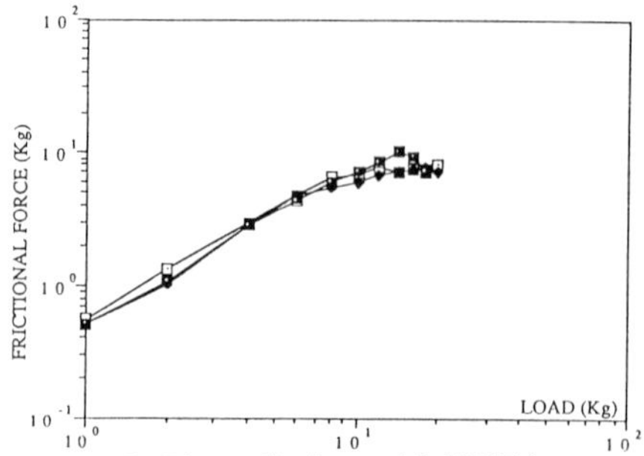

Figure 2. Phenolic/aramid (35%) composite experimental results.

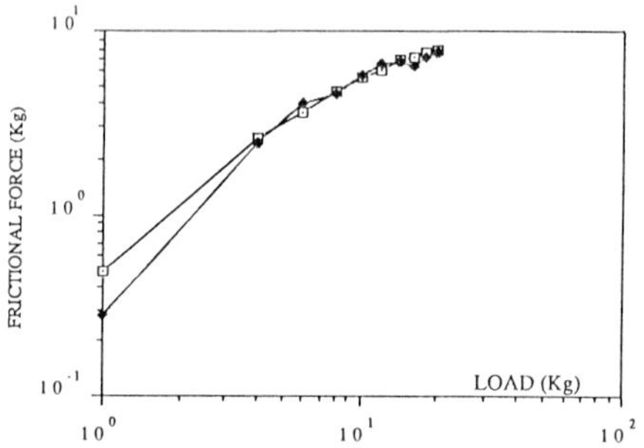

Figure 3. Phenolic/aramid (50%) composite experimental results.

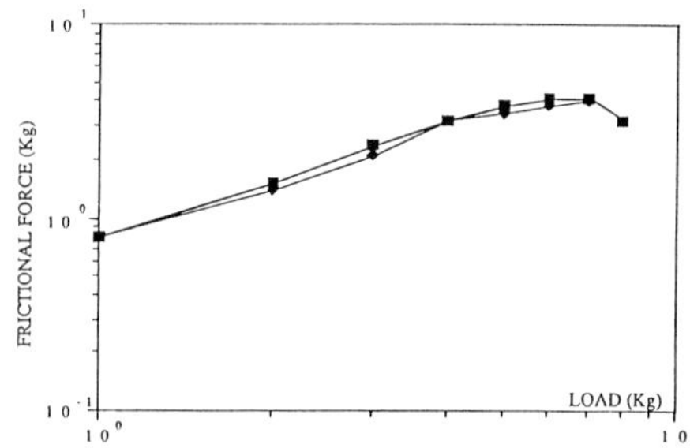

Figure 4. Phenolic/carbon (35%) composite experimental results.

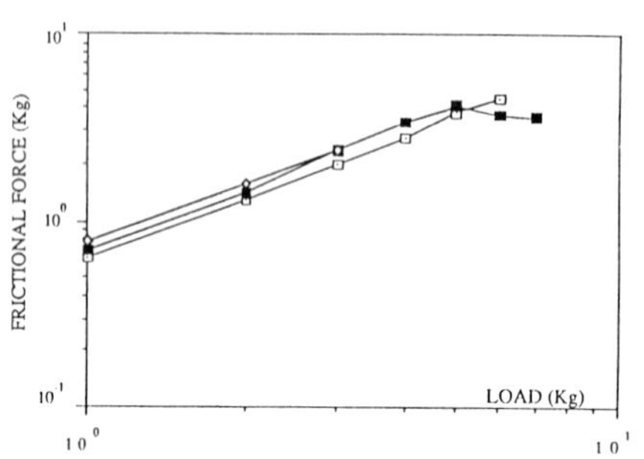

Figure 5. Phenolic/glass (35%) composite experimental results.

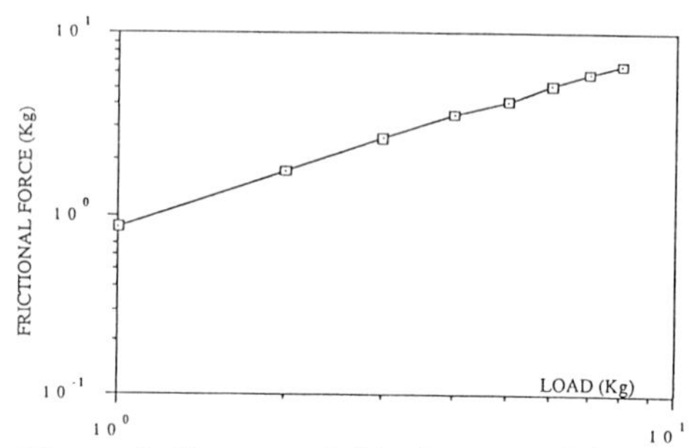

Figure 6. Commercial brake material experimental results.

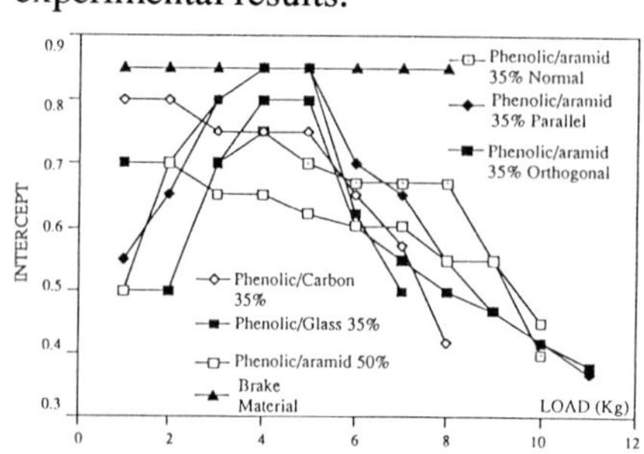

Figure 7. Intercepts of individual points of experimental results for each material plotted against load.

transition to a certain elastic asperity deformation model from a plastic response, which produces contact areas which decrease at a rate less than the increase in the normal load. A more reasonable explanation is that this is due to subtle thermally induced changes in the interface rheology and surface mechanical properties. This view may be supported by using the following analysis.

Using:- $F = W / k\, W^{n'} = W^{(1 - n')} / k$ (From:- $F = W / A$ and $A = k'' W^{n'}$)

and describing the friction using the Adhesion Model of Friction

$$\mu = \tau / P + \alpha \text{ then } \quad \mu = \tau k / W^{(1 - n')} + \alpha$$

and therefore $F = \tau k W^{n'} + \alpha W$

The material parameters may be expressed as exponential functions of temperature of the form $k = k_o \exp(bT)$ and $\tau = \tau_o \exp(aT)$, being terms related to the bulk and surface rheology terms. Then:-

$$F = k_2 \exp(+CT) W^{n'} + \alpha W$$

where $C = (b - a)$ when the temperatures of the bulk and surface are the same. When they differ by δT then the term for the exponential becomes:-

$$\exp(bT_b - aT_s) = \exp(bT_b - a(T_b + \delta T) = \exp(CT_b - a\,\delta T)$$

substituting this above and taking logarithms of both sides then:-

$$\ln F = \ln(k_2 \exp(CT_b - a\,\delta T)W^{n'} + \alpha W$$

for $T = 0$ then:- $\ln F = \ln(k_3 W^{n'} + \alpha W)$

Factorising this by removing W^n where $n' < n < 1$; since n' is the range 2/3 to 1, and the experimental data shows the initial load index to be around 1 in general, then the error in doing this will be small. So

$$\ln F = \ln(W^n(k_3 + \alpha)$$

and therefore $\ln F = n \ln W + \ln(k_3 + \alpha)$

Table 1. Load index analyses of the simulations overall and the initial values, as well as the intercepts and loads of deviation from linearity for different materials and fibre orientations studied.

MATERIAL	ORIENTATION	LOAD INDEX OVERALL	LOAD INDEX INITIAL	INTERCEPT	DEVIATION LOAD
EPOXY/ARAMID	PARALLEL	0.672	1.054	0.86	8
PHENOLIC	NORMAL	1.035	1.163	0.7	8
ARAMID(35%)	PARALLEL	0.885	1.080	0.75	5
	ORTHOGONAL	0.926	1.103	0.7	5
PHENOLIC	PARALLEL	0.921	1.029	0.75	10
ARAMID(50%)	ORTHOGONAL	1.064	1.279	0.75	10
PHENOLIC	PARALLEL	0.783	1.004	0.8	6
CARBON	ORTHOGONAL	0.833	0.970	0.8	6
PHENOLIC	NORMAL	1.010	1.010	0.78	6
GLASS	PARALLEL	1.095	1.095	0.65	6
	ORTHOGONAL	0.922	1.129	0.78	5
BRAKE		0.990	0.990	0.86	>8

This analysis predicts what is observed experimentally that the double logarithmic plots of F against W, shown in Figures 1-6, will initially be linear and have a gradient of n and intercept ($k_3 + \alpha$). This remains so until there is some significant change in k_3 which occurs when a significant temperature difference, δT, occurs between the bulk and the surface when deviation from linearity is expected to evident. The initial gradients, corresponding intercepts and loads at which the plots first deviate from initial load index linearity are also shown in Table 1.

There is some evidence from some of the systems that the load of deviation is correlated with the level of friction and therefore the temperature of the counterface. The phenolic composites are all similar in their intercept values emphasizing the matrix dominant nature of this parameter. However they do differ illustrating the contribution of the reinforcement, which is further shown in the load and extent to which the systems begin to diverge from linear behaviour. This latter point is demonstrated by Figure 7 where the intercept of a line of gradient unity has been drawn through each of the experimental points and plotted as a function of the load. This emphasizes the surface rheology term by removing the bulk deformation term. Both the glass reinforced composite and the brake material show the intercept invariant whilst the aramid and carbon systems show rates of decrease that mirror the effectiveness of the 'lubricating' ability of the interfacial 'third body' i.e. carbon > aramid(35%) > aramid(50%).

The temperature in the contact is proportional to the frictional work since $F = \mu W^n$ then the exponential term, exp (CT), is in reality a load dependent term, exp (CW^n). It is to be expected that the these terms are sensitive to the thermal environment, i.e.

when then thermal mass of the counterface has been raised significantly, at higher loads and frictional levels. The curvature in the load index plots of the carbon and aramid systems result from the adiabatic shear softening as the temperature difference between the surface and bulk of the composite increases. Carbon systems are known to exhibit this effect (Lancaster,1968: Fusaro,1986) and their is some evidence that aramid systems do also (Freidrich,1987). For the glass composite and the brake material this mechanism is does not take place, with the particulate debris from the glass system abrading the counterface and the thermal conductivity of the brake material reducing the build up to temperature in the contact.

Thus a general picture of the processes operating in the contact emerges; with all the load indices starting off around 1. The systems heat up as a result of the frictional work increasing due to growth in the contact area due to decreasing flow stress at increasing temperature, this effect is included in the T_b term. The term a.δT then operates as the localised surface heating takes effect, depending on the material this can cause an apparent reduction in the load index, which Figure 7 demonstrates. The glass composite and brake material have similar constant intercept values; that is they do not 'fade': whilst the other systems do, as shown by the rate of reduction in the intercept value, which is in the relative order carbon > aramid(35%) > aramid(50%) > glass.

The trends in load index may therefore be tentatively assigned to the generation of a transferred layer and the modified composite surface produced at high energy dissipation, or after protracted sliding in the current contact configuration. The aramid and carbon third body has an interfacial rheology that is very temperature dependent. This in effect produces a layer that acts as a high temperature lubricant. The glass composite shows a much reduced effect, indicating that the base resin may undergo surface degradation to form less efficient lubricating species.

5. CONCLUSIONS

The main purpose has been to evaluate the effectiveness of the simulation and associated analysis in assessing the frictional characteristics of potential brake material constituents. This appears to be relatively novel and economical way of achieving this. Two responses are observed for the simulation studied here:-

- For the epoxy/aramid, phenolic/aramid and phenolic/carbon composites, evidence of third body formation and lubrication of the contact, demonstrated by the trends in the load index.
- For the phenolic/glass composites and commercial brake material, show a direct dependency of the friction force on the load and considerable evidence of material damage to the sample and the counterface.

These fundamentally different responses are generated by two separate mechanisms, the first by a thermally induced 'lubrication' of the contact causing a marked reduction of friction with load, while in the latter the lubricant is not produced or is abraded away during sliding. The former is a mechanism of 'fade' and is not desired in a brake material, therefore aramid fibres alone do not possess the fundamental type of response required in a brake material. However the composites used commercially are consist of many components and the characteristics shown here may be masked by more complex interactions. Indeed practically the interest in using aramid fibres commercially seems to arise from their ability to substitute for asbestos in brake material mixes without requiring reformulation.

REFERENCES

Bowden F P, Tabor D 1951 *Friction and Lubrication of Solids, Part 1* (Clarendon Press Oxford)
Bowden F P, Tabor D 1964 *Friction and Lubrication of Solids, Part 2* (Clarendon Press Oxford)
Briscoe B J 1981 *Phil Mag* **A43** 511-27
Briscoe B J, Stolarski T A, Davies G J 1984 *Trib Intl* **17** 129
Briscoe B J, Yao L H, Tweedale P J 1985 *Proc Intl Conf EUROTRIB 85* **II** Paper 5-1.7
Briscoe B J, Tweedale P J 1987 *Proc Intl Conf 'ICCM VI'* **5** 282
Chang H W 1983 *Wear* **85** 81-91
Clerico M, Patierno 1979 *Wear* **53** 279
Freidrich K, Cirino M, Pipes R B 1987 *Proc Intl Conf 'ICCM VI'* **5** 303
Jain V K 1983 *Wear* **92** 279-92
Lancaster J K 1968 *J Appl Phys* **1** 549
Lancaster J K, Giltrow J P 1970 *Wear* **16** 359-74
Lancaster J K 1972a *Tribology* **5** 249-55
Lancaster J K 1972b *Wear* **20** 315
Lancaster J K 1972c *Wear* **20** 335
Lodge J F, Howell A S 1954 *Proc Phys Soc* **B67** 89-98
Rhee S K 1970 *Wear* **16** 431-445
Rhee S K 1971 *Wear* **18** 471-477
Rhee S K 1974 *Wear* **28** 277-281
Rhee S K, Tarr W R 1975 *Wear* **33** 373-375
Rhee S K 1976 *Wear* **37** 291-297
Rhee S K, Tsang P H S, Jacko M G 1985 *Proc Intl Conf 'Wear of Materials'* 129
Rhee S K, Tsang P H S, Wang Y S 1989a *Wear* **133** 39-45
Rhee S K, Tsang P H S, Jacko M G 1989b *Wear* **133** 23-37
Tanaka K 1977 *Proc Intl Conf 'Wear of Materials'* 510-27

Paper presented at Int. Conf. on New Materials and their Applications, Univ. of Warwick, 1990

Wear of aluminium-alumina metal-matrix composites by abrasion and erosion

I M Hutchings and A Wang*

Department of Materials Science and Metallurgy, University of Cambridge, Pembroke Street, Cambridge CB2 3QZ, UK

ABSTRACT: The wear by abrasion and erosion has been investigated of composites containing up to 30% by volume of discontinuous δ-alumina fibres in a planar random array in an aluminium alloy matrix. The composites were more resistant than the matrix alloy to two-body abrasive wear by SiC particles, but showed a transition in behaviour with the size of the abrasive particles. With the smallest particles the wear resistance increased linearly with fibre volume fraction up to 30%, while with larger particles the wear resistance passed through a maximum at an intermediate volume fraction. For erosion by SiC particles, the composites showed decreasing wear resistance with increasing reinforcement at all impact angles; similar behaviour occurred with SiO_2 erodent, except at the lowest impact angle where the trend was reversed. These observations can be understood by considering the extent of fibre fracture associated with the different conditions of wear.

1. INTRODUCTION

Metal-matrix composites (MMCs) form one class of 'new materials' with attractive mechanical properties. In particular, they offer the potential of high values of specific stiffness and strength, in combination with relatively high fracture toughness (Feest 1986). Recent surveys have predicted a major increase in the use of MMC materials in the near future (e.g. Anon. 1989); it is forecast, for example, that in twenty years MMCs may constitute 30% of the mass of a typical aircraft gas turbine engine (Jeal 1988).

In many applications, MMC materials may experience friction and wear, but few studies have so far been made of their tribological behaviour. In a recent review it was pointed out that composite materials as a whole appear to have received relatively little attention from the tribological community, with a small decline over the past ten years in the proportion of papers concentrating on these materials among those presented at the biennial International Conferences on Wear of Materials (Hutchings 1989).

Limited studies suggest that MMCs under conditions of lubricated or unlubricated sliding against a smooth counterface tend to show lower wear rates than the unreinforced matrix alloy (Hosking et al. 1982, Surappa et al. 1982, Kishore 1983), although wear of the counterface may on the other hand be enhanced by the abrasive action of the hard reinforcing phase. The resistance of MMC materials to abrasive wear by hard particles has consistently been reported to be better than that of the unreinforced matrix. For example, aluminium and titanium alloys containing TiC or WC particles showed appreciably greater resistance to abrasion (by the dry-sand rubber wheel method) than the plain matrix alloys (Ayers 1984), and aluminium alloy MMCs reinforced with Al_2O_3 particles have shown lower abrasive wear rates in tests against a

* Present address: Dept. of Mechanical Engineering, Clemson University, Clemson SC 29634, USA

fixed alumina abrasive than the unreinforced alloy (Surappa et al. 1983). However, the latter investigators also noted an increase in abrasive wear rate as the volume fraction of reinforcement was raised from 3% to 5%. Bhansali and Mehrabian (1982), in contrast, found that the resistance of aluminium-alumina particulate composites to abrasion in the dry sand rubber wheel test was greatest for a high volume fraction of particles. The tribological behaviour of aluminium alloy particulate MMCs has been reviewed by Prasad and Rohatgi (1987).

Little work has been reported on the response of MMCs to erosive wear by solid particle impact. Turenne et al. (1988) studied the erosion by an aqueous slurry of silica sand of aluminium alloy MMCs containing short alumina-silica fibres, and found very little benefit in terms of increased wear resistance to be provided by fibre volume fractions of up to 20%, except at shallow angles of incidence. At an impact angle of 15°, however, the effect of the reinforcement was substantial, with a strong increase in wear resistance associated with increasing fibre volume fraction.

The purpose of this paper is to present the results of an investigation of the wear rates and mechanisms associated with abrasion and erosion of MMCs containing short alumina fibres in an aluminium alloy matrix.

2. EXPERIMENTAL METHODS

The MMCs studied in this work contained discontinuous δ-alumina fibres (Saffil RF, ICI plc) in an aluminium alloy matrix (Al6061: 1% Mg; 0.8% Si; 0.25% Cu; 0.2% Cr). The fibres had a mean diameter of 3 μm and length of ~200 μm, with a tensile strength of ~2 GPa and Young modulus of ~300 GPa (Dinwoodie et al. 1985). The composites were made by liquid metal squeeze-infiltration of fibre preforms, and were therefore anisotropic, containing fibres distributed through the matrix in a planar random array as illustrated in Figure 1. Specimens were prepared with volume fractions of fibres ranging from 5% to 30%; tests were also performed with the unreinforced matrix alloy. All materials were tested in the as-cast condition.

Microhardness measurements, made with a Vickers indenter at a load of 500 grams force (4.9 N) on polished sections through the composites, are plotted in Figure 2. The materials showed a linear increase in hardness with volume fraction of fibres, and also significant anisotropy; the hardness measured by indenting a polished surface perpendicular to the fibre plane was in all cases higher than that measured on a surface parallel to the fibre plane.

Abrasive wear tests were performed with a pin-on-disc two-body abrasion apparatus (Mercer and Hutchings 1988, Wang and Hutchings 1989), in which the flat end of a cylindrical specimen 5 mm in diameter was loaded against the plane surface of a rotating disc carrying silicon carbide bonded abrasive paper. In these tests the applied normal load was 1.9 N, giving a nominal contact pressure of 0.1 MPa. All tests were performed in air at a humidity of 50 to 70%. The specimen ran on a single track during the test at a speed of 0.63 ms^{-1}, and the vertical displacement of the specimen was continuously monitored and recorded.

Erosive wear tests were performed with a gas-blast rig (Hutchings and Deuchar 1987), in which the erodent particles were introduced into air flowing along a cylindrical nozzle 300 mm long and 3 mm inside diameter; a constant differential pressure of 0.5 bar was maintained between the ends of the nozzle. Both silicon carbide and silica erodents were used, sieved to a mean particle size of 100 μm. Representative samples of these two types of particle are illustrated in Figure 3. The silicon carbide particles were slightly more angular than the silica particles. Under the conditions employed in these experiments the velocity of the particles leaving the nozzle was 65 ± 5 m s^{-1}, as measured by the double rotating disc method (Ruff and Ives 1975). The test specimen, with dimensions $20 \times 10 \times 5$ mm, cut from a cast billet,

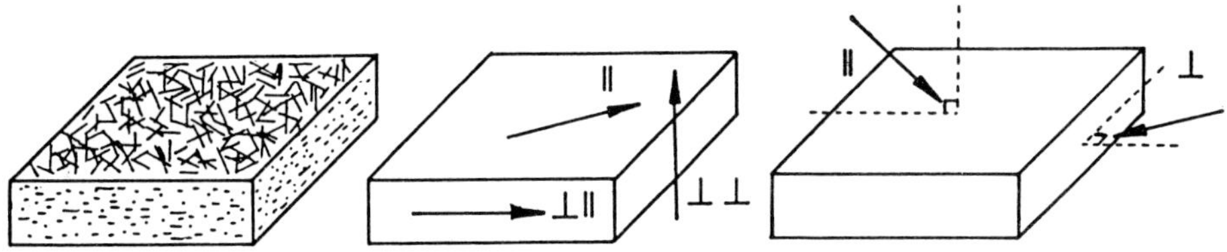

Figure 1.Schematic diagrams showing (a) the distribution of fibres in the MMC samples; (b) the orientations used in the abrasion experiments; (c) the orientations used in the erosion tests.

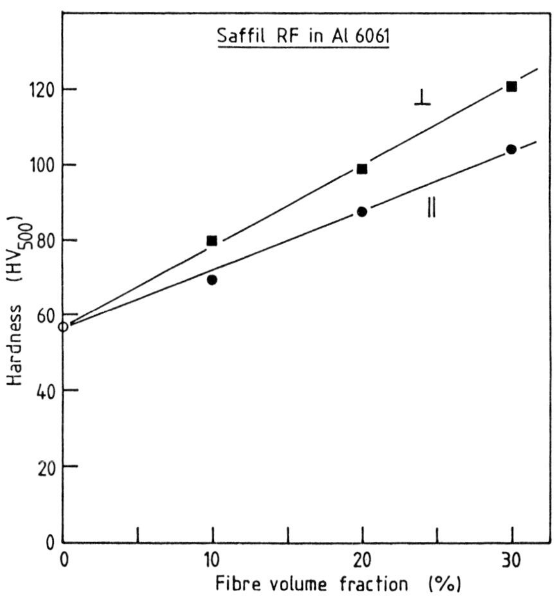

Figure 2. Microhardness values for indentations made on polished surfaces parallel and perpendicular to the fibre plane.

(a)

(b) 100 μm

Figure 3. Examples of the abrasive particles used in the erosive wear tests: (a) silicon carbide (b) silica.

was mounted at a measured angle to the axis of the nozzle. The wear rate was determined by weighing the specimen before and after erosion by 25 gram increments of erodent particles. Before each weighing, loose particles were removed from the specimen by an air blast.

3. RESULTS

3.1 Effect of Specimen orientation

Abrasion and erosion experiments were performed with specimens cut from the billets in different orientations. In the case of abrasion, three orientations were used: with the sliding direction oriented perpendicularly to the fibre plane, and parallel to the fibre plane with the plane of the abrasive paper either parallel to, or perpendicular to, the fibre plane, as shown in Figure 1(b). Erosion tests were made in two different orientations: with the eroded surface parallel to the fibre plane, and with it perpendicular to the fibre plane with the impact direction in the fibre plane (Figure 1c).

In all cases there was little influence of specimen orientation on the measured wear rate. In the abrasion tests, sliding normal to the fibre plane often resulted in a slightly higher wear rate than the other two orientations, although this was not consistently the case. In the erosion tests no systematic effect of orientation could be discerned. No account is, therefore, taken of fibre orientation in the results reported in the following sections; the data represent the mean results for all the different orientations used.

3.2 Wear by abrasion

Since the length of the pin specimen was continuously monitored, the wear process could be followed closely. Fresh abrasive paper was used for each experiment. In all cases it was observed that an initial high wear rate, lasting for some 5 to 10 seconds (i.e. about 10 to 20 traversals of the track) was followed by a lower and effectively constant wear rate for at least the next 60 seconds. A typical record of wear over this period is shown in Figure 4. The reduction from the higher initial to the lower 'steady-state' wear rate was associated with some blunting of the abrasive particles and the removal of a few large and weakly bonded grit particles from the abrasive paper.

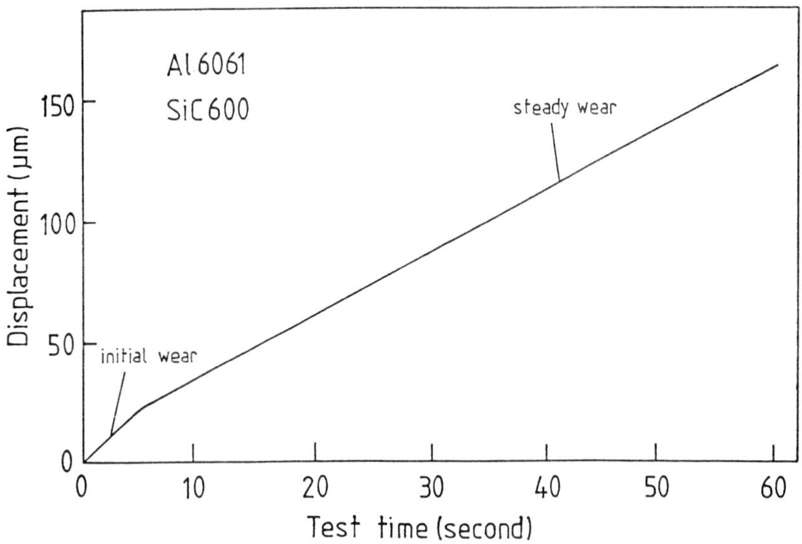

Figure 4. Typical record of specimen length change against time for the duration of an abrasion test.

From the slope of the wear trace (as shown in Figure 4) values of wear rate, expressed in dimensionless form as the reduction in specimen length per unit sliding distance, were derived for both the initial and steady-state regimes. Although the initial wear rates were always greater than those measured in the corresponding steady-state, the trends observed with grit particle size and volume fraction of reinforcement were similar for both; values of steady-state wear rate only are therefore reported here.

Figure 5 shows steady-state wear rates for the composite materials abraded against silicon carbide abrasive papers with different particle sizes, ranging from 600 mesh (20 μm) to 240 mesh (60 μm) mean particle diameter. All the materials showed a strong variation of wear rate with abrasive particle size.

The variation of wear rate with the volume fraction of reinforcement in the composites, however, followed a different pattern for different abrasive particle sizes. With the smallest abrasive particles (600 mesh, 20 μm), a monotonic decrease in wear rate was associated with increasing fibre volume fraction. Figure 6 shows the wear rate data from Figure 5 plotted in terms of wear resistance (1 / wear rate) relative to that of the unreinforced matrix; the effect of the reinforcement can thus be clearly seen. For abrasion by the 600 mesh particles, the wear resistance increased linearly with volume fraction of fibres. With larger abrasive particles the dependence was no longer linear, and at the highest fibre contents the wear resistance was lower than that for intermediate proportions of reinforcement.

Examination of the worn surfaces by SEM indicated significant differences in the response of the fibres to abrasion. As Figure 7 shows, with the smallest abrasive particles there was little evidence of fibre fracture, even in the composites with the highest volume fraction. With larger abrasive particles, however, fibre fracture occurred leading to the removal of fibre fragments and associated regions of matrix.

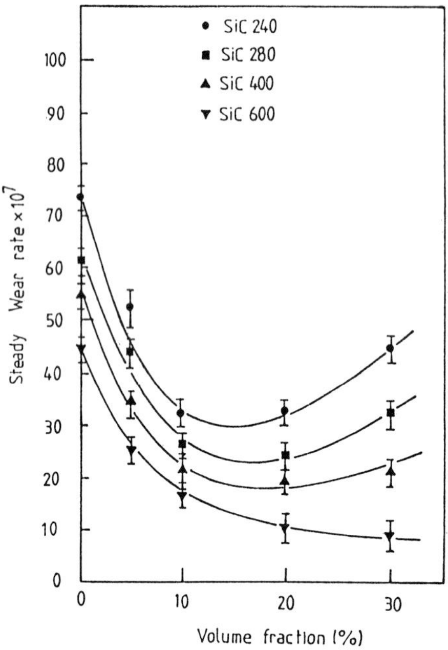

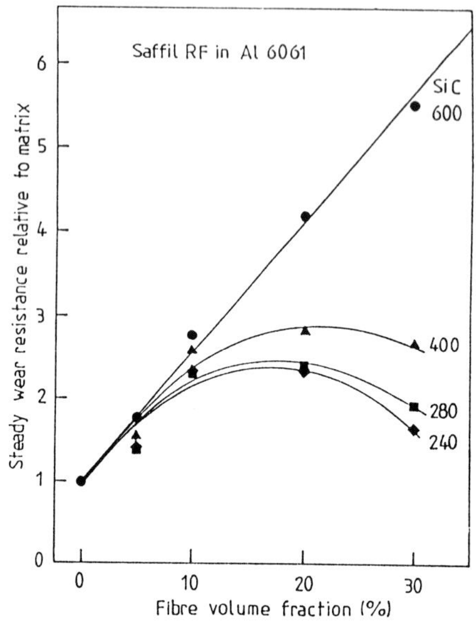

Figure 5. Values of steady state wear rate (reduction in specimen length per unit sliding distance) for MMCs with different fibre volume fractions abraded against silicon carbide particles of different sizes.

Figure 6. Data from Figure 5 expressed in terms of wear resistance (1 / wear rate) relative to that of the unreinforced alloy.

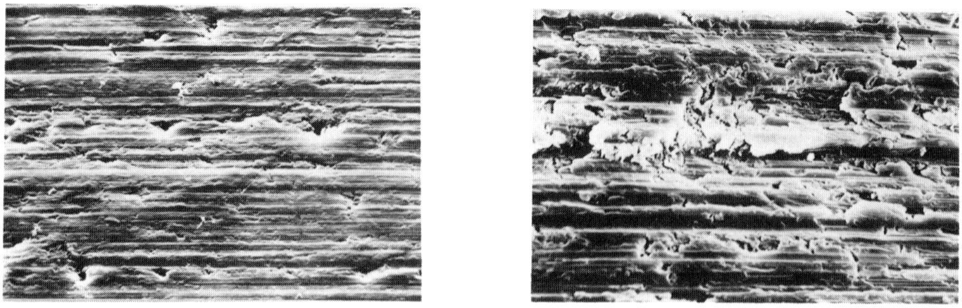

(a) (b) 10μm

Figure 7. Surfaces of composites with 20% volume fraction of fibres, abraded against (a) 600 mesh (20 μm) silicon carbide grit; (b) 240 mesh (60 μm) silicon carbide grit.

3.3 Wear by erosion

The specimens were weighed after every 25 grams of erodent had struck the surface. In each case, after a very small initial transient, plots of mass loss against the cumulative mass of erodent were linear, and values of steady-state erosion rate were therefore expressed as the (dimensionless) gradients of these plots. The erosion rates due to silicon carbide were considerably greater, by a factor of nearly two, than those caused by the silica particles. Figure 8 shows the dependence of erosion rate on impact angle, for composites with different volume fractions of reinforcement eroded by these two types of particle. In each case maximum wear was observed at an oblique angle.

For normal incidence (90° impact angle) the erosion rate increased with increasing fibre content. This was also true at lower impact angles, except at 20° for erosion by silica particles when the trend was reversed.

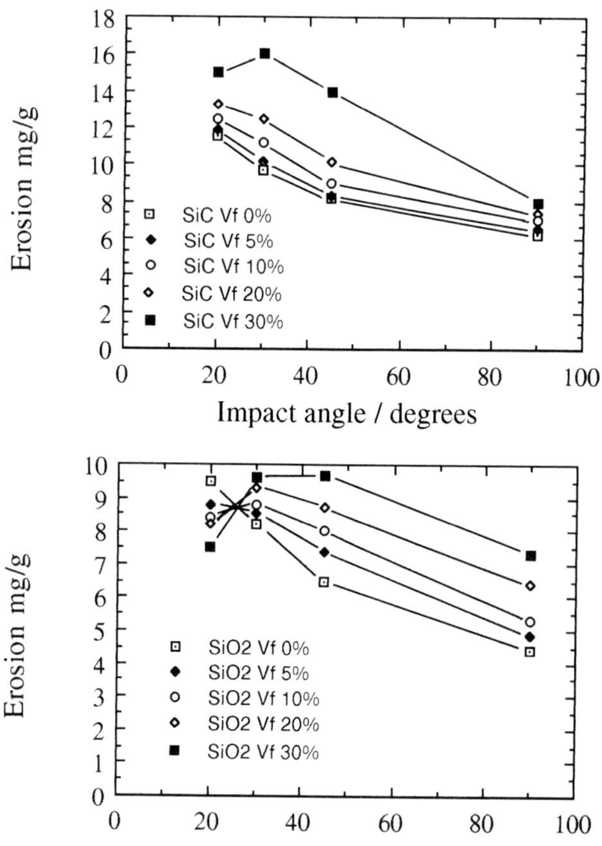

Figure 8. Variation of erosion rate (mass lost per unit mass of imparting particles) with impact angle for the composites with different volume fractions of fibres, eroded by (a) silicon carbide particles and (b) silica particles.

Figure 9 shows the surfaces of composites containing 20% volume fraction of fibres, after erosion at impact angles of 90° and 20°. At 90° the differences between samples eroded with silica and silicon carbide were not pronounced, although the indentations in the surface are more sharply defined in Figure 9(a) where they have been made by the more angular silicon carbide particles. Surfaces eroded by both types of particle showed narrow pits and steep-sided indentations where fibres had been fractured and removed. At 20° a greater difference was noted, with more evidence of fibre fracture (as illustrated also in the cross-section in Figure 10) on the surfaces eroded by silicon carbide particles.

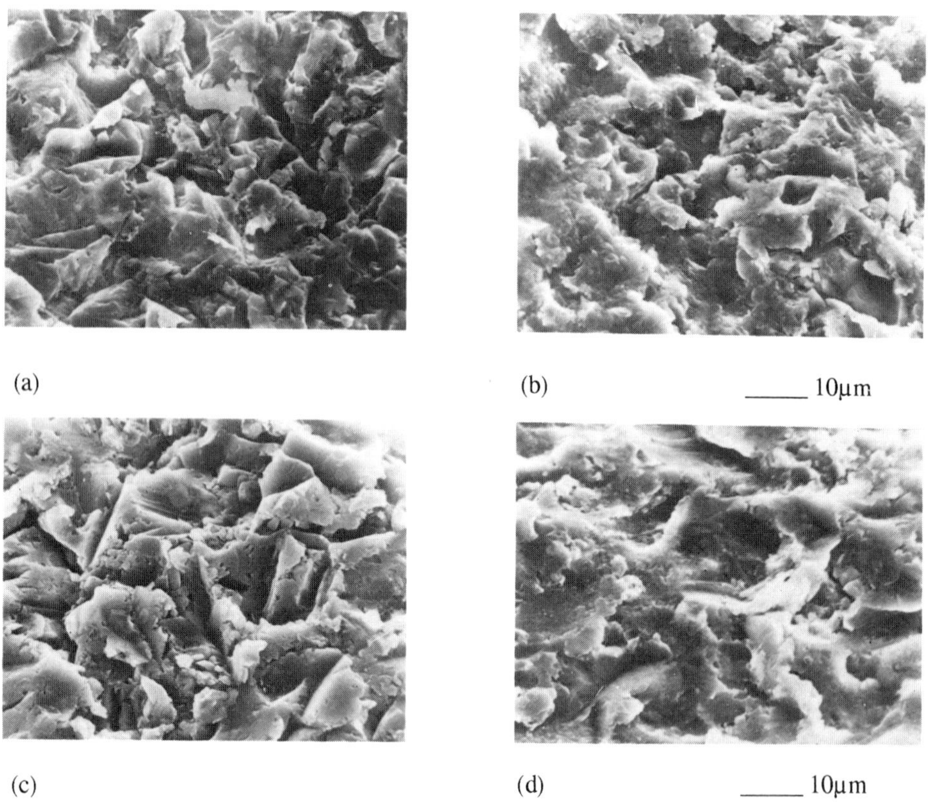

Figure 9. Surfaces of composites with 20% volume fraction of fibres eroded at (a and b) 90° and (c and d) 20°. The erodents were (a and c) silicon carbide and (b and d) silica particles.

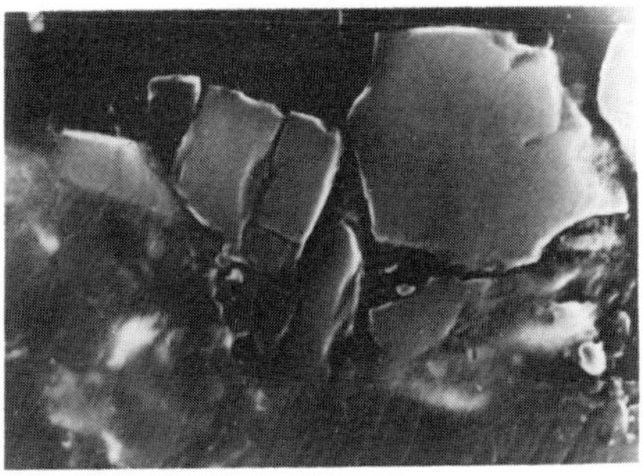

_______ 3μm

Figure 10. Cross-section of composite with 20% volume fraction of fibres eroded at 90° impact angle by silicon carbide particles, showing fracture of the reinforcing fibres.

4. DISCUSSION

The observations reported above highlight the strong dependence of the wear behaviour of these MMC materials on the applied conditions. The results of the abrasion experiments show that, provided fibre fracture can be avoided, the abrasive wear resistance of a short-fibre reinforced MMC can be substantially greater (in this case by up to six times) than that of the unreinforced matrix. The factor by which the wear resistance is enhanced can be several times that by which the bulk indentation hardness is increased. However, when fibre fracture occurs the benefits of reinforcement are much less and high fibre volume fractions can be positively detrimental to abrasion resistance, since broken fibres may not only form a source of wear debris themselves but may also enhance the removal of the matrix material. The strength of the interface between the fibres and the matrix, although not investigated in this work, would also be expected to influence the extent to which fibre breakage contributes to material loss.

In this work the transition from plasticity-dominated wear (involving plastic ploughing and cutting of both matrix and fibres) to extensive fibre fracture was brought about by an increase in abrasive particle size. It could also arise from an increase in the load carried by each particle, as has been shown for the abrasive wear of the brittle boride layer on a boronized steel (Wang and Hutchings 1989). The conditions for fibre fracture to occur during abrasive wear of a MMC can be identified by calculating the load carried by each grit particle, which in the present experiments depends on the particle size, and employing the theory developed by Hutchings and Wang (1988). For the MMCs studied here the conditions for fibre fracture predicted in this way show good agreement with those actually observed (Wang and Hutchings 1989b).

Fibre fracture also plays a central role in the wear of these MMCs by erosion. Only in the case of erosion at shallow impact angles (20°) by the less angular and weaker silica particles was widespread breakage avoided. It was only under these conditions that fibre reinforcement was beneficial in reducing the wear rate; for higher impact angles, and under all conditions for the impact of the hard and angular silicon carbide particles, the effect of fibre reinforcement was positively detrimental to erosion resistance. The impact conditions and the nature of the erosive particles are clearly very important in determining the extent of fracture; high angles and impact velocities will tend to favour fracture, as will large (and therefore more massive) particles of angular shape and high strength.

Further work is needed to establish more clearly the range of conditions, for both abrasion and erosion, over which fracture of the reinforcing phase in MMCs can be avoided, and the influence of the strength of the reinforcement, its shape and distribution, and the strength of the interface, on the resulting wear rate. It is, however, clear that careful attention must be paid in the use of MMCs to the possibility of changes in the wear mechanisms, and associated wear rates, due to fracture of the reinforcing phase. For some applications protective coatings may be necessary in order to provide acceptable lifetimes.

5. CONCLUSIONS

From these investigations of the abrasive and erosive wear of MMCs containing discontinuous alumina fibres in an aluminium alloy matrix, it can be concluded that the extent of fracture of the reinforcing phase plays a critical role in determining the wear behaviour. If widespread fibre fracture can be avoided, then wear rates by erosion or abrasion can be reduced by increasing the fibre content of the composite. The benefits of fibre reinforcement in increasing wear resistance depend strongly on the precise conditions of wear; under the most favourable conditions examined, the increase in wear resistance was much greater than the corresponding increase in bulk hardness. On the other hand, if extensive fibre breakage occurs then composites containing high volume fractions of reinforcement may show higher wear rates than those with lower proportions. The tendency for fibre breakage will increase with abrasive particle size, angularity and hardness, and in the case of two-body abrasion with the applied normal load; for erosive wear, it will increase with impact angle and velocity.

REFERENCES

Anand K and Kishore 1983 *Wear* **85** 163-169
Anon. 1989 *Advanced Materials and Processes* February 13
Ayers J D 1984 *Wear* **97** 249-266
Bhansali K J and Mehrabian R 1982 *J. Metals* **34** 30-34
Dinwoodie J, Moore E, Langman C A J and Symes W R 1985 *Proc. 5th Int. Conf. on Composites ICCM V* eds W C Harrigan et al. (New York: AIME) 671-685
Feest E A 1986 *Materials and Design* **7** 58-64
Hosking F M, Folgar Portillo F, Wunderlin R and Mehrabian R 1982 *J. Mat. Sci.* **17** 477-498
Hutchings I M 1989 *Proc. Conf. Friction and Wear* Oct. 1989 Bad Nauheim FRG ed K-H Zum Gahr (Oberursel: Deutsche Gesellschaft für Materialkunde) (in press)
Hutchings I M and Deuchar D W T 1987 *J. Mat. Sci.* **22** 4071-4076
Hutchings I M and Wang A G 1988 *Phil. Mag.* **A57** 197-206
Jeal R H 1988 *Metals and Materials* **4** 539-542
Mercer A P and Hutchings I M 1988 *Wear* **124** 165-176
Prasad S V and Rohatgi P K 1987 *J. Metals* **39** 22-26
Ruff A W and Ives L K 1975 *Wear* **35** 195-199
Surappa M K, Prasad S V and Rohatgi P K 1982 *Wear* **77** 295-302
Turenne S, LeDéoré C and Masounave J 1988 *Wear Resistance of Metals and Alloys* ed G R Kingsbury (Metals Park: ASM International) 39-45
Wang A and Hutchings I M 1989a *Mat. Sci. Tech.* **5** 71-76
Wang A and Hutchings I M 1989b *Wear* **124** 149-163

Inst. Phys. Conf. Ser. No 111
Paper presented at Int. Conf. on New Materials and their Applications, Univ. of Warwick, 1990

Tribology of hard coatings

R. Kingswell, D.S. Rickerby* and S.J. Bull

AEA Industrial Technology, Harwell Laboratory, Oxfordshire. OX11 ORA
*Rolls-Royce plc, Derby.

ABSTRACT: Hard ceramic and metal coatings are used successfully in a wide range of applications. However the inter-relationships between microstructure, properties and wear behaviour are poorly understood. The control of coating microstructure by variations in deposition conditions, to a certain extent, offers the potential for optimising a coating for a given application. In this paper we discuss the abrasive, erosive and sliding wear behaviour of hard coatings and relate these to the microstructural changes (and hence physical properties) produced by variations in either process parameters or process type.

1. INTRODUCTION

Coatings are being used increasingly in applications where wear behaviour is a primary or important secondary requirement. For example physical vapour deposition (PVD) and chemical vapour deposition (CVD) has been used to deposit TiN and TiN/TiC multilayers on drill bits, Hintermann (1984), cemented carbide lathe tools, Bart and Woerner (1986) and gear cutting hobs, Niemi et al. (1986). Thick ceramic layers deposited by thermal spraying are often used in applications where their electrical, corrosion or thermal properties are of primary importance but wear resistance is an important secondary requirement, for example thermal barrier coatings on diesel engine and gas turbine components. The use of hard coatings offers several advantages over bulk materials since surface properties can be engineered to be independent of the bulk to produce a composite system whose overall performance exceeds that of a ceramic monolith, also product costs can be reduced by using cheaper substrate materials.

The basic wear mechanisms of adhesion, abrasion and erosion are poorly understood for most coating systems. In addition it must be recognised that the unique microstructures developed in coatings processes results in mechanical properties and wear behaviour significantly different from that of similar bulk materials. In this paper the wear behaviour of PVD and thermal spray coatings are discussed and related to their unique microstructures and properties. Factors effecting wear behaviour under abrasive, adhesive and erosive wear conditions are given together with examples for specific coating systems.

2. EXPERIMENTAL

Sliding wear tests were performed on a pin-on-disc tester using a 3 mm diameter steel sphere counterface. Tests were performed unlubricated using a load of 2.5 N. Wear depth was measured by means of a displacement transducer attached to the moving arm holding the indenter.

Abrasion tests were performed using a Taber abrasive test machine (Teledyne, Taber, North Tonawanda, N.Y., USA). Two abrasive wheels were caused to rotate by the samples rotating beneath them at a rate of 1 rev s^{-1}. Weight loss is an average of four identical samples and is expressed as a function of the number of revolutions. During each test wear debris was removed by a suction device, in addition the abrasive wheels were dressed between each test, to remove adhered debris.

Erosion tests were performed using a modified bead blaster (Guyson Ltd., UK) through which predetermined quantities of alumina grit were passed. Compressed air was used to control the erodent velocity, which was varied in the range 50 to 180 m/s.

3. DEPOSITION TECHNIQUES, MICROSTRUCTURES AND PROPERTIES

3.1 Deposition Techniques

The growing importance of ceramic coating technology is evident from the wide variety of techniques which have been developed for coating production, Figure 1.

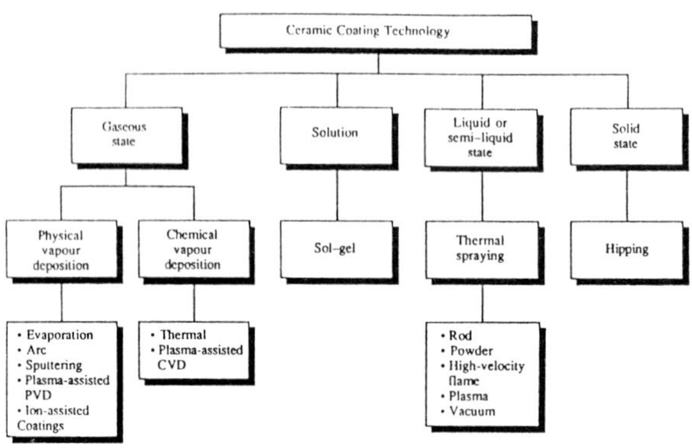

Fig. 1 Classification of coating processes.

Of these physical vapour deposition (PVD), chemical vapour deposition (CVD) and thermal spraying are used commercially to produce adherent, high density deposits. These techniques cover coating production from less than 1 μm to several millimetres as shown in Figure 2. PVD coatings are applied using evaporation dependent or sputtering dependent techniques. The former takes place under a low pressure (< 10^{-5} torr) and the coating material follows a collision free line of site path to the substrate. Therefore, difficulties can arise when coating components of complex geometry. This can be overcome to some extent by using multiple sources as in the Mulli-Arc system. Alternatively, a sputtering process can be employed, where the higher pressure, larger cathode and good 'throwing' power enables different component geometries to be coated simultaneously. In PVD processes the coating quality is largely dictated by the substrate cleanliness, its temperature during deposition, the deposition pressure and application of a substrate bias; Rickerby and Newbery (1988).

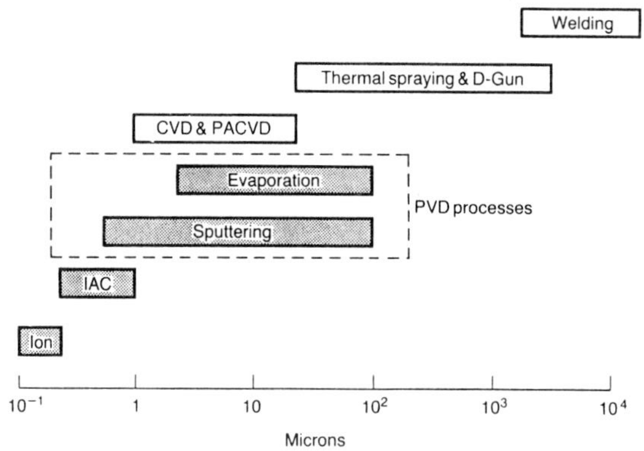

Fig. 2 Effective depth of treatment for a range of coating technologies.

Thermal deposition is used for production of coatings with thickness in excess of 50 μm. This lower limit is largely dictated by the deposition process, which involves heating and accelerating particles in a gaseous medium onto a substrate, where in a molten or semi-molten state, they spread to form lenticular 'splats'. The splats are typically 5 μm thick and to ensure a continuous layer a coating must be several splat layers thick. The most successful method of thermal spraying has been plasma spraying, where a DC arc is used to provide temperatures in excess of 20,000°C and nozzle exit velocities of 200-500 ms^{-1}. There are two basic techniques: air plasma spraying (APS) and vacuum plasma spraying (VPS). The latter can be used to eliminate the deleterious effect of oxide entrapment at intersplat boundaries in metal coating systems. The higher velocities obtainable in VPS also leads to reduced porosity and improved intersplat cohesion; Wolf and Longo (1980).

3.2 Microstructures and Properties

PVD and thermal spray coatings are anisotropic as shown in Figure 3. PVD coatings consist of columnar grains perpendicular to the substrate surface, while thermal spray coatings consist of lenticular splats with the splat boundaries parallel to the substrate surface. In both PVD and thermal spraying the grain/splat boundaries have a significant influence on the mechanical properties and wear behaviour.

The columnar structure present in PVD coatings has been attributed by Dirks (1977) and Henderson (1974) to a combination of atomic shadowing and limited surface mobility. The structures associated with PVD deposition have been classified by Movchan and Demchishin (1969). Three main structures were identified. Zone 1 consisting of tapered columns with domed tops. Zone 2 having parallel columns with a planar surface and Zone 3 which consists of equiaxed grains. Later Thorton (1977) identified a Zone T consisting of poorly defined fibrous grains. The type of microstructure produced will depend on the deposition conditions.

Increasing the substrate temperature and/or substrate bias during deposition tends to alter the structure from a Zone 1 type to that of a Zone T or 2.

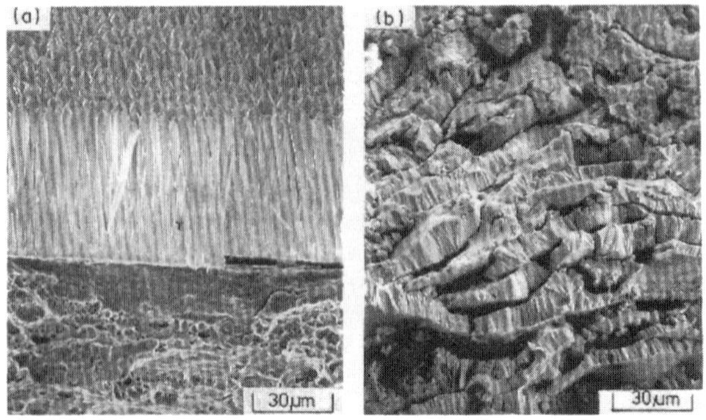

Fig. 3 Fracture surface of a PVD and an APS coating showing the anisotropic grain structure.

Thermal spray deposits consist of lenticular splats, porosity and spherical or angular unmelted particles. Intersplat cohesion is dependent on the degree of melting, particle velocity on impact, coating material and surface temperature during deposition. Porosity tends to decrease with increasing particle velocity. However, high velocities reduces the dwell time of particles in the plasma, increasing the likelihood of entrapped unmelted particles in the final structure leading to higher porosity levels and reduced particle cohesion. Therefore, obtaining the required structure needs careful control of the numerous process parameters.

The properties that are influenced by the degree of grain cohesion are also important in determining wear behaviour. These properties are summarised in Figure 4.

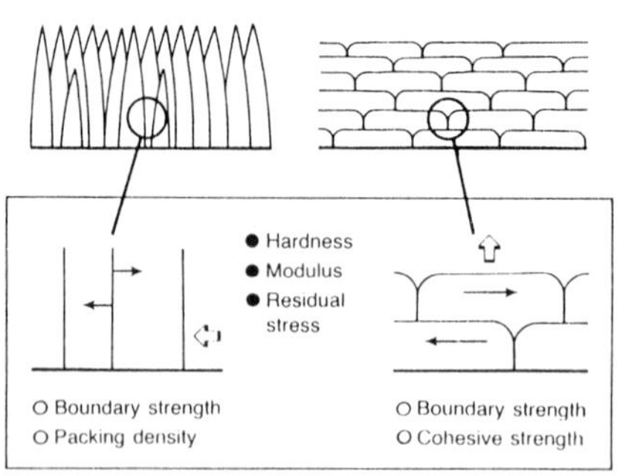

Fig. 4 Schematic diagram of PVD and thermal spray coating microstructures.

In PVD coatings high packing density and improved intercolumnar cohesion leads to increased hardness but also increases the residual stress levels presesnt in the coating, reducing the critical load for coating detachment. These effects are illustrated in Figure 5, where substrate bias has been used to control coating density.

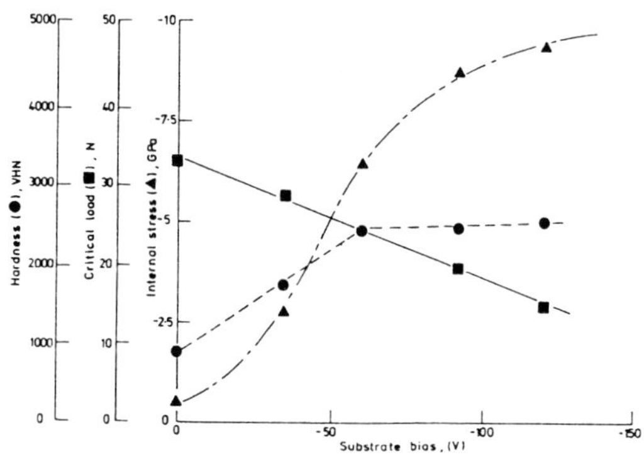

Fig. 5 Variation in coating hardness, critical load for coating removal and residual stress level as a function of substrate bias during deposition.

The properties of thermally sprayed coatings can also vary significantly, for example, intersplat cohesion has a significant effect on the apparent Young's modulus of a coating, as shown in Figure 6, which contrasts the behaviour of VPS, APS and bulk tungsten. Air plasma spray tungsten has considerable levels of oxide at the grain boundaries and under an applied load splat slippage occurs along these weak planes resulting in a low Young's modulus and a low stress to failure. Reducing the oxide level by vacuum plasma spraying more than doubles the apparent Young's modulus and has a significant affect on fracture behaviour of the coating. The APS coating fails by fracture along the splat boundaries, giving a stepped fracture appearance, while the VPS coating fails by fracturing of splats, resulting in a planar fracture surface. Residual stresses are higher in the VPS coatings partly as a result of the higher deposition temperatures which result during spraying, but also because of the lack of stress relief from splat slippage.

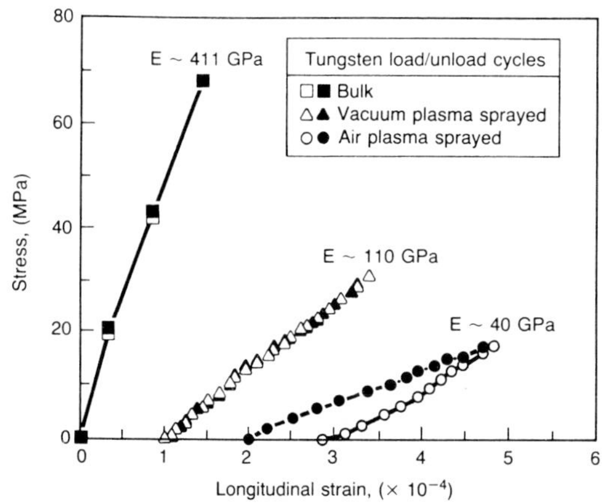

Fig. 6 Typical load extension curves for APS, VPS and bulk tungsten materials.

4. THE WEAR BEHAVIOUR OF HARD COATINGS

In common with bulk materials coatings can wear by a number of different mechanisms, and these can be divided into four categories:

(i) Abrasive wear - involving the cutting of one material by particles of another during sliding wear.

(ii) Adhesive wear - involving the formation of junctions between materials and the removal of debris by plastic deformation or oxidation during sliding wear.

(iii) Chemical wear - involving the chemical interaction between sliding surfaces to produce new, weakly bonded phases which can subsequently be removed (e.g. oxidative wear).

(iv) Erosive wear - involving the removal of material by impacting particles either by plastic or brittle fracture.

The mechanisms operating in any situation will depend on the test conditions, the microstructure and properties of the coating and counterface material, be it coated or uncoated.

4.1 Sliding Wear of PVD Coatings

The mechanisms of material removal from PVD coatings that can operate during sliding wear are summarised in Figure 7.

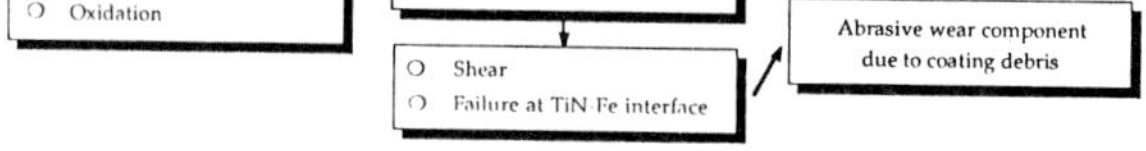

Fig. 7 Important wear mechanisms acting during sliding wear of hard PVD coatings.

Surface adhesion is high for uncoated steel running against a steel counterface resulting in high contact shear stresses and material loss by delamination and/or oxidation. In contrast, the mechanisms operating when coated samples are running against a steel counterface are dependent on the coating microstructure. Under sliding wear conditions PVD coatings of TiN with open, Zone 1 type microstructures (see Figure 8(a)) wear via an adhesive mechanism. The pyramidal tops of the open columns act as abrasive particles cutting into the steel sphere removing material from the sphere which becomes embedded into the rough surface of the coated disc. Adhesion then occurs between the steel particles on the disc and the steel sphere, resulting in high shear stresses within the sphere and TiN coating, similar to that observed in tests on uncoated steel discs. The high stress levels developed leads to failure at the coating/substrate interface and relatively high wear rates. As the coating density is increased there is a move towards the denser Zone T, 2 microstructure (see Figure 8(b)) and as the columns of TiN are forced into close contact the likelihood of sphere material becoming transferred is reduced. In this situation adhesion is no longer the dominant mechanism and gross failure of the coating or sphere is not observed. Instead the mechanism is mild, oxidative wear and the debris produced is a brown powder of Fe_3O_4; Bull and Rickerby (1989).

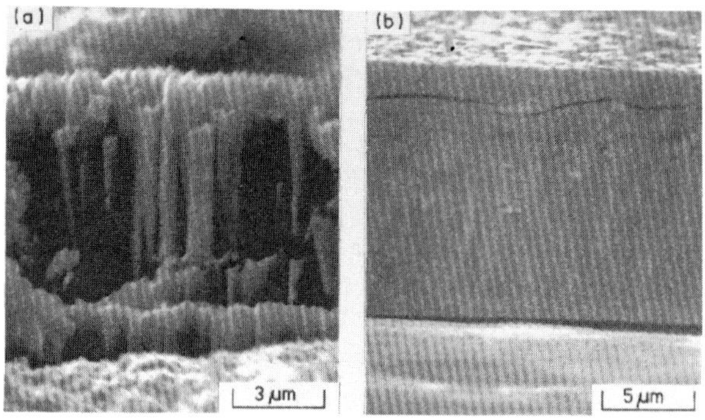

Fig. 8 Fracture surfaces of PVD coatings
(a) open Zone 1-type microstructure,
(b) a dense Zone T-type microstructure.

4.2 Abrasive Wear of PVD Coatings

The abrasive wear behaviour of thin PVD coatings is also dependent on the coating microstructure and hence on the coating technique and deposition conditions. These effects are summarised in Figure 9.

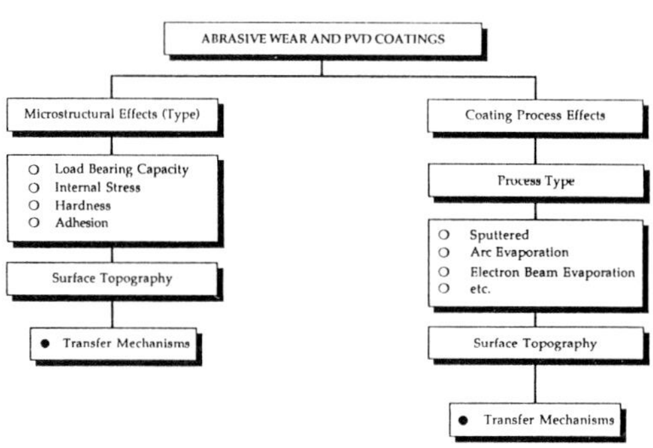

Fig. 9 Important process and microstructural effects in the abrasive wear of PVD coating.

As was the case for sliding wear, the behaviour of the coating is significantly influenced by the 'openness' of the columnar structure. The results of abrasive wear tests performed on TiN coatings possessing the various types of microstructures are shown in Figure 10.

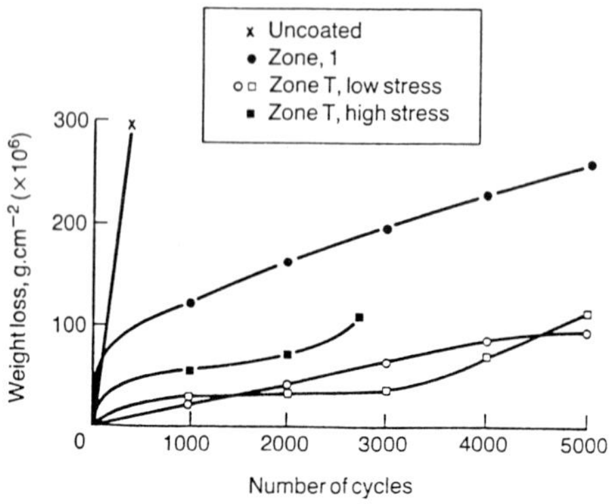

Fig. 10 Variation in weight loss with number of cycles for uncoated and TiN coated steel tested under mild abrasive conditions.

The counterface material was a coarse abrasive wheel (CS17). For uncoated steel specimens wear proceeds via plastic ploughing with material loss occurring at the intersection of scratch tracks. The Zone 1 sample possessed an open columnar microstructure which terminated with pyramidal shaped asperities. Abrasion tests on this type of sample produced an initially rapid wear rate, which can be explained by the removal of the loosely bonded titanium nitride columns.

As the substrate bias is increased the structure becomes denser and the coating surface becomes planar, as shown in Figure 8(b). The higher hardness resulting from the increase in density offers improved load support and thus the degree of ploughing is reduced. However, as shown in Figure 5, increases in the substrate bias can also produce an increase in the residual stress levels present within the coating which results in a reduction in the critical load for coating detachment. In such situations the stress levels developed ahead of an abrading particle can lead to debonding and spallation in a similar manner to failure ahead of a single pass scratch test. Similarly stresses developed behind an abrasive particle can lead to fracture and material loss after one pass or several passes of abrasive particles. As the number of cycles increases the regions where coating loss has occurred grow by a microchipping mechanism. With the highly stressed Zone T microstructure, pit formation, resulting from debonding at the coating/substrate interface, becomes more severe, a direct consequence of the high residual stress levels developed during coating formation. Thus an optimum level of residual stress appears to give the best combination of high hardness and adequate coating/substrate bonding which in turn leads to maximum wear resistance for Zone T microstructures.

4.3 Erosive Wear of Thermally Sprayed Tungsten

Erosion tests performed on APS and VPS tungsten have highlighted the importance of microstructure in determining wear behaviour. The important parameters affecting coating behaviour are summarised in Figure 11, together with the erosion mechanisms observed.

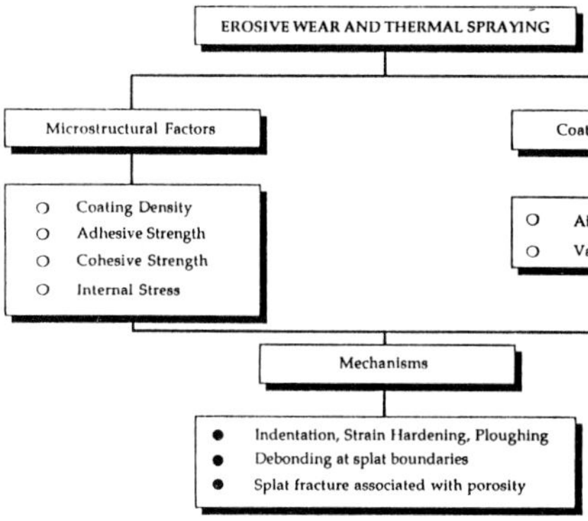

Fig. 11 Important process and microstructural effects in the erosive wear of thermally sprayed materials.

The poor bonding between splats in APS metal coatings resulting from the oxide formation at intersplat boundaries has been discussed previously. In such cases significantly higher erosive wear rates are observed when compared with VPS and bulk materials, as shown in Figure 12 where wear rate is given as a function of erodent velocity.

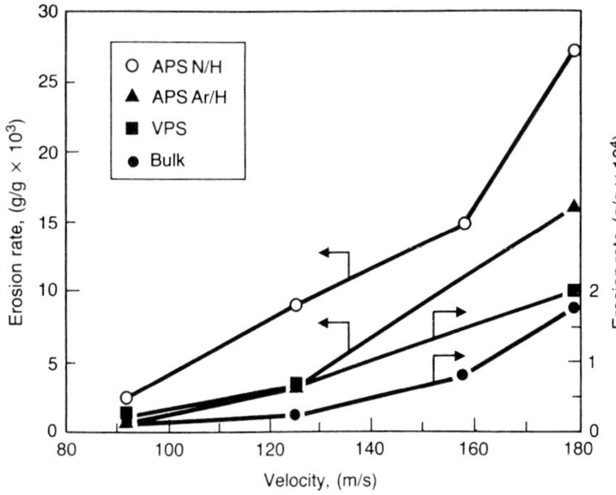

Fig. 12 Variation in erosion rate with erodent velocity for bulk and plasma sprayed tungsten.

Tests on all three classes of material (APS, VPS and bulk) gave rise to an exponential relationship between steady state wear rate and erodent velocity. The shape of the curves were indicative of erosive wear occurring via brittle mechanisms. For APS tungsten, the high erosion rates observed result from material loss occurring via fracture along the weak splat boundaries. The energy required for removal of splats is very low and there is little evidence of damaged splats because the removal rate is so high, i.e. removal occurs after very few impact events. In contrast, the well-bonded splats in VPS coatings are not removed as a whole during erosion testing. The main mechanism of material removal is microchipping, often associated with porosity. Ductile behaviour was also observed in tests on VPS and bulk tungsten, as wear scores were present on the eroded surface. This was particularly noticeable at low impingement angles; Kingswell et al. (1990).

5. CONCLUSIONS

The microstructure and properties of hard coatings produced by physical vapour deposition and thermal spraying techniques are significantly different from those of the bulk. The wear properties are dependent on the microstructure developed during deposition and particularly the strength of the grain/splat boundaries. Optimum wear properties can only be achieved if the wear mechanisms operating in a given application are known and the process type and conditions used have been optimised to produce the required coating structure with the desired mechanical, properties. The results given here indicate that with proper control the benefits of hard coatings in wear applications can be significant.

6. REFERENCES

Bhat, D.G. and Woerner, P.F., 1986, J. Metals, **38** 68.

Bull, S.J. and Rickerby, D.S., 1989, Proc. 16th Int. Conf. on Metallurgical Coatings, San Diego, California.

Dirks, A.G. and Leamy, H.J., 1977, Thin Solid Films, **47** 219.

Henderson, D., Brodsky, M.H. and Chaudhari, P., 1974, Appl. Phys. Lett., **25** 641.

Hinterman, H.E., 1984, Wear, **100** 381.

Kingswell, R., Rickerby, D.S., Bull, S.J. and Scott, K.T., 1990, Proc. 17th Int. Conf. on Metallurgical Coatings, San Diego, California.

Movchan, B.A. and Demchishin, , 1969, Fizika Metal., **28** 653.

Niemi, E., Korhonen, A.S., Harju, E. and Kauppinen, V., 1986, J. Vac. Sci. Tech., **A4** 2763.

Rickerby, D.S. and Newbery, R.B., 1988, Vacuum, **38** 161.

Thorton, J.A., Ann. Rev. Mater. Sci., 1977, **7** 239.

Wolf, P.C. and Longon, F.N., 1980, Proc. 11th Int. Thermal Spray Conf., The Hague.

Inst. Phys. Conf. Ser. No 111
Paper presented at Int. Conf. on New Materials and their Applications, Univ. of Warwick, 1990

High temperature fretting damage to silicon nitride ceramics: Palliation by surface treatment

Z.P. Wang*, C. Ruiz
Oxford University, Dept. Engineering Science, Parks Road, Oxford, OX1 3PJ
(* now at sheffield University)

ABSTRACT: The palliation of fretting damage between silicon nitride and a nickel alloy at 600^0C by means of surface treating the ceramic has been studied. Of the four treatments considered, ion implantation offers the greater potential while sputtering with Zr/ZrN and gold coating are least effective.

1. INTRODUCTION

Silicon nitride is regarded as a candidate for the fabrication of gas turbine components, such as blades. These components may be joined to metals through interlocking joints. A typical case it that of the turbine blades, held by means of dovetail joints to nickel alloy disks. The contact conditions in such joints consist of a contact pressure of about 150 MPa and relative displacements of $\pm$ 0.5 mm at temperatures equal to or just above 600^0C. These conditions have been reproduced in a number of tests in which a 50x3x5 mm beam of the ceramic is loaded in 3-point bending under a fixed force. A pad, fixed through a connecting rod to a crankshaft with an adjustable stroke, is pressed against the face of the beam under tensile stress, causing the damage. The residual strength in the beam after a given number of oscillations of the pad is measured by removing the pad and increasing the main force until fracture occurs. A detailed description of the experiment and a discussion of the results obtained with as-ground and ground and polished beams has been presented elsewhere [1,2,3].

In all cases, the surface damage was sufficient to initiate small cracks that caused a severe reduction in the residual strength of the beam. To alleviate this damage, four actions have been considered : a) to prevent the direct contact by means of a soft interlayer, b) to decrease the local contact stress introducing a relative thick compliant layer, c) to increase the fracture toughness of the ceramic and d) to increase the surface hardness. Of these, only the first and the last have been studied in this investigation.

2. Zr-ZrN SPUTTERING

Sputtered zirconium nitrides are used to improve the wear resistance of tools [4] A nitride film is brittle and tends to flake. When sputtering Zr on a Si_3N_4 surface, the strong affinity of Zr to oxygen means that a zirconium silicide, giving an improved bond between the surface and the coating, is formed. ZrN is an effective oxidation barrier that gives substantial protection in a wear environment [5]. The thermal expansion coefficients of Zr and ZrN are similar to that of Si_3N_4, thus avoiding

the problem of flaking off due to thermal stressing.
Thirty specimens have been treated by R.G. Duckworth, in accordance with the following procedure:

1- Sputtering Zr in vacuum of 2.5×10^{-3} torr of argon at 3kw for 6 cathodes and 15 minutes, giving a 0.25 μm layer.

2- Sputtering Zr in a vacuum of 2.0×10^{-3} torr of argon and 0.8×10^{-3} torr of nitrogen at 3kw for 6 cathodes and 45 minutes, giving a 0.25 μm layer of ZrN.

3- Bombarding Ne in a vacuum of 10^{-6} torr at 500 keV (dose 10^{16} ions/cm^2). The ions, scattered following a gaussian distribution, produce a maximum damage at depth of about 0.5 μm, improving on the other hand the bond strength.
The test results are summarised in Table 1.

Table 1-Zr/ZrN treatment

Set number	Number per set	M.o.R. (MPa)	Standard Deviation	Coeff.of Variation	Cycles	Weibull modulus (m)
9	10	556 (560)	70	0.127	0	9.7 (10)
10	10	452 (418)	67	0.147	15,000	7.3 (4.5)
11	10	394 (374)	46	0.118	30,000	10 (10)

In Table 1, the numbers in brackets correspond to the as-ground (untreated) specimens, tested for comparison. Each set consisted of 10 specimens, set 9 was tested as treated. In sets 10 and 11 the modulus of rupture and Weibull modulus (M.o.R.) were determined after 15,000 and 30,000 wear cycles respectively. The results show that there is virtually no difference between the treated and the untreated specimens before wear but that surface treatment does improve the wear resistance as reflected by the residual M.o.R. and Weibull modulus.

3. ION IMPLANTATION + Zr/ZrN SPUTTERING

Ion implantation is currently attracting interest as a possible surface treatment process for improving wear-resistance in a wide range of materials. The majority of the investigations performed to date have concentrated on the effect of ion implantation in metals. Modifications to the mechanical properties of ceramic surfaces have been reported for sapphire [6] and SiC [7]. The intention was to increase the wear resistance of the Si_3N_4 by improving hardness and indentation fracture toughness and thus resist the formation and propagation of surface cracks while the Zr/ZrN sputtering would reduce contact damage during the initial period by preventing direct contact.
Fifty specimens were treated by R.G. Duckworth, in accordance with the following procedure:

1-Sputtering Zr in a vacuum of 2.5×10^{-3} torr of argon at 3kw for 6 cathodes and 2 minutes to give 10 nm layer.

2-Implanting N^+ at 100 keV (dose 1.7 to 3×10^{17} ions/cm^2)

3- Sputtering Zr in a vacuum of 2.5×10^{-3} torr of argon at 3kw for 6 cathodes and 45 minutes

4- Sputtering Zr in a vacuum of $2.0x10^{-3}$ torr of argon and $0.8x10^{-3}$torr of nitrogen at a power of 3kw for 6 cathodes for 45 minutes
5- Repeating 3,4 twice
6- Repeating 3.
The final coating had a thickness of about 5 μm.
The test results are summarised in Table 2, where the figures in brackets correspond to the untreated specimens, as before.

Table 2. Ion implantation + Zr/ZrN sputtering

Set number	Number per set	M.o.R. (MPa)	Standard Deviation	Coeff.of variation	Cycles	Weibull modulus (m)
12	10	564 (560)	77	0.137	0	8.2 (10)
13	20	464 (418)	82	0.176	15,000	7.0 (4.5)
14	20	426 (374)	58	0.137	30,000	10 (10)

As in the case of the plain sputtering, the as-treated specimens do not show any improvement over the plain as-ground specimens, but there is now a more pronounced improvement in the residual M.o.R. after 15,000 cycles and after 30,000 cycles of wear by 11% and 14% respectively. The corresponding figures for the sputtering without ion implantation are 8% and 5%. It may be assumed that improvement at the early stages of wear is due to the presence of the interlayer while at the late stages it is an increase in the fatigue resistance, i.e., as a result of the resistance to crack propagation imparted by the ion implantation. The Zr/ZrN coating is easily detached from the surface as shown in Fig.1 and the drop in strength that occurs after the first few wear cycles is due to the direct contact between ceramic and metal. The only reason why the second batch of specimens-sets 12,13,14- is slightly better than the first-set-9,10,11-is the thicker coating.
Once that wears off, there is little difference between the untreated and the specimens treated by sputtering Zr/ZrN.
Only the ion implantation makes any difference in the long run.
It is interesting to note that the Weibull modulus is virtually the same for set 9 and for the as-ground specimens and drops slightly, from 10 to 8.2 in set 12.

Fig.1. Worn surface (ion implant.+ Zr/ZN)

A possible reason is that ion implatation introduces compressive residual stresses around fracture origins near the surface while it also causes surface damage that could act as new fracture starters.
These two factors, having reverse effects on the strength, result in an increase in the scatter of the M.o.R. around the same average value. In the as-ground specimens,m drops to a very low value - 4.5 - when wear damage is sufficiently severe to present a new risk of failure in addition to the machining damage and internal defects that were present originally.
As wear damage becomes more intense,it overshadows the two potential causes and there is less scatter, m increases while the M.o.R. continues to drop.
The most significant effect of the surface treatment is seen to increase the Weibull modulus at the early stage of wear, improving the reliability of the material.

4. GOLD COATING

Gold, used as a solid lubricant, was sputtered to produce a 28 nm layer on a specimen. In another specimen, Ne was bombarded onto the surface after sputtering a 14 nm layer and a second coating of the same thickness was then sputtered. The two specimens were worn for 15,000 cycles and the wear process was observed by removing the specimens from the testing rig every few cycles.

Fig.2. Gold coated specimen after 1000 cycles

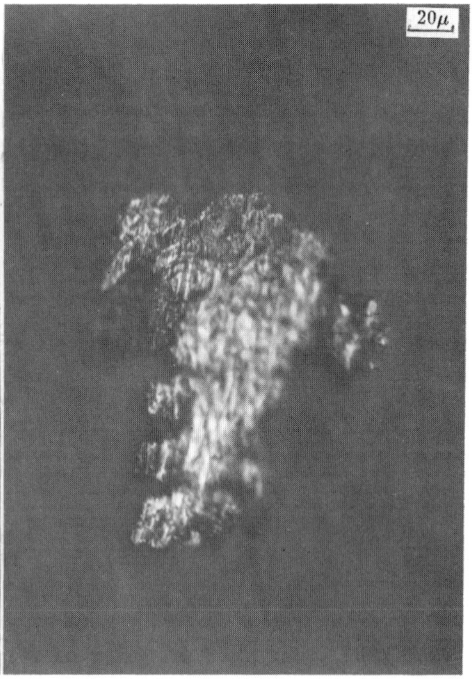

Fig.3 Particle detached from the coating

In the specimen without bombardment, the coating rubbed off in a few cycles- Fig.2.- but piled up between the contact surfaces, flaking off as shown in Fig.3., indicating the presence of high local contact stresses and large plastic deformation. There was no reduction of the contact damage although it might be expected that the gold particles would act, as it was intended, as a lubricant. The possible reason is the presence of ceramic particles, abrasive and hard. Bombardment improved adhesion but had no effect on the resistance to wear damage.

5. NICKEL COATING

Nickel coatings are widely used to increase resistance to corrosion and wear in metals and plastics but have not been applied to ceramics due to the mismatch between the thermal expansions of the base and the coating. Several routes were followed unsuccessfully until a technique was finally developed to produce a bond between the nickel and the Si_3N_4 that was Bregarded as satisfactory.

The technique consisted in:-

1- Sputtering nickel in a vacuum of 0.1 torr of argon for 2.5 hours to give a 30 nm layer

2- Bombarding Ne in a vacuum of 10^{-6} torr at 500 keV (dose 10^{16} ions/cm^2)

3- Repeat 1 for about 40 hours to obtain a layer 470 nm thick. The total thickness of the coating is thus 0.5 μm.

Only one specimen was treated. After heating to 600^0C, numerous cracks

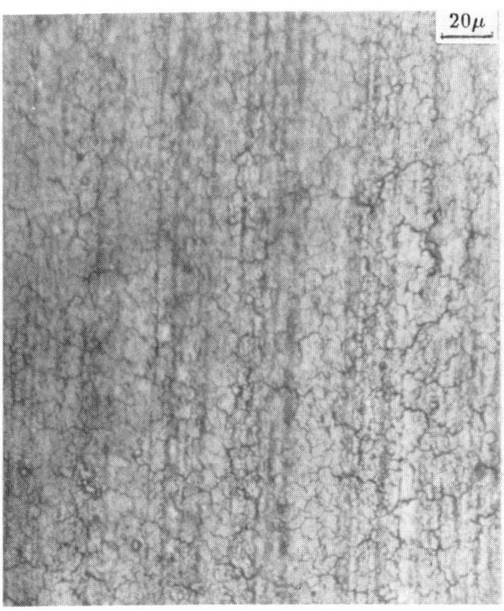

Fig.4. Nickel coating after heating to 600^oC

Fig.5. Worn nickel coating after 15000 cycles

appeared -Fig.4- due to thermal expansion mismatch. During the wear test, the progressive damage to the coating was monitored. After 15000 cycles- Fig.5- deep grooves were apparent, corresponding to similar markings in the pad. The nickel coating, even then, continued to act as a soft interlayer, deforming due to the high contact stresses but remaining fixed to the ceramic base.

6 CONCLUSIONS

Ion implantation, combined with a sputtered coating of Zr-ZrN, does improve the resistance to fretting fatigue to the ceramic to a greater extent than plain Zr-ZrN coating. The poor mechanical strength of gold and the weakness of the bonding with the ceramic make coating with gold quite ineffectual but nickel coating has considerable potential and is worth exploring further.

REFERENCES

1- Z.P. Wang and C. Ruiz, 'Effect of wear damage on the strength of HIP-Si_3N_4', J.Am. Ceram. Soc. 72,6 (1989) 1076

2- Z.P. Wang and C. Ruiz, 'Characterization of contact damage of Syalon in contact with Waspalloy', Wear, to be published

3- Z.P. Wang and C. Ruiz, 'Characterization of contact damage of Si_3N_4 in contact with Waspalloy', Trans. and Journal, Brit. Institute Ceramics, to be published.

4- R.G. Duckworth, 'High purity sputtered tribological coatings', Thin Solid Films, 86 (1981) 213

5- R.G. Duckworth, R.E. Happer and C. Jeyness, 'Back scattering analysis of ZrN alloys', Nuclear Inst. & Methods in Phys. Res., 15B (1986) 272

6- P.J. Bunnett and T.F. Page, 'Criteria for mechanical property modifications of ceramic surfaces by ion implantation', Radiation effects in insulators, Univ. Surrey, Guildford, July 15-19, 1985

7- S.G. Roberts and T.F. Page, 'The effect of N_2^+ and B^+ ion implantation on the hardness behaviour and near surface structure of SiC', J. Mater. Sci., 21 (1988) 457.

ACKNOWLEDGEMENTS

The advice and assistance of Dr. R.G. Duckworth, who did the surface treatment and the numerous helpful discussions with Dr.G. Syers, whose enthusiastic support for this project has proved invaluable, are gratefully acknowledged.

Inst. Phys. Conf. Ser. No 111
Paper presented at Int. Conf. on New Materials and their Applications, Univ. of Warwick, 1990

Ceramic rolling elements for machine tool spindle bearings

F P Wardle S W Siu A T Synnott
RHP Bearing Research Centre, P O Box 18, Newark, Notts., NG24 2JF

ABSTRACT: Machine tool spindle bearings are required to operate over a wide range of speeds and yet exhibit high rigidity, accurate running and low friction. As speeds increase above 0.5×10^6 DN (DN = bearing bore in mm x speed in rpm) centrifugal force acting on the rolling elements leads to a progressive increase in bearing friction and loss in rigidity. Ceramic rolling elements, due to their light weight reduce centrifugal force effects and enable significant improvements in bearing performance to be achieved relative to bearings fitted with steel rolling elements. In spring preloaded bearings friction levels are reduced by 30 - 50%, rigidity is increased by 12 - 100% and maximum bearing speeds may be increased by at least 30%. For back-back bearing arrangements induced preload at high speed is reduced by 70% and the limiting speed increased by 13%.

1. INTRODUCTION

Machine tool spindles represent a critical application for rolling element bearings. They are required to have extreme accuracy of rotation, high rigidity, running temperature little above ambient and in some applications, run at maximum speeds approaching those of Aero engines. Lubrication conditions for bearings are also minimal, either air-oil or grease and bearing loads are light (Hertzian stresses are less than 1500 N/mm^2). Under these conditions the parasitic load generated by ball centrifugal forces degrades bearing performance and ultimately limits bearing speed.

Ceramic rolling elements are attractive because of their light weight. Hot pressed silicon nitride has a density about 40% that of steel which, when translated to rolling element centrifugal forces represents a substantial reduction. Light weight though is not the only requirement of rolling element material. It must not degrade the operational reliability of the bearing in any way and it must be capable of being machined to the standards of precision and finish normally expected of machine tool spindle bearings. Hot pressed silicon nitride rolling elements are available in grade 5 precision with surface finishes better than 0.01 μm Ra. In this respect they are at least as good as their steel counterparts.

The technical advantages of ceramic rolling elements and their operational reliability in the machine tool bearing application is the main subject of this paper. The discussion is based upon theoretical analyses and experimental observations of bearing performance in machine tool spindle test rigs. Hybrid test bearings were not optimised in design but retained the same internal geometry as bearings with conventional steel rolling elements. Experimental results obtained so far show them to have an advantageous performance in terms of speed capability, reduced friction, increased rigidity and reduced thermal distortion.

2. RELEVANT PROPERTIES OF CERAMIC

Hot pressed silicon nitride (Si_3N_4) has so far appeared the most promising ceramic material available for rolling bearing applications. Apart from low density it has a number of properties which might be considered advantageous over steel. Table 1 compares the basic material properties of silicon nitride with that of bearing steel. Its lower thermal conductivity, lower thermal expansion, higher electrical insulation, higher hardness and good dimensional stability are beneficial in the high speed machine tool bearing application whilst its retention of hardness and strength at high temperatures is an advantage in other bearing applications.

Fatigue life is not an important parameter in machine tool spindle bearings, however, for it to remain so a new material should not result in a substantial life reduction. Fatigue testing has so far shown hot pressed silicon nitride to have an L_{10} life at least as good as bearing steel. Experimental data has however exhibited more scatter in individual results and occasional infant mortalities have been experienced. Coupled with the fact that a reliable non-destructive test technique is not yet available to detect sub-surface material flaws, the fatigue performance of hot pressed silicon nitride may currently be considered a disadvantage in safety critical applications.

Its modulus of elasticity is also higher than that of bearing steel and in a hybrid bearing the steel raceways may be subjected to higher Hertzian contact stress levels for the same level of loading than in a conventional bearing. This is not a disadvantage in the lightly loaded spindle bearing application but may be so in other bearing applications where fatigue life is a performance limitation.

Finally, the failure modes of silicon nitride are satisfactory for rolling bearing applications. It does not fracture as might be expected but spalls when subjected to high cyclic stresses or wears under low stress - high sliding speed conditions in a manner similar to bearing steel.

MATERIAL PROPERTY	ROLLING BEARING STEEL	SILICON NITRIDE
Density (g/cm^3)	7.85	3.19
Modulus of elasticity (N/mm^2)	208 000	315 000
Poisson's Ratio	0.3	0.26
Coefficient of thermal expansion (k^{-1})	11×10^{-6}	3.2×10^{-6}
Thermal conductivity (W/mK)	40 - 50	24 - 40
Specific Resistance (Ω mm^2/m)	0.1 - 1	$10^{17} - 10^{18}$
Hardness HV10	700 (20 °C)	1700 (<800°C)
Bending strength (N/mm^2)	2400 (20 °C)	700 (<1000 °C)
Fracture toughness ($MN/m^{3/2}$)	25	6 - 8

TABLE 1 COMPARISON OF BEARING STEEL AND CERAMIC MATERIAL PROPERTIES

3. PERFORMANCE LIMITATIONS OF HIGH SPEED MACHINE TOOL SPINDLE BEARINGS

Bore grinding, milling of aluminium spars and CNC turning are examples of applications where high speed angular contact ball bearings are used. The applications are characterised by intermittent or light cutting forces and resulting bearing stresses are low enough for material fatigue not to be a serious limitation to bearing performance. The parameters bearing stiffness, running temperature, torque and maximum speed are more important as they enable improvements in machine performance to be obtained in respect of cutting accuracy, throughput power and range of cutting conditions. Currently maximum speed is perhaps the most important bearing limitation as new tool materials are enabling higher machining speeds or specific cutting power levels to be used. Higher bearing speeds can be exploited either directly to achieve higher machining speeds or indirectly to achieve larger diameter, more rigid spindle assemblies.

In the machine tool spindle application maximum bearing speed, running temperature and rigidity are functions of mounting arrangement and/or lubrication conditions. Spindle bearing arrangements embrace a wide variety of configurations, mainly to support complex loading situations. In this paper the advantages of ceramic rolling elements are discussed relative to two bearing arrangements: the spring loaded and the back-back arrangement, and for two types of lubrication: air-oil and grease. These situations cover the different limitations which arise from a wide variety of spindle designs.

3.1 Spring Preloaded Air-Oil Lubricated Bearings

Spring preloading is intended to allow bearing-shaft assemblies take up any axial thermal expansions that may occur and so maintain constant bearing load. In this arrangement air-oil lubricated angular contact bearings exhibit minimum values of running temperature and torque at a specific oil flowrate, Figure 1. Reducing oil flow much below this optimum value results in bearing seizure whilst increasing oil flow above the optimum increases bearing torque and temperature significantly. In practice, a margin of safety is allowed and oil flows are set within a tolerance band just to the high side of the optimum.

Under these lubrication conditions bearing speed is limited by wear of ball, inner and occasionally outer raceway surfaces. The wear takes the form of banding; it is progressive and results in a slow deterioration in the running accuracy of the bearing and eventually failure through rough running, cage break-up or overheating. Figure 2 shows a micrograph of a steel ball surface after running at 1.35 x 10^6 DN for 1000 hours. The wear band is typical; gross material removal has occurred within the contact zone and a deposit has built up alongside it, comprising wear debris and a dark brown polymeric layer attributed to lubricant degradation.

At high speeds, the ball - inner ring and ball - outer ring contact angles are different as a result of ball centrifugal forces and pure rolling cannot occur (Fig.3a). The difference in inner and outer contact angles gives rise to a ball spinning motion with most of the spinning occurring at the inner raceway. It is this spinning motion which contributes most bearing wear. The inner raceway spin/roll ratio as defined in Figure 3b has been found to be useful in indicating the likelihood of wear. Duration tests between 100 and 1000 hours have been performed for bearings subjected to different axial loads and speeds and wear rates of the balls measured using a profilometer. The results are shown plotted against ball-inner raceway spin/roll ratio in Figure 4. The

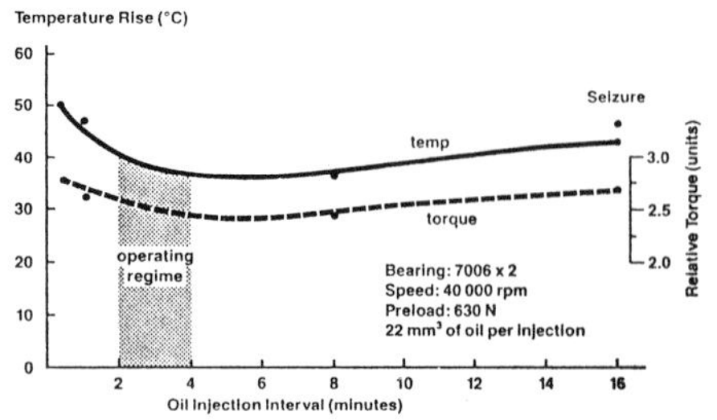

Fig. 1 Dependence of bearing temperature and torque on oil flow rate

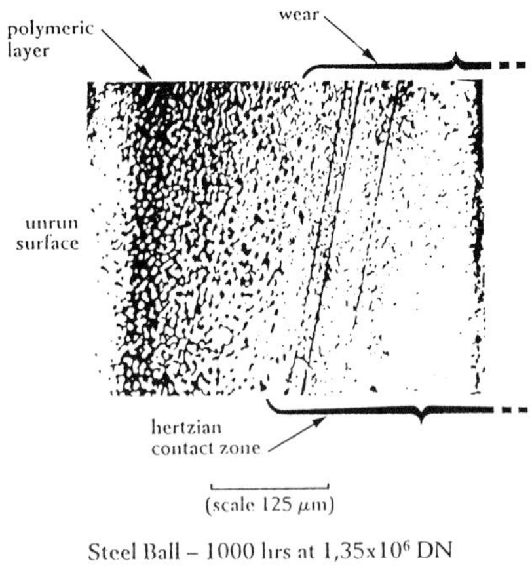

Fig. 2 Typical Spindle Bearing Wear Scar

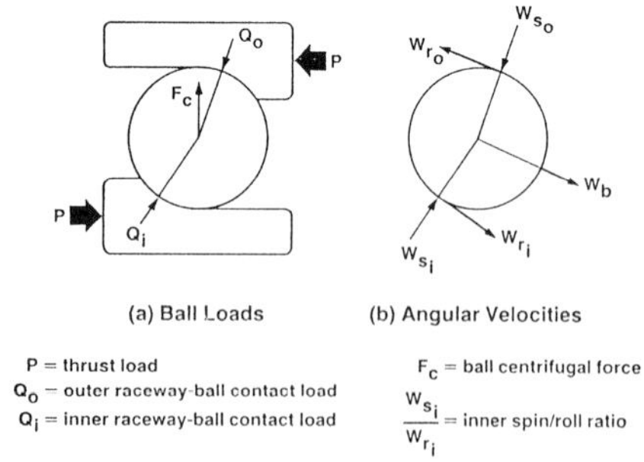

Fig. 3 Ball loads and angular velocities in high speed Angular Contact Ceramic Ball Bearings

wear rate is the linear wear depth of the ball per rolling distance. Although there is some experimental scatter for spin/roll ratios above 0.47, below this value wear rates are consistent and negligible. Figure 4 therefore supports a maximum spin/roll ratio criterion of 0.47 above which significant wear may be expected.

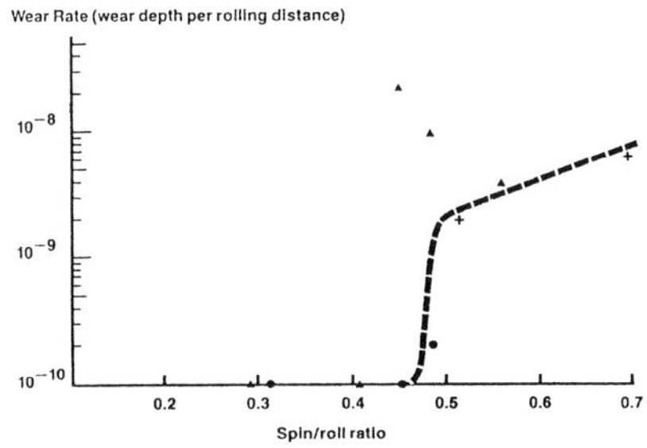

Fig. 4 Wear rate versus spin/roll ratio for machine tool spindle angular contact ball bearings

3.2 Spring Preloaded Grease Lubricated Bearings

Grease lubricated high speed angular contact bearings exhibit many of the characteristics of air-oil lubricated bearings but are subjected to the added limitations of grease life and no forced cooling.

The quantity of grease used to lubricate a bearing has a similar effect on running temperature and torque as oil flow-rate. There is an optimum grease quantity for which both parameters are a minimum. Reducing grease pack below this optimum results in bearing seizure whereas increasing grease pack above the optimum leads to churning and hence excess torque and temperature.

Grease type is also important. To date testing has shown a barium complex grease to exhibit the best high speed lubricating properties. Limiting speed is again governed by bearing surface wear and lubricant degradation. Figure 5 shows a micrograph of an inner raceway taken from a bearing run at 1.02 x 10^6 DN for 500 hours lubricated by Kluber LDS 18 Super. The wear pattern exhibits the same characteristics as that for air-oil lubrication shown in Figure 2. However, a basic difference between the wear process for grease and oil is that for the latter wear rates are stable whereas for grease, wear will accelerate during running as the grease loses its lubricity. The interval to initial wear, i.e. grease life, exhibits a speed dependence with a disproportionately low grease life being obtained at high speed.

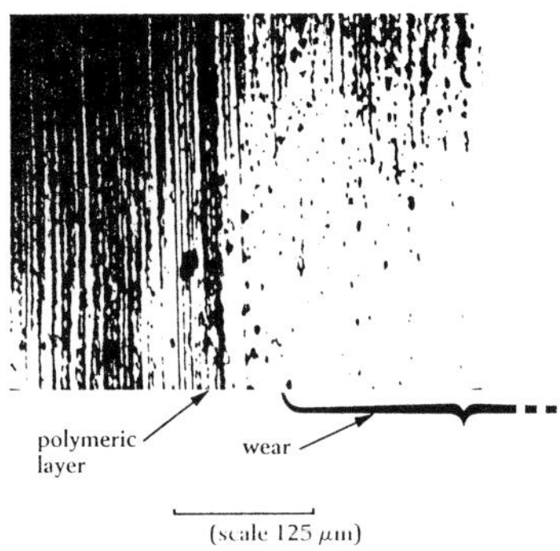

Fig. 5 Inner Race - 500 hrs at 1,02 x 10^6 DN Grease Lubricated

Maximum bearing speeds for grease lubricated bearings are determined from duration tests lasting at least 500 hours. On completion of the test the bearing and grease pack should not exhibit any signs of deterioration, thus inferring a substantially longer life could be achieved. Maximum bearing speeds determined in this way are approximately 70% of those for air-oil lubrication.

3.3 Back-Back Bearing Arrangements

In the back-back bearing arrangement inner and outer rings of both bearings are rigidly fixed in position and bearing preload is set by making the axial distance between inner track centres less than that between outer track centres, Figure 6. This arrangement gives good bearing stiffness and can take load in any direction but is sensitive to frictional heating. In a majority of applications heat generated by bearings is conducted more efficiently through the outer ring and housing than through inner ring and shaft, also more heat is generated at the inner than that at the outer so inner ring temperatures are usually higher than outer ring temperatures. In the back-back arrangement the difference between inner and outer ring expansions produces an increase in ball loads which is termed 'thermal preload'. For grease and air-oil lubricated back-back bearing arrangements thermal preload provides the main limitation to maximum bearing speed. At high speeds bearing power losses, frictional heating and temperature gradients are also high and a speed can be reached where thermal preload feeds back and increases bearing power loss still further. If this happens both bearing power loss and thermal preload continue to increase until seizure results.

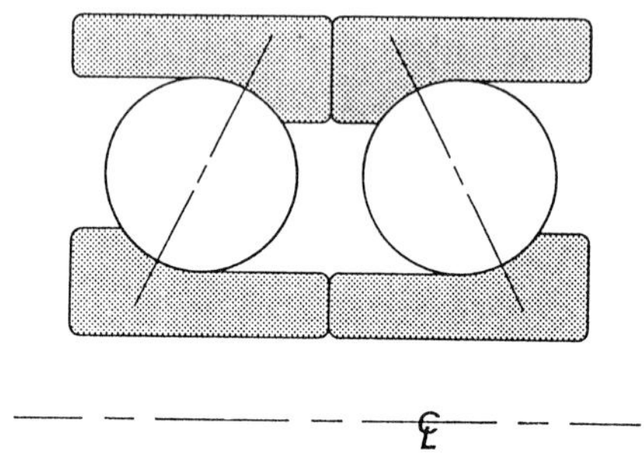

Fig. 6 Back to back angular contact ball bearing arrangement

The maximum speed of a back-back bearing arrangement depends to an extent upon the application as heat conduction rates and thermal growths are a function of shaft and housing design. Bearing power loss though, is an important parameter the reduction of which would enable increases in speed to be obtained in any given application.

4. THE ADVANTAGES OF CERAMIC ROLLING ELEMENTS IN MACHINE TOOL SPINDLE BEARINGS

4.1 Spin/Roll Ratio

As explained in section 3.1 maximum speed of spring preloaded, air-oil lubricated bearings is largely influenced by spinning friction and a limiting value of ball inner spin/roll ratio of 0.47 has been found to apply to the operational conditions encountered in machine tool spindle bearings. High spin/roll ratios are produced at the ball-inner raceway contacts when speeds and hence ball centrifugal forces are high enough to create differences in contact angle between ball-inner and ball-outer contacts. The benefit of the lighter weight of ceramic rolling elements can be seen from the theoretical comparisons of Figures 7 and 8 which shows how ball loads (Fig. 7) and spin/roll ratios (Fig. 8) vary with speed for steel balled and ceramic balled bearings of the same internal geometry subject to the same level of preload. Over a range of preloads the theoretical maximum speeds which can be attained by bearings with ceramic rolling elements, without exceeding a spin/roll ratio of 0.47, are about 40% higher than that for conventional bearings.

Examination of bearing surfaces after duration running have been consistent with theoretical results. For example Figure 9 shows a micrograph of a wear band on a ceramic ball run in a 7006 bearing at a speed of 45,000 rpm and 226 N preload for a period of 1000 hours. The test condition generated a theoretical spin/roll ratio of 0.47 and negligible wear was

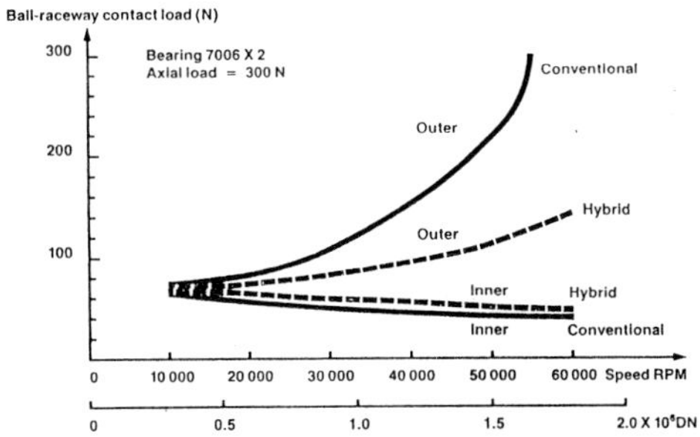

Fig. 7 Contact load versus speed for conventional and hybrid bearings

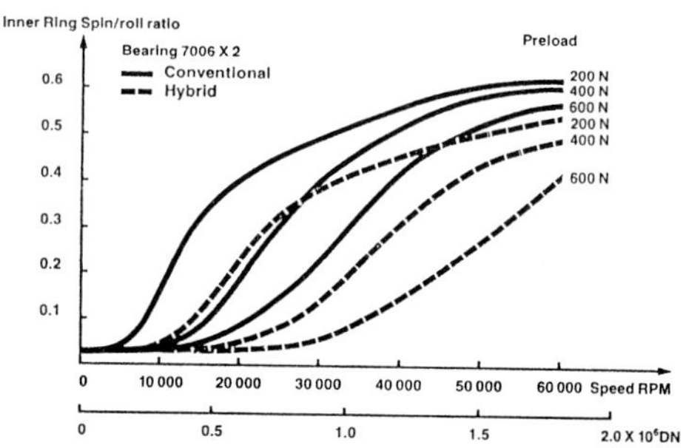

Fig. 8 Spin/roll ratio versus speed for conventional and hybrid bearings

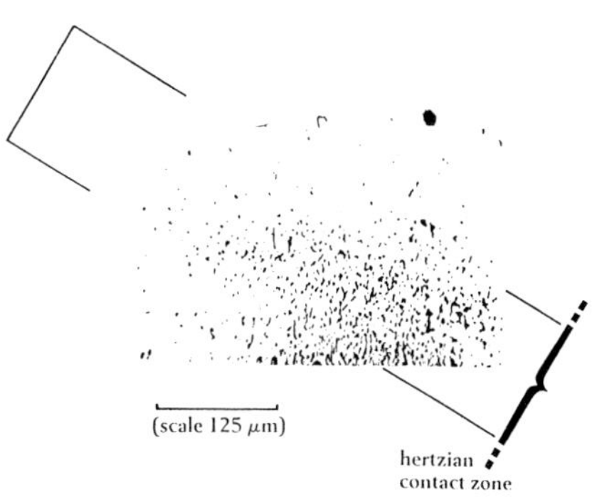

Fig. 9 Ceramic Ball - 1000 hrs at 1,35 x 10^6 DN

observed. Inner and outer raceway surfaces also remained in good condition. So far experimental work has shown the limiting speed of spring preloaded hybrid bearings to be at least 30% higher than for conventional bearings. The ultimate limiting speed has not yet been reached and further work needs to be performed.

The 0.47 spin/roll ratio criterion can be used to define a minimum control preload necessary for machine tool bearings to run at a given maximum speed without incurring wear, Figure 10. This enables preload to be matched to bearing application conditions and to be set at a practical minimum level in order to reduce friction and running temperature. Control preloads for ceramic balled bearings are about 70% less than that for conventional bearings.

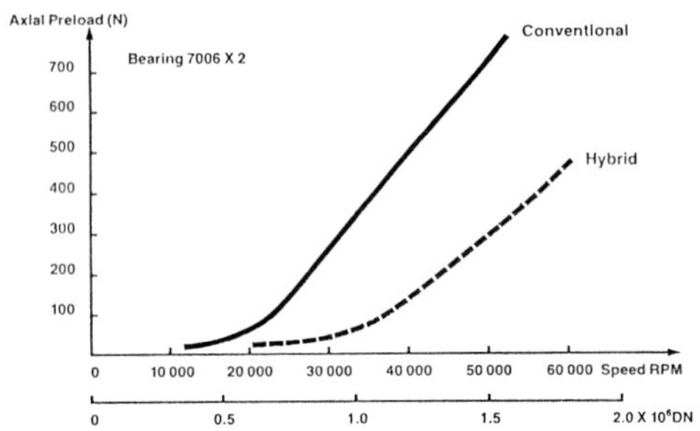

Fig. 10 Axial preload required to maintain a Spin/roll ratio of 0.47

4.2 Bearing Friction

The relative improvement in bearing friction levels produced by ceramic rolling elements was determined from temperature and supply power measurements taken on spring preloaded angular contact ball bearings. Both air-oil and grease lubrication conditions were investigated. The test rig comprised of a motorised spindle fitted with two 7006 size angular contact ball bearings each having a nominal contact angle of 15°. A water jacket built into the spindle housing enabled cooling of the stator.

Comparative tests were performed on bearings fitted with steel rolling elements and with ceramic rolling elements. Bearing temperature measurements are shown as a function of speed and load in Figure 11. Temperature rise above ambient recorded for the bearing fitted with ceramic rolling elements was found to be about half of that for the conventional bearing. Improvements in running temperature were also supported by supply power measurements. The bearings with ceramic rolling elements requiring only about 70% the supply power of the conventional bearings.

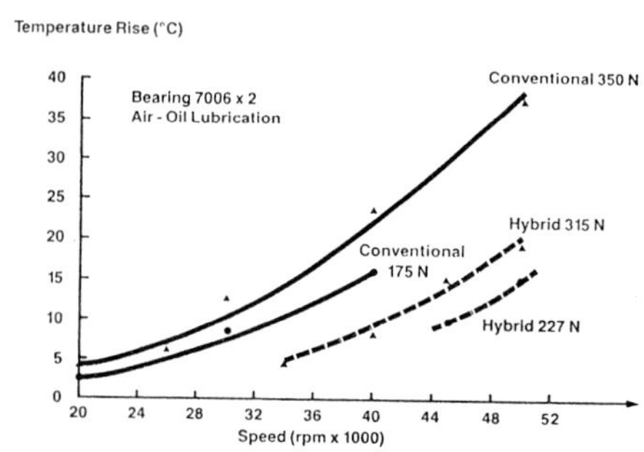

Fig. 11 Temperature versus speed for conventional and Hybrid Bearings

Similar results were obtained for grease lubrication. At a test speed of 34,000 rpm and a preload of 318 N the bearing with ceramic rolling elements was found to run at 10°C above ambient compared to 16°C for the conventional bearing, Figure 12. Supply power measurements for the bearing with ceramic rolling were about 75% of those for the conventional bearing.

4.3 Thermal Preload in Back-Back Bearing Arrangements

A pair of RHP 7013 x 2 (65 mm bore) angular contact ball bearings were mounted back-back at one end of a spindle. They were closely spaced (27 mm between centres) and face adjusted to give a total axial preload deflection of 12.5 um. This resulted in a preload of 125 N for the conventional bearings and 145 N for the hybrid bearings. Bearing preload could be determined from dynamic measurements of the amplitude of the ball pass stress wave on the outer ring. The stress wave amplitude - bearing load relationship was determined by calibration for known axial loads applied to the bearings.

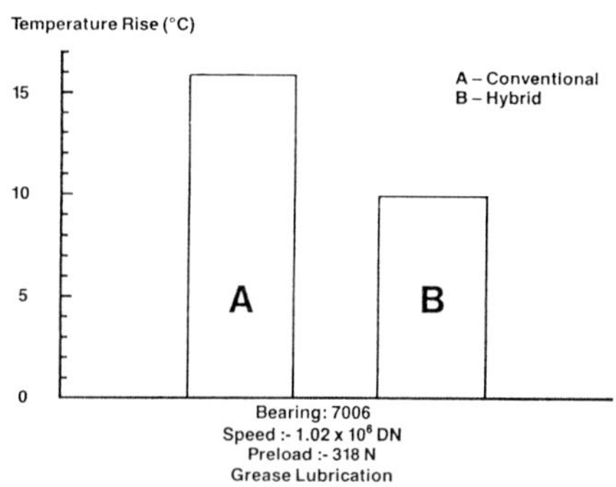

Fig. 12 Temperature rise for conventional and Hybrid Bearings

Thermal preload tests consisted of running the bearings at a set speed until outer ring temperatures stabilised. At this point preload and outer ring temperature measurements were recorded. The procedure was repeated at speed increments until a limiting speed was reached at which stable running conditions could not be obtained. Thermal preload was determined mainly for grease lubrication conditions. Figure 13 shows the results. For the hybrid bearings induced preload increased gradually with speed to 930 N at 17,000 rpm. Whilst for the conventional bearings preload increased at a higher rate to 2,150 N at 15,000 rpm. Thus in terms of induced preload at a given speed the hybrid bearings shows a 70%

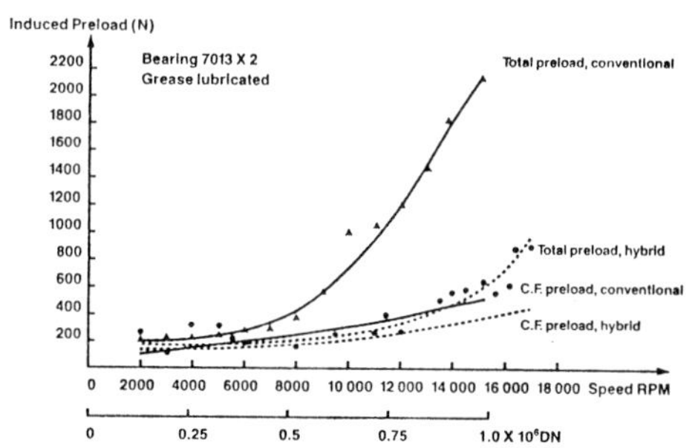

Fig. 13 Induced preload versus speed for back-back angular contact ball bearings

improvement. The induced preload has two components, one due to rolling element centrifugal force, the other due to thermal effects. Figure 13 indicates the component due to the centrifugal force for both steel and ceramic rolling elements. Thermally induced preload is dominant for steel rolling elements whereas for ceramic rolling elements the thermal preload is much lower in magnitude.

Running temperatures, Figure 14, correlate with induced preload. Temperature rise above ambient for the hybrid bearing being almost half that for the conventional bearing.

Similar results were obtained with oil mist lubrication. The hybrid bearings again producing about 70% less induced preload at a given speed than the conventional bearings.

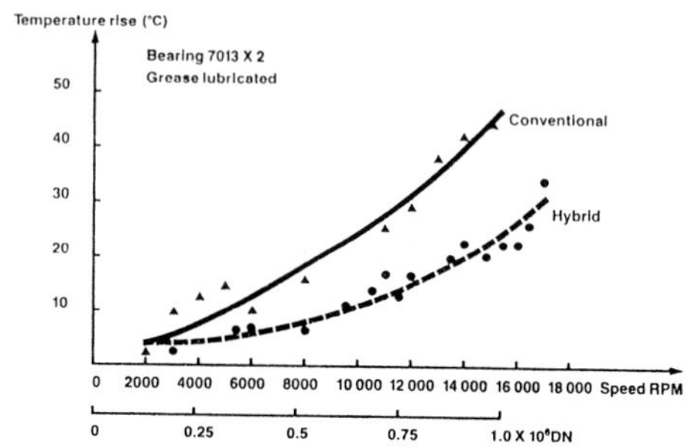

Fig. 14 Temperature rise versus speed for bearings arranged back-back

The results show thermal distortion to be substantially less in back-back bearing arrangements using hybrid bearings compared to conventional bearings mounted in the same arrangement. Thermal preloads at a given speed are reduced by almost 70% and speeds may be increased by at least 12%.

4.4 **Rigidity**

For a given bearing geometry the higher Young's modulus of ceramic results in hybrid bearing stiffness approximately 12% higher than that for conventional bearings under non-rotating conditions. As bearing speed increases the differences become more pronounced. Both axial and radial components of bearing stiffness reduce with speed due to ball centrifugal force creating a difference between inner and outer contact angles. The reductions can be as high as 70% of the static stiffness values, Figure 15, for both ceramic and steel rolling elements. The main advantage of ceramic rolling elements is that bearing stiffness is maintained to a much higher speed so that at a given speed the stiffness of a hybrid bearing may be up to twice that of a conventional bearing.

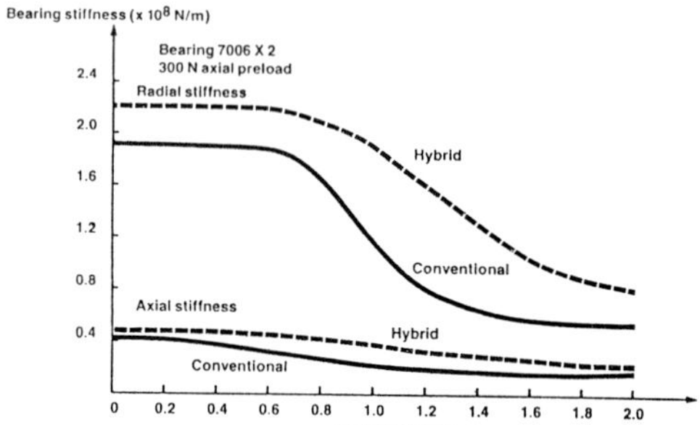

Fig. 15 Bearing stiffness versus speed for conventional and hybrid bearings

5. DISCUSSION

Ceramic rolling elements are a relatively new development in ball bearings and relative to steel rolling elements some of their current disadvantages will improve. Costs may be expected to reduce as improvements to manufacturing techniques are introduced and the difficulty in detecting subsurface flaws may be resolved as the results of research into inspection methods are made available. However, a disadvantage that cannot be resolved and which will limit their use in hybrid bearing forms is their high rigidity; under a given bearing load this results in higher Hertzian stresses than for steel rolling elements and for some high load bearing applications a theoretical reduction in fatigue life.

Ceramic rolling elements are therefore likely to be confined to specialist bearing applications and the machine tool spindle bearing is a good initial application. Material fatigue and the need for flaw detection are unimportant as Hertzian stresses are characteristically low. Higher speeds, higher rigidity, lower power losses and running temperatures are the technical improvements required of spindle bearings and in this respect the ceramic rolling element enables substantial gains to be made. Experimental work and theoretical analysis undertaken by RHP has shown hybrid spindle bearings can run at least 30% faster than their conventional equivalents under both grease and air-oil lubrication conditions. Bearing friction and running temperatures have also been shown to be reduced by 30 - 50% and rigidity to be increased by 12 - 100% depending on bearing speed. Induced preload in back-back bearing arrangements is also reduced by up to 70% whilst maximum speed is increased by 13%.

A point which should be made is that although the above improvements are substantial they were achieved on test rigs which have no other limitations and which allow the full benefits of ceramic rolling elements to be attained. Simply retrofitting hybrid bearings into existing spindle designs may not achieve such large improvements as spindle performance may be limited by other factors such as poor dynamic stability. Hybrid spindle bearings should however enable radical changes in spindle design to be made with the overall benefits of better rigidity and operating speed range.

ACKNOWLEDGMENTS

The authors would like to thank RHP Precision for funding the work reported in this paper and Spheric Engineering for the supply of silicon nitride balls.

Inst. Phys. Conf. Ser. No 111
Paper presented at Int. Conf. on New Materials and their Applications, Univ. of Warwick, 1990

Development of antifriction materials of new generation

V Nikolayev, V Domoratsky

State University, Gagarin Av. 72, Dniepropetrovsk-10, GSP 320625, The Ukraine, USSR.

ABSTRACT: Conventional self lubricating bearing materials have inadequate wear resistance, carrying capacity and resistance to corrosion in aggressive media. Various options for improving the structure of the materials to overcome these limitations are discussed. Cr-based coatings on the surface of the bronze skeleton were made by chemical vapour deposition of chlorides, carbonyls and other compounds. Application of 6-10% by volume of Cr to the bronze improves operating characteristics.

1. INTRODUCTION

Currently a great variety of types and sizes of self lubricating bearings are made of flouroplastic impregnated bronze for use in many industries. These bearings can stand large loads and work unlubricated over a wide temperature range (from -200 to +300 deg C). They work well in air or other gases, in vacuo, and in non-lubricating liquids. They do not give any problems with static electricity nor is fretting corrosion observed. These bearings are made from a strong backing (usually steel) plated with copper with a porous layer of antifriction alloy powder (usually bronze) sintered onto it and impregnated with a mixture of flouroplastic with a filler (e.g. PTFE D and molybdenum disulphide).

However application of materials of this type is currently restricted by problems of durability and capacity and by corrosion in aggressive environments (e.g. marine fogs).

To overcome these problems the authors have investigated methods of making these materials with a Cr-based protective coating on the bronze to increase corrosion and wear resistance. Care was taken to avoid the coating process altering the porosity characteristics of the structure.

2. POSSIBLE TECHNIQUES

The use of electroplated coatings and of solid state diffusion coating techniques is found to close the porosity. This prevents coating of the bronze particles beneath the surface and also prevents impregnation with solid lubricants. As a result the bearing properties are damaged and the number of scrap bearings rises.

The selected approach was the CVD of a mixture of Chromium and Chromium carbides on the surface of the sintered bronze skeleton. Chemical Vapour Deposition over a wide temperature range was carried out using various Cr containing compounds such as chromium chloride and organometallic compounds such as bisethylbenzylchromium. The compounds tested have decomposition temperatures ranging from 300 to 800 degC and various thermal decay rates in vacuo with either oxygen excess or deficiency. The first of these compounds, chromium chloride, has been previously used to treat other materials, but is not widely used.

The second variant is the use of organometallic compounds (OMC) or chromium carbonyls; these materials have low decomposition temperatures of 300 degC or less. The presence of residual oxygen in the reactor reduces the decomposition temperature of carbonyls even further. The deposition rate is high enough even at low temperatures. Two different methods are used to carry out the process. The first is the feeding of the OMC vapours into the chamber by saturation of a gas stream outside the reaction chamber. The second is the supply of the OMC as a liquid directly into the deposition zone so that it evaporates close to the articles being coated.

Currently these methods are not applied industrially for coating materials but our results show that they are potentially valuable.

3. EXPERIMENTAL PROCEDURE

Porous bronze was made by loose sintering tin bronze onto 0.3% carbon steel or stainless steel strip with 18Cr,9Ni. The bronze powders were water atomised. As is shown in fig. 1a the particle shape is spheroidal with a dendritic surface structure (fig 1b). The powders used were screened -200 +63 microns and the thickness of the sintered layer was 300-400 microns.

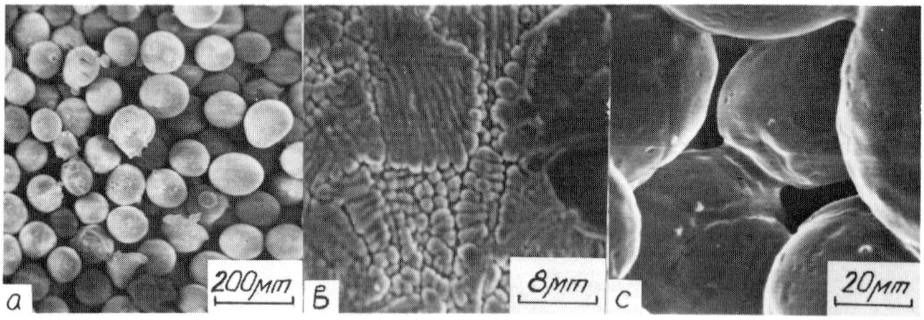

Fig. 1. Atomised bronze powder and sintered layer.

CVD using chromium chlorides (method 1) was carried out in the range 750-850 degC, for times of 1.5-3 hours using an external vapour supply to the chamber. External vapour supply of OMCs give difficulties in the consistency of the concentration of the vapour in the deposition zone due to dilution of the OMC vapours by reaction products.

When using the second method the vapour is created in the chamber by evaporation of liquid OMC by the radiated heat near to the articles being coated. High deposition rates and coating continuity make this method preferable for CVD using OMCs.

The process of thermal decomposition and coating using chromium carbonyls and bisethylbenzylchromium was carried out in a vacuum furnace with a special hot zone designed to protect the elements from the reagent stream. The parts were hung on a jig which could be rotated about a vertical axis within the deposition chamber. Liquid nitrogen traps were used to condense decay products outside the chamber.

The process was investigated at substrate temperatures from 300 to 800 degC. OMC was supplied through a metering device at rates of 10-12 cc/hr and the initial vacuum in the chamber was 8 Pa.

4. RESULTS AND DISCUSSION

The coatings formed consist of a mixture of chromium and complex chromium carbides. Metallographic investigation of the samples by taper section shows that the coating thickness is relatively uniform and does not vary by more than about 3 microns for a coating thickness of 15-20 microns (fig 2) and 5 microns for a coating thickness of 30 microns. The microhardness of the coatings is up to 20 GPa, while the cores of the bronze particles are only 5.5-5.7 GPa. The coating has good adhesion to the bronze substrate and provides a continuous coating of the entire surface of the skeleton throughout the sintered layer (fig 3), although one finds signs of coating spalling on individual particles in the case of deposition at low substrate temperatures. Coating porosity can also be found in areas of surface defects (fig 3c).

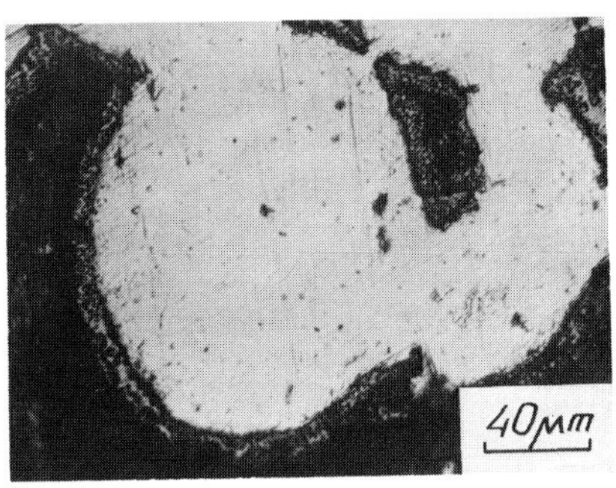

Fig. 2. Coating on bronze particles, on the backing from structural steel

These phenomena are probably due to very low concentrations of OMC vapour over the surface being treated.

The rates of decomposition and deposition rise with temperature in the active zone; however, in the case of chromium chlorides etc economically viable rates of deposition are only reached at temperatures exceeding 0.8 of the melting point of the bronze skeleton. At these temperatures the bronze layer sinters rapidly and porosity is reduced with areas of closed pores or zero porosity.

Consequently the amount of lubricant which can be introduced into the sintered layer is reduced. This causes poor bearing performance. These disadvantages and defects are not found using the second method i.e. OMC evaporation in the immediate vicinity of the article being treated (fig 3a).

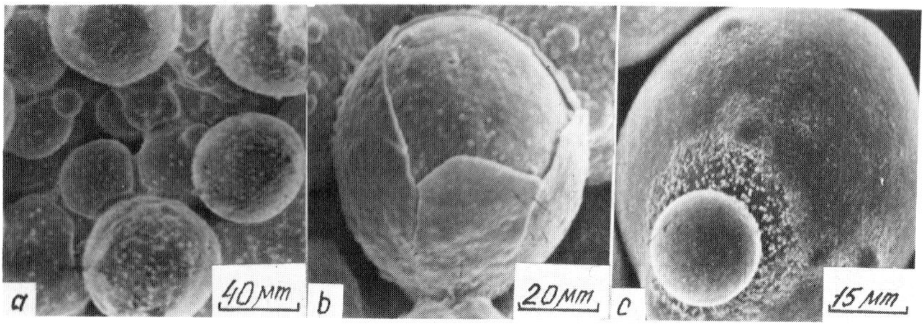

Fig. 3. Sintered bronze skeleton with coating

Here the deposition rate is high enough even at low temperatures and ranges from 10-35 microns/hour for different temperatures. The bronze layer retains it's interconnected porosity.

5. CONCLUSIONS

Samples of bearings were made by both the methods described. Tests were carried out on a friction pair with a mating surface with a hardness of 45-50 HRc. Using the second method stable operation up to a load of 60-70 MPa was possible before dripping occurred at sliding velocities up to 1 m/sec. Without the coating the load to cause dripping was 20 MPa and using the first method it rose to only 30-35 MPa. Thus the use of the coating method of introducing liquid OMC to be evaporated in the vicinity of the article being coated is considered to be industrially applicable for improved operating capacity. In addition, the wear and corrosion resistance of bearings tested in a marine fog was improved by a factor of 3-5.

The production of bearings of this type will certainly require special equipment or the improvement of existing furnace types. However, this should not be difficult as the requirements are not demanding. The plants are absolutely safe and need not meet any special requirements. Operating experience of the equipment for CVD of bronze bearings shows that it has a high efficiency.

Note: the authors of this paper would like to thank A F Sanin for his help in discussing the results of their experiments, and J J Dunkley for reading the manuscript.

Paper presented at Int. Conf. on New Materials and their Applications, Univ. of Warwick, 1990

Study of stress relief patterns in diamondlike carbon films

Suman B.Iyer, K.S.Harshavardhan[a] and Vikram Kumar[b]

Materials Science Division,National Aeronautical Laboratory,Bangalore-560 017, INDIA, a) Materials Research Centre,Indian Institute of Science Bangalore- 560 012, INDIA,b)Department of Physics,Indian Institute of Science Bangalore- 560 012, INDIA.

ABSTRACT: An optical microscopy study of stress relief patterns in diamondlike carbon films is presented. Interesting stress relief patterns are observed which include the well known sinusoidal type, branching pattern and string of beads pattern.The last one is shown to relieve stresses under marginal conditions. Two new stress relief patterns are noted in the present study. One of them is of a sinusoidal shape with two extra branches at every peak position. The distribution of different stress relief forms from the outer edge of the films towards the interior is markedly dependent on film thickness. Our new patterns support the approach in which the stress relief forms have been analysed earlier using the thoery of plate buckling.

1. INTRODUCTION:

Recently, there has been considerable interest in the study of diamond like carbon (DLC)films which have a wide range of applications (Angus and Jansen 1988, Bachman and Messier 1989, Deutchman and Partyka 1989,Rabalais and Kasi 1988, Spear 1989,Sundgren and Hentzell 1986,Tsai and Bogy 1987). For DLC films used as protective coatings in various tribological and optical applications as well as overcoats on magnetic storage media, their mechanical behavior and stability are of primary importance. High compressive stresses have been found by many research workers in carbon films deposited onto glass, silicon, stainless steel and alumina substrates (Matuda et al 1981, Gille and Rau 1984, Nir 1984). In many cases,these compressive stresses caused buckling of the films.Some very interesting buckling phenomena have been reported in carbon films deposited by electron beam and ion beam techniques. vuckling forms such as wavy wrinkles with definite shapes and widths are observed. Although buckling in these films is undesirable, study of buckling is quite useful in estimating the mechanical properties such as Youngs modulus, adhesion energy etc,. Also in view of the very high figure of merit of the DLC films as wear resistace coatings, a fundamental understanding of this wrinkling and associated stress relief patterns is very important. This forms the basis for the present investigations.

The present report is organised as follows. Section 2 describes the method of preparation of DLC films used for this study. The general characterization of the films including the measurement of the internal stress are presented in section 3. The various stress relief patterns are qualitatively discussed in section 4, bringing out the new features. Section 5 is devoted to modelling of the observed relief patterns and estimating the mechanical properties of the films using them. Section 6 describes our initial results on hydrogenated films and Section 7 summarises the results of the present investigations.

2. EXPERIMENTAL :

Diamondlike carbon films were prepared by radio frequency plasma assisted chemical vapour deposition technique. A 13.56 MHz Materials Research Corporation RF sputtering system, model SEM-8622, was used for this purpose. The gas used was a mixture of propane and n-butane. The films were deposited at an RF power density of ~ 0.25 W cm^{-2}. The negative substrate bias during the deposition was optimised to 1000 V. A constant sputtering pressure of 25 mTorr was maintained during deposition. Substrates used for deposition include glass, crystalline silicon and germanium,GaAs, quartz and stainless steel, kept at ambient temperature (30°C). Prior to deposition, the system was pumped down to a vacuum $\sim 10^{-6}$ Torr. The substrates were pre etched in an Argon plasma at an RF bias voltage of 750 V for 10 minutes.

3.CHARACTERIZATION:

Transmission high resolution electron microscopy and x-ray diffraction studies show that the films are amorphous in nature. The films are transparent for wavelengths 2 - 15 μm and are characterised by high electrical resistivity (10^{8}-10^{10} Ohm cm) and excellent chemical etch (corrosion) resistance. Film density was measured by microweighing of the films deposited on thin glass substrates. The average density is found to be 2- 2.5 gm cm^{-3} which corresponds favourably with earlier reports.The hardness ofthe films is not measured quantitatively,but the extreme hardness of the films can be demonstrated by the scratch test. The films contain high internal stresses giving rise to buckling beyond a critical thickness, which depends on the deposition conditions. When deposited on a thin substrate, the internal stress in the film caused a spherical deformation of the substrate. The sign of the curvature indicates that the internal stress is compressive in nature and its magnitude is given by the modified Stoney equation (Jaccodine and Schlegel 1966)

$$\sigma_i = \frac{E_s}{6\,(1-\nu_s)}\;\frac{t_s^2}{t_f}\;\frac{1}{R}$$

Where E_s and ν_s are the Youngs modulus and Poissons ratio of the substrate respectively. σ_i is the residual stress in the film and R is the radius of curvature of the substrate. It has been measured experimentally in the present study by Newton rings method. The internal compressive stress in the film evaluated this way is 3 x 10^{10} dynes/cm^2.

4.INTERNAL STRESS RELIEF PATTERNS :

Our DLC films deposited on glass substrates contain large compressive internal stresses and display characteristic stress relief patterns.

These include the well known (Matuda et.al 1981, Gille and Rau 1984, Nir 1984)sinusoidal type, string of beads pattern etc,.We have used an optical microscope, Neophot 2 ,for these studies. As also observed by earlier workers,in the present study, buckling began after exposure to air at atmospheric pressure and was accelerated greatly by high humidity. In most cases, the buckling started at the film edge. Considerable effort has been made to understand the stress relief patterns particularly in hard carbon films and there have been attempts to study the thickness dependence of the buckling patterns by Matuda et.al(1984) and Nir(1984) . Figures 1 and 2 display the thickness dependence of the profile of the wrinkles from the outer edge of the substrate towards the interior of our DLC films.

The following points have been noted in our present investigations.There are overall six kinds of stress relief patterns as mentioned below.

1.(i) Peeling (ii) flower like buckling with cracks (iii) flower lie buckling with waves (iv)dense buckled up regions with neither cracks nor waves (v) wavy or sinusoidal patterns and (vi)string of beads pattern. Patterns (ii) and (iii) are shown in figures 3a and 3b.

2.For the thickest film (2500 Å - 20 min.) peeling of the film has been observed at the outer edge.

3.The stress relief pattern at the inner edge of a 2500 Å (20 min)thick films is similar to that at the outer edge of a 1500 Å (12 min.)thick film whose inner edge pattern is again similar to the outer edge pattern of a 1250 Å(10 min.)film.

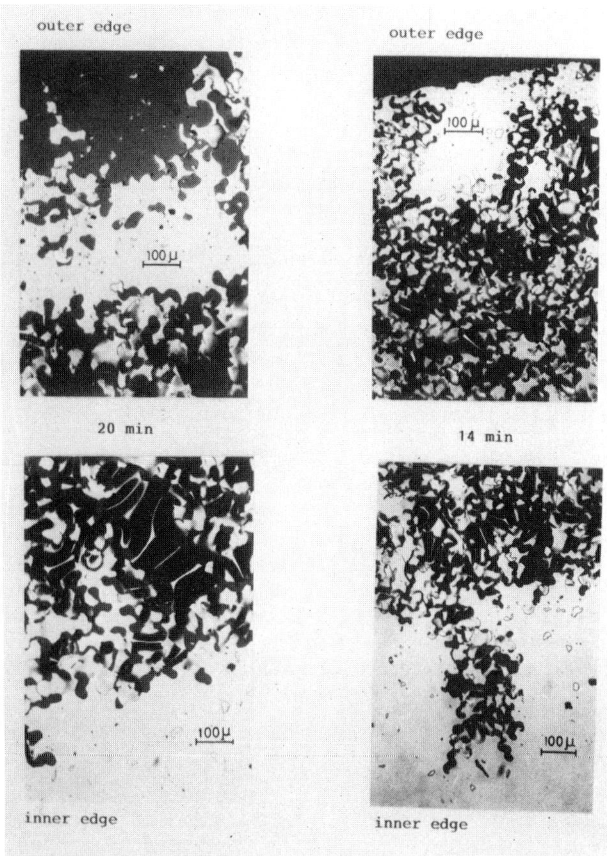

Fig.1 Thickness dependence of the profile of the wrinkles from the outer edge of the substrate towards the interior of the film.

20 min = 2500 Å ; 14 min = 1750 Å

4.Sinusoidal buckling waves travelling over some distance and string of beads pattern [these are absent in 2500 Å (20 min) thick and 1500 Å (12 min.)thick films].

5.The thinnest sample (750 Å -6 min.) we have studied has only string of beads pattern at the outer edge of the film .

From this comparative study it is clear that the residual internal stress in the film is markedly dependent on the film thickness, increasing with the film thickness. The stress relief starts at the outer edge of the substrate through peeling or buckling of the film. It proceeds towards the centre of the film there by decreasing the residual stress progressively. With adhesion remaining the same the six

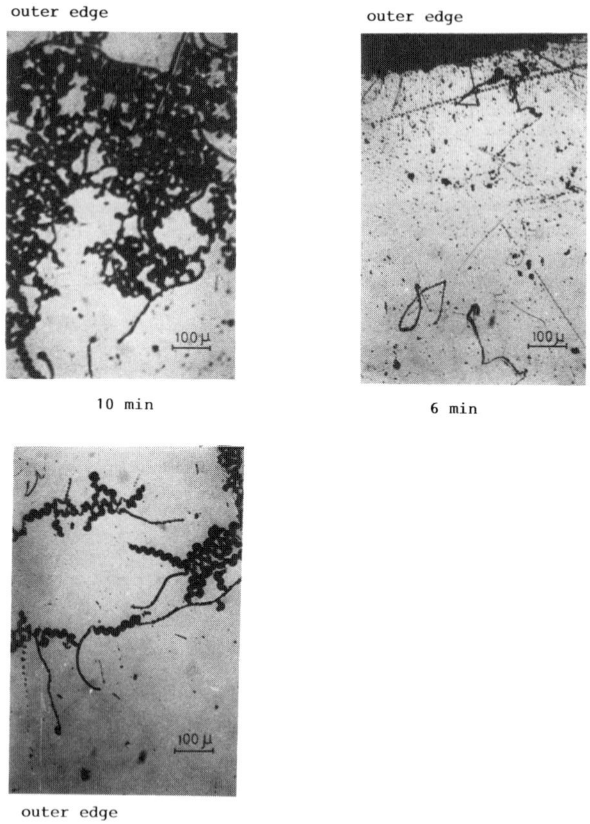

Fig.2 Thickness dependence of the profile of the wrinkles from the outer edge of the substrate towards the interior of the film
10 min = 1250 Å ; 6 min = 750 Å

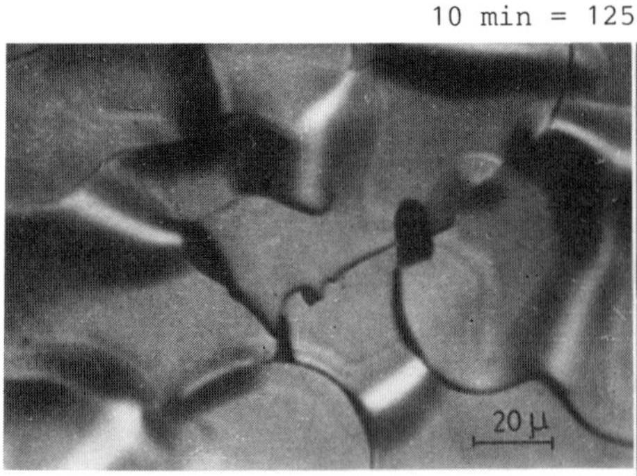

Flowery pattern with cracked edges

Fig. 3a

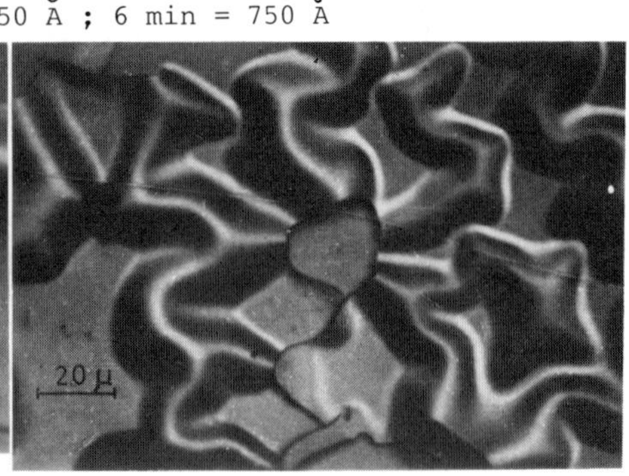

Flowery pattern with waves

Fig. 3b

stress relief forms listed above correspond to a decreasing magnitude of the residual stress value. It is clearly seen that sinusoidal wavy pattern and the string of beads pattern occur under marginal stress conditions, which are of the order of 10^9 dynes/cm^2.

Figure 4 shows that films which are grown under similar conditions and thus have the same mechanical properties, show wavy stress relief patterns with dimensions directly proportional to the film thickness.

We now report two new stress relief patterns. The first one is basically of a sinusoidal shape with two extra branches at every peak position, as seen in figure 5.The second one has features which look like dotted parallel lines (figure 6)with a slight buckling out of the film at each dot. Several workers have attempted to theoretically understand these stress relief patterns. As discussed in the next section, these new patterns,particularly the one with features corresponding to the sinusoidal shape having two extra branches at each of the peak,strengthen the validity of the models put forward by Nir (1984)which use the general theory of buckling of plates.

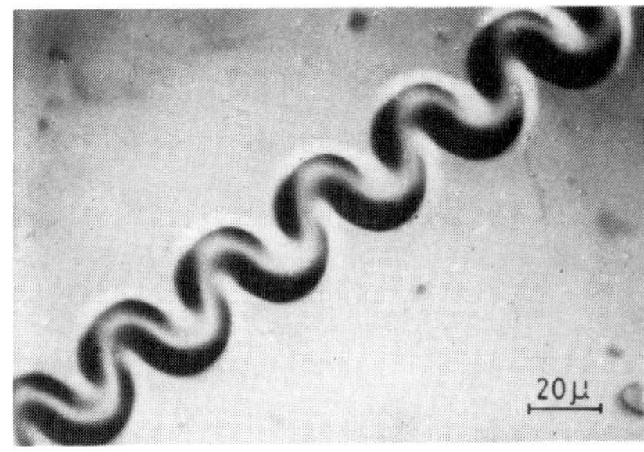

14 min

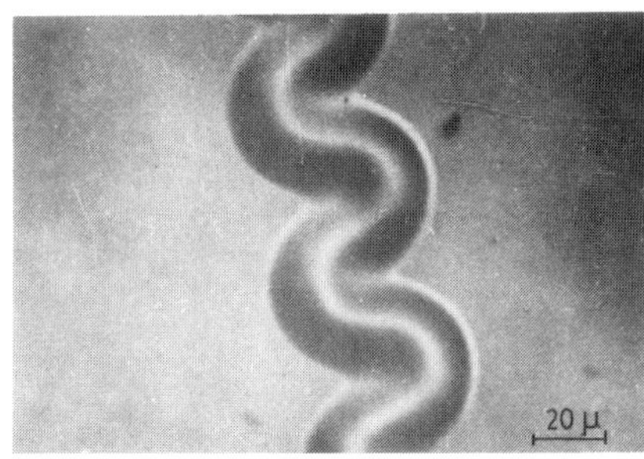

20 min

Fig.4 The figure indicates that the wavy stress relief patterns have dimensions which are directly proportional to the film thickness.
14 min = 1750 Å ;20 min = 2500 Å

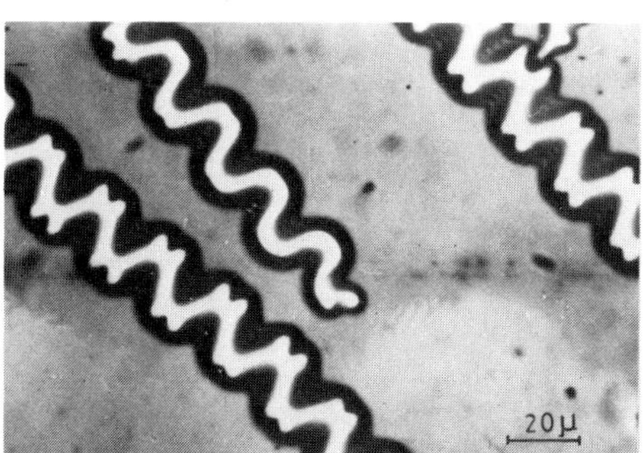

Wavy relief pattern with two spikes at each peak

Fig. 5

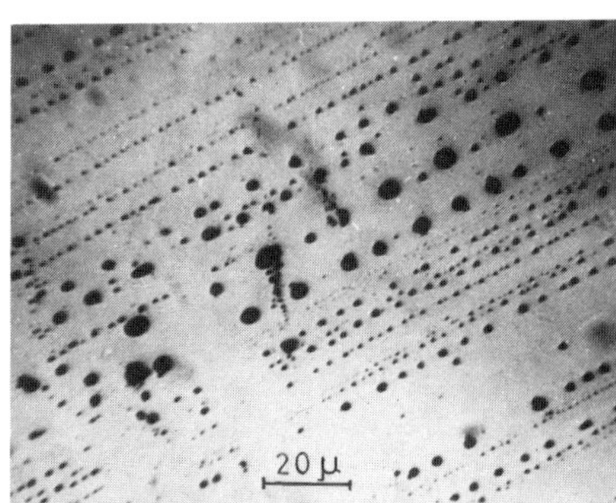

Fig.6 Stress relief patterns-Dotted parallel lines with a slight buckling out of the film at each dot.

5. COMPARISON WITH THOERY :

Earlier, Nir (1984) has applied a general theory of buckling of shells to analyse the stress relief patterns in hard carbon films. In brief, the thin film is treated as a thin plate attached by adhesive forces to the substrate. We obtain the buckling equation as

$$D\left(\frac{\partial^4 w}{\partial x^4} + 2\frac{\partial^4 w}{\partial x^2 \partial y^2} + \frac{\partial^4 w}{\partial y^4}\right) + \sigma_x h \frac{\partial^2 w}{\partial x^2} + \sigma_y h \frac{\partial^2 w}{\partial y^2}$$

$$+2\tau_{xy} h \frac{\partial^2 w}{\partial x \partial y} + f = 0 \quad \ldots\ldots\ldots\ldots\ldots\ldots (1)$$

where

D	- moment of inertia of the film
E	- Youngs modulus
h	- film thickness
u	- Poissons ratio
x,y,z	- coordinates relative to the substrate
u,v,w	- film coordinates as defined in the elastic theory
σ_x and σ_y	- internal compressive stresses
τ_{xy}	- shear stress
f	- external force.

As discussed by Nir (1984), the solution of equation (1) has a physical meaning as the deflection of the film from the substrate in the normal direction. One type of the solution of equation (1), which is physically acceptable is

$$w = 1 + \cos(kx + qy) \quad \ldots\ldots\ldots\ldots\ldots\ldots\ldots\ldots\ldots (2)$$

This solution implies that the maximum displacement of the ridge formed by the film buckling is along a straight line given by

$$kx + qy = 2n\pi \qquad n = 0, \pm 1, \pm 2, \ldots \quad \ldots\ldots\ldots (3)$$

The solution consists of two families of parallel ridges crossing each other as described by equation (3) and shown as a sketch on the right hand side. Once the buckling process begins, it proceeds in a certain direction. When it reaches a point where the ridges cross each other, it may turn into either of them or branch into both of them. The two extra branches which appear at the peak positions in wavy patterns of

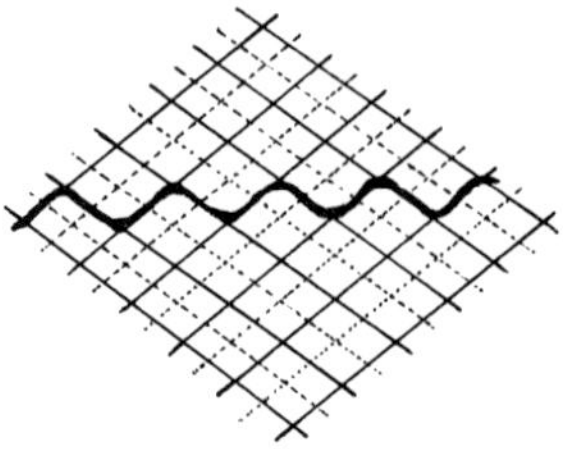

Two families of parallel ridges crossing each other

figure 5 seem to be manifesting such a branching.The wavy pattern at the centre of figure 5 suggests that, beyond a certain point the stress is not enough for bukling to branch out and a normal wavy pattern without branching seem to be a more favourable situation.

The pattern in figure 6 is not very interesting in itselt. But it seems to correspond to another solution suggested by Nir (1984),viz., a continuous solution which does not vanish in isolated regions only.

Matuda et.al(1981) have studied the dependence of the size of wrinkles on the film thickness. The cross sectional shape of the wrinkle is approximated by

$$w(x) = \frac{h}{2}\left\{1 + \cos\left(\frac{2\pi}{l}x\right)\right\} \quad \ldots\ldots(4).$$

where l is the length of the lifted up region and h is the height at the mid point. If $h \ll l$, Young's modulus E of the film is given by

$$E = \frac{1 + \pi^2 h^2/4l}{\pi^2 h^2/4l}\,\sigma \quad \ldots\ldots.(5).$$

where σ is the average internal stress in the film. We have used eq. 5 to calculate Young's modulus for the 2500 Å thick film (20 min) for which the measured average internal stress is 3×10^9 N m^{-2} and h/l is estimated as 0.15 with the result $E = 1.6 \times 10^{10}$ N m^{-2} which agrees favourably with the earlier reported value (Matuda 1981).

Matuda et al (1981) have also derived the expression for the adhesion energy U_a per unit area

$$U_a = \frac{\pi^4 (h/l)^2 E d^3}{12\,(1-\gamma^2)}\;\frac{L}{l_0^2\,(L - l_0)}$$

where

d - thickness

γ - Poisson's ratio

Two orthogonal wrinkles of width l extend in the x and y directions over a square of width L.

Figure 7 shows crosses consisting of orthogonal wrinkles. Using these the adhesion energy was estimated to be 1000 ergs / cm^2 which is also in close agreement with the values calculated by Matuda (1981).

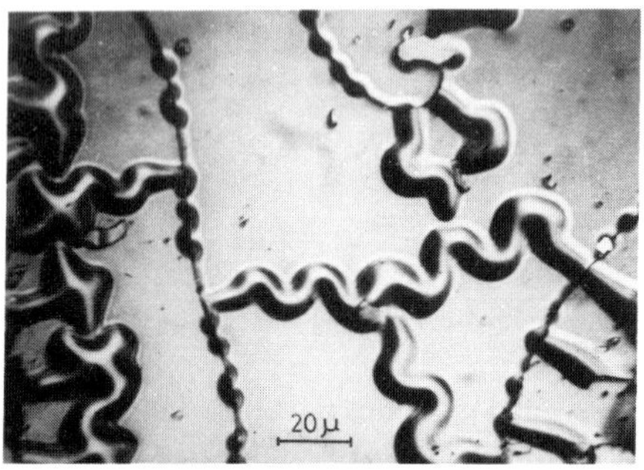

Stress relief patterns
String of beads
Sinusoidal shape & crosses

Fig.7

6.HYDROGENATION :

For practical applications of DLC films as hard coatings in tribology, optics and magnetic storage media, buckling and peeling are highly undesirable. In order to reduce the residual compressive stress in the films, during deposition, hydrogen has been added to the active hydrocarbon precursor gas upto 40 volume percent. The internal stress has been found to decrease as function of increasing volume percent of hydrogen as shown in figure 8.

We speculate that addition of hydrogen provides structural flexibility to the otherwise elastically rigid amorphous network as also suggested by Angus and Jansen (1988). Strain is relieved in these networks by the appearance of some medium range ordering process in which clusters of sp^2 bonded carbon atoms can form. The low optical gap exhibited by these films (hydrogenated) is also suggestive of such a clustering process. Our initial experiments (such as scrath tests) indicate that the hardness of the films is decreasd relative to the unhydrogenated films. We are at present investigating these aspects by a detailed analysis of the changes occuring in the chemical bonding (sp^2 and

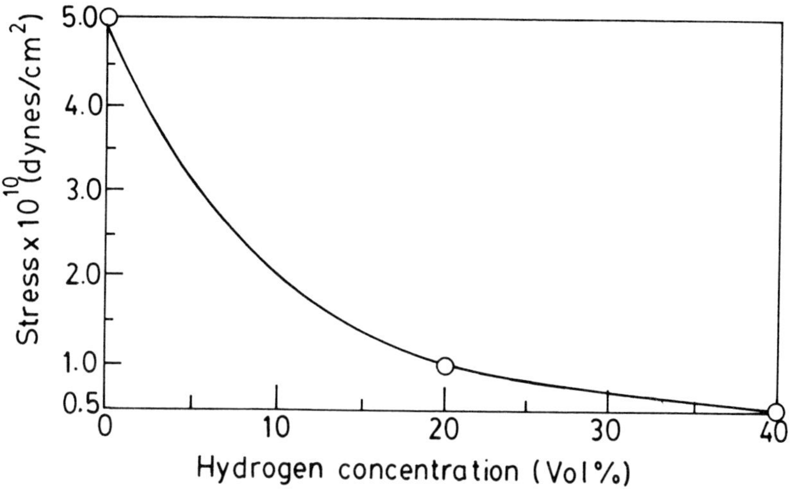

Fig.8 Internal Strain of the film as a function of hydrogen concentration

sp^3) of the network by various spectroscopic techniques (infra red and Raman).

7. SUMMARY :

By a systematic optical microscopy study, we have analysed stress relief patterns in diamond like carbon films prepared by plasma assisted chemical vapour deposition. Comparative study of the thickness dependence of the relief patterns clearly shows that string of beads occurs under very marginal conditions preceded by sinusoidal shape at slightly higher stress conditions. Two new stress relief patterrns have been reported one of which is a sinusoidal pattern with two extra branches at each peak and the other consists of dotted parallel lines. These new pattens support the earlier approach in which the stress relief forms have been analysed using theory of plate buckling. The stress relief patterns have been used to estimate Young's modulus and adhesion energy of the films.

Angus JC and Jansen F 1988 J.Vac.Sci.Technol A6 1778
Bachman PK and Messier R 1989 Chem.and Eng. News 67 24
Deuctchman AH and Partyka JH 1989 Advanced Materials and Processes 6 29
Gille B and Rau B 1984 Thin Solid Films 120 109
Jaccodine RJ and Schlegel WA 1966 J.Appl.Phys. 37 2429

Matuda N, Baba S and Kinbara A 1981 Thin Solid Films 81 301
Nir D 1984 Thin Solid Films 112 41
Rabalais JW and Kasi S 1988 Science 239 623
Spear KE 1989 J.Am.Ceram.Soc. 72 171
Sundgren JE and Hentzell 1986 J.Vac.Sci.TechnolA 4 2259
Tsai H and Bogy DB 1987 J.Vac.Sci.Technol.A 5 3287

Inst. Phys. Conf. Ser. No 111
Paper presented at Int. Conf. on New Materials and their Applications, Univ. of Warwick, 1990

Processing and properties of ceramic fibres, with emphasis on silicon carbide

V. Kalyvas[1], A.P. Kemp[2], B. Rand[2], G.C. East[1], J.E. McIntyre[1], F.L. Riley[2], L.F. Chen[2] and J.C. Ko[2].

University of Leeds, Department of Textile Industries[1] and School of Materials (Division of Ceramics)[2], Leeds LS2 9JT, U.K.

ABSTRACT: The properties and thermal stability of high strength ceramic fibres are reviewed. Alumina and silicon carbide fibres show loss of strength at temperatures above 1000°C due to grain growth phenomena combined with internal degradative reactions in some cases. Routes to silicon-based fibres involving powder sintering and pyrolytic conversion of polymeric precursors are discussed and illustrated, where appropriate, with examples.

1. INTRODUCTION

The successful development of fibre-reinforced ceramics for high temperature applications poses severe experimental difficulties, such as control of the fibre-matrix interface at elevated temperature, and the fabrication of composites containing continuous fibres. However, one of the greatest handicaps to the more rapid development of this class of materials is the lack of an adequately oxidation-resistant ceramic fibre with high strength and stiffness, maintained at temperatures greater than about 1100°C. This article will first compare the properties of existing ceramic fibres and then highlight some of the difficulties which are encountered in the fabrication of new continuous fibres based on silicon carbide.

2. HIGH TEMPERATURE PROPERTIES OF HIGH STRENGTH CERAMIC FIBRES

Ceramic fibres can be classified into three main groups: carbon, oxide and non-oxide fibres.

2.1 Carbon

A wide range of carbon fibres is presently available commercially, and the properties of these fibres are well documented (Watt and Perov, 1985; Johnson, 1990). The Young's modulus can be controlled with values up to about 800 GPa, tensile strengths up to 7 GPa, and strains to failure as high as 2.4%. However, many fibres are weak under compressive loading (Johnson, 1990; Fitzer, 1990; Rand, 1985). Because fabrication involves temperatures approaching 3000°C, fibre microstructures and associated mechanical properties are stable to temperatures in excess of 2000°C, but only under non-oxidising conditions. Thus, these fibres are the only stable fibres available for the production of ceramic matrix composites intended for high temperature applications, but they will require protective coatings to prevent fibre degradation during prolonged service in oxidative environments.

2.2 Oxide Fibres

The most important oxide fibres are based on alumina and aluminosilicate systems. The production and properties of these fibres have recently been reviewed by Birchall (1985) and Stacey (1988). Aluminosilicate fibres, with silica contents greater than ~35%, are usually spun from the melt and are mostly used in thermal insulation at temperatures up to ~1400°C. The fibres more suitable for composite fabrication have higher alumina contents and cannot be produced from the melt because of the high temperatures involved (> 1850°C, depending on composition). Slurry and solution spinning processes have therefore been devised. In the slurry process, alumina and hydrated alumina particles, dispersed in liquids containing organic binders, are extruded, dried and sintered to form continuous polycrystalline filaments. The solution route makes use of viscous solutions of aluminium salts, which can be gelled after extrusion, then dried and sintered. A number of fibres produced in such a way is available (see Table 1), but most fibres are development products, available only in small quantities and very costly.

Table 1: Properties of High Performance Ceramic Fibres.

Fibre	Composition w/o	E /GPa	UTS /GPa	Dia. /µm	CTE $/10^{-6}\ K^{-1}$	Density $/g\ cm^{-3}$
CARBON						
Carbon (range)	100% C	200-700	1.4-5.5	7-10	-0.4 to -1.8	1.7-2.0
SILICON CARBIDE						
SiC, AVCO (SCS-6)	CVD SiC on C core.	406	3.9	143		3.0
SiC, Nicalon	59% Si, 31% C 10% O	190	2.5-3.3	10-20	3.1	2.55
Tyrrano	Si, C, O, Ti	210	3.0	10		2.4
SiC, VLSI	whisker	580	8.4	6		3.3
ALUMINA						
Nextel 480	70% Al_2O_3, 28% SiO_2, 2% B_2O_3	224	2.3	11		3.05
Nextel 312	62% Al_2O_3, 24% SiO_2, 14% B_2O_3	154	1.75	11		2.7
Sumitomo	85% Al_2O_3, 15% SiO_2	230	2.0	9-17		3.2
Du Pont FP	>99% α-Al_2O_3	380	1.38	20	5.7	3.9
Du Pont PRD-166	α-Al_2O_3 + 20% ZrO_2					

Fibres from the slurry route generally have diameters in the region of 20 μm. The grain size of the stable α-alumina phase is of the order 0.5μm. Control of grain size is achieved through the additives in the precursor slurry, the most common of which is silica which stabilizes the low temperature stable γ-phases of alumina at higher temperatures. Some fibres contain B_2O_3, and Du Pont's new PRD166 fibre contains approximately 20% ZrO_2, claimed to provide a measure of transformation toughening. The presence of silica reduces the room temperature modulus, but not the strength. According to Stacey (1988), this is due to the grain size, fibres with larger grain sizes have greater stiffness. The room temperature strengths and moduli of alumina-based fibres are high (Table 1), but because of their high density, ~ 4000 kgm^{-3}, their specific values do not compare well with those of carbon. Alumina-based fibres show advantages over carbon and carbide fibres, however, in that they are stable in oxidising atmospheres up to high temperatures. Under reducing conditions they are stable to temperatures of 1200°C or so, this stability limit being largely determined by the presence of SiO_2 in the fibre, which can be reduced to the volatile SiO. Alumina is compatible with non-oxides, such as silicon nitride (to ~1500°C) and carbide ceramics. Creep becomes significant at temperatures above 1000°C and is also controlled by grain boundary impurities. Generally, the smaller the grain size in the fibre the lower the temperature at which grain growth begins to be appreciable and degradation of properties starts. Thus, a small grain size might be desirable for low temperature use, but a larger grain size would confer good retention of mechanical properties at high temperatures. Stacey (1988) has measured the relative stabilities of alumina-based fibres, in terms of the temperature to which they maintain 50% of their room temperature strength, and found:

Fibre	Critical temperature / °C
Nextel 312	1200
Nextel 480	1430 (ox); 1200 (red)
Du Pont FP	1400
Du Pont PRD166	> 1400 (75% strength retention at 1400°C).

The control of grain growth in these fibres and its dependence on the intermediate alumina phases produced is not yet fully understood. Improved understanding would be expected to lead to better control, with a consequent improvement in composite performance.

2.3 Non-oxide fibre

Of the many non-oxide monolithic ceramics developed for use at high temperature, only materials based on SiC, Si_3N_4 and BN appear to have been fabricated in fibrous form. By far the most widely investigated are the SiC-based fibres for which three main production routes have been reported. In the first a silicon carbide layer is built up by chemical vapour deposition (CVD) on a core fibre of carbon or tungsten. The second comprises spinning an organosilicon polymer which is subsequently pyrolysed to SiC in a manner analogous to the formation of carbon fibre from PAN. The third route is to form a fibre from a dispersion of SiC powder in an organosilicon polymer solution, which after pyrolysis is sintered at high temperature. Only the first two of these routes have led to commercial products.

A range of silanes have been used in the vapour deposition process. The resultant fibre consists of a relatively unreacted core covered by the characteristic columnar morphology of silicon carbide. In the AVCO fibres, based on a carbon fibre core, there are two distinct silicon carbide layers with differing morphology (Nutt and Wawner, 1985), possibly the result of two sets of deposition conditions. The outer layer has a relatively coarse microstructure, with angular surface grains acting as stress raising flaws, rendering the fibre sensitive to surface defects and abrasion. To counteract this effect, a third, outer, graded, carbon/SiC layer can be deposited to reduce the stress concentration at the filament surface. These coatings are of considerable significance in determining the micromechanical properties of composites incorporating the fibres. Fibres produced by this process are of the order 100-150 μm in diameter. Strength and modulus values, although high at room temperature (Table 1), decrease at temperatures in excess of ~600°C (Yajima, 1985). The most important application of these monofilaments is in metal-matrix composites.

The organosilicon polymer precursor route to silicon carbide fibre has produced many candidate products from a wide range of monomeric base units.

The first successful SiC-based fibre was developed by Yajima et al. in the early 1970's from polycarbosilanes (Yajima, 1985). Fibres obtained from melt spinning were crosslinked in the presence of oxygen by formation of Si-O-Si and Si-O-C bonds (Taki et al. 1989), and then pyrolysed under inert gas or in vacuum to give the ceramic fibre. Fibres made by a similar process are sold by the Nippon Carbon Co. under the trade name "Nicalon". Extensive research on the fibre and on its performance in composites at high temperature, has shown that degradation occurs above ~1000°C. TEM of Nicalon fibre shows ~2 nm SiC grains (Chaim et al., 1988) within a Si-O-C amorphous phase (Porte et al., 1989), together with paracrystalline carbon. Heating leads to reaction of oxygen with the excess carbon and the formation of porosity. The silicon carbide grains simultaneously grow to micrometer dimension. Both processes lead to the formation of crack nucleation sites, and thereby reduced strength.

Sintering of silicon carbide particles to form fibres (Boecker et al., 1987) provides possibly the simplest and most direct route, loosely following the route to monolithic silicon carbide. High specific surface area SiC powder is mixed with a polysilane carrier and sintering aids, and spun. The polysilane is converted to a SiC-based matrix by heating, and conglutinates the fibre during sintering at ~ 2000°C to form the final product. Polycrystalline filaments of diameter ~20-150 μm can be obtained with a density > 96% theoretical and good thermal stability. However, the room temperature strength is low compared with that of fibres prepared from a purely polymeric route, < 1.4 GPa, a value comparable with monolithic silicon carbide produced under high quality processing conditions (Kendall et al., 1989).

2.4 Summary of Fibre Properties

Figure 1 compares the strengths of various fibres at elevated temperatures. Excessive grain growth is the main weakening factor in alumina fibres of low SiO_2 and B_2O_3 content. Both the polymer precursor and the CVD routes to silicon carbide fibre give materials with high strength, but of limited thermal stability. The CVD route is slow and expensive, and the large diameter fibres cannot readily be wound and woven into fabric for continuous fibre reinforcement. The production routes to smaller diameter

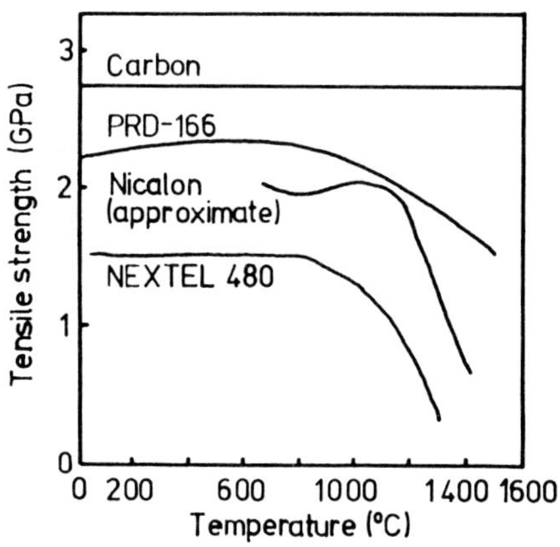

Fig. 1. Temperature dependence of the tensile strength of some ceramic fibres, compared with that of a carbon fibre of modest strength.

fibres need to be refined to give high strength and stiffness at higher temperatures than is at present possible. In the case of the polymeric precursor route the ceramic fibres are microcrystalline and it is desired to maintain this microcrystallinity to higher temperatures and to eliminate oxygen to avoid the disrupting degradation reactions. Control of the fibre composition is essential to achieve both of these objectives. Elements additional to Si and C may be required to stabilise the microcrystalline structure (e.g. in Si-C-N systems). An alternative is to fabricate sintered SiC products with small grain size.

The production of sintered and polymer-based fibres is discussed below. Many publications have dealt with the control of the polymer structure and chemistry in the attempt to eliminate residual oxygen. However, while polymer chemistry is important it is essential also that the polymer has the correct sets of characteristics for spinning into fibre of required diameter and cross-sectional shape. This is not a trivial matter and the requirements of chemistry and spinnability may not coincide, especially in the area of melt spinning.

3. SPINNING TECHNIQUES

The essential features of basic fibre spinning techniques are briefly outlined here to illustrate the difficulties and solutions encountered when spinning materials from previously untried polymeric systems.

Spinning involves extruding a liquid or suspension of controlled viscosity through narrow apertures of chosen diameter (40-300 μm, in a spinneret), drawing the fibre by a process of extensional flow, and then rendering the strands solid to maintain the structure and morphology before collecting the formed fibre on a bobbin.

For a polymer material to be suitable for spinning, it must be capable of being transformed into a "low" viscosity medium, either by fusing or by dissolving in a low viscosity solvent. On a more operational level, it is desirable to use high molecular weight (MW) materials to facilitate fibre handling, and materials which are solid under ambient conditions following spinning. These latter "constraints", however, can be sometimes relaxed if necessary.

There are three types of spinning:

1) Melt spinning, where the melt and the spun fibre have the same composition, and the polymer is thermoplastic in character. It is extruded in the liquid state and solidified by cooling. The technique is suitable for polymers and glasses, and generally provides a fibre of circular cross-section.

2) Dry spinning. The spinning fluid, or dope, consists of the required material dissolved or dispersed in a volatile solvent. After extrusion, the solvent is evaporated by hot gas, leaving a polymer fibre, normally of "elliptical" or "dog-bone" cross-section, although circular cross-sections can be obtained by careful control of the appropriate parameters.

3) Wet spinning. This differs from dry spinning in that the solvent is removed from the fibre by passing the extruded filaments through a liquid in a coagulation bath. The bath contains a solution of the dope solvent in a liquid which is a non-solvent for the polymer. A concentration gradient exists across the fibre-liquid interface, causing the solvent to diffuse out from the fibre. Normally a product of approximately circular cross-section results.

One of the most important process variables is the rate of solidification. If the rate is too high, the differential contraction between the inner ("fluid") material and the outer ("solid") skin may result in a poor quality fibre. In the case of melt spun materials this contraction is not significant; however in dry spinning, the outer skin may collapse giving a fibre of non-circular cross-section. The situation is worse for wet spinning, where a porous, low strength fibre may result. This is illustrated in Figure 2 for an early attempt at forming a PAN/silica fibre in this laboratory (see Section 4).

If, however, the solidification rate is too low, the thread-line may remain incompletely solidified throughout spinning and winding, and thereby stick to other fibres on the bobbin. This problem may be avoided by appropriate adjustment of the relevant parameters, namely extrusion rate, winding up speed, and the solidification process directly, i.e. control of gas flow rate and temperature in melt and dry spinning, and composition and temperature of the coagulation bath in wet spinning.

The effects of fibre sticking and mechanical damage from subsequent handling may be significantly reduced by the application of a surface coating ("size") before final take-up and winding.

4. ROUTES TO NON-OXIDE CERAMIC FIBRES

In this section, selected routes to non-oxide ceramic fibres are discussed. Emphasis is given to SiC-based fibres and the overview is highlighted with examples from current work in our laboratories.

4.1 Carbothermal Reduction

One possible route to the in situ generation of SiC in the fibrous form is by carbothermal reduction of silicon dioxide, SiO_2. This reaction involves silicon monoxide vapour, i.e.

$$SiO_2(s) + C(s) \longleftrightarrow CO(g) + SiO(g)$$

$$SiO(g) + 2C(s) \longleftrightarrow 2CO(g) + SiC(s)$$

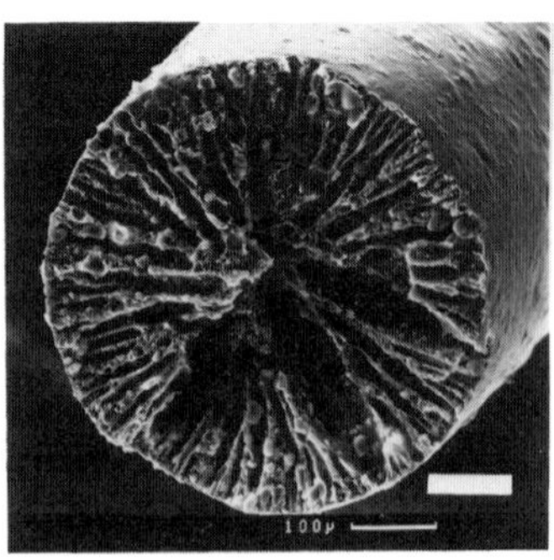

Fig. 2. PAN/SiO_2 as spun fibres with fast solidification rate.

The basic requirement for the precursor fibre is a fine uniform dispersion of SiO_2 in a carbon matrix.

The initial choice of a carbon source, pitch, whilst having the highest carbon yield, was abandoned due to the difficulties in dispersing both tetraethylorthosilicate (TEOS) and silica. Similarly, TEOS was found to be immiscible with PAN dissolved in dimethylformamide (DMF). However, uniform dispersions of fine amorphous silica powder (specific surface area 380 m^2/g) with PAN/DMF were obtained with the use of silane coupling agents and high shear mixing. Standard ceramic dispersing agents were found to be ineffective. These dispersions were wet spun into a H_2O/DMF coagulation bath and the PAN subsequently converted to carbon using standard carbon fibre preparation procedures (Watt, 1985). This stage gave a fibre with a porous microstructure. Subsequent high temperature treatment to form SiC led to a further increase in porosity, due to the volume change associated with the above reactions. There was a concomitant decrease in tensile strength, from ~ 0.7 GPa for the C/SiO_2 "intermediate" fibre, to ~ 0.1 GPa for the final SiC fibre.

Formation of SiC in the heat treated samples was conveniently monitored by X-ray diffraction (XRD) (Figure 3), and found to begin between 1300°C and 1400°C.

The poor mechanical properties, obtained as a result of the high porosity, may be improved by densifying the fibre by sintering, after incorporation of sintering aids, for example, an organoboron compound (Riedel et al.). This work is currently being undertaken. Sintering could also be facilitated by impregnation with an organosilicon polymeric precursor. A possible advantage of this route to sintered SiC fibre is that lower diameter filaments (< 40 μm) are obtained as standard, compared with the nominal 100 μm from the "direct" method.

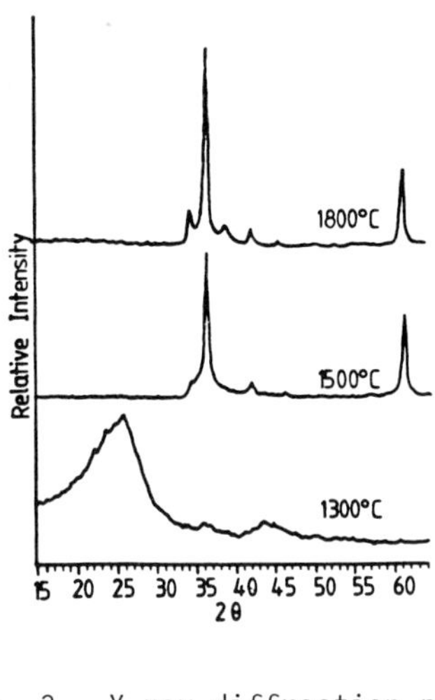

Fig. 3. X-ray diffraction patterns for PAN/SiO_2 fibre heat treated at different temperatures.

In order to improve the dispersion of silica in the carbon fibre to a microscopic level prior to reduction, uniformity is required on a molecular scale. This might decrease the porosity, and should reduce the pore size. Alternative systems are therefore under investigation. For example, TEOS is miscible with a cellulose acetate/acetone solution, giving mixing on a molecular scale of carbon and silica species. However, some separation may occur because in order to control the viscosity of the spinning dope, it is necessary to partially hydrolyse the TEOS to SiO_2 using H_2O and acid catalyst.

The dope was dry spun to yield the precursor fibre. Conversion to the intermediate and final ceramic fibre is achieved by the same method as for the PAN/SiO_2 system, but in this instance higher densities resulted in the final product.

4.2 Polymer Precursors

The polymer precursor route to ceramic fibres has produced many candidate experimental fibres due to the variety of monomers and polymerisation reactions available and the diversity of the copolymers that can be produced from them. Important contributions have been made by Yajima (1985), West (1986), Schilling (1986) and Baney (Baney et al., 1984).

The requirements for an effective polymer precursor are:

1) Spinnability - the polymer should be soluble and/or fusible; it should have a controlled molecular weight distribution and be solid at room temperature.

2) Stabilisation - it should have the ability to cross-link through reactive groups after spinning.

3) Ceramic yield - the polymer should contain rings and cages, especially at backbone silicon atoms, and be reactive enough to form a high cross-link density after stabilisation.

4) Stoichiometry - the polymer should be of such a composition to provide the required stoichiometry in the ceramic fibre.

Various approaches have been adopted in attempts to satisfy these requirements.

The general route to SiC-based fibres from polymeric precursors is shown in Figure 4. The similarity to the carbon fibre fabrication process is striking. The first successful fibre was developed by Yajima et al. (1986) from polycarbosilanes (PCS). These polymers were obtained from thermolysis of polydimethylsilane (PDMS) or tetramethylsilane in an autoclave or by "catalytic" treatment at 700°C in Ar at 1 atmosphere pressure. A variety of redistribution reactions take place through the free radicals produced by chain scission of PDMS, followed by dehydrogenation and demethylation reactions. The resultant complex mixture of polycarbosilanes with branched cyclic and cage units is fractionated to provide the spinnable material by dissolving in benzene or hexane, filtering to remove the insoluble fraction and by distillation at high temperature in vacuo to remove the low MW volatiles. The soluble/fusible "high" MW material thus obtained is far from a conventional fibre-forming polymer. Our own work shows how the yield of various fractions changes with thermolysis temperature, in the presence of $AlCl_3$ catalyst. Temperatures in excess of 400°C are required for the maximum yield of ~ 20%. However, yields of up to 65% can be obtained under different thermolysis conditions (Hasegawa et al., 1986). Yajima used melt spinning to form the fibres from this fraction at unusually high spinning temperatures (up to 350°C) in air. Subsequent cross-linking of the spun fibres in air led to

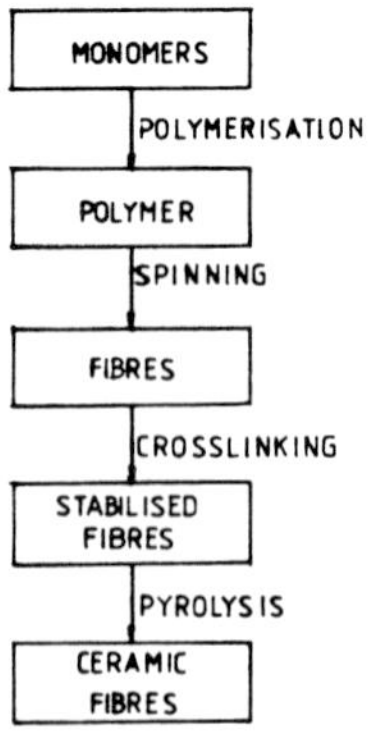

Fig. 4. Polymer route to ceramic fibre.

the desired infusible material. In this process, oxygen reacts mainly with Si-H groups to form Si-O-Si and Si-O-C linking bonds, and Si-O-H groups, as revealed by Infrared (IR), Nuclear Magnetic Resonance (NMR), thermal analysis and gas evolution measurements during the subsequent pyrolysis. Thus, oxidation increases the cross-link density, but it has the disadvantage of introducing oxygen into the structure which is only fully eliminated at high temperatures at the expense of the mechanical properties of the ceramic fibre. PCS fibres crosslinked by electron irradiation have been shown to exhibit superior thermal properties to those crosslinked in air (Okamura, 1987).

From work in our laboratories, changes in the main functional groups of a thin PCS film were monitored using Fourier Transform Infra-Red (FTIR) spectroscopy after heating in air at 100°C for different lengths of time (Figure 5). The intensities and areas of the bands at about 3300 (O-H), 1720 (C=O), and 1000-1100 cm^{-1} (Si-O-Si, Si-O-C and Si-C-Si) are increased,

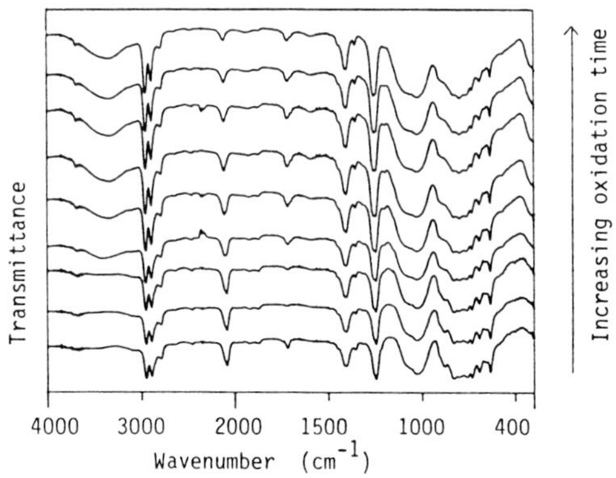

Fig. 5. FTIR of a PCS film as a function of the duration of the oxidative treatment.

whereas the reverse is true for the 2100 and 840 cm^{-1} Si-H bands (relative to Si-C (which at this temperature is assumed invariant)). Although the basic mechanism of these changes has been proposed (Hasegawa, 1989), some details remain uncertain. Hasegawa et al. (Hasegawa et al., 1983; Hasegawa, 1989) have shown how the composition of evolved gases during pyrolysis of the polymer changes after oxidative cross-linking. The un-cross-linked polymer evolves low MW oligomers at temperatures above 200°C with H_2 and hydrocarbons being evolved at ~400°C. Between 600 and 800°C the gaseous products are mainly H_2 and CH_4. After oxidation the amount of low MW oligomers evolved is not significant. Figure 6 from our work shows the effect of increasing oxidative stabilisation on the ceramic yield of PCS via thermogravimetric analysis. The marked increase in ceramic yield is evident.

Through variations in the starting materials, the thermolysis conditions and the extent of fractionation of the products, a wide range of polycarbosilanes can be produced. In these, the nature and density of reactive sites can be made to vary considerably. With a high density of reactive sites it is possible to cross-link the polymer by thermal treatment alone. However, if melt spinning is at all possible in this case, the product is strongly dependent on the rate of the cross-linking reaction. If the rate is low, then the fibre microstructure will worsen with increasing time as cross-linked "particulate" material with changing chemical and rheological characteristics is supplied from the reacting melt reservoir. At high cross-linking rates the polymer does not melt. In both cases, depending on the solubility of the polymer, dry or wet spinning could be possible.

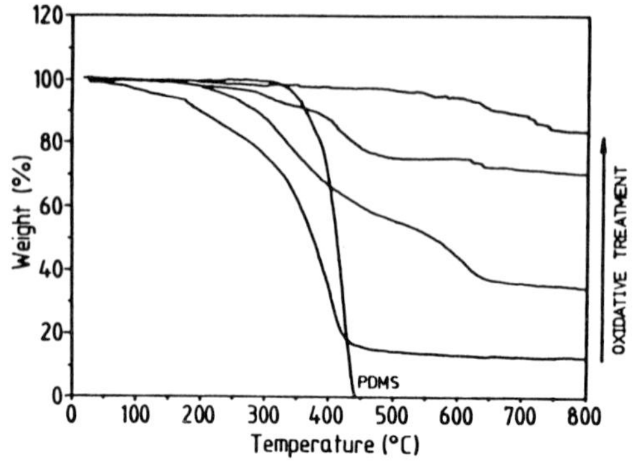

Fig. 6. TGA traces of PCS as a function of oxidative treatment at 180°C in air.

Other polymer precursors that have been studied in some detail are polysilastyrenes (PSS), polyvinylsilanes (PVS) and polysilazanes. West et al. (West, 1986) approached the problem of oxygen reduction and simplification of the production route to SiC-based fibres with the development of polysilastyrene, belonging to the polysilane group of polymers by virtue of its silicon backbone. A series of such soluble/fusible polymers has been synthesised and characterised. The polymers are degraded completely during pyrolysis via the formation of cyclics. However, polymers with aromatic groups attached to the silicon backbone leave behind ceramic residues which are increased after exposure to ultra-violet (UV) radiation. For example, the polymer $-(Me\text{-}Si\text{-}Me)_x.(Me\text{-}Si\text{-}Ph)_y-$, termed polysilastyrene due to the structural similarities with polystyrene, has been melt spun into fibres and cross-linked by UV irradiation to partially transform it into a polycarbosilane, and pyrolysed to SiC fibre. Depending on the irradiation conditions, hollow or solid ceramic fibres have been obtained (Sinclair, 1984). The simplicity of the process and the reduction in oxygen content are advantages over the Yajima route but the low ceramic yield (theoretically 45%) is a disadvantage. Problems associated with complete stabilisation of the irradiated fibre have been addressed via the inclusion of polyfunctional unsaturated silanes into the polymer and this enhances the cross-link density, but reported work in this direction is very limited.

Polysilastyrenes can also be converted to polycarbosilanes (PCSS) by thermolysis. The product can be highly reactive at low temperatures and is used as a binder in monolithic ceramic fabrication. PCSS has also been melt spun into fibres, cross-linked by UV irradiation and pyrolysed to high performance ceramic fibres (Teijin Ltd., 1986). Work in this laboratory has followed the above route with some modifications and continuous ceramic fibres have been produced (Figure 7).

Schilling (1986) developed polycarbosilanes (of typical formula $-(Me_3Si)_x(Me\text{-}Si\text{-}Me)_y(CH_2CH\text{-}Si\text{-}Me)_z-$), branched polysilahydrocarbons ($(=(Me\text{-}Si\text{-}CH_2)\text{-}(Me\text{-}C{=}CHCH_2)\text{-})$) and sodium derived vinylic polysilanes (see following example) based on various mono- and poly-functional vinyl-substituted chlorosilanes and unsaturated hydrocarbons. They employed the concept of the molar functionality, F, of the mixture of the chlorosilane monomers, i.e.

$$F = \frac{\sum_{i=1}^{n} x_i f_i}{\sum_{i=1}^{n} x_i}$$

where f_i and x_i are respectively the functionality and concentration of monomeric species i. The higher the value of F the more cross-linked are the polymeric structures. As a result of this the ceramic yields are higher, but yields of the useful high molecular weight fraction of the reaction mixture are lower. Linear polycarbosilanes, polysilanes and polysilahydrocarbons give negligible ceramic yields. The latter can be significantly improved when branching at backbone silicon atoms is introduced.

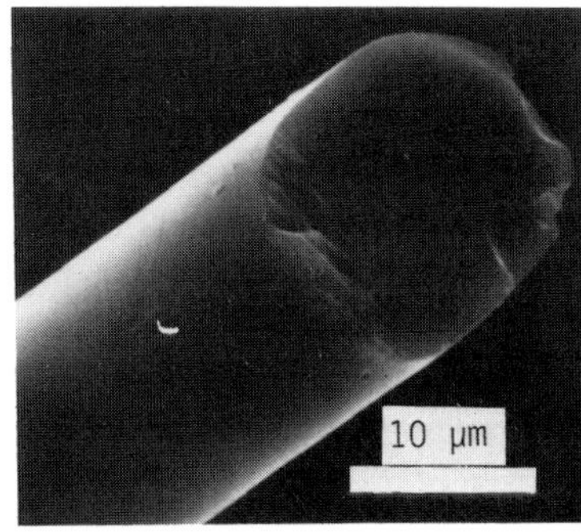

Fig. 7. Ceramic fibre from a polysilastyrene-based polycarbosilane.

Schilling (1986) also found that the reactivity of Si-H and Si-CH=CH_2 groups are higher in K/tetrahydrofuran than in Na-toluene environments. Thus, by using Na-toluene, soluble self curable vinyl polysilanes were obtained in useful yields, which gave ceramic yields of up to 60%, as illustrated in the following example:

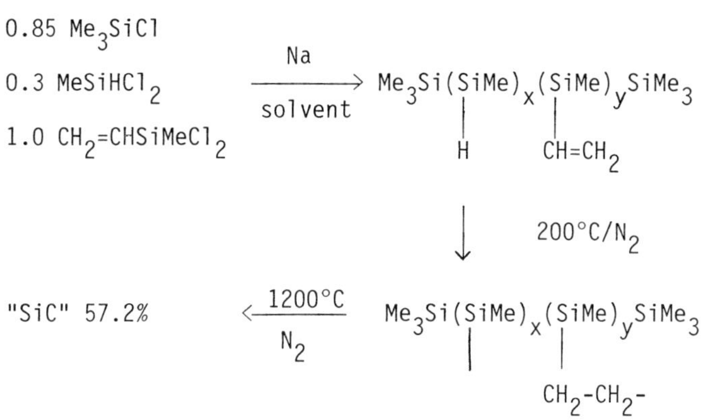

Although certain polyvinylsilane precursors can be thermally cross-linked, they are difficult to melt spin by virtue of their high reactivity. For example, Figure 8 shows differential scanning calorimetry traces for the thermal treatment in N_2 of a PVS polymer precursor synthesised in our studies. The unoxidised sample shows two distinct exothermic peaks. The initial peak at ~150°C is due to the thermal cross-linking reactions, the higher temperature peak being due to thermal degradation processes. On prior cross-linking in air for different periods the low temperature peak is gradually eliminated. This is a useful approximate indication of the onset of cross-linking. Figure 9 shows a PVS fibre cross-linked in air and pyrolysed to 1200°C in argon.

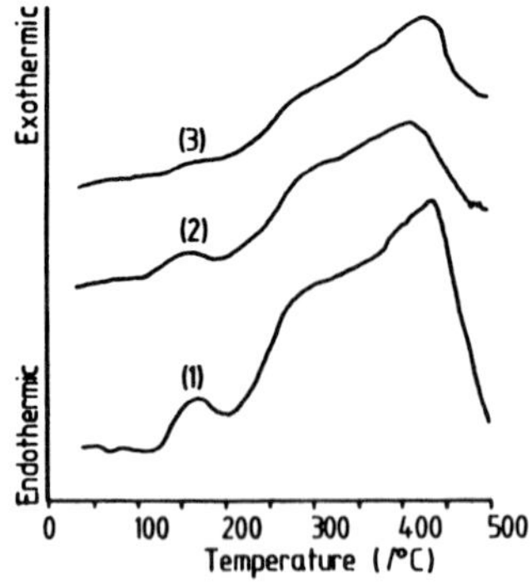

Fig. 8. DSC traces for the heat-treatment of polyvinylsilane in air: (1) as synthesised; (2) and (3) after increasing oxidative treatment.

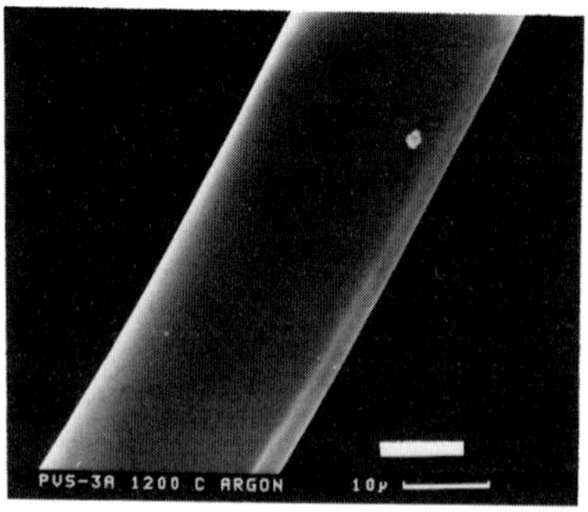

Fig. 9. Ceramic fibre from polyvinylsilane precursor.

The final development in this brief review of the polymer route to SiC comes from Baney et al. (1984) at Dow Corning, who showed that organodisilane materials (obtained as "waste" from the industrial synthesis of chloromethylsilanes) can be converted, through redistribution of Si-Cl and Si-Si bonds, to heavily branched polymethylchlorosilanes by heating in the presence of a tetrabutylphosphonium chloride catalyst. Further modification is necessary in order to reduce the high reactivity of the Si-Cl groups and consequently to improve the handling properties. The hydrido-derived polymer gives the highest ceramic yield of 90%, but is pyrophoric. Some of these chemically modified polymers have been spun into fibres, stabilised by oxidation in air and pyrolysed into ceramic fibres with Si-C-O composition.

Silicon nitride-based fibres have also received considerable attention. They can be produced from polycarbosilane fibres pyrolysed in an ammonia atmosphere (Okamura, 1987) or from precursors based on polysilazanes (Wynne et al., 1984). There are three main routes to polysilazane precursors. The first is thermolysis of methylsilazane compounds to a low MW polysilazane of complex structure which can be melt spun to Si-C-N-O fibre (Penn et al., 1986).

$$CH_3SiCl_3 + CH_3NH_2 \longrightarrow CH_3Si(NHCH_3)_3$$

$$\downarrow \text{thermolysis}$$

$$[CH_3(CH_3NH)Si(CH_3N)]_x \cdot [CH_3Si(CH_3N)_{1.5}]_y$$

$$\downarrow$$

SiCNO ceramic fibres, 65% yield

The second was developed by Seyferth et al. (Seyferth, 1987) and consists of reacting chlorosilanes with ammonia. Recent work by this team has produced near stoichiometric, white Si_3N_4 fibres from dry spinning of a particular polysilazane followed by UV curing and pyrolysis in a stream of ammonia.

$$RSiHCl_2 + NH_3 \longrightarrow (RSiNH.)_x \longrightarrow SiNC \text{ or } Si_3N_4 \text{ ceramic fibres, } < 90\% \text{ yield.}$$

In the final route, the mixed chlorodisilane by-product used as a SiC precursor as outlined above, is reacted with hexamethyldisilazane under thermolysis conditions to form tractable polysilazanes.

$$\text{chlorodisilanes} + [(CH_3)Si]_2NH \longrightarrow \underset{\text{(MPDZ)}}{(CH_3SiSiCH_3NH)_x} \longrightarrow \text{SCNO fibre } < 70\% \text{ yield}$$

$$HSiCl_3 + [(CH_3)_3Si]_2NH \longrightarrow \underset{\text{(HPZ)}}{(HSiNH)_{3/2}} \longrightarrow \text{SiNCO fibres } \leqslant 70\% \text{ yield.}$$

4.3 Conclusion

Sintering of silicon carbide powder is a promising route for the fabrication of SiC fibres. Carbothermal reduction of SiO_2 can be used to generate the fine, polycrystalline SiC in fibrous form, but the fibres require high temperature processing and the incorporation of sintering aids to ensure adequate densification.

A wide variety of routes involving polymer precursors are also under investigation. To avoid the use of oxygen in the stabilisation of the fibres prior to pyrolysis, it is necessary to synthesise new polymers. These polymers might contain groups that can be readily eliminated to reveal reactive sites on the Si-C polymer backbone, to allow cross-linking reactions to take place. However, the polymers must also be spinnable and pyrolysable to stoichiometric ceramic product. The simultaneous attainment of all these requirements is a daunting task.

The elucidation of the mechanisms of structural development during pyrolysis, of the pyrolysis reactions themselves and how these are determined by the structure and composition of the cured polymer are essential to the development of fibres with the required thermal stability.

REFERENCES

Baney R.H., Gaul J.H., Hilty T.H. 1984, in "Emergent process methods for high technology ceramics", pp. 253 ff, Davis R.F., Palmer H., Porter R.L. (eds), Plenum Press, New York.

Birchall J.D., Bradbury J.A.A., Dinwoodie J. 1985, "Alumina fibres: Preparation, properties and applications", Chapter 4 in "Strong Fibres", Watt W., Perov B.V. (eds) Handbook of Composites Vol. 1, Kelly A., Rabotnov N. (Series eds), North-Holland, Amsterdam.

Boecker W.D.G., Chwastiak S., Frechette F., Lau S-K. 1987, "Single-phase alpha-SiC reinforcements for composites", Proc. Silicon Carbide Symp., Columbus, Ohio, 407 ff.

Chaim R., Heuer A.H., Chen R.T. 1988, "Microstructural and micromechanical characterisation of SiC and Si-C-N ceramic fibres produced from polymer precursors", J. Am. Ceram. Soc. 71, 960 ff.

Fitzer E. 1990, "Carbon fibres and composites", in "Carbon Fibres, Filaments and Composites", 169 ff, Figueiredo J.L. et al. (eds), Kluwer Academic Publishers.

Hasegawa Y., Okamura K. 1983, "Synthesis of continuous silicon carbide fibre. Part 3: Pyrolysis process of polycarbosilane and structure of the products", J. Mat. Sci. 18, 3633 ff.
Hasegawa Y., Okamura K. 1989, "Synthesis of continuous silicon carbide fibre. Part 6: Pyrolysis process of cured polycarbosilane fibre and structure of silicon carbide fibre", J. Mat. Sci. 24, 1177 ff.
Johnson D.J. 1990, "Structure and properties of carbon fibres", in "Carbon Fibres, Filaments and Composites", 119 ff, Figueiredo J.L. et al. (eds), Kluwer Academic Publishers.
Kendall K., Alford N.Mc.N., Clegg W.J. and Birchall J.D. 1989, "High strength and Weibull Modulus from colloidal processing of ceramics", Br. Ceram. Proc. 45, pp. 79-91.
Nutt S.R., Wawner F.E. 1985, "Silicon carbide filaments: microstructure", J. Mat. Sci. 20, 1953 ff.
Okamura K. 1987, "Ceramic fibres from polymeric precursors", Composites 18 [2], 107 ff.
Penn B.G., Daniels J.G., Ledbetter F.E., Clemons J.M. 1986, "Preparation of SiC-Si_3N_4 fibres by the pyrolysis of polycarbosilazane precursors: A review", Pol. Eng. Sci. 26 [17], 1191 ff.
Porte L., Sartre A. 1989, "Evidence for a silicon oxycarbide phase in the Nicalon silicon carbide fibre", J. Mat. Sci. 24, 271 ff.
Rand B. 1985, "Carbon fibres from mesophase pitch", Chapter 6 in "Strong Fibres", Watt W., Perov B.V. (eds) Handbook of Composites Vol. 1, Kelly A., Rabotnov N. (Series eds), North-Holland, Amsterdam.
Riedel R., Passing G., Petzow G. "Pyrolysis of a boron-containing compound for doping of SiC", to be published in "Proceedings of the 2nd International Conference of Powder Processing Science", Deutsche Keramische Gesellschaft, Hausner H. (ed).
Schilling C.L. Jr. 1986, "Polymeric routes to silicon carbide", Br. Pol. J. 18 [6], 355 ff.
Seyferth D. 1987, "Organosilicon polymers as precursors for Si-containing ceramics: recent developments", Pol. Preprints 28, 389 ff.
Sinclair R.A. 1984, "Pyrolysis of polysilanes", Chapter 21 in "Ultrastructure processing of ceramics, glasses and polymers", Hench L., Ulrich D.L. (eds), John Wiley and Sons, New York.
Stacey M.H. 1988, "Developments in continuous alumina-based fibres", Br. Ceram. Trans. J. 87, 168 ff.
Taki T., Okamura K., Sato M. 1989, "A study of the oxidation curing mechanism of polycarbosilane fibre by solid-state high-resolution nuclear magnetic resonance", J. Mat. Sci. 24, 1263 ff.
Teigin Ltd. 1986, "Polycarbosilastyrene copolymer, method of producing the same, and process for producing shaped SiC articles from the same", European Patent 212 485 A2.
Watt W. 1985, "Chemistry and physics of the conversion of polyacrylonitrile fibres into high modulus carbon fibres", Chapter 9 in "Strong Fibres", Watt W., Perov B.V. (eds), Handbook of Composites Vol. 1, Kelly A., Rabotnov N. (Series eds), North-Holland, Amsterdam.
West R. 1986, "The polysilane high polymers", J. Organometall. Chem. 300, 327 ff.
Wynne K.J., Rice R.W. 1984, "Ceramics via polymer pyrolysis", Ann. Rev. Mater. Sci. 14, 297 ff.
Yajima S., Hayashi J., Omori M., Okamura K. 1976, "Development of silicon carbide fibre with high tensile strength", Nature 261, 683 ff.
Yajima S. 1985, "Silicon carbide fibres", Chapter 6 in "Strong Fibres", Watt W., Perov B.V. (eds), Handbook of Composites Vol. 1, Kelly A., Rabotnov N. (Series eds), North-Holland, Amsterdam.

Fiberamic: A new silicon carbonitride ceramic fiber with high thermal stability

Abstract

RHÔNE-POULENC is developing a new silicon CarboNitride fiber named FIBERAMIC®.

The main feature of that fiber is its good thermomechanical behavior up to 1300°C - 1400°C.

With a homogeneous amorphous structure based on a continuum of silicon-centered tetrahedrons, this fiber can be used at elevated temperatures because of its structural and chemical stability.

Mechanical and thermomechanical investigations made with FIBERAMIC® have demonstrated good high temperature durability such as a retention of 80% of the room temperature strength at 1400°C as well as a high temperature stability after ageing in oxidizing atmosphere.

Current efforts are devoted to the production of FIBERAMIC® at the pilot plant scale and the evaluation of thermomechanical properties of FIBERAMIC®-based composites.

1. INTRODUCTION

In recent years, high temperature materials like ceramic and metal matrix composites are gaining considerable attention for specific components used in space, aeronautics and automotive industries. One of the keys of this development is the avaibility of a stable ceramic fiber, even in oxidizing atmosphere at high temperature.

A wide range of ceramic reinforcements have been commercially available such as alumina-silica fiber or silicon oxycarbide fiber. However, they lack oxidative stability at elevated temperatures (above 1000°C - 1200°C) mainly because of their chemical and structural evolutions during heating.

Pure Silicon Carbide monofilaments can also be prepared by chemical vapor deposition on a Carbon substrate but their very high diameters (140μm) don't allow the fiber to be woven into fabrics commonly used.

RHÔNE-POULENC is developing a Silicon CarboNitride fiber named FIBERAMIC® which is a continuous textile yarn of 250 or 500 filaments obtained through pyrolysis of a polysilazane precursor.This concept can be readily understood with reference to the synthesis of carbon fibers which are industrially used today. This route permits the production of continuous ceramic filaments with small diameters.

The aim of this Paper is to give information on the preparation of FIBERAMIC® and to describe its structure and mechanical properties.

2. EXPERIMENTAL

Room temperature mechanical properties are measured by single filament test methods using a gauge length of 10 mm and an extension rate of 2 mm/min. Strength and diameter correspond to the average from about 30 individual tests. Elastic Moduli are determinated by the compliance method.

Thermomechanical properties are measured also by single filament tests method using a gauge length of 25 mm.This equipment was developed in collaboration with Société Européenne de Propulsion.Temperature is controlled along the filament at +/– 5°C. Tests are performed under air after a stabilization of the filament in the hot furnace for 3 min.

3. PREPARATION OF FIBERAMIC

Figure 1 shows the four steps necessary to produce FIBERAMIC®. In the first one, polysilazane precursors are obtained by reacting ammonia or amine with chlorosilanes :

$$2 \equiv Si - Cl + 3\,NH_3 \longrightarrow \equiv Si - NH - Si \equiv + 2NH_4Cl$$

With this type of polymer we can adjust the average molecular weight in the range of 1,000 to 20,000 in order to select a polymer with good spinnability.

Green fibers can be prepared by melt or dry spinning. A continuous green yarn with 250 or 500 filaments is produced. In order to avoid the melting of green fiber during pyrolysis, RHÔNE-POULENC has patented thermal and chemical curing processes suitable for polysilazanes such as:

$$\equiv Si - H + \equiv Si - CH = CH_2 \xrightarrow{Pt,\ h\nu} \equiv Si - CH_2 - CH_2 - Si \equiv$$

The development of the new curing processes leads to the preparation of ceramic fibers with a low oxygen content as compared to that obtained through oxidation or hydrolysis.

The cured fiber is then converted into ceramic Silicon CarboNitride fiber by pyrolyzing in an inert atmosphere in the range of 1200°C to 1400°C ; this produces flexible, black fibers. Their optimum properties and stability are related to the densification of the structure (see fig. 2).

4. STRUCTURE AND PROPERTIES OF FIBERAMIC®

Structural Description

The typical composition of FIBERAMIC® is Si : 57 wt %, N : 22 wt%, C : 13 wt % and O : 8 wt %, in phases it is equivalent to about only 5 to 10 % of free carbon and 90 to 95 % of $SiN_XC_YO_Z$.

Figure 3 shows X-RAY diffraction patterns of FIBERAMIC® after a 5 hour'heat treatment : it remains amorphous up to 1400 C ; therefore no defect appears by local excess grain growth.

In fact, ^{29}Si-NMR measurements (see fig. 4) show that FIBERAMIC® can be described like a continuum of silicon-centered tetrahedrons with Nitrogen,Carbon and Oxygen as first neighbours. The structure is homogeneous and the free Carbon is organized in small clusters as shown by bright field Transmission Electronic Microscopy (see fig.5).

Mechanical and thermomechanical properties

Results of mechanical testing on FIBERAMIC® are as follows : Figure 6 shows the distribution of diameters; the average value is about 14 μm and the distribution is narrow.

FIBERAMIC® exhibits typical brittle failure with an average tensile strength in the range of 1500 - 2200 MPa and Weibull parameter of about 7 (see fig.7).The failure is due to small surface defects as shown on figure 8.

Measurement of FIBERAMIC® toughness, based on Griffith equation, gives an average k_{1C} value of 2.5 $MPa.m^{-1/2}$ despite an amorphous structure,as compared to monolithic Silicon Carbide (4.5 $MPa.m^{-1/2}$).It is established that the control of the shape and size of surface flaws is the way to reach high failure strength.

Elastic moduli of FIBERAMIC® is about 200 GPa, related to the density and composition of Silicon CarboNitride.

As expected from its thermal stability, FIBERAMIC® has an excellent thermomechanical behavior in an oxidizing atmosphere as shown on figure 9, Silicon CarboNitride retains 80 % of its room temperature strength when tests are performed at 1400°C.

After ageing in oxidizing atmosphere, FIBERAMIC® tensile strength decreases slowly with time as shown on figure 10, the fiber retains more than 85% of the initial stress after 100 hours at 1000°C or after 10 hours at 1200°C.

Those results show that the chemical and structural stabilities of the amorphous Silicon CarboNitride fiber up to 1400°C provide interesting thermomechanical properties in oxidizing atmosphere.

5. CONCLUSIONS

The new ceramic fiber FIBERAMIC® described in the present paper offers good mechanical properties at room temperature in comparison to the other ceramic reinforcements.

Amorphous Silicon CarboNitride presents very interesting intrinsic properties in terms of thermomechanical behavior up to 1400°C in oxidative atmosphere.

The material and the product line present a large potential for further improvement and optimization and should find an interesting place among ceramic, metal or glass composites.

Various grades of FIBERAMIC® are under development in order to optimize the thermomechanical properties of the fibers and composites.

6. ACKNOWLEDGEMENTS

The authors wish to thank the team of Mrs OBERLIN (Laboratoire M. Mathieu - PAU - France) for its intensive work and microscopic studies as well as Mr TAULELLE (Laboratoire de spectroscopie du solide, Jussieu - PARIS VI - France) for the NMR analysis and interesting discussions.

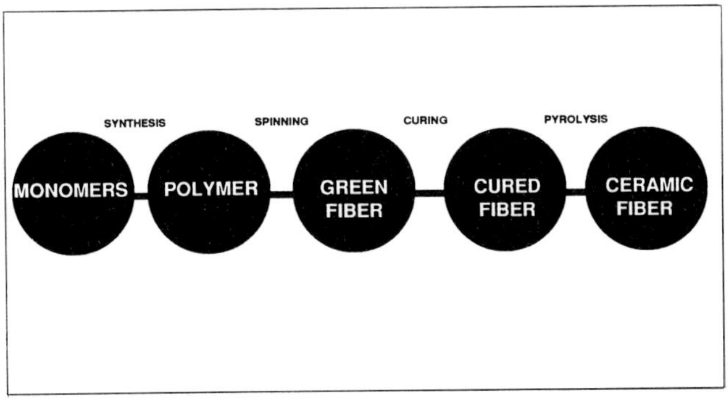

Figure 1 **Preparation of SiNC Fibei**

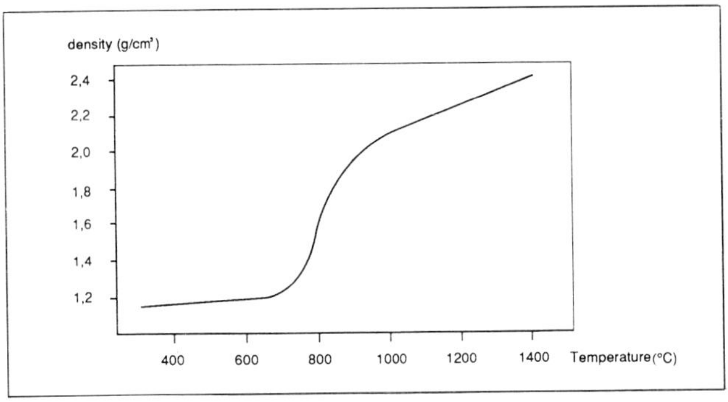

Figure 2 **Density versus temperature**

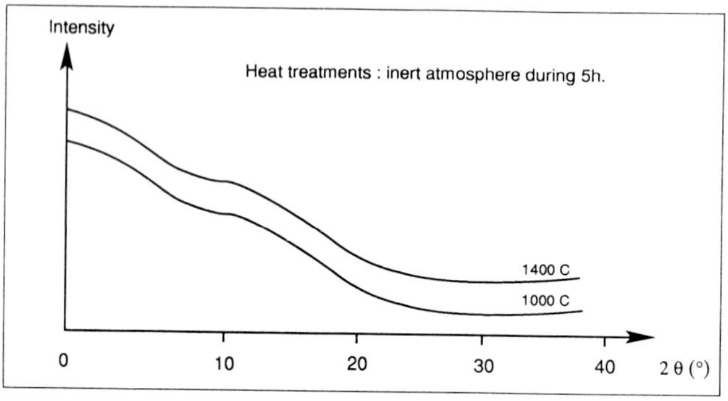

Figure 3 **X-RAY diffraction patterns**

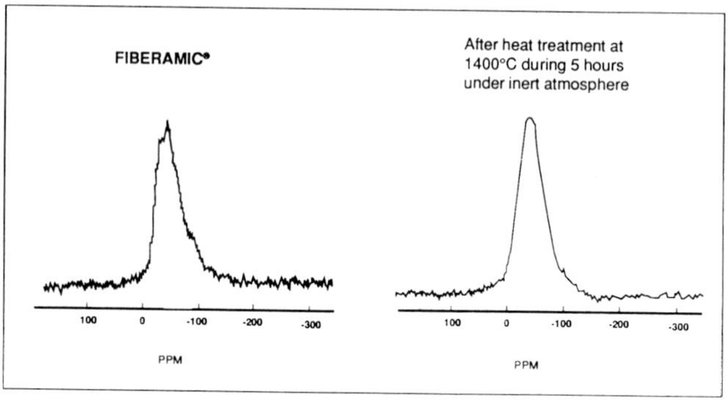

Figure 4
^{29}Si – NMR measurements of FIBERAMIC®

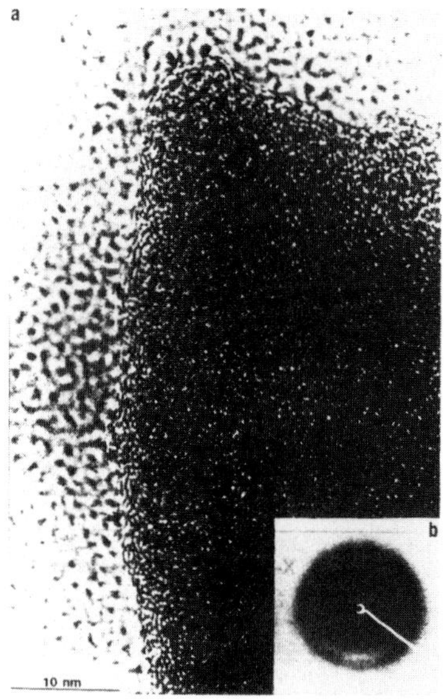

Figure 5
Bright field Transmission Electronic Microscopy of FIBERAMIC®

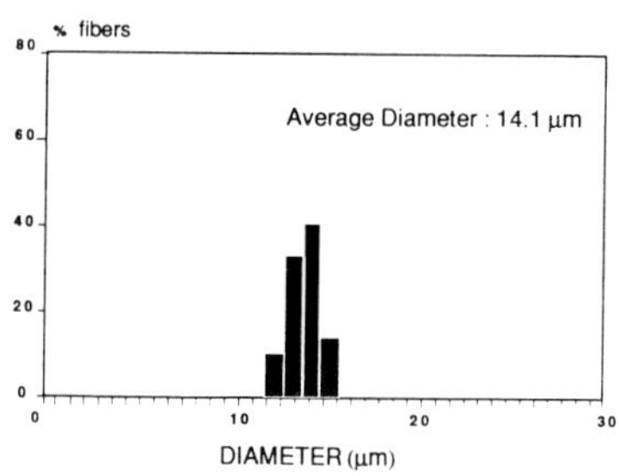

Figure 6
Diameter distribution

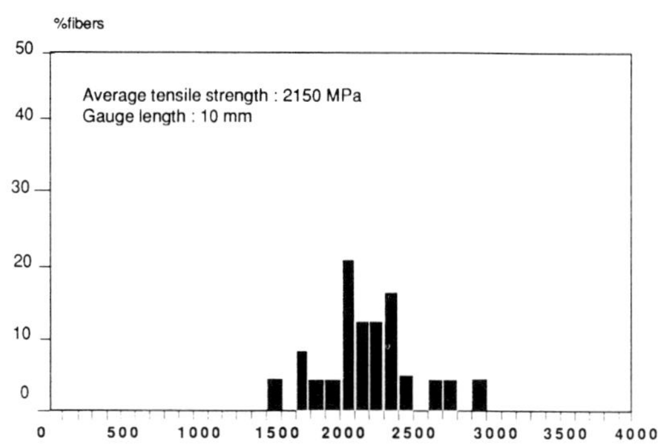

Figure 7
Strength distribution of FIBERAMIC®

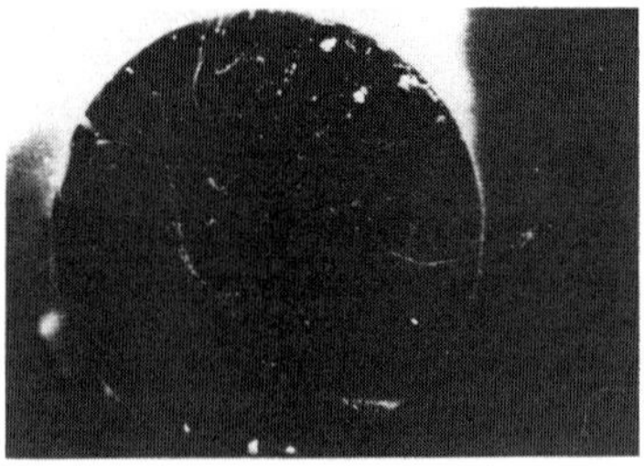

Figure 8
SEM analysis of failure surface

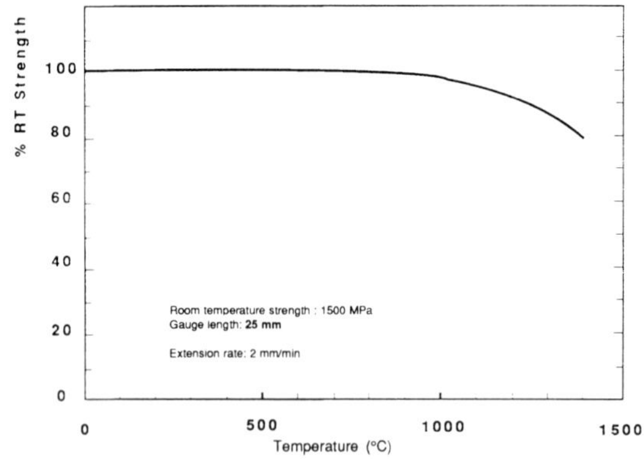

Figure 9
Strength retention versus temperature in air
(after 3 min. of stabilization time)

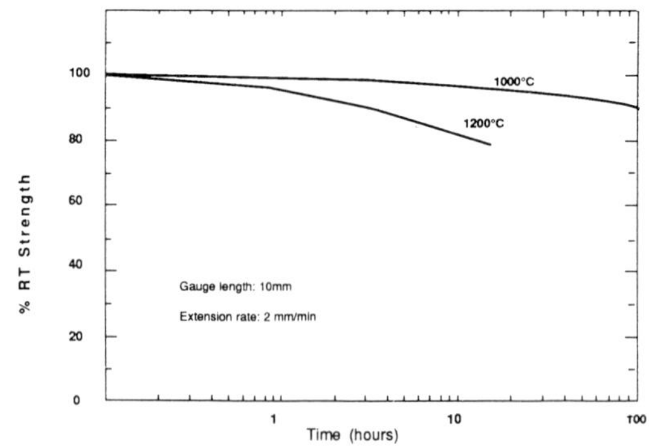

Figure 10
Strength retention after ageing in air
versus time

Inst. Phys. Conf. Ser. No 111
Paper presented at Int. Conf. on New Materials and their Applications, Univ. of Warwick, 1990

Coatings for composites—a theoretical assessment

C F KNIGHTS

AEA Technology, Harwell, Didcot, OXON, OX11 ORA, UK

ABSTRACT: Materials to act as interlayers to prevent sintering between ceramic fibres and ceramic matrices during fabrication and use in vacuo or inert environments have been identified and temperature limits indicated. Service use in oxidising atmospheres and under continual temperature cycling that causes cracking severely limits service temperatures. Such limitations will exist for oxide composites, when non-oxide interlayers are required to stop sintering between fibre and matrix, and for non-oxide composites such as those based upon carbon and/or silicon carbide. The current approaches to provide limited high temperature oxidation protection are considered.

1. INTRODUCTION

Successful incorporation of fibres in a ceramic matrix requires prior consideration of how the fibres and matrix will interact during fabrication and subsequent use. This paper considers uses of chemical thermodynamics to establish the extent of chemical interaction between fibre and matrix and the need for interlayers to alter the interaction. Consideration will also be given to the nature of external coatings to protect the composite from aggressive in-service environments such as air and combustion gases.

The common fibres and matrices are carbon, silicon carbide,silicon nitride, alumina, alumino-silicates (mullite), silica and glasses; boron is only used as a fibre. Presently, carbon fibre can retain its strength to 2770K (short term), while silicon carbide and alumina are limited to temperatures $\leq$1370K (Phillips 1988). Boron and silica fibres are limited to 600°C (Fitzer and Gadow 1986). Fibre reinforced concretes and metal matrices will not be considered.

2. FABRICATION

Generally, the temperatures required for fabrication will be higher than the in-service temperatures but the environment can be chosen to be non-aggressive.

For carbon-carbon (C-C) and SiC-SiC composites it may be necessary to coat the fibres in order to reduce sintering and bonding with the matrix. This can be achieved by using weakly bonded sacrificial coatings of carbon or SiC on fibres of carbon, SiC or Si_3N_4. Provided there is no free silicon in the SiC there will be no chemical interaction or dissolution between C and SiC up to 2545°C (Massalski 1988). Silicon carbide will probably possess a thin surface layer of silica. In closed pores, gaseous reaction products could build up to substantial pressures (Table 6). Silicon nitride at 1800K will create a nitrogen pressure of 0.9 bar.

The use of carbon or SiC fibres in oxide matrices will be limited by the pressures of carbon monoxide (Table 1). In vacuum, the escape of gaseous products from interacting interfaces could displace the reactions significantly. In the absence of pumping or in the presence of a backfill of carbon monoxide the continuing reaction could be inhibited. Carbides of titanium and zirconium as interlayers in an alumina system would be inferior to the use of SiC; reactions would produce liquid titanium-aluminides and zirconia respectively.

Fabrication of oxide fibre/oxide matrix composites also requires interlayers to reduce fibre/matrix sintering. Such inter-layers could be SiC and, at lower temperatures, carbon (Table 1). Zirconia fibre and matrix composites could employ ZrC, ZrB_2 or SiC interlayers. The presence in glasses of oxides with a less negative Gibbs energy of formation ($\Delta_f G°_T$) than silica would impose lower temperatures on the fabrication procedure. At 1273 and 1573K, TiN and TiB_2 are more stable than C and SiC in contact with glass such as mullite ($3Al_2O_3.2SiO_2$). In general oxide interlayers will dissolve in an oxide matrix or an oxide fibre. Fabrication seems relatively straightforward compared to the problems of fibre/matrix/environment interactions during in-service use.

3. IN-SERVICE USE

For composites containing non-oxide components appropriate external coatings will be required to protect the composite overall and appropriate internal interlayers will be required between the fibre and matrix. In-service, thermal cycling and imposed strains will cause cracks. These must not be allowed to propagate through the fibres. Thus the matrix-fibre interface should have the ability to stop cracks. Weak bonding or a separated interface is necessary. The non-reactive, non-sintering interlayers required to fabricate the composite are ideal for this purpose. A mismatch in coefficients of thermal expansion (cte, Strife and Sheehan 1988) will ensure some physical separation between fibre and matrix.

Table 1 Interaction of oxides with carbon and carbides*

Fibre or Coating	Matrix	Temp K	Gaseous Products CO	Al	Al_2O bar	SiO	Other Products
C	Al_2O_3(1)	2500	0.04	0.06	0.15	-	
C	SiO_2(c)	1800	1.1	-	-	0.007	-
SiC	SiO_2(1)	2000	0.12	-	-	0.34	-
SiC	Al_2O_3(1)	2000	0.006	0.003	0.004	0.006	-
SiC	Al_2O_3(1)	2500	0.08	0.03	0.05	0.14	-
TiC	Al_2O_3(1)	2500	1	-	-	-	TiAl(1)
ZrC	Al_2O_3(1)	2500	0.01	0.02	-	-	ZrO_2(c)

* Data from Chase et al 1985

3.1 Inert Gas or Vacuum Environments

Such environments will impose restrictions similar to those already considered in relation to fabrication. An additional criterion is the life expected from the component. Hillig (1986) suggests that for long term use pressures of gaseous decomposition products at the external surface of the composite should be $\leq 10^{-6}$ bar; internally the pressures could be up to 0.1 bar provided the gases cannot escape readily.
Thus ZrO_2 and Y_2O_3 satisfy this external criterion for surface temperatures up to 2500K in vacuo or in air; Al_2O_3 could be used only up to 2300K in vacuo while silica is restricted to 1900K (Table 2).

Table 2 Volatility of oxides in vacuo and in oxygen#

Oxide M_xO_y	Temp* K	Pressure of gaseous products/bar MO_2	MO	M	O_2	Conditions⁺
ZrO_2(c)	2500	3.5E-7	4.2E-7	1E-12	2E-7	vacuum
	2500	3.5E-7	2 E-10	1E-18	0.1	O_2 imposed
Y_2O_3(c)	2500	-	5 E-7	2E-10	1E-7	vacuum
Al_2O_3(c)	2300	2.7E-9	2E-7	6E-7	4E-8	vacuum Al_2O_3 = 4E-8 Al_2O_2 = 1E-9
Al_2O_3(l)	2500	1.0E-6	0.2E-6	8E-10	0.1	O_2 imposed Al_2O_2(g)= 5E-11 Al_2O(g) = 5E-12
SiO_2(c)	1900	4.4E-8	2.0E-6	3E-15	9E-7	vacuum
SiO_2(l)	2100	1.4E-6	0.7E-6	9E-17	0.09	O_2 imposed

Data from Barry and Rand 1981 and Chase et al 1985
* Temperature for a total pressure of gaseous products of $\sim 10^{-6}$ bar.
\+ and additional product pressures (bar)

When oxygen is present in the environment the pressures of oxygen species with low values of O/M are greatly suppressed and the oxide can be used at a higher temperature before the total pressure of gaseous products exceeds 10^{-6} bar.

In vacuo the non-oxide ceramics can be used at higher surface temperatures than the oxides except for SiC, which is restricted to $\leq$1950K because of the vapour pressure of silicon, and for silicon nitride which is restricted to $\leq$1300K because of the nitrogen pressure (Table 3). In an imposed nitrogen atmosphere the nitrides can have a higher surface temperature before reaching the 10^{-6} bar limit due to metal vapour.

Table 3 Volatility of non-oxide ceramics

Material (Cryst)	Temp* K	Vapour pressure of gaseous species/bar					Other phases formed	Ref
		Metal	Carbide	C_1	C_2	C_3		
C	2550	-	-	1.4E-6	7E-8	1.1E-6	-	M
SiC	1950	1.2E-6	1E-11	1E-11	5E-13	1E-11	C(c)	J
TiC	2400	4.7E-7	-	4.4E-8	7E-9	1.1E-7	C(c)	J
ZrC	2500	7.E-10	-	1.8E-7	4E-8	5.4E-7	C(c)	J
			B					
TaB_2	2600	5E-9	2.2E-6				Ta(c)	S
HfB_2	2600	0.2E-6	0.5E-6					S
			N_2					
Ta_2N	1800	1E-16	9.6E-7				Ta(c)	K,H
HfN	2400	4.3E-7	3.6-E7				Hf(c)	S,H
Si_3N_4	1300	4E-11	4.6E-7				Si(c)	M

* = Temperature for a total pressure of gaseous products of 10^{-6} bar
M = Barry and Rand 1981 H = Hultgren et al 1973. J = Chase et al 1985
K = Kubaschewski and Alcock 1979
S = Schick 1966

3.2 Oxidising Environments

In oxidizing environments the surface must meet the volatility criterion, but in addition the effects of oxygen permeability through the structure and the effects on the permeability of temperature cycling must be considered.

Under isothermal conditions, silicon carbide, surpassing all other non-oxide ceramics, can be used in air up to 1920K (1650°C, Gnesin 1974). Thus apart from the fact that current SiC fibres lose their strength at >1370K (Phillips 1988) the SiC-SiC composite could be used isothermally up to 1920K. A carbon fibre/SiC matrix composite could in principle maintain the fibre contribution to its strength up to the oxidation limit for SiC (1920K).

Unfortunately, few applications involve isothermal conditions. Usually a composite has to withstand temperatures cycling between stand-down and service conditions and during service. Silicon carbide loses its oxidation resistance under these conditions due to cracking of the protective silica layer as a result of the cte mismatch.

Oxide matrices and fibres are inherently oxidation resistant. Creep strength, however, is relatively low and currently alumina fibres lose strength above 1370K (Phillips 1988) and silica loses half its room temperature strength at 870K (Fitzer and Gadow 1986).

Sintering between oxide matrix and fibre can be prevented during fabrication by interlayers (Table 1). If sintering at service temperatures can occur, the crack resistance of the composite will be impaired unless the non-oxide interlayer present to allow fabrication can be maintained during service.

Alumina and stabilised zirconia have rather high oxygen permeabilities (Figure 1) and can provide little protection of non-oxide interlayers or fibres. For example, a 3mm layer of ZrO_2 - CaO would allow complete oxidation of a 5μm thick layer of silicon carbide beneath it in 70 days at 1350K. Alumina is better (5 years at 1350K) but rapidly deteriorates at higher temperatures (70 days at 1560K).

Silica has a particularly low permeability which accounts for its protectiveness on SiC on which it forms naturally during oxidation. Silica as an outer coating on alumina and zirconia is limited by the formation of alumino-silicate and zircon ($ZrSiO_4$). The changes in thermal expansion coefficients and subsequent temperature cycling will undoubtedly cause complete break-up of the surface protection.

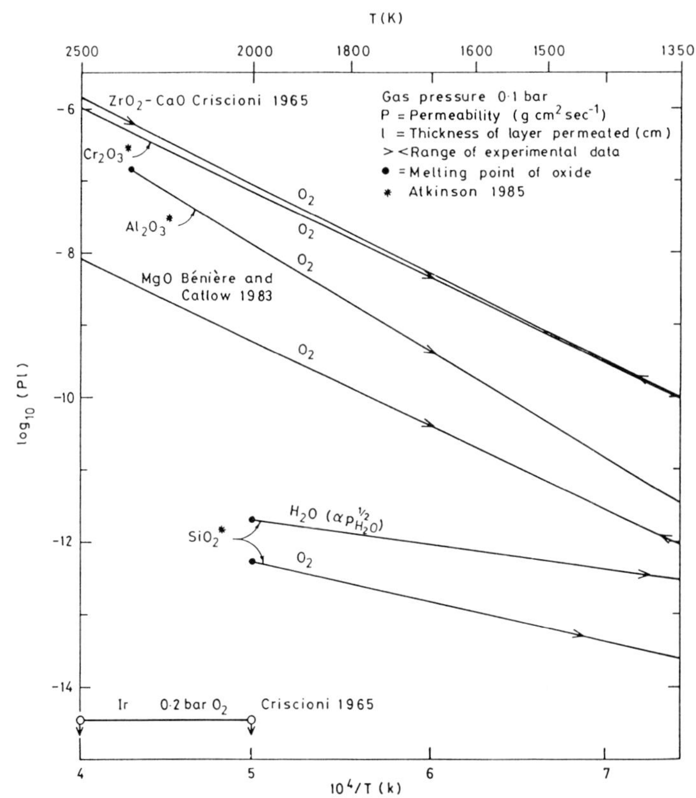

Figure 1. Permeability of oxides to oxygen

The external carbon-carbon composites used on the US space shuttle are protected by siliconising the surface with Si and SiO gases to form SiC (Strife and Sheehan 1988 and Lewis 1989). On cooling the SiC cracked due to its higher thermal expansion coefficient compared with that of the carbon matrix. These cracks were filled with a tetraethyl ortho silicate and sodium ethoxide. This provided protection at temperatures below the point at which, with rising temperature, the cracks would close because of the cte mismatch and expel some of the liquid phases. The treatment lasted for 75h, adequate for 70-80 missions. This approach would not be satisfactory for longer service or one in which temperatures vary more frequently.

Carbon starts to oxidise at 640K (Strife and Sheehan 1988). Oxidation protection must therefore be available at these low temperatures. Boron additions have proved to be effective since they oxidize to give B_2O_3 [melting point 723K (450°C)] which can fill cracks and reduce the permeation of oxidants. Boric oxide has a relatively low volatility 1 x 10^{-7} bar pressure at 1300K. Unfortunately water vapour enhances volatility dramatically due to the formation of gaseous HBO_2 and $(HBO_2)_3$ species. Thus borate is inadequate as an external protection (Jawed and Nagle 1986, Table 4).

Table 4 Borate Volatility

$$B_2O_3(l) + H_2O(g) = 2\,HBO_2(g) = (2/3)(HBO_2)_3(g)$$

Temp K	P_{H_2O}	P_{HBO_2}	$P_{(HBO_2)_3}$
		bar	
900	0.1	8.0E-7	1.5E-3
	0.01	2.5E-7	4.8E-5
1000	0.1	8.3E-6	8.9E-4
	0.01	2.6E-6	2.8E-5

Thus the problem common to most composites containing non-oxide components and used in oxidising, high temperature, non-isothermal environments is how to maintain the presence of a liquid to block cracks which form in an outer coating. Cracks can form because of thermal expansion mismatch or external impacts. The conceptual structure for a self-healing external coating is

Outer oxide low volatility	Inner oxide low oxygen permeability	liquid oxide eg B_2O_3 to seal cracks	Fibre and matrix

Unfortunately the boric oxide and, to a lesser extent, silica form low melting point eutectics with the other oxides (Table 5), and both will react with non-oxide fibres and matrices (Table 6). These reactions will seriously undermine the integrity of the composite and gaseous products will permeate throughout its open porosity.

Table 5 Attack of oxides by B_2O_3 and SiO_2

System	Liquid phases*	
	Temp/K	Composition/mole%
B_2O_3 - SiO_2	913	20 wt% SiO_2
	1013	40 wt% SiO_2
B_2O_3 - Al_2O_3	1773	9 Al_2O_3
	2173	20 Al_2O_3
B_2O_3 - ZrO_2		no data found
SiO_2 - Al_2O_3	1868	5 Al_2O_3
SiO_2 - ZrO_2	1949	5 ZrO_2 + $ZrSiO_4$
Al_2O_3 - ZrO_2	1983	43 wt% ZrO_2 eutectic
B_2O_3 - Al_2O_3 - SiO_2	1773	40 wt% Al_2O_3 Mullite and SiO_2 formed
Al_2O_3 - ZrO_2 - SiO_2	1882	eutectic with Mullite, $ZrSiO_4$ and SiO_2

* Levin et al 1964

JTA graphite incorporates these principles to allow the graphite to be used for a few hours at 1700°C (Criscione 1965). The graphite is impregnated with an intimate mixture of SiC (12 vol %), ZrB_2 and HfB_2 (21 vol %). During oxidation and subsequent volatilisation of boron and silicon components a surface is left comprising an outer layer of mainly $(ZrHf)O_2$, inner layers of SiO_2 and silicon carbide, and liquid B_2O_3 filling cracks; the structure is unstable (Table 5). Other studies have examined graphite impregnated with zirconium diboride (ZrB_2) and boron (Gray 1984) and a coating on graphite of Al_2O_3, SiC, Si and B (9,60,30,1 wt% levels) (Veltri and Galasso 1982). Improvements to silicon carbide for use above 1650°C have been sought with an outer layer of ZrO_2 stabilised with Y_2O_3 and an inner layer of alumina (Patten, Moss and Forcht 1981). SiC-Al_2O_3 is reasonably stable but eutectic liquids of Al_2O_3-ZrO_2 can form at 1710°C and Al_2O_3-Y_2O_3 liquids can form at 1800°C (Levin et al 1964). Patten, Moss and Forcht (1981) claimed that the coating showed no attack in 90-minutes at 1925°C (2200K).

Table 6 Attack of non-oxides by B_2O_3 and SiO_2*

Non-Oxide	Oxide	Temp K	Equilibrium products					
			Solids	Gases (Bar)				
				B_2O_3	B_2O_2	CO	SiO	SiO_2
C	B_2O_3(l)	1600	B_4C	1E-4	3E-4	0.007		
		1800	B_4C	3E-3	6E-3	0.7		
SiC	B_2O_3(l)	1800	B,SiO_2	3E-3	0.02	0.01	0.033	6E-9
C	SiO_2	1400	SiC			4E-3	6E-6	8E-14
		1600				0.096	3E-4	4E-11
		1800				1.1	7E-3	6E-9
								Si(g)
SiC	SiO_2	1600				6E-4	2E-3	3E-8
		1800				0.011	0.033	2E-6
		2000				0.12	0.34	4E-5

*Data from Barry and Rand 1981 and Chase et al 1985

The presence of other oxides in silica increases the oxygen permeation rate considerably. (At 900°C additions of alumina or TiO_2 at 20 wt% increase permeability by x 2.3; 20% Al_2O_3 + 5% Na_2O by x8.6 and 30% GeO_2 or 30% B_2O_3 by x67 and x406 respectively (Schlichting 1984). Silicon carbide fibres in an alumina matrix showed a x10 increase in the oxidation rate at 1300°C due to alumina incorporation into the silica film growing on the SiC (Jakus and Nair 1990). Silicon carbide can in turn enhance the rate of crystallisation of Pyrex glass even during hot pressing (Murty and Lewis 1989). Long term oxidation protection of composites at high temperature is a major design problem.

4. CONCLUSIONS

1. In terms of minimising unwanted chemical reactions, a number of satisfactory options for interlayers are available to allow the fabrication of oxide and non-oxide ceramic composites and their use in vacuo or in inert atmospheres.

2. For isothermal use in oxidising atmospheres, external coatings of silicon carbide can in principle be used up to 1650°C to protect composites based upon C-C and SiC

and to protect internal non-oxide interlayers in Al_2O_3, alumino silicates and, at lower temperatures, glasses containing oxides less stable than silica.

3. Use of entirely non-oxide composites in oxidising atmospheres with fluctuating temperatures poses severe problems due to cracking of the protective outer coatings. Borate protection for even a few hours is only possible up to about 1100°C for C-C. Oxide-oxide composites could be used provided sintering between matrix and fibre is not a problem. Otherwise, non-oxide interlayers have to be used which need to be protected from oxygen permeating through cracks within the oxide matrix and through the oxide lattice itself. The protection methods presently being examined are chemically unstable.

REFERENCES

Atkinson A 1985 Rev. Modern Phys. **57** 437

Barry T I and Rand M H 1981 Provision of the Database of SGTE (Scientific Group for Thermodata in Europe), see Ansara I and Sundman B 1981 Computer Handling and Dissemination of Data ed P S Glaeser (Oxford: Elsevier Science) p 154

Bénière F and Catlow C R A 1983 Mass Transport in Solids (New York: Plenum Press) p 360.

Chase M W, Davies C A, Downey J R, Frurip D J, McDonald R A and Syverud A N 1985 J Phys. Chem. Ref. Data **14** Suppl.1.

Criscioni J M 1965 USA Report ML-TDR-64-173 Part 2

Fitzer E and Gadow R 1986 Amer. Ceram. Soc. Bull. **65** 326

Gnesin G G 1974 Refractory Carbides ed G V Samsonov (London: Consultants Bureau)

Gray P E 1984 US Patent Application 676 985 (Chem. Abs. **105** 101694)

Hillig W B 1986 Tailoring Multiphase and Composite Materials ed R E Tressler and G L Messing (New York: Plenum) p 697

Hultgren R, Desai P D, Hawkins D T, Gleiser M and Kelly K K 1973 Selected Values of the Thermodynamic Properties of Metals and Alloys (Ohio: ASM)

Jakus K and Nair S V 1990 Composite Sci. Technol. **37** 279

Jawed I and Nagle D C 1986 Mat. Res. Bull. **21** 1391

Kubaschewski O and Alcock C B 1979 Metallurgical Thermochemistry (Oxford: Pergamon)

Levin E M, Robbins C R, McMurdie H F, Roth R S, Negas T, Cook L P and Dennis J R, 1964, 1969, 1975, 1981, 1983, 1987 Phase Diagrams for Ceramicists (Ohio: Amer. Ceramic Soc.)

Lewis C F (ed) 1989 Mat. Eng. January 1989, p27

Massalki T B 1988 Binary Alloy Phase Diagrams (Ohio: ASM)

Murty V S R and Lewis M H 1989 J. Mat. Sci. Lett. **8** 571

Patten J, Moss R W and Forcht B A 1981 US Patent 4,500,602

Phillips D C 1988 Risφ Conf. Mechanical and Physical Behaviour of Metallic and Ceramic Composites.

Schick H L 1966 Thermodynamics of Certain Refractory Compounds (London: Academic Press)

Schlichting J 1984 J. Non Cryst. Solids **63** 173

Strife J R and Sheehan J E 1988 Ceramics Bull. **67** 369

Veltri R D and Galasso F S 1982 US Patents 4,472,476; 4,476,164 and 4,476,178

Inst. Phys. Conf. Ser. No 111
Paper presented at Int. Conf. on New Materials and their Applications, Univ. of Warwick, 1990

Fabrication aspects of glass matrix composites for gas turbine applications

Dr D A Clarke

Rolls-Royce plc
Advanced Technology Centre
University of Warwick
Coventry, UK.
CV4 7AL

Summary

Fibre reinforced glasses and glass-ceramics have mechanical and thermal properties which make them suitable for use in gas turbines as alternatives to metals such as titanium and nickel alloys. Although initially developed nearly 20 years ago it is only recently that they have come to prominence as 'real' demonstration parts.
This article presents a brief background to the development of glass matrix composites (GMC's) including the basic fabrication methods. Some of the key problems involved in manufacturing production quantities of GMC gas turbine components are described and the need for highly repeatable, and hence largely automated, processing methods highlighted.

1. Background to Development of GMC's for Gas Turbines

The development of fibre reinforced glasses and glass-ceramics in the UK was initiated in the late 1960's/early 1970's in a programme at UKAEA Harwell. This work was partly sponsored by Rolls-Royce and was aimed at developing a high temperature analogue to a polymer matrix composite using the then newly developed carbon fibre materials and a borosilicate glass matrix [1].
This initial work demonstrated both the high strength and toughness of the material at elevated temperatures and the fundamental concepts of fabricating continuous fibre reinforced glass into various shapes including flat plates and washers and circular hoops.

The primary limitation to this material was the poor oxidation resistance of the carbon fibre, resulting in degradation of the composite properties when heated in air to temperatures above about 400C. This problem still presents a restriction on the use of nearly all carbon fibre reinforced composites and caused the initial work at

Harwell to remain largely unused except for limited applications in space structures and low expansion structures for optical/electronic use [2,3].

In the late 70's/early 80's there occurred a resurgence of interest in ceramic matrix composites (CMC's) following the development in Japan of a silicon carbide fibre yarn, 'Nicalon'. This material suffers little degradation in air at temperatures upto 900-1000 C and is reasonably handleable by conventional fibre/composite processing methods. It has a modulus of about 250-300 GPa and an average filament diameter of 10-20um [4,5,6].

The majority of CMC work with this fibre was initially based on glass matrix systems rather than higher temperature materials such as silicon carbide or silicon nitride, principally due to the largely proven fabrication techniques suitable for use with glass matrices.
In their work on C/glass, Harwell had developed the now widely used 'pre-preg' / hot-press system for material fabrication, (figure 1), a system which has the dual advantages of being relatively inexpensive to set up on a lab scale and also relatively simple and easy to characterise.
With the availability of a reasonably stable 'high'-temperature fibre, research groups throughout the world thus began a new series of work programs aimed at developing a patentable CMC system based around a SiC yarn reinforcement.

In the UK a program was set-up at UKAEA Harwell to utilise their C/borosilicate technology on a SiC/borosilicate material. This was again influenced by Rolls-Royce requirements and was subsequently extended to examine higher temperature matrix materials including magnesium-alumino-silicate (MAS) matrices. This latter work was a joint undertaking between Rolls-Royce, Harwell, Pilkington Glass and the Department for Trade and Industry under the Advanced Ceramics for Turbines Program [7].
In the USA two main activities were undertaken, one at Corning Glass, developing a calcium-alumino-silicate (CAS) system and one at United Technologies Research Corporation developing CAS and lithium-alumino-silicate (LAS) matrix systems. All these materials are crystalline glass-ceramic materials rather than amorphous glasses and as such are suitable for use under load at quite high temperatures (1000 C). The results of these programs have been widely reported and the key material properties are summarised in table 1 [8,9].

Table 1 clearly shows the potential weight saving which is possible when the glass-ceramic matrix composites are compared to equivalent 1000 C capable metal alloys, (generally nickel based materials).

The CMC's have approximately half the density of most nickel systems resulting in similar specific mechanical properties. Among the additional advantages of the oxide glasses and glass-ceramics for gas turbine applications are their inability to burn, (being fully oxidised), and their essentially non-strategic availability.

The development programs of the 1980's were almost universally successful in proving the material capabilities of glass and glass-ceramic matrix composites. Following on from this work most of the major gas turbine manufacturers now appear to have embarked on programs to develop the manufacturing technology required to fabricate generic component shapes. If successful, such work should lead to composite parts for ground based trials and ultimately full production.

2. Fabrication Principles of Glass Matrix Composites

There are two main routes to fabrication of glass and glass-ceramic matrix composites, both based on principles used in manufacture of polymer matrix systems.

The first route is the most widely used method, consisting of the preparation of a 'pre-preg' fibre/glass-powder material which is then hot-pressed at temperatures of 800-1400 C to cause melting/sintering/compaction of the matrix phase. This process is shown schematically in figure 1 [1].
The 'pre-preg' material is produced by drawing a desized fibre tow through a slurry, (which contains the matrix phase as a ground powder), followed by winding of the impregnated yarn onto a drum to give a sheet of material. Once dry the sheet can be removed from the mandrel and cut to suitable shapes for hot-pressing. Pressing is normally in a uniaxial system using a graphite or refractory metal tool. This is heated via an induction system to minimise heat-up times and avoid uncontrolled matrix crystallisation effects and fibre degradation.

The second route is apparently simpler but requires a more detailed understanding of chemical synthesis techniques to generate the matrix phase and may require specialised technology to produce a (complex) fibre preform. Using a sol-gel method it is possible to form a low viscosity aqueous based liquid with the basic composition of the required oxide system. This can be used to infiltrate a 1, 2 or 3D fibre preform using simple vacuum/pressure infiltration methods, (schematic figure 2), at (or near to) room temperature. The preform is then dried to remove the aqueous content of the matrix, typically leaving 5-50% solids within the fibrous structure. By numerous repetitions of this infiltration/drying process it is possible to gradually build up a dense oxide matrix within the fibre preform. The entire structure is then put through an elevated temperature cycle in an ambient pressure furnace to sinter the matrix phase together. In some cases it may be applicable to use overpressure sintering techniques at this stage to ensure full densification. In either case the processing technology is relatively straightforward and reproducible although potentially very time consuming.

3. Manufacturing Technology Limitations

The key manufacturing limitations of these technologies are as follows:

3.1 Pre-preg/Hot-press

a) The refractory glass compositions of primary interest have relatively high viscosities (>10^5 poise [11]) unless heated to temperatures of 1100-1300 C when fibre degradation becomes rapid [6].
Even at these temperatures the relatively high viscosity of the melt will result in fibre movement ('wash') within the tool as the glass is forced around the preform thus making prediction of the final fibre positions difficult. Consequently, to achieve a genuinely uniform final product, the starting material, (the pre-preg), must also be totally uniform with a completely even fibre matrix distribution and a constant thickness. In an ideal material the glass particles would be sufficiently small and numerous to form a near continuous phase in the pre-preg and during pressing would simply sinter together, giving virtually no fibre movement from the initial lay-up.
Allied to fibre/matrix movement is the problem of achieving full densification in a complex shape tool cavity where the thickness of pre-preg in the pressing axis varies from point to point, (figure 3). Under these circumstances unless the material is accurately formed to the geometry of the cavity, initially thick sections will compact before the thinner ones, again leading to distortion of the fibre lay-up.
Finally, to mould a complex 3D shaped part to low tolerances whilst accurately controlling fibre position essentially requires the use of a closed-tool moulding process. Unless the pre-preg is completely uniform in compaction behaviour then when the tool is shut there may still be some sections which have not achieved full densification. Equally, if some sections of pre-preg do not compact as much

as required the tool will not shut, almost regardless of applied pressure (figure 3).

b) When using induction heating systems to heat large blocks of material (such as a toolset), the majority of the tool is actually heated by conduction of heat generated in the skin of the material. Depending on the tool material and the frequency of the induction unit the thickness of this inductively heated surface layer can be varied from fractions of a millimetre to a few centimetres [10].
The problem with this system is that, particularly for a simple part such as a large flat panel, there is an appreciable temperature gradient across the part, (figure 4). To allow all parts of the section to reach thermal equilibrium thus requires a dwell period close to the pressing temperature. The result of this is that the edges of the panel will be at temperature for longer than the centre and properties such as the fibre/matrix interface structure and the level of matrix devitrification will vary across the panel.
Removing this temperature gradient, (and any associated vertical gradient, figure 4), presents a considerable problem if a rapid heating rate is to be maintained over the whole tool.

3.2 Sol-Gel infiltration methods

There are two main problems in using this type of system, firstly that of producing a suitable fibre preform and secondly reducing the time and effort involved in infiltrating this with a matrix to an acceptable level.

a) The types of preforms suitable for use range from a 1D lay-up similar to the pre-preg material described above but infiltrated with a sol-gel matrix rather than a ground powder, through to complex 3D woven, braided and knitted structures. It is these latter 3D parts which offer the most interest owing to their potentially much improved interlaminar shear properties compared to a more conventional 1 or 2D lay-up [12].
For the fabricator, manufacture of 3D structures poses considerable problems in materials such as silicon carbide. Firstly the material is inherently brittle and the available yarns, (principally the Nicalon material), cannot be bent around radii of less than 1-2mm without fracturing filaments and causing general yarn damage. Since most weaving, and particularly knitting, technologies require the bending of yarns around very small radii, (close to the yarn diameter), it is difficult to produce a

preform which does not contain large quantities of broken fibre debris as short (mm) lengths of material. In fact, this may not significantly influence the final material properties. It does however, impose handling problems for manufacturers in areas such as health and safety control, in preventing any of these filament fragments from entering electrical equipment where their conducting nature can cause severe damage, and also simply in keeping a continuous weaving system operating with a yarn which is likely to break during manipulation.

b) Infiltrating a high density fibre preform with a liquid presents little problem and can be carried out either under pressure or vacuum to achieve optimum filling.
Using a sol-gel system requires the infiltrated preform to then be dried, and probably calcined at a reasonably elevated temperature, (400-900 C), to prevent the dried solids from redissolving when the next infiltration is carried out.
Once the first infiltration stage is complete it is almost inevitable that some sections of the fibre preform will contain isolated voids formed as the matrix dried and shrank. Filling these voids may be impossible on further infiltration.
The fabricator is thus presented with the tasks of firstly minimising matrix shrinkage on drying to limit the overall number of (time-consuming) infiltrations required, and also varying the viscosity of the liquid matrix system at each infiltration to try and ensure that as much void space as possible is reached and filled each time.
A further consideration which must be addressed is the question of what level of residual void space is acceptable in a given component. This obviously will vary from design to design, but unless residual voidage can be closely controlled this may ultimately limit the use of sol-gel fabricated components to certain areas.

4. Conclusion

In conclusion, it can be seen that although glass/glass-ceramic matrix composites have been proved to have mechanical properties which make them suitable for use in gas turbines there is considerable manufacturing research required before they can be said to be ready for even pilot production.
A small number of key deficiencies have been highlighted, most of which are under investigation. A considerable

quantity of highly innovative development work is almost certainly required to achieve acceptable solutions.
The overall requirement of a true production process for these materials must be 'complete' repeatability to ensure high-quality parts with a low overall scrap rate and (ideally), the minimum of post-fabrication inspection. This is the goal which manufacturing technology groups are now trying to achieve.

References

1. The technology of carbon fibre reinforced glasses and ceramics. Sambell R A, Phillips D C, Bowen D H, J Mat Sci 7 (1972) p.1454

2. Industrial applications and markets for CMC's Clarke D A Adv. Cer. and Met. composites, short course, Univ. Cambridge March 1990

3. Composite laser mirror Stalcup R K, Sigman W R, U.S. Pat No. 4451118 May 1984

4. Manufacturers data sheets - 'Nicalon' Silicon Carbide fibres Nippon Carbon Co. Japan

5. Effect of curing conditions on mechanical properties of SiC fibre (Nicalon) Ichikawa H, Teranishi H, Ishikawa T J Mat Sci Letts 6 (1987) pp 420-422

6. Mechanical and structural characterisation of Nicalon SiC fibres up to 1300°C Bunsell A R, Simon G, Comp Sci tech 27 (1986) pp 157-171

7. Overview of the ACT programme Bennett A Inst. of Cers - Adv Eng with Ceramics, Dec 89 London

8. High Strength silicon carbide fibre reinforced glass matrix composites. Prewo K M, Brennan J J, J Mat Sci 15 (1980) pp 463-468

9. Silicon carbide yarn reinforced glass matrix composites. Prewo K M, Brennan J J, J Mat Sci 17 (1982) pp 1201 - 1206

10. Induction heating handbook Davies J, Simpson P, McGraw-Hill 1979

11. Handbook of glass properties Bansal N, Doremus R, Academic Press 1986

12. Preform fibre architecture for ceramic matrix composites Ko F K, Am. Cer. Bull 68 (2) 1989 pp 401-414

TABLE 1

	T_{max}	σ_T	E	P	K_{IC}	α	$\sigma_{T/P}$	E/P
	°C	GPa	GPa	g/cm^3	$MPa/m^{0.5}$	$x10^{-6}$	$GPa/kg/cm^3$	
Nicalon/borosilicate								
UD-0°, VF-45%	550	0.7	120	2.3	26	3.3	0.3	52
$[0/90]_s$, VF-45%	550	0.4	60	2.3	26	3.0	0.2	26
Nicalon/LAS								
UD-0°, VF-45%	1000	0.7	150	2.5	16	2.2	0.3	60
Titanium (IMI 829)	700	0.9	120	4.5	-	8.0	0.2	27
Nickel (C1023)	950	1.0	211	8.0	-	11.0	0.1	26

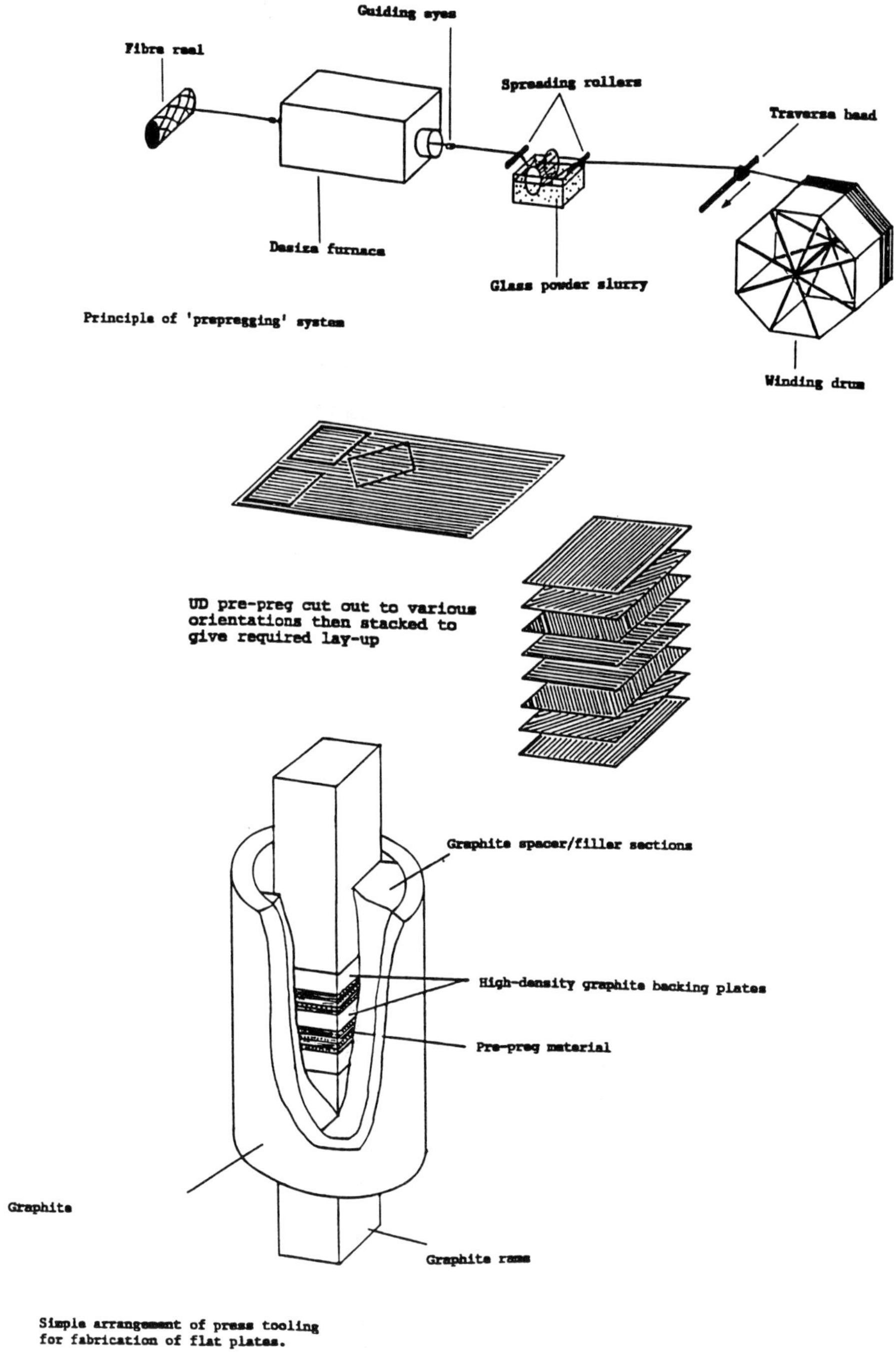

Figure 1. Stages in fabrication of a glass matrix composite panel by 'pre-pregging' and hot-pressing.

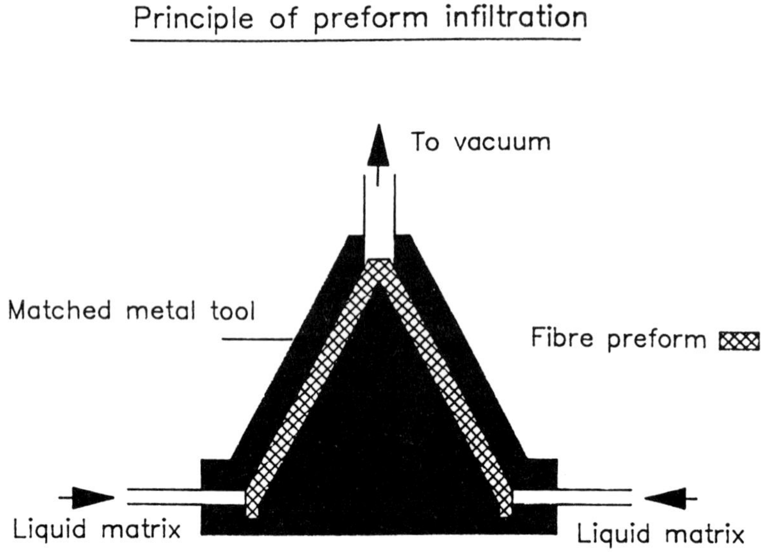

Figure 2. Schematic of vacuum infiltration route to impregnate a fibre preform with a liquid matrix.

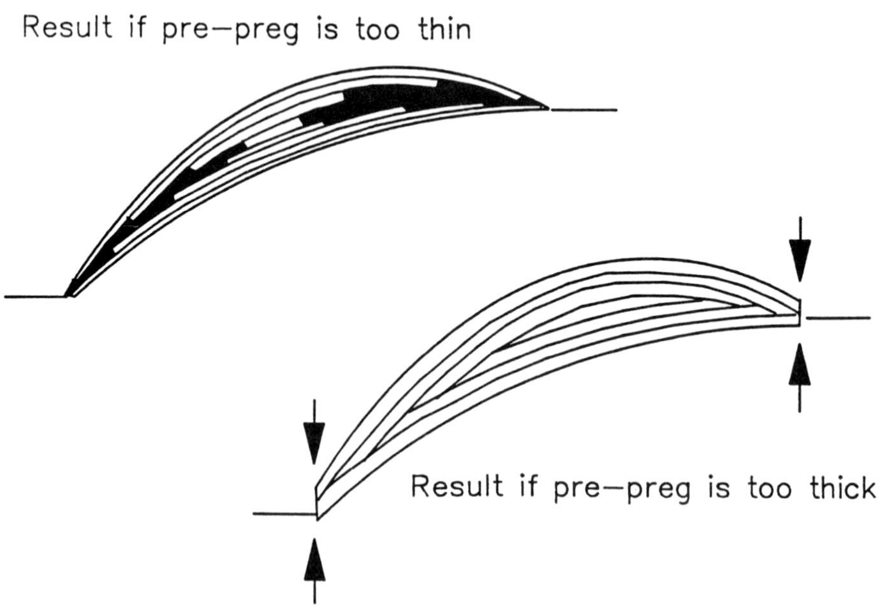

Figure 3. Effect of incorrect pre-preg thickness on compaction of a complex section part.

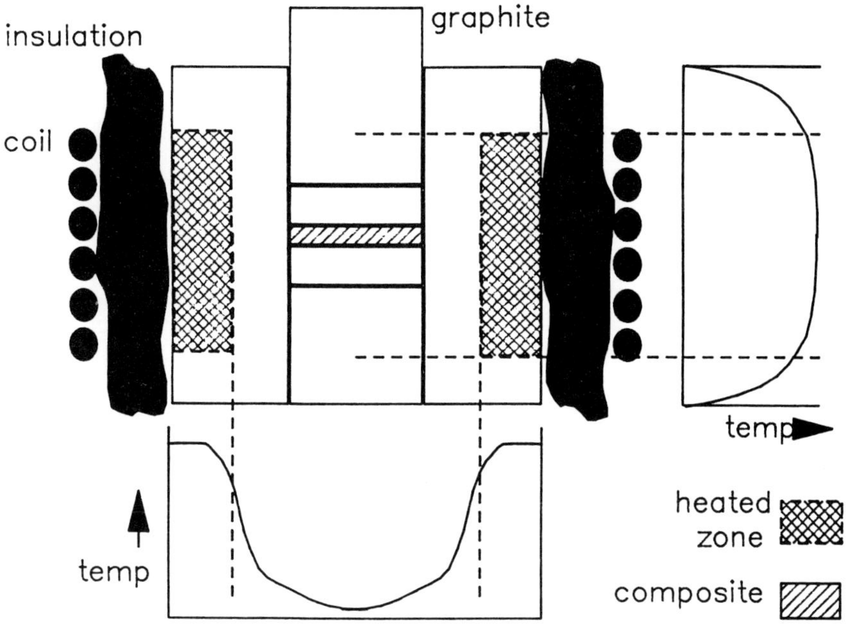

Figure 4. Schematic of temperature gradients present in a large inductivily heated tool-set.

Inst. Phys. Conf. Ser. No 111
Paper presented at Int. Conf. on New Materials and their Applications, Univ. of Warwick, 1990

Interface, matrix and fibre microstructure in SiC-reinforced glass-ceramic composites

V.S.R.Murthy, M.W.Pharaoh and M.H.Lewis

Centre for Advanced Materials Technology, Department of Physics, University of Warwick, Coventry. CV4 7AL. U.K.

ABSTRACT: Matrix crystallisation and the mechanism and kinetics of 'C' rich interface development have been studied for various silicate compositions and pressing temperatures. In some systems the hot pressing cycle neccessary for densification and critical interface development is incompatible with optimum matrix microstructure. The development of refractory composites, currently limited by the fibre stability in processing or in service conditions, has been explored using sol-gel stoichiometric silicate glasses and diphasic eutectic glasses (conventially melted) followed by crystallisation of low thermal expansion phases from the MAS and BAS systems.

1. INTRODUCTION

1.1 Principles of Microstructure Development

The microstructural requirements for high-performance fibre-reinforced ceramic matrix composites are moderately well established, at least for uniaxial stressing, but attainment of the required fibre, matrix and ineterface properties consistent with realistic fabrication procedures remains an elusive goal. Silicate glass and glass-ceramic matrices have played a key role in the experimental modelling of composite behaviour and are attractive from the viewpoint of low density, ease of fabrication and the control of thermal expansion. The recent resurgence of interest in ceramic-matrix composites has been due not only to inadequate fracture-toughness in monolithic ceramics but also to the availability of fine SiC fibres from polymer precursors which, fortuitously, develop low-cohesion graphitic interfaces by 'in-situ' reaction with silicate matrices.

As a prelude to a review of microstructural studies of glass and glass-ceramic matrix composites at Warwick some general principles will be surveyed concerning the interaction between mechanisms which occur during composite fabrication using Fig.1 as a framework for discussion.

During the essential time period for matrix densification, preferably by viscous flow, the matrix is susceptible to crystallisation at temperatures below the liquidus (Te). Densification above Te is possible if the silicate glass is near to a low-melting composition where fibre degradation is negligible and 'in-situ' interface reaction kinetics are adequate. Ideal conditions are those in which densification and interface development precede a controlled matrix crystallisation via 'nucleation' and 'growth' thermal cycles (Fig.1). This is rarely possible because currently available fibres, such as Nicalon, are susceptible to structural and strength degradation above ≈ 1000°C and most glass-ceramics of interest from the viewpoint of thermal expansion and high-temperature capability have liquidus temperatures above 1200°C.

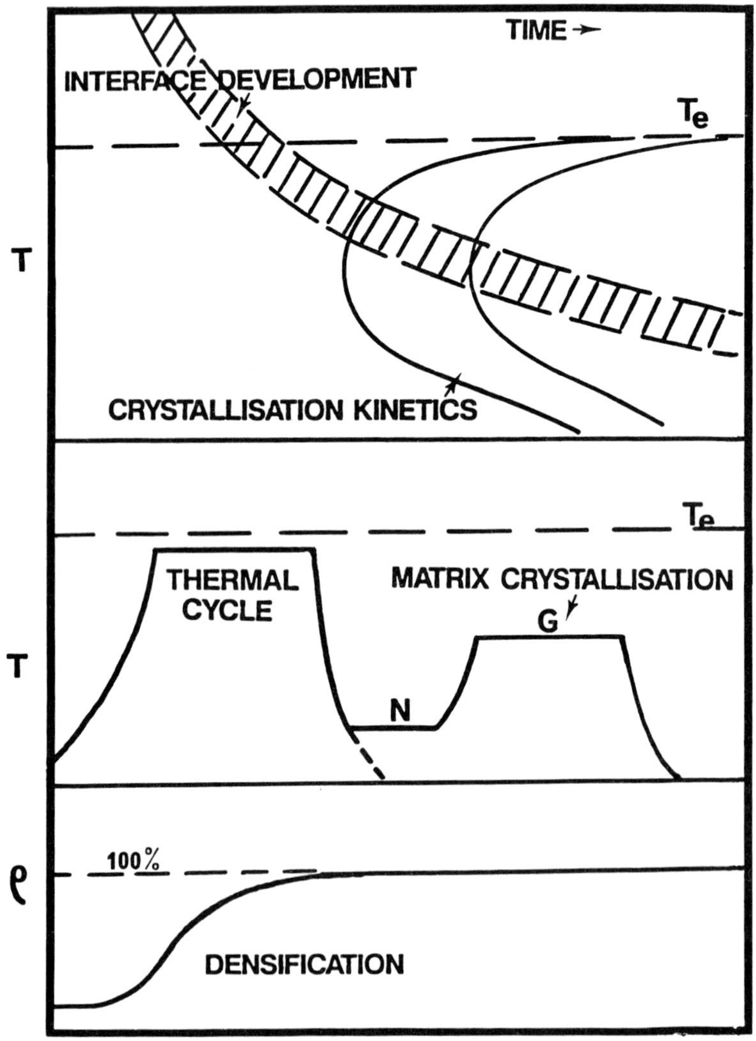

Figure 1. Schematic illustration of matrix crystallisation and interface development in CMC's.

1.2 Silicate System Selection

Glass and glass-ceramic compositions which have so far been studied as CMC matrices are generally based on low thermal expansion phases, developed as monolithic microstructures. Examples are the Li_2O - Al_2O_3 - SiO_2 (LAS) and MgO - Al_2O_3 - SiO_2 (MAS) glass ceramics containing the β-spodumene and cordierite respectively. In this programme we have studied mono and diphasic matrices derived from MAS and BAS (BaO - Al_2O_3 - SiO_2) together with low thermal expansion borosilicate based glasses (Table 1).

System	Processing Method	Hot-Pressing Temperature, °C	Major Crystalline Phase (s)
sodium borosilicate	conventional melting	$\geq$750	α-cristobalite
MAS	"	$\geq$950	μ, α-cordierite
BAS	sol-gel	1650	Hexacelsian
BAS 1	conventional melting	1350	Monoclinic celsian
BAS 2	sol-gel	1650	Hexacelsian, Mullite

Table 1: Glass and glass-ceramics examined for CMC development.

The selection of specific composition is normally a compromise between high temperature potential and ease of fabrication or microstructure control. Stoichiometric glasses which crystallize to near-monophase silicates present high temperature melting problems and crystallize rapidly either during melt cooling or matrix sintering. Sol-gel preparative routes are not a solution to the latter problem. Di or tri-phasic microstructures evolve from eutectic glasses which are more readily melted or sintered but the ratio of phase volumes may not provide the required mean thermal expansion and there is a greater probability of glassy residues.

One objective of the Warwick programme has been to identify silicate compositions with a range of potential application temperatures which may be controllably crystallized during or after hot-pressing (or, in the case of borosilicates, are stable glasses). A second objective has been the identity of reactions between matrix and fibres resulting either in the required interface microstructures or in fibre degradation. The studies have been conducted largely on small composite tiles propared by the hot-pressing of silicate glass containing a random mixture of chopped fibres. This is an efficient method for studying matrix crystallization together with interface microstructure.

2. FIBRE STRUCTURE AND STABILITY

To date, two different types of continuous silicon carbide fibres are available as a reinforcing medium for CMC's. (i) fibres obtained by the melt spinning of organic precursors e.g. Nicalon[1] or Tyranno[2] and (ii) fibres produced via chemical vapour deposition (CVD) e.g AVCO[3] or sigma[4] monofilaments.

2.1 Monofilaments

AVCO SiC monofilaments (SCS-6) produced via CVD are large in diameter ($\approx 142\mu m$), possess very high tensile properties (tensile strength $\approx$4GPa and elastic modulus $\approx$400GPa) and are stable to 1200°C. The fibre is produced by thermal decomposition of controlled hydrogen - silane gas mixture onto a carbon core ($\approx 33\mu m$). Initially, pyrolitic graphite ($\approx 1\mu m$) is deposited onto the carbon core which is claimed to improve the surface for silicon carbide deposition. Following this a sheath of silicon carbide is deposited in a columnar crystal morphology (Fig.2a). The sheath consist of β-SiC grains extending in the radial direction with $\langle 111 \rangle$ preferred orientation. the columnar grains within the sheath vary in size; near the substrate 40-50nm followed by 90-100nm at the surface. Such variation in grain size is attributed to the temperature difference within the production chamber (1300°C at the entrance and 1150°C at the exit). Magic angle spinning nuclear magnetic resonance (MAS NMR) studies on the monofilaments revealed that the SiC is mainly β-SiC (3C) with other polytypic forms (6H and 15R) (Fig.2b). Further, the fibre is coated with twin layers of amorphous carbon (by varying the ratio of Si to C) towards the surface. The carbon rich layer is claimed to heal surface microcracks on the SiC and acts as a low-cohesion layer at the fibre-matrix interface following composite fabrication.

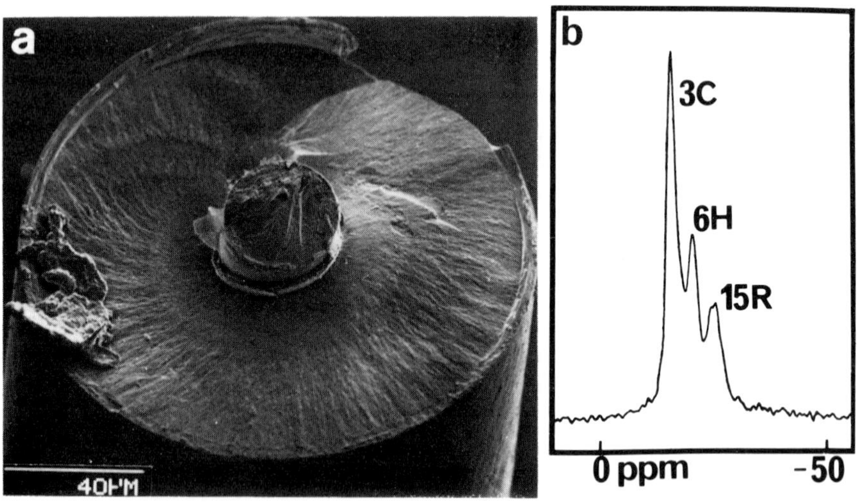

Figure 2:(a)AVCO (SCS-6) SiC monofilament. (b) ^{29}Si MAS-NMR spectrum indicating different polytypes of SiC.

[1]Nippon Carbon Co. Ltd, Tokyo, Japan.
[2]Ube Industries Ltd, Tokyo, Japan.
[3]AVCO speciality Materials, USA.
[4]B.P. Research, UK.

Although the fibre is stable to 1200°C, the large fibre diameter is a major limitation during fabrication of CMC's. Above 1200°C, a dispersion of fine silicon particles (produced during non-stoichiometric CVD) is detrimental to high temperature strength due to grain boundary sliding (Dicarlo 1986, Brun and Borom 1989).

2.2 Nicalon and Tyranno Fibres

Continuous, non-stoichiometric silicon - oxy - carbide fibres (of size range 8-15μm) developed from organic precursors have a limited temperature capability, up to ≈1000°C. Nicalon and Tyranno are commercially available silicon-oxy-carbide fibres synthesised from polycarbosilane and polytitanocarbosilane respectively. While the C/Si ratio is similar for both fibres, Tyranno additionally contains 2-4wt% titanium to improve the high temperature stability. In the as recieved condition, both are non-crystalline. ^{29}Si MAS NMR spectra from the fibres consist of four peaks (Fig.3) representing mixed local silicon tetrahedra (SiO_xC_{4-x}; $0 \leq x \leq 4$). In Tyranno these peaks are on a much broader resonance indicating the highly distorted structure associated with greater oxygen and titanium. High temperature annealing (1300°C for 75 hrs) induces a progressive segregation of C and O to their respective tetrahedra with an accompanying increase in the long range order, evidenced by spectral line narrowing. Further, titanium within the fibre reacts with the carbon to form TiC precipitates. With further high temperature annealing (1400°C for 170 hours) Tyranno fibres develop well defined crystals of SiC (mainly β) and SiO_2.

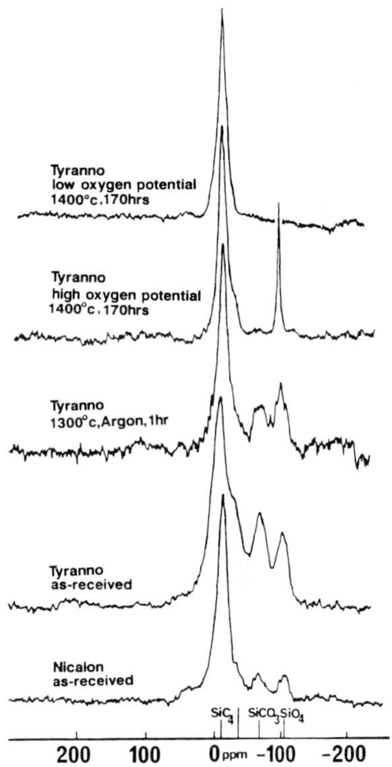

Figure 3:^{29}Si spectra from Nicalon and Tyranno fibres

At this stage, structural changes are sensitive to oxygen potential. At lower oxygen potential (i.e argon pressure) β SiC is the only crystallisation product, while oxygen in the fibre combines with Si and C forming SiO and CO respectively (Fig.3). In spite of C and Si loss, the fibre retains its shape but strength is low due to porosity within the fibre (Murthy et.al 1989). Fibres developed from methylpolysilane, containing nano-crystalline stoichiometric SiC (80wt%) with low oxygen (2wt%) exhibit promising elevated temperature properties (Lipowitz et.al 1988).

3. SILICATE MATRICES

3.1 Glasses

Sodium borosilicate glass with low coefficient of thermal expansion (3.3×10^{-6}°C^{-1}) is a potential matrix material for lower temperature ($\leq$500°C) applications (Dawson et al 1987) However, during densification the normally stable borosilicate glass crystallises into cristobalite (Fig.5a), a polymorph of silica. The matrix devitrification is due to heterogeneous nucleation of crystallisation on the surface of glass powder particles and on fibre-matrix interfaces. Further, the minor constituents of the glass diffuse into the fibre, modifying the localised glass to a silica rich composition. Silica enrichment at the interface increases the driving force for crystallisation (Murthy and Lewis 1989b). Such crystallisation is undesirable in composite formation since cristobalite induces microcracking in the matrix which is detrimental to mechanical properties and can hinder the kinetics of interface development.

The kinetics of cristobalite formation can be controlled by incrementing the processing temperature to near the liquidus (Te).

3.2 Glass-Ceramics

In the majority of glass-ceramic systems the infiltrated glass partially crystallises during the densification. However, in some cases, a post heat treatment is essential to complete the 'ceraming' process. A low thermal expansion matrix is a prerequisite for interface decohesion and prevent clamping of the fibres during the cooling cycle.

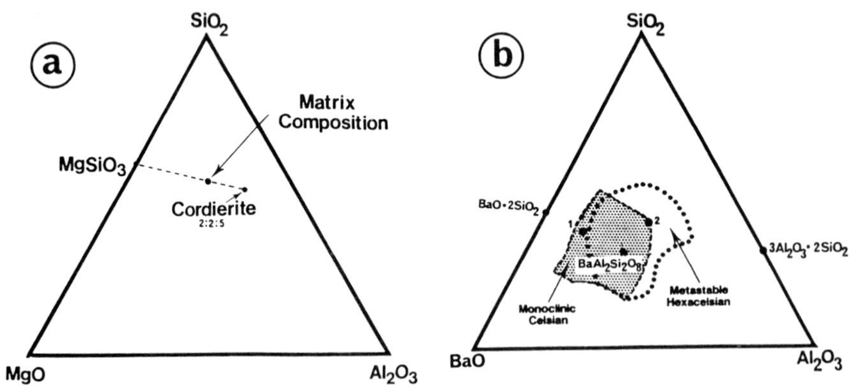

Figure 4: (a) Cordierite composition relative to stoichiometric MAS and enstatite. (b) Diagram indicating phase fields for monoclinic and hexagonal celsians and compositions studied in this investigation.

For example, in the MAS system, stoichiometric cordierite is not suitable as a matrix due to fibre degradation. However, by choosing an appropriate composition, glass can be hot-pressed at lower temperatures and further can be crystallised to α or μ cordierite and enstatite having markedly different thermal expansions. A composition selected on the stoichoimetric cordierite ($2MgO.2Al_2O_3.5SiO_2$) - enstatite ($MgO.SiO_2$) join line (Fig.4a) crystallises to cordierite (mainly α) with small proportions of enstatite which moderates the thermal expansion.

Further, controlled crystallisation has a greater significance in refractory systems such as barium alumino silicate (BAS). In the BAS, refractory celsian crystallises to two crystallographic forms: (a) celsian with hexagonal symmetry (referred to as hexacelsian) having higher thermal expansion (8×10^{-6}) and (b) monoclinic celsian with lower thermal expansion (3×10^{-6}). The primary phase field for monoclinic celsian compared with the experimental limits of hexacelsian in monoclinic form is delineated in Fig.4b.

In refractory systems, conventional glass preparation via melting and quenching is difficult for obtaining stoichiometric and near stoichiometric compositions due to high melting temperatures and inability to suppress crystallisation on cooling. In stoichiometric celsian powders, prepared via a sol-gel route, crystallisation of hexacelsian could not be suppressed. Additionally, sol-gel powders crystallise at lower temperatures ($\approx$ 800°C), which is undesirable from the view point of hot-pressing and interface development. To suppress the crystallisation of hexacelsian, the following di or tri phasic compositions have been examined.

A composition on the celsian - sanbornite tie line (Fig.4b, composition 1) within the phase field of primary celsian, has crystallised to a lower thermal expansion monoclinic phase. Eutectic reaction between celsian and sanbornite, further decreases the melting point enabling densification at lower temperature. However, the residual glass is resistant to crystallisation as sanbornite ($BaO.2SiO_2$). An alternative composition (Fig.4b, composition 2) crystallises predominantly as mullite, with low proportion of hexacelsian and aluminosilicate (Fig.7c) and decreases the thermal expansion to $\approx 6.0 \times 10^{-6}$. Such a glass composition is beneficial not only for lowering thermal expansion but also decreases the average density, which is an advantage in several applications.

4. FIBRE - MATRIX INTERACTIONS

4.1 'In-situ' Interface Development

In silicon-oxy-carbide fibre reinforced silicate matrices, a low cohesion carbon rich layer forms initiated by solid state cross diffusion at the interface. This carbon rich layer, which has a layered structure, weakly bonds to the matrix and controls the micromechanical properties of the interface. While the primary driving force for the cross-diffusion reaction is the silica concentration gradient across the interface, the kinetics of C-layer formation is controlled by the fibre and matrix chemistry, temperature, time and other minor chemical constituents (Cooper and Chyung 1989). During the solid state reaction, graphite inherently present in the fibre precipitates at the fibre/matrix interface and grows into large colonies, forming a thick layer, by the deposition of carbon formed via the oxidative reaction; $SiC + O_2 \rightarrow SiO_2 + C$.

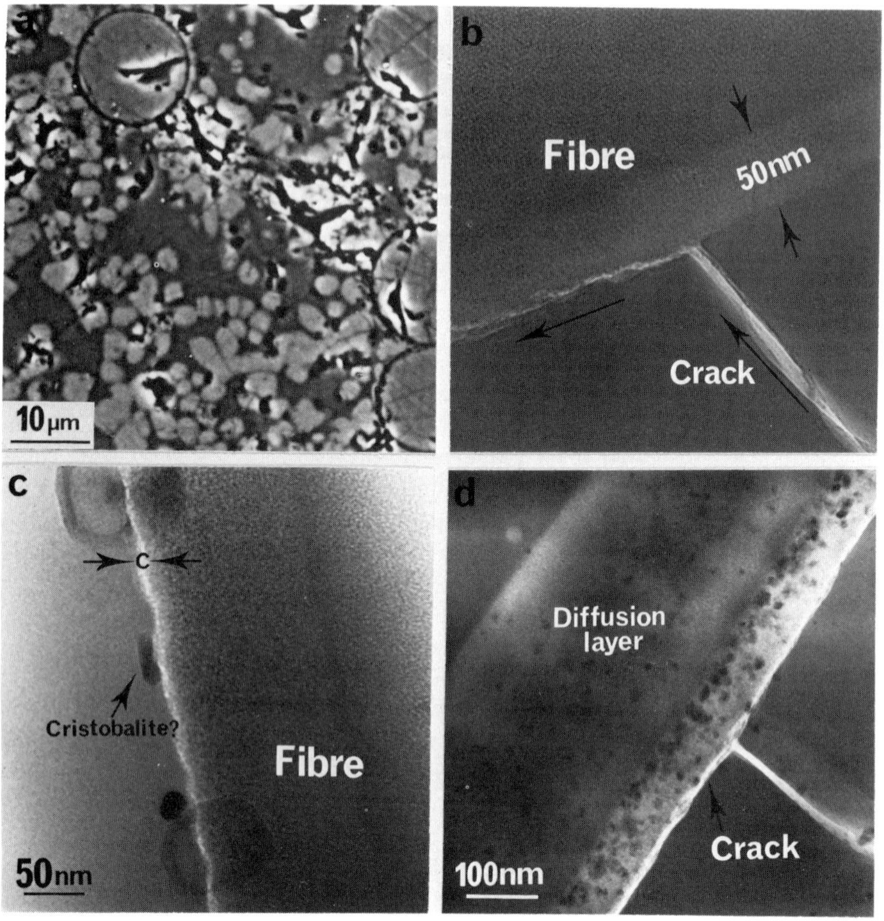

Figure 5:Matrix crystallisation and interface development in SiC - borosilicate composites. (a) cristobalite formation at 950°C. (b) Interface in Nicalon at 950°C. (c) Interface in Nicalon at 1100°C and (d) interface in Tyranno at 1100°C.

In SiC - borosilicate glass composites which are produced at temperatures about 950°C the interface is observed with a ≈50nm diffusion band extending inwards from the original fibre surface (Fig.5b). The diffusion band formed is due to transport of minor elements: (Na, Al and probably B) into the fibre coupled with cross diffusion of Si and O. At this stage carbon precipitation is not detectable at the fibre - matrix interface, however, it is evident from the presence of an efficient crack deflection in thin sections that this type of interface has a low cohesion and is responsible for the impressive macroscopic mechanical behaviour. At higher temperatures, near Te a thin carbon layer (of the order of 10-20nm) forms together with an extensive diffusion band

(≈200nm) (Fig.5c). The fibre chemistry also effects the carbon precipitation at the interface together with the extent of the diffusion band. Tyranno's more disordered structure and lower carbon content produce a wider diffusion band but a thinner carbon rich layer (Fig.5d).

In a MAS glass ceramic having a higher basicity the 'C' rich layer is well developed and is of the order of 50nm (Fig.6a). The 'C' rich layer has further grown to 150nm in BAS-SiC composites with higher basicity and processing temperature (Fig.7a). Tyranno SiC fibres undergo severe degradation during hot pressing due to the indiffusion of matrix elements (Fig.7b).

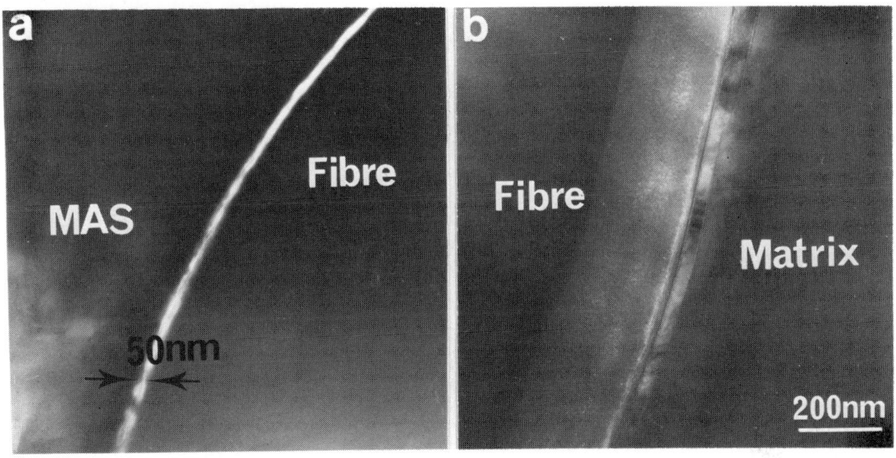

Figure 6:(a)C-rich layer formation and (b) silica bridging in SiC-MAS system.

4.2 Degradation

In various silicon carbide fibre reinforced glass and glass ceramics the in-situ formation of a 'C' rich layer, having a low cohesion and interfacial shear stress ($\tau \simeq$ 2MPa), satisfies the debonding requirements for pull out during transient overload. In a high temperature oxidising environment the interfacial carbon oxidises to carbon monoxide (CO) and SiC forms a passive silica layer (Fig.6b) which increses the bond strength between the fibre and matrix. Due to the fibre - matrix bonding the interfacial shear stress increase by an order of magnitude and reverses the 'tough' composite behaviour to that characteristic of brittle monolithic ceramics.

5. INTERFACE DEVELOPMENT VIA FIBRE PRECOATING

To overcome problems, such as extensive diffusion of matrix elements and oxidation of interfacial carbon followed by silica bridging, the fibre matrix interface has to be modified. This can be done either by additions to the matrix which segregate to the interface or by coating the fibre with a diffusion barrier layer and a low cohesion layer. In-situ interface modification via the addition of carbide forming transition elements was unsuccessful because of discontinous carbide precipitation (Brennan 1986). Hence a coating proir to composite fabrication is essential. Chemical vapour deposition (CVD) is the most widely used technique for deposition of such layers, though several other methods are available.

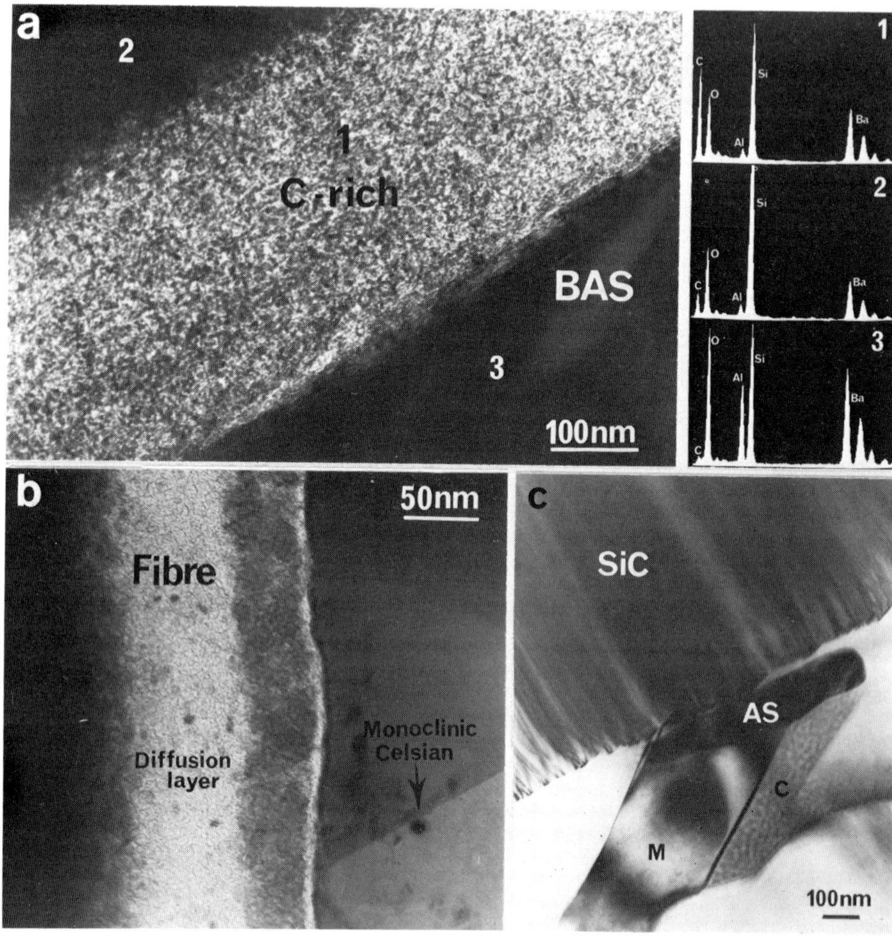

Figure 7:(a) Thick C rich interface layer in Nicalon containing BAS (comp.1) and corresponding analytical traces (b) extensive diffusion in Tyranno fibres (comp.1) (c) tri-phasic structure in comp.2 (C-celsian, m-mullite, as-aluminosilicate).

For example, plasma assisted CVD has been employed to deposit TiN onto Nicalon fibres (Fig.8a) as an effective diffusion barrier layer for matrix elements, but it is unstable at high temperatures. For effective fibre pullout a combination of a low cohesive layer followed by a protective, stable, refractory layer is desireable.

The primary coating should possess a layered structure (Van der Waal type) with low cohesion and shear stress coupled with stability and non reactivity at elevated temperatures. The second layer is essentially a refractory layer to protect the low cohesion layer and fibre during processing and from oxygen induced degradation during service. Carbon or BN and SiC are potential coating materials for primary and secondary layers respectively. To assess the reaction between the matrix and future stoichiometric SiC fibres or coatings, SiC whiskers have been added to the stoichiometric BAS glass (Fig.8b). The SiC-glass interface is stable and non-reactive.

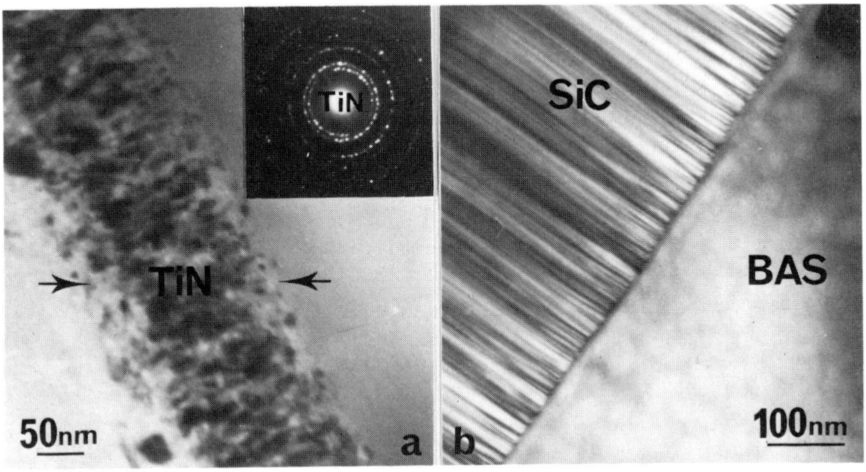

Figure 8:(a) TiN interlayer via plasma assisted CVD (b) non-reactive interface in SiC whisker containing BAS.

6. SUMMARY

(i) The interaction between kinetics of densification, interface-reaction and matrix crystallisation is critical for optimum microstructure development. Low thermal-expansion matrices are essential.

(ii) 'Stable' glass compositions may be susceptible to crystallisation when processed from the powdered or sol-gel state or in the presence of fibres. An example is that of conventional Pyrex

(iii) Sub-1000°C fabrication is preferred for Nicalon/Tyranno fibre composites but densification of refractory silicates is not normally possible at such low temperature.

For $\geq$1000°C fabrication or service conditions, intrinsically-stable stoichiometric SiC fibres are required.

(iv) 'In-situ' graphitic interfaces provide good models for optimum mechanical response (the detail of which remains to be studied) but restrict application temperatures to 600-700°C due to oxidation/SiO_2 bridging.

(v) In-situ graphitic interfaces do not form with stoichiometric SiC and for high temperature composites fibre precoating with 'Van der Waal' compounds or low yield-stress metallics is required. Diffusion/reaction barriers may be neccessary between fibre/interface layer and/or interface/matrix.

REFERENCES

J.J.Brennen 1986 'Tailoring of multiphase and ceramic composites' (New York:plenum) pp. 549.

Brun M.K and Borom M.P 1989. *J.Am.Ceram.Soc.* **72** 1993.

Cooper R.F and Chyung k 1987. *J.Mater.Sci* **22** 3148.

Dawson D.M, Preston R.F and Purser A. 1987. *Brit.Ceram.Proc.* **39** 215

Dicarlo J.A 1986 *J.Mater.Sci.* **21** pp.217.

Lipowitz J, LeGrow G.E, Lim T.F and Langley N 1988. *Ceram.Eng.Sci.Proc* **9** 931

Murthy V.S.R, Lewis M.H, Smith M.E and Dupree R 1989a *Mater.Lett* **8** 263.

Murthy V.S.R and Lewis M.H 1989b. *J.Mater.Sci.Lett* **8** 571.

Pharaoh M.W, Murthy V.S.R and Lewis M.H 1989. (Rolls Royce Report)

ACKNOWLEDGEMENTS

Financial support for the research programme on composites has been gratefully recieved from the SERC, MOD and Rolls-Royce plc.

Paper presented at Int. Conf. on New Materials and their Applications, Univ. of Warwick, 1990

Fabrication and mechanical properties of improved fibre reinforced ceramics

J.J.R. Davies, R.F. Preston, R.J. Lee, and K.N. Walls*.

Advanced Engineering Materials Department,
AEA Technology Harwell,
Oxon OX11 ORA.

*Dept. of Engineering Science,
University of Oxford.

Abstract

The quality and performance of unidirectional carbon fibre reinforced borosilicate glass has been evaluated after production by two different hot-pressing conditions. This combination of fibres and matrix can be regarded as a model system for high temperature applications, as it overcomes certain fabrication difficulties associated with fibre-matrix interfacial reactions in silicon carbide fibre reinforced ceramics. Attempts have been made to reconcile differences between the tensile and flexural behaviour of this carbon fibre borosilicate material, and to address problems associated with matrix crystallinity, residual thermal strains and anomalous Poisson's ratio behaviour.

In addition a candidate high temperature structural composite based on Tyranno (silicon-titanium-oxy-carbide) fibres in a glass ceramic matrix was examined, particularly to further define the non-uniform Poisson's ratio effect.

1. Introduction.

Ceramic composites are potentially useful engineering materials, combining brittle fibres and brittle matrices to produce tough, strong high temperature composites. Early work on reinforced glasses and glass ceramics was performed by workers at Harwell [Sambell, Phillips, and Bowen, 1974], and concentrated on carbon fibre reinforced glass and glass ceramics. The performance of these materials was limited by rapid fibre oxidation above 400°C however, and they were not used commercially [Linger and Pratchett, 1972]. Further work into ceramic composites was stimulated by the development of oxidation resistant silicon-oxy-carbide type fibres by S.Yajima and co-workers in the mid seventies. The most common brands of these fibres are Nicalon silicon-oxy-carbide fibre, (Nippon Carbon Co. Ltd) and Tyranno silicon-titanium-oxy-carbide fibre, (Ube Industries Ltd). Both of these fibres are currently used in several experimental ceramic composites. Fabrication of most of these composites involves the densification of a mixture of powder and fibres, together with formation of the correct fibre-matrix interface. This interface is formed by oxidation-reduction reactions between the fibres and matrix [Cooper and Chyung, 1987], and the conditions necessary for this may sometimes be incompatible with matrix consolidation requirements. Recently during an investigation of the ballistic impact resistance of stressed ceramic matrix composites, some unexpected crystallinity was observed in carbon fibre reinforced borosilicates, illustrating the production problems which can arise even in simple composite systems [Phillips, Park, and Lee, 1990].

This paper presents a discussion of the mechanical properties of two fibre reinforced glass and glass ceramic systems based on carbon fibre reinforced borosilicate and silicon-titanium-oxy-carbide fibre reinforced glass ceramic. Fine silicon carbide-type fibres and carbon fibres are produced by controlled pyrolysis of polymer precursors and have similar tensile strengths. The ceramic fibre and matrices differ significantly in their expansion coefficients and maximum usable temperatures. Basic property data are given in Table 1. Both composites rely on a weak fibre-matrix bond in order to achieve high strength and toughness. The weak bond in the carbon fibre composite is due to non-reaction between the fibre and matrix, while the silicon carbide type fibre composite relies on a fibre-matrix reaction to form a weak carbonaceous fibre-matrix bond during fabrication.

Table 1: Thermomechanical Properties of Fibres and Matrices.

	Young's Modulus (GPa)	Strength (MPa)	Expansion Coefficient (10^{-6} °C^{-1})	Temperature Limit (air) (°C)
Fibres				
Hercules HMU	380	2760	8.5 (rad) 0.54 (ax)	400
Tyranno	206	2740	3.10	1000
Matrices				
Borosilicate Glass	60	150	3.2	550
Glass Ceramic	119	150	2.5-5.5	1200

2. Experimental Procedure.

2.1. Production of Carbon Fibre-Borosilicate Model Materials.

The slurry impregnation method for composite production was first developed at Harwell, and involves production and hot-pressing of a fibre-powder prepreg. Prepreg is produced by spreading the received carbon fibre tow, then infiltrating with a slurry which contains matrix powder, organic binder and a solvent. The impregnated tow is then wound onto a drum and dried to give a flat prepreg sheet. This sheet is strong and flexible enough to be handled in preparation for subsequent shaping and hot-pressing.

Carbon-fibre borosilicate prepregs were produced, and hot pressing was carried out under two composite fabrication routes; one referred to as a high temperature-low pressure route, and the other a high pressure-low temperature route. The former method was preferred in earlier work [Sambell, Phillips, and Bowen, 1974], whereas the latter was applied to materials produced for a recent testing programme [Phillips, Park, and Lee, 1990]. In the high temperature route preforms were slowly heated to 350°C in a graphite die in order to ensure binder removal. The preform was then heated rapidly to 1100°C, when 4 MPa pressure was applied to compact the matrix. Heating was maintained to 1200°C, whereupon the material was then allowed to cool. In the low temperature route a maximum temperature of only 950°C was used, with pressure of 12 MPa for compaction. These methods were compared by means of the following experimental procedure:

- Optimisation of the prepreg production process for carbon fibre.
- Production of a batch of prepreg.
- Hot-pressing of prepreg into 15 x 50 mm bars under the following conditions:
  i). Low temperature pressing: 850°C under 12 MPa for 10 mins.
  ii). High temperature pressing: 1200°C under 4 MPa for ≈1 min.
- 3-point flexural testing at 40:1 span to depth ratio.
- X-ray diffraction of composite to identify the crystalline condition of the matrix.
- Examination of polished sections.

The optimum fabrication conditions were then employed to produce 100 mm square plates of carbon fibre-borosilicate composite for subsequent mechanical testing.

2.2. Production of Fibre Reinforced Glass Ceramic.

Harwell Laboratory currently uses a reinforced glass ceramic as a model material. This has a fully crystalline matrix surrounding Tyranno (silicon titanium oxy-carbide) fibres, and has a 3-point flexural strength greater than 1100 MPa [Davies and Preston, 1989]. This material was also manufactured in the form of 100 mm square plates, and specimens were taken from two such plates for mechanical testing.

2.3. Production of Mechanical Test Specimens.

The set of testing criteria recommended by the Composites Research Advisory Group (CRAG) were adhered to as closely as possible for this testing programme [CRAG, 1988]. Although these rules were originally formulated for the testing of polymer composite materials they have been adopted for this work. The carbon fibre-borosilicate CRAG test specimens are described in Tables 2 and 3.

The flexural strength of the Tyranno fibre-glass ceramic was measured at 40 mm span, with a specimen width of 5 mm, and did not follow CRAG guidelines. CRAG tensile testing standards could not be fully followed, as the maximum width of composite plate pressed was 100 mm. This allowed a maximum tensile gauge length of only 60 mm, with 20 mm end tabs, both of which values are around 50% of the minimum CRAG specifications. The Tyranno fibre composite was tested in tension with a slightly reduced gauge length of 50 mm, and correspondingly lengthened end-tabs. The surface finishes employed on specimens were:

Carbon-borosilicate:	As-pressed
Tyranno-glass ceramic:	Machined to 50 μm diamond equivalent.

Tests were conducted on an Instron 1185 machine with a cross-head rate of 0.5 mm min^{-1}. All tests were monitored with 0-90 strain gauges, with some specimens mounted with extra gauges to verify equality of strain on both sides of the bars. Strain gauge data were gathered on a computer.

Table 2: Standards for 3-Point Flexural Testing.

Flexural Parameter	CRAG Recommendation	Value Used.
Bar thickness-	2 mm (or nearest possible)	2 mm
Bar length-	80 mm (L/d=40)	80 mm
Bar width-	10 mm (±0.25 mm)	10 mm
Overhang at end rollers-	≥5 mm	10 mm
Support rollers diameter-	10 mm (reducible to 6 mm)	10 mm
Loading roller diameter-	25 mm (reducible to 10 mm)	25 mm

Table 3: Standards for Tensile Testing.

Tensile Parameter	CRAG Recommendation	Value Used.
Specimen gauge length	100-150 mm	60 mm
Length of end-tabs	>50 mm	28 mm
Specimen width	10 - 20 mm constant	Uniform 10 mm
Thickness profile	Uniform along length	Nominally 2 mm

3. Experimental Results.

3.1. Carbon-Borosilicate Fabrication Trials.

Carbon fibre composites fabricated according to the high temperature fabrication route densified more satisfactorily than by the low temperature route, with fewer handling difficulties caused by friability of the specimen edges. The full fabrication data set is presented in Table 4.

Mechanical testing revealed that the high temperature route material was much stronger than the lower temperature material, with a flexural strength of 1230±50 MPa.

X-Ray Diffraction showed that the high temperature route material had a glassy borosilicate matrix, whereas the low temperature material contained cristobalite phase in the matrix. Examination of polished sections revealed that large quantities of this cristobalite phase were present in the low temperature composite (Figure 1). The high temperature pressed material had very little evidence of crystalline material, with only small traces occasionally visible on the surface of fibres. 100 mm square plates of carbon- borosilicate composite were therefore produced via the high temperature production route.

Table 4: Fabrication Results for Carbon Fibre-Borosilicate Composite.

Fabrication Route	Flexural Strength (40 mm Span)	Matrix Condition
850°C, 12 MPa, 5 mins.	533 ± 110 MPa	Microcracked, cristobalite present
1200°C, 4 MPa, 1 min.	1226 ± 52 MPa.	Glassy

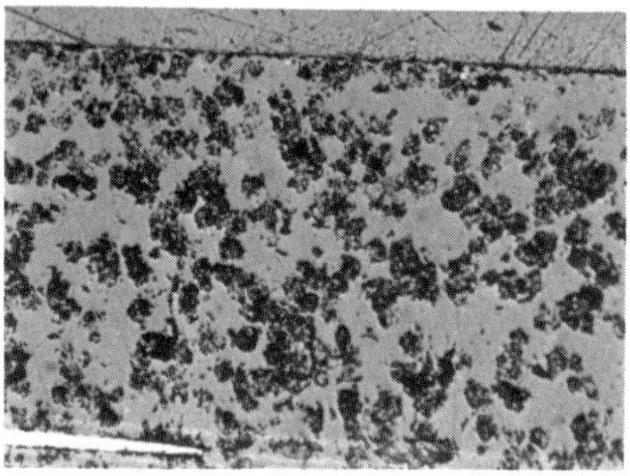

Figure 1: Cristobalite Particles in Matrix of Carbon Fibre-Borosilicate Composite. The fibre in the lower left-hand corner is approximately 8 μm in diameter

3.2. Mechanical Testing.

Results of the mechanical testing of carbon-borosilicate and Tyranno-glass ceramic composites obtained from large plates of material are presented in Table 5. The materials displayed much higher strengths in flexure than in tension. Flexural and tensile strength data for the carbon fibre-borosilicate composite were plotted on Weibull curves shown in Figures 2a and 2b. A straight line fitted through the points gave a Weibull modulus of 9-13 for the material in both types of test, with errors of fit due to the small number of tensile samples.

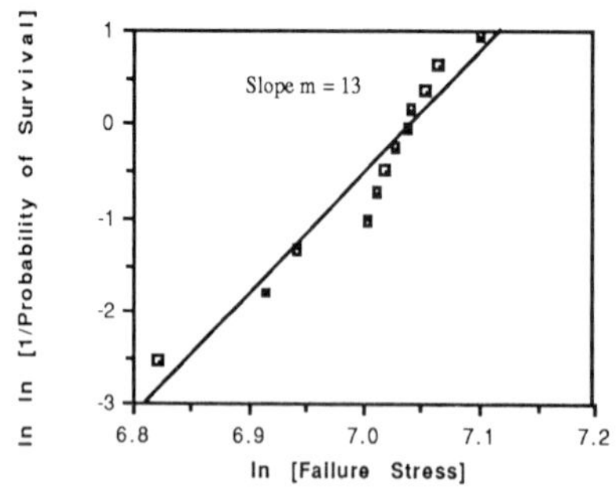

Figure 2a: Weibull Plot of Flexural Strength of Carbon Fibre-Borosilicate (MPa).

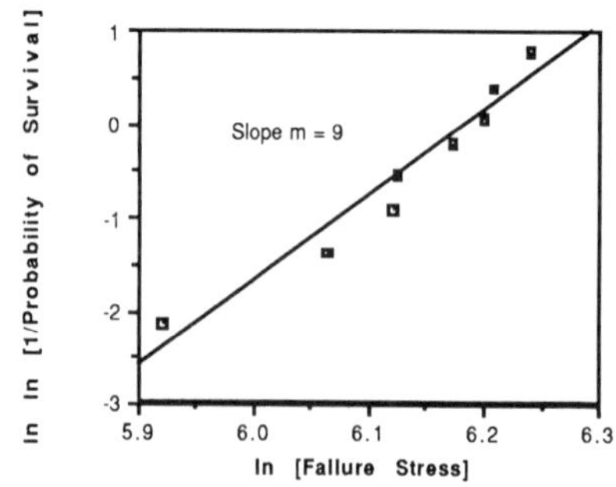

Figure 2b: Weibull Plot of Tensile Strength of Carbon Fibre-Borosilicate (MPa).

Ignoring the slight difference in tensile test configurations, the tensile strengths of the two composites were significantly different, with a strength of 462±42 MPa for the carbon-borosilicate composite, and 709±26 MPa for the Tyranno-glass ceramic. Additionally, the tensile fracture surface of the carbon-borosilicate had only short fibre pull-out, with a "woody" appearance overall, while the Tyranno-glass ceramic failed in a very fibrous manner, with much splitting along the gauge section prior to failure.

Both the glass and glass ceramic composites displayed the transverse strain reversal effect first noted by Kim and Katz [1988]. Under monotonic loading, the carbon-borosilicate expanded transversely to exceed its maximum dimensions by a maximum of approximately 0.05 %, and the Tyranno fibre-glass ceramic by a strain of ≈0.02 %, (Figures 3 and 4). An extensometer was also used on the Tyranno-glass ceramic composite, and this showed a lower expansion than the strain gauge results, with a maximum strain of 0.016 %. This may be a function of interaction between the gauges and the splitting damage associated with these materials. Transverse expansion for the Tyranno material was enhanced by cycling (Figure 4)

Table 5: Mechanical Testing Strength Data.

Composite	Flexural Strength (MPa)	Flexural Modulus (GPa)	Tensile Strength (MPa)	Tensile Modulus (GPa)
Carbon-Borosilicate	1104 ± 77	115 ± 7	462 ± 42	144 ± 2
Tyranno-Glass Ceramic	1150 ± 150	-	707 ± 26	-

Table 6: Summary of Transverse Strain Behaviour.

Composite	Loading Conditions	Position	Minimum Strain	Maximum Strain
Carbon-borosilicate	(monotonic tensile)	(central gauge)	-0.0060%	+0.050%
Tyranno-glass ceramic	(monotonic tensile)	(central gauge)	-0.0230%	+0.005%
Tyranno-glass ceramic	(monotonic tensile)	(end gauge)	-0.0230%	+0.020%
Tyranno-glass ceramic	(cycled tensile)	(central gauge)	-0.0240%	+0.035%
Tyranno-glass ceramic	(cycled tensile)	(end gauge)	-0.0260%	+0.054%
Tyranno-glass ceramic	(monotonic flexural)	(central gauge)	-0.0430%	+0.050%

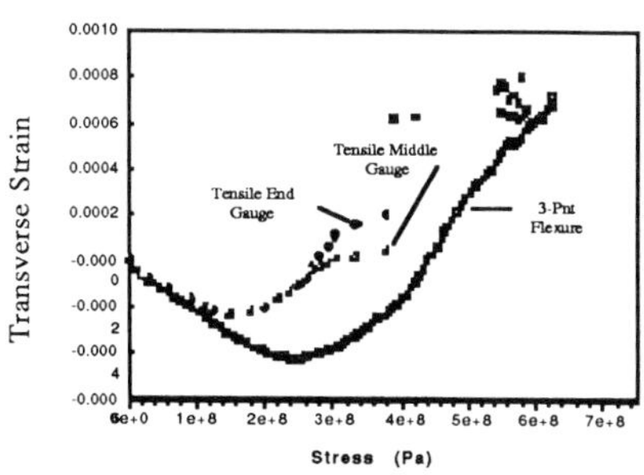

Figure 3: Transverse Strain in Flexure and Tension for Tyranno Fibre-Glass Ceramic.

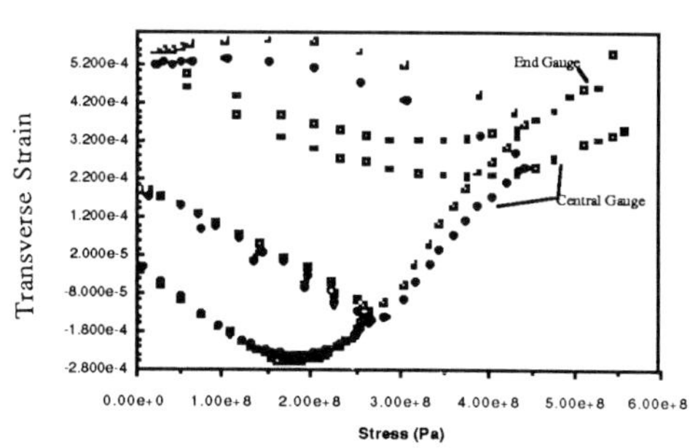

Figure 4: Transverse Strain in Tension as Function of Gauge Position.

4. Discussion of Results.

4.1. Discussion of Carbon-Borosilicate Fabrication Results.

The results of this study show that cristobalite-induced matrix cracking occurs in carbon fibre-borosilicate composites as has also been observed in silicon carbide fibre materials [Murty and Lewis, 1989]. Production of this phase is therefore a phenomenon relating to the form and temperature treatment of the matrix as well as to the presence of fibres. The surface nucleation and growth of crystalline phases in sintered glass compacts is well documented, occurring rapidly for fine borosilicate powders heated to about 900°C [Shin and Knowles, 1990]. Heterogeneous nucleation of the phase occurs on particle surfaces, and for borosilicates this phase is cristobalite, a polymorph of silica. This has a higher thermal expansivity than the parent glass, (27×10^{-6} versus 3.25×10^{-6}), and undergoes a β-α transformation with a 5.7% volume decrease on cooling through 230°C. The resulting strain mismatch on cooling of the crystallised glass causes severe cracking around the cristobalite particles. The hot pressing temperatures of borosilicate composites often lie in the 850-950° C range, and the matrix therefore can suffer from cristobalite cracking damage. The physical coherence of the matrix is important for protecting the ceramic composite against internal oxidation at high temperatures, and production of cristobalite during fabrication is a major problem.

The improved matrix structure of the high temperature route composite material was due to the lower glass viscosity, ($\approx 2 \times 10^4$ poises), which gave good consolidation, and dissolution of crystalline phases. During the pressing dwell, any cristobalite which formed on heating was redissolved, and efficient removal of relic particle surfaces also prevented cristobalite nucleation during cooling. Therefore the only areas available were fibre surfaces, which were obviously not such efficient nucleation sites.

The difference in strength for the composites is probably a function of the low glass viscosity at the higher temperatures. Softer matrix particles give less chance of contact damage to the fibres, and this effect clearly more than compensates for any thermal degradation that the fibres may undergo.

With many glassy-matrix composite systems, it is not possible to alter fabrication conditions sufficiently to inhibit phase nucleation, as the formation of the necessary weak fibre-matrix interface may be hindered. This is the case for some silicon carbide fibre containing materials examined at Harwell. In order to circumvent this problem the fibre-matrix interface must be pre-existing as a fibre coating.

4.2. The Flexural and Tensile Strength Ratio for Carbon Fibre-Borosilicate Composite.

Strength testing results in this programme show good agreement between the flexural and tensile Weibull moduli of the carbon fibre-borosilicate composite, giving confidence in the quality of the test procedures. Other workers have not found this similarity between the Weibull moduli in the two types of test, probable due to the use of inappropriate experimental techniques.

There is a larger ratio between the flexural and tensile strengths of carbon-borosilicate than is generally found even in other ceramic composites. This ratio is generally around 1.6 to 2, but for this carbon fibre composite the value of σ_f/σ_t is 2.39. There are several possible explanations for this including shift of the neutral axis in flexure due to matrix microcracking, Weibull stressed volume differences, and simple fibre bundle effects. These are compared in **(i).** to **(iii).** following:

(i). Neutral Axis Shift.

Matrix microcracking of the tensile face of a 3-point flexural bar shifts the neutral axis, causing a strength overestimation of approximately 10%. This implies a true flexural strength of 1004 MPa [Davidge and Briggs 1988]. The true experimental ratio for the strengths would then be 2.17. This effect is therefore not responsible for the discrepancies observed, but may be applied as a correction to the experimental data.

(ii). Weibull Stressed Volume.

The Weibull modulus of the composite, "m", which is about 10 for the carbon fibre-borosilicate tested here, can be substituted into a conventional equation which relates the relative stressed volumes in tension and in flexure, and ultimate strengths for the material:

$$\frac{\sigma_f}{\sigma_t} = \left[2.\ (m+1)^2 \cdot \frac{V_t}{V_b} \right]^{\frac{1}{m}} \qquad (1)$$

This equation gives a ratio of only 1.68, compared with 2.39, or the ratio of 2.17 from treatment **(i)**. Inapplicability of the simple Weibull theory may be because failure of the composite does not involve selection of a single critical flaw, but instead involves the bundle-like failure of many fibres, each with their own distinct flaw population.

(iii). Simple Fibre Bundle Theory.

This was first applied to ceramic composites by Davidge and Briggs, [1988], and relates the Weibull modulus of the reinforcement fibre to the composite strength. This invokes simple theory for the failure of bundles of fibres, assuming that at ultimate failure the heavily microcracked matrix makes no contribution to the behaviour of the composite. The composite is therefore treated as an array of fibres dead loaded at the both ends, with transfer of load from a broken fibre to the entire length of the remaining fibres. Equation (1) is modified to relate the length of the fibre under stress to the stress distribution in the tensile and flexural specimens:

$$\frac{\sigma_f}{\sigma_t} = \left[(m+1) \cdot \frac{L_t}{L_b} \right]^{\frac{1}{m}} \qquad (2)$$

Evaluation of this equation with an assumed fibre Weibull modulus of 4 [Hitchon, McCausland and Phillips, 1975] gives a strength ratio for the composite of 1.39. Simple fibre bundle theory probably fails because of load transfer from fractured fibres to their nearest neighbours via the microcracked matrix, and pure bundle end loading conditions do not exist [Davies, 1989].

These simple theories currently available for comparing flexural and tensile strengths therefore do not give agreement with experimental data. The most appropriate approach may be to apply fibre bundle theory, but the results show that there must be substantial modifications to the simple case. This will be necessary due to the presence of the microcracked matrix, which due to the fibre-matrix interfacial friction stress will still allow stress transfer between broken fibres and the rest of the bundle.

4.3. Fracture Behaviour of Carbon Fibre-Borosilicate and Tyranno Fibre-Glass Ceramic.

The Tyranno reinforced glass ceramic underwent a more fibrous fracture than the carbon-borosilicate material (Figure 6). This isolated difference in behaviour between ceramic composite systems is not unusual, but if the difference in the fibre-matrix interfacial shear strengths is taken into account, then some confusion is apparent.

There is a large radial expansion mismatch between the carbon fibres and the surrounding matrix on cooling, which pulls the fibres away from the surrounding matrix. This should give low fibre-matrix shear strengths, and should allow much greater fibre pull-out lengths for the carbon fibre-borosilicate composite than for the Tyranno-glass ceramic, where the fibre-matrix bond is stronger. Simple analysis of the matrix microcrack spacing in polished sections of failed composite [after Aveston, Cooper, and Kelly, 1971] shows that the carbon fibre material has an interface bond of approximately 5 MPa, and the Tyranno fibre composite a bond of up to 35 MPa. Using a simple force balance, the fibre pull-out length in the borosilicate composite should be a maximum of around 2 mm, much greater than observed (approximately 100 μm). The Tyranno material correspondingly should have maximum fibre pull-out lengths of around 0.3 mm, and this is similar to the lengths observed. The cause of this must be the nature of the fibre-matrix bond or frictional resistance which gives a maximum matrix failure stress in conjunction with high composite strength and toughness. Analysis of the mechanics may be relatively straightforward for composites containing smooth fibres, such as Tyranno, but may be more complicated for systems containing carbon fibres. This is because carbon fibres often have rough surfaces, and complex mechanical keying between fibres and matrix may occur, causing the fibre-matrix interfacial shear stress to vary during composite failure and preventing the application of simple fibre pull-out models

4.4. Anomalous Stress-Strain Behaviour of Ceramic Composites.

Both unidirectional materials tested in this programme displayed so-called "ductile" ceramic composite stress-strain curves, with marked reductions in elastic moduli due to matrix failure. The transverse expansion noticed is not easily explained however. There are two plausible explanations for the expansion: Relief of thermal stresses and axial cracking:

(i). **Tensile matrix stresses.** The greater cooling contraction of the matrix compared with the fibres causes tensile matrix stresses in the material at room temperature [Sambell, Phillips, and Bowen, 1974]. This thermal stress is given by Equation (3),

$$\sigma_m = \frac{(\alpha_m - \alpha_f) \,.\, \Delta T \,.\, E_f \,.\, V_f}{\left[V_f \,.\, \left\{\frac{E_f}{E_m} - 1\right\} + 1\right]} \qquad (3)$$

where ΔT is the fabrication cooling interval over which no strain relief occurs, and α denotes a thermal expansivity. The tensile thermal stress in the borosilicate matrix was consequently 54 MPa, and in the glass ceramic matrix 83 MPa. Relief of this strain with the composite still under stress could only give a positive strain increase of 0.023% for the carbon fibre-borosilicate composite, and .020% for the Tyranno fibre-glass ceramic, and the mechanism is therefore not significant.

(ii). Axial Cracking. A cited explanation for transverse cracking is the formation of axial matrix cracks in the material which are wedged open by debris or relative motion during stressing [Nardone and Prewo, 1988, Harris,1990]. This the most likely cause of the transverse expansion, but is also the most difficult to quantify. Results show that the occurrence of the effect seems to be independent of the test technique. Tyranno composite specimens tested in tension and monitored by gauges at both the centres and tab ends show differences in behaviour between the two monitoring positions (Figures 3 and 4). At low imposed stresses strain reversal occurred, and the strain was clearly similar at the centre of the specimen to near the gripped ends. As the material was loaded further however, the material showed much greater expansion near the end tabs than at the centre of the gauge length, indicating that the gripping of the composite affects the transverse strain. Examination of specimens has shown that stress concentration causes enhanced damage near the end-tabs (Figure 5). This would enhance axial cracking near the grips, and would tend to cause greater crack opening. The transverse strain effect seen in these materials is therefore a combination of a materials cracking effect which occurs evenly at low stresses, combined with the influence of the gripping conditions at higher stresses.

The Tyranno fibre composite also showed set strain effects when unloaded, and these results are shown in Figure 6 for both the transverse and longitudinal strains in a different sample of composite. Clearly the rate of longitudinal set strain accumulation levels above 450 MPa, while the transverse strain continues to rise, which implies that the transverse microcracking in the matrix is not directly connected with the transverse expansion, giving further indications that the continuation of axial cracking causes transverse expansion in ceramic matrix composites.

Figure 5: End Tab Damage of Tyranno Fibre-Glass Ceramic Composite

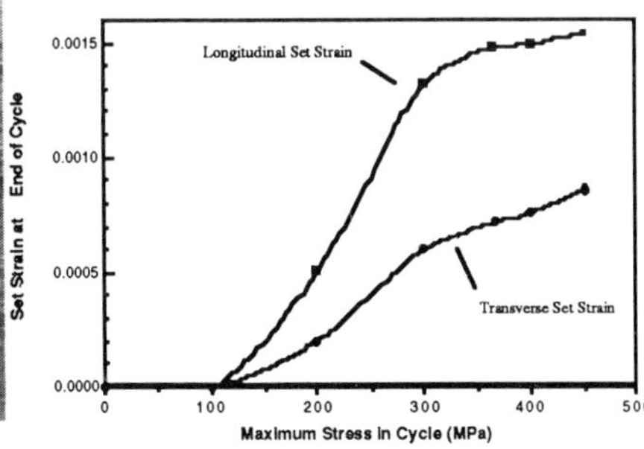

Figure 6: Set Strains in Tyranno Fibre-Glass Ceramic Composite

5. Conclusions.

i). Carbon fibre reinforced borosilicates have improved matrices and strength properties when fabricated at temperatures near the working point of the matrix. The material is free from cristobalite, and the higher pressing temperatures do not appear to adversely affect the strength of the composite when measured in flexure.

ii). There is a high ratio between the flexural and tensile strengths of the carbon-borosilicate material tested in this programme. Current (simple) relationships for describing composite strength are inadequate, and further work is needed on modelling. Although the flexural strength of this carbon fibre composite is high and compares favourably with other composites studied, the tensile strength is lower than for other strong composites, and is also lower than the tensile strengths presented by previous workers using similar fibre and matrix materials. It is therefore prudent for mechanical test data to be accompanied by details of the fabrication process used in the provision of testing materials.

iii). The fracture surfaces of carbon-borosilicate composites do not display as great a degree of fibre pull-out when compared with silicon carbide reinforced ceramics, despite the lower fibre-matrix interfacial strengths registered by matrix microcracking. There is no proven explanation for this at present.

iv). The transverse expansion of ceramic composites now appears to be a general phenomenon, and occurs in both flexural and tensile tests. Evidence indicates that the effect is due to axial cracking, and that enhancement of cracking near tensile testing grips will cause an apparent increase in the transverse expansion.

6. References.

1. Aveston J , Cooper GA and Kelly A, **"Single and Multiple Fracture"**, Paper 2, Properties of Fibre Composites, NPL Conference 1971.
2. Cooper RF and Chyung K, **"Structure and Chemistry of Fibre-Matrix Interfaces in Silicon Carbide Fibre-Reinforced Glass Ceramics: an Electron Microscopy Study"**, Journal of Materials Science, **22** (1987), pp. 3148-3160.
3. **"CRAG Test Methods for the Measurement of the Mechanical Properties of Fibre Reinforced Plastics"**, Composites Research Advisory Group, edited by Curtis PT , Royal Aircraft Establishment Technical Report 88012, February 1988.
4. Davies JJR, and Preston RF, AEA Technology Harwell, 1989, to be published.
5. Davies JJR, **"The Production and Detailed Mechanical Behaviour of Ceramic Composites for Advanced Applications"**, Internal Report AERE G5362, 1989.
6. Harris B, **Private Communication**, Materials Science Dept., University of Bath, 1990.
7. Hitchon JW, McCausland WH, and Phillips DC, **"The Variability of the Tensile Strength of Unidirectional Carbon Fibre Reinforced Plastics"**, Internal Report AERE R8217, November 1975.
8. Kim RY and Katz AP, **"Mechanical Behaviour of Unidirectional SiC/BMAS Ceramic Composites"**, Ceramic Engineering Society Proceedings, 9[7-8], pp. 853-860, 1988.
9. Linger KR and Pratchett AG, **"The Thermal Degradation of Carbon Fibre Reinforced Ceramics"**, Internal Report AERE R7142, 1972.
10. Murty VSR and Lewis MH, **"Interface structure and matrix crystallisation in SiC (Nicalon)-Pyrex composites"**, Journal of Materials Science Letters 8 (1989) 571-572.
11 Nardone V and Prewo KM, Journal of Materials Science 23 (1988) 168.
12. Phillips DC, Park N, and Lee RJ, **"The Impact behaviour of High Performance Ceramic matrix Composites"**, Composites Science and Technology 37 (1990) 249-265.
13. Sambell RAJ, Phillips DC, and Bowen DH, **"The Technology of Carbon-Fibre-Reinforced Glasses and Ceramics"**, Proc. of "Carbon Fibres: Their place in Modern Technology", Plastics Institute, London, 1974.
14. Shin DW and Knowles KM , **"Microstructure-Property Relationships in SiC Fibre-Reinforced Borosilicate Glass Composites"**, Paper C400/049 I Mech E 1990, 163-168.

7. Acknowledgements.

The authors of this paper would like to acknowledge the help and support given to them by the following people: Dr D.C. Phillips for comments on the manuscript, Dr R.W. Davidge for advice on data interpretation, and Dr. D.T. Livey and the AEA Technology Underlying Research Programme for financial support.

Inst. Phys. Conf. Ser. No 111
Paper presented at Int. Conf. on New Materials and their Applications, Univ. of Warwick, 1990

Metal filament and ceramic fibre reinforced glass-ceramic matrix composites

I W Donald, B L Metcalfe and D J Bradley

MOD(PE), Atomic Weapons Establishment, Aldermaston, RG7 4PR, UK

ABSTRACT: The preparation of glass-ceramic matrix composites with matrix thermal expansions, α, matched approximately to those of a variety of metallic filaments and ceramic fibres of diameter < 25 μm is reported. Preliminary data are presented giving details of the glass-ceramic compositions employed, the process conditions required in order to produce matched thermal expansion systems, and some of the properties of the resultant composites. For comparison, data are also given for some systems where α is not matched. The best composites exhibit mean strengths in excess of 1GPa, coupled with work of fracture values of the order of $30kJm^{-2}$.

1. INTRODUCTION

Reinforcement of ceramic materials by filaments and fibres produced some very encouraging results in the 1960s and 1970s in terms of providing tough composites with high values for work of fracture (Donald and McMillan, 1976). For example, values of up to ≈ $20kJm^{-2}$ were reported for work of fracture of some continuous uni-directionally aligned ceramic fibre systems (Aveston, 1972), and values up to $16kJm^{-2}$ were found for some discontinuous, randomly-orientated metal filament reinforced composites (Donald and McMillan, 1977). It was soon appreciated, however, that the factors which promote toughness in such materials do not necessarily lead to corresponding increases in strength. This was due in part to the fact that, with the exception of carbon fibres, the fibres available commercially at that time were of relatively large diameter, >50μm, and could act as powerful stress concentration sites; this was particularly true when the thermal expansion characteristics of the fibres did not match those of the matrix. Some of the carbon fibre systems did achieve simultaneous increases in both fracture toughness and strength (see Phillips 1974), but could not be used in oxidising environments at temperatures > 400°C. Due to these disadvantages, interest in ceramic matrix composites declined during the 1970s.

Over the last few years, however, there has been a significant revival of interest in these materials (see Donald, 1989). This has been due to the advent of a whole new range of small diameter, high temperature ceramic fibres, available commercially at realistic prices. Using the new silicon carbide fibres in glass or glass-ceramic matrices, for example, composites with strengths in excess of 1GPa coupled with high values for work of fracture, of the order of 50 kJm^{-2}, have been prepared (Brennan and Prewo, 1982; Dawson et al, 1988). Glasses and glass-ceramics offer the best scope for accomplishing simultaneous increases in both strength and toughness because these materials exhibit relatively low elastic moduli. In addition, they offer the advantage that composite consolidation can be carried out at relatively low temperatures, thereby minimizing matrix-fibre interaction, and matrix thermal expansions can, in principle, be matched to those of the reinforcing phases employed.

Recent reported work has concentrated on the use of the new ceramic fibres, metal filaments having been neglected. Metal reinforcement does, however, offer a number of distinct advantages over ceramic fibres, in particular a lower susceptibility to damage during composite fabrication. Inexpensive metal filaments are now also available commercially in the form of multi-filament tows with individual filament diameters down to 2 μm (Bekaert metal fibre), although they are limited to only a few specific metals and alloys at the present time. Additionally, although still in the development stage, metal filaments in the form of glass-coated Taylor-microwire are becoming available in a wider range of materials (see Donald 1987). The fact that Taylor-microwire is produced directly with a glass coating makes these materials ideal candidates for the production of glass or glass-ceramic composites directly by the hot-pressing of filament tows; and this method has indeed been shown recently to be a viable route for producing composites (Donald et. al. 1988).

Table 1 - Metal filaments and ceramic fibres

Filament or fibre	Diameter (μm)
316L stainless steel	8 and 22
Inconel 601	12 and 22
Alumina (Nextel)	10
Silicon carbide (Nicalon)	15
Taylor-wires:-	
a) Cu / pyrex	6 / 16
b) Ni-Si-B alloy / pyrex	6 / 13

In the present paper, following from earlier work (Metcalfe and Donald, 1989), we report some preliminary results outlining the preparation of glass-ceramic materials with thermal expansions approximately matched to those of a variety of metal filaments and ceramic fibres. A number of composites have been prepared using these glass-ceramics; the individual process conditions are described, and the properties of the materials are compared and discussed.

Table 2 - Compositions of glass-ceramics employed (mole %)

Glass	Li_2O	Na_2O	MgO	ZnO	B_2O_3	Al_2O_3	SiO_2	P_2O_5	TiO_2	ZrO_2
LZS	18.0	4.8	-	17.9	4.3	-	53.7	1.3	-	-
LMAS-1	17.7	4.3	10.9	-	3.8	11.5	50.8	1.0	-	-
LMAS-2	15.9	3.8	18.9	-	3.4	10.4	41.6	1.0	3.8	1.2
MAS-1	-	-	23.4	-	-	19.8	48.1	1.4	5.1	2.2
MAS-2	-	-	22.5	-	-	19.1	46.5	4.9	4.9	2.1

2. MATERIALS EMPLOYED

A number of different matrix and reinforcing phases were employed in the present work, as summarized in Tables 1 and 2. Micrographs of some of the starting filaments are shown in Figure 1.

3. EXPERIMENTAL PROCEDURES

3.1 Preparation and characterization of glass-ceramics

Glasses were generally prepared by melting the constituent oxides, carbonates and phosphates in a Pt-Rh crucible for ≈ 3 hours at 1450-1500°C. The melt was poured into distilled water and the resulting glass frit was remelted, this procedure being carried out twice to give a homogeneous product. The thermal properties of the as-quenched powder 0.6-1.0 mm in size were determined by differential thermal analysis employing a Stanton Redcroft Model 674 DTA operating at a heating rate of 10°C min^{-1}. A preliminary assessment of the efficiency of the nucleating agents used was made by DTA employing different particle sizes of glass, following the method of Thakur and Thiagarajan (1966). Optimum nucleating temperature régimes were found using the method described by Marotta et al (1981, 1982). Glass rods ≈ 8mm diameter were also cast, annealed and subjected to various heat-treatment schedules. The thermal expansion characteristics of the resultant glass-ceramics were monitored, at a heating rate of 5 °C min^{-1}, over the range ambient to 700 °C, using a dilatometer supplied by Ceramic Developments Ltd. A minimum of three measurements was made for each expansion determination. The crystal phases present in the glass-ceramics were found using standard X-ray diffraction methods.

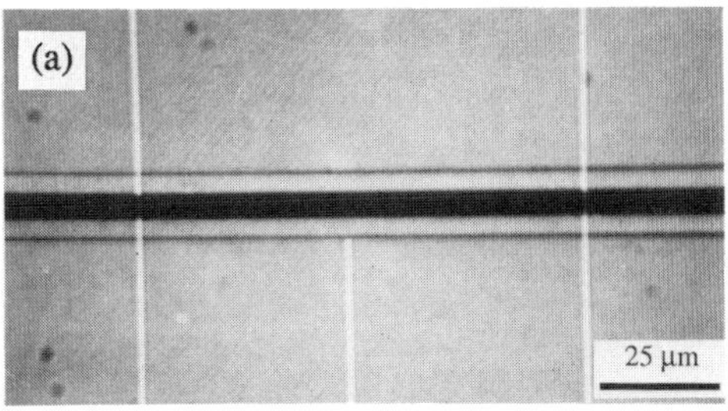

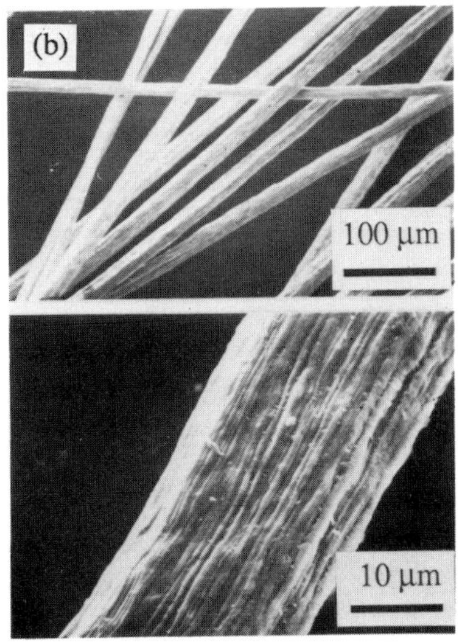

Figure 1 - Micrographs of metal filaments
a) Cu microwire, b) stainless steel

3.2 Composite fabrication

Glass powders <44 µm in size were prepared by ball-milling glass frit in acetone for severel hours. Composite plates were subsequently prepared at AERE(Harwell) by impregnating

Table 3 - Process conditions employed for fabrication of composites

Composite system	Processing/heat-treatment schedule
A) Stainless steel or Inconel / LZS	(a) Heat at 50 °$Cmin^{-1}$ to 585°C and apply pressure (7MPa); heat at 15°$Cmin^{-1}$ to 750 °C and hold for 30 min; cool to 400°C and release pressure; continue cooling to room temperature.
	(b) Heat to 465°C, apply pressure (7MPa) and hold for 60 min; heat to 750°C and hold for 60 min; cool to room temperature.
B) Taylor-wires	Heat at 50°$Cmin^{-1}$ to 930°C, apply pressure (1MPa) and hold for 5 min; release pressure and cool to room temperature.
C) SiC / LMAS-2	(a) Heat to 650°C, apply pressure (13MPa) and hold for 30 min; heat to 900°C and hold for 30 min; cool to room temperature.
	(b) Heat to 1200°C, apply pressure (13MPa) and hold for 30 min; cool to 650°C and hold for 30 min; heat to 950°C and hold for 30 min; cool to room temperature.
D) Al_2O_3 / LMAS-2	Heat to 650°C, apply pressure (13MPa) and hold for 30 min; heat to 900°C and hold for 30 min; cool to room temperature.
E) SiC / MAS-2	Heat to 810°C, apply pressure (13MPa) and hold for 30 min; heat to 1100°C and hold for 30 mins; cool to room temperature.

ceramic fibre or metal filament tows with a slurry of glass powder in a methyl ethyl ketone and poly vinyl acetate base, followed by hot-pressing in nitrogen in a graphite die. The process conditions employed are summarized in Table 3. These conditions were derived on the basis of the thermal and expansion data noted for the glass-ceramics as described in section 4.

Table 4 - Thermal properties of starting glasses

Glass	T_g (°C)	T_{xp1} (°C)	T_{xp2} (°C)	T_{ms} (°C)	T_{liq} (°C)
LZS	459	650	737	845	926
LMAS-1	507	714	-	950	1069
LMAS-2	547	763	-	960	1163
MAS-1	796	1137	-	1315	1463
MAS-2	763	1097	-	915	1357

Table 5 - Nucleation efficiency

Glass	ΔT (°C)
LZS	0
LMAS-1	38
LMAS-2	6
MAS-2	4

Consolidation of the SiC fibre composites was carried out both at low (≈50-100°C >T_g) and high (≈T_{liq}) temperatures in order to assess the effect of consolidation temperature on overall composite performance. In addition, these composites were prepared using two different glass-ceramic matrices in order to determine the effect of thermal expansion mismatch on the resultant properties. Some composites were also prepared directly by hot-pressing tows of Taylor-wire; further details relating specifically to the preparation of these materials are given elsewhere (Donald et al, 1988; Donald and Metcalfe, 1990).

3.3 Mechanical testing of composites

Composite plates ≈ 1.5 mm thick were sliced into bars 50 mm x 5 mm in size and tested in 3-point bending using a span of 40 mm at a crosshead speed of 0.5 mm min^{-1}. For comparison, some samples were also tested using a span of 16 mm. Values of flexural strength were calculated from a knowledge of the specimen dimensions and maximum load, whilst work of fracture values were found, where appropriate, by integration of the area under suitable load-displacement curves. Integration was normally performed only up to a maximum cross-head displacement corresponding to four times that at the maximum load. A minimum of 5 samples was used for each volume fraction band for the metal reinforced systems, whilst 30 samples were normally used for each of the ceramic fibre systems.

3.4 Examination of composites

Composite microstructures were examined optically using sectioned and polished samples. Fracture surfaces were examined by scanning electron microscopy.

4. RESULTS AND DISCUSSION

4.1 Characterization of glass-ceramics

The thermal properties of the glasses employed in this investigation are summarized in Tables 4 and 5, and Figure 2. Data for determining the optimum nucleation temperature régimes are summarized in Figure 3. Some data for the lithium zinc silicate, LZS, systems have been reported previously in more detail (Donald et al, 1989; Metcalfe and Donald, 1990). Hot-pressing consolidation temperatures were chosen to correspond to the second, higher temperature nucleating régime. In principle, the thermal expansion of the LZS can be matched both to the Ni-based superalloys and to stainless steel, depending on the precise heat-treatment parameters employed. In the present work, composites containing stainless steel filaments were prepared under the same conditions as those containing the Inconel filaments. This gave

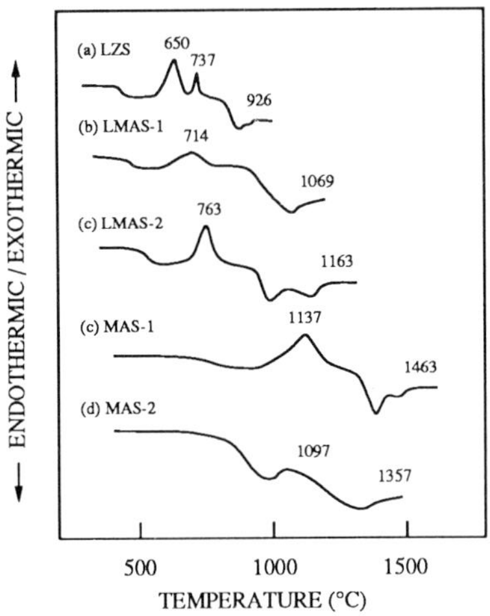

Figure 2 - DTA traces of glasses

Table 6 - Thermal expansion data

Glass-ceramic	Thermal expansion (10^{-6}°C^{-1}) {20-460°C}	{20-700°C}
LZS	14.6	17.2
LMAS-2(a)	5.8	8.6
LMAS-2b)	6.4	9.1
MAS-2(a)	4.9	4.9
MAS-2(b)	10.0	8.8
MAS-2(c)	8.2	7.6
MAS-2(d)	9.6	8.5
MAS-2(e)	4.7	5.0

a system with approximately matched thermal expansion when using the Inconel reinforcement. Thermal expansion data are shown in Table 6 and Figure 4. Two base glass compositions were chosen for possible use with the ceramic fibres; a lithium magnesium alumino-silicate, LMAS, and an alkali-free magnesium alumino-silicate, MAS. The base LMAS glass (LMAS-1) exhibited only a diffuse crystallization exotherm as resolved by DTA. Crystallization was, however, improved (LMAS-2) by the addition of TiO_2 and ZrO_2 (Figure 2). The MAS glass (MAS-1), based close to the cordierite composition, exhibited a very high liquidus temperature; this glass was subsequently modified (MAS-2) in order that the fabrication temperature could be lowered. The crystal phases present in the glass-ceramics are summarized in Table 7. The precise amount of each phase is dependent on the overall heat-treatment conditions employed. Reference to Table 6 illustrates that the thermal expansion of both the LMAS and MAS glass-ceramics can be approximately matched to that of alumina, depending on the precise heat-treatment schedule adopted. The expansion of the MAS glass-ceramic can also be matched to that of SiC.

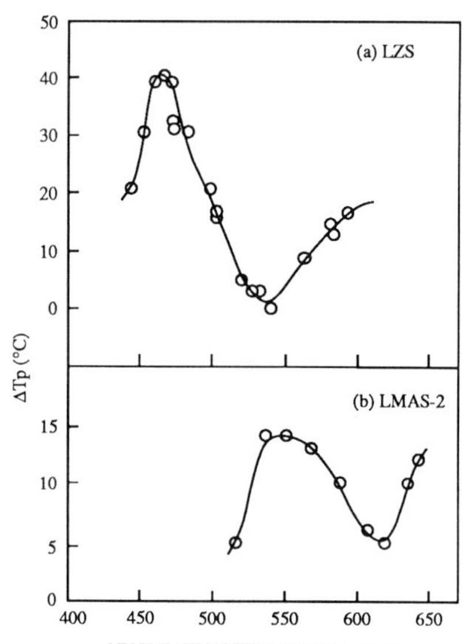

Figure 3 - Optimum nucleation temperature régimes

4.2 Microstructures and mechanical properties of composites

The microstructures of a number of the composites are shown in Figure 5. Mechanical properties of the composites are summarized in Table 8, representative apparent stress-displacement curves are illustrated in Figure 6, and fracture surfaces are shown in Figure 7.

Table 7 - Crystal phases present in glass-ceramics

Glass-ceramic	Heat treatment (°C)	Crystal phases
LZS	585° + 750°	α-crist.+ α-quartz + tridymite + β_1-Li_2ZnSiO_4 + γ_o-Li_2ZnSiO_4
LMAS-2(a)	650° + 900°	$Li_2Al_2Si_3O_{10}$ + $LiAlSi_2O_6$ + Mg_2SiO_4
LMAS-2(b)	1200° + 950°	$Li_2Al_2Si_3O_{10}$ + $LiAlSi_2O_6$ + Mg_2SiO_4
MAS-2(a)	810° + 1100°	$MgAl_2O_4$ + $ZrTiO_4$ + $Mg_3(PO_4)_2$ + SiO_2 + glass
MAS-2(b)	810° + 1150°	$Al_6Si_2O_{13}$ + $Mg_2Al_4Si_5O_{18}$ + $Mg_3(PO_4)_2$ + SiO_2
MAS-2(d)	860° + 1150°	$Al_6Si_2O_{13}$ + $Mg_2Al_4Si_5O_{18}$ + $Mg_3(PO_4)_2$ + $ZrTiO_4$
MAS-2(e)	860° + 1400°	$Al_6Si_2O_{13}$+ $Mg_2Al_4Si_5O_{18}$ + $ZrSiO_4$ + glass

The degree of scatter associated with the mechanical property data for the metal filament reinforced systems is high. This scatter may be associated with the non- uniform filament distribution noted in Figure 5(a). Broad trends in behaviour can, nevertheless, be resolved within the various families of data. For example, individual flexural strength data are plotted against volume fraction reinforcement, V_f, for the 22µm stainless steel and Inconel filaments in Figure 8. It can be seen that a broad maximum in flexural strength is found at $V_f \approx 0.4$. The decrease in strength at higher V_f most probably arises due to the effects of an increasing proportion of filament-filament

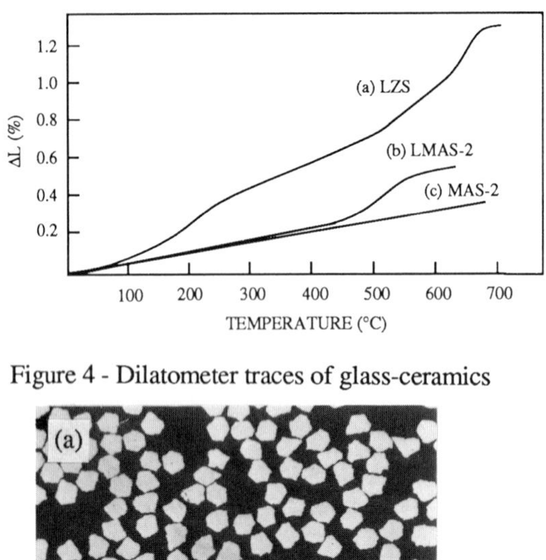

Figure 4 - Dilatometer traces of glass-ceramics

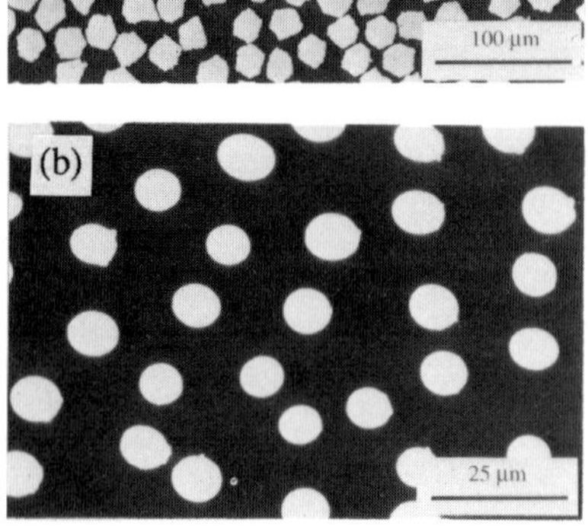

Figure 5 - Microstructures of composites a) LZS/40v% stainless steel, b) pyrex/Cu

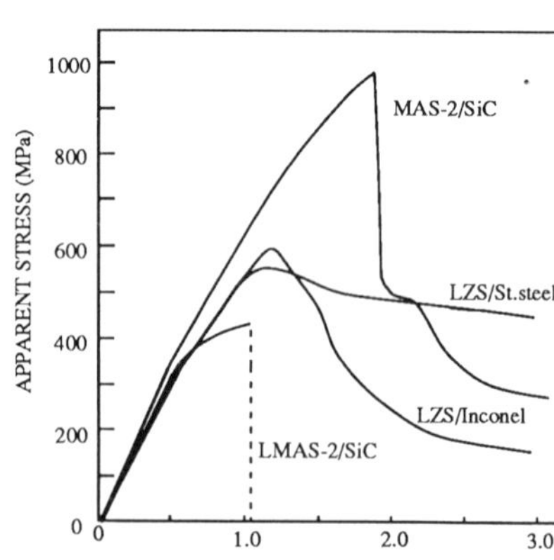

Figure 6 - Representative apparent stress-displacement curves

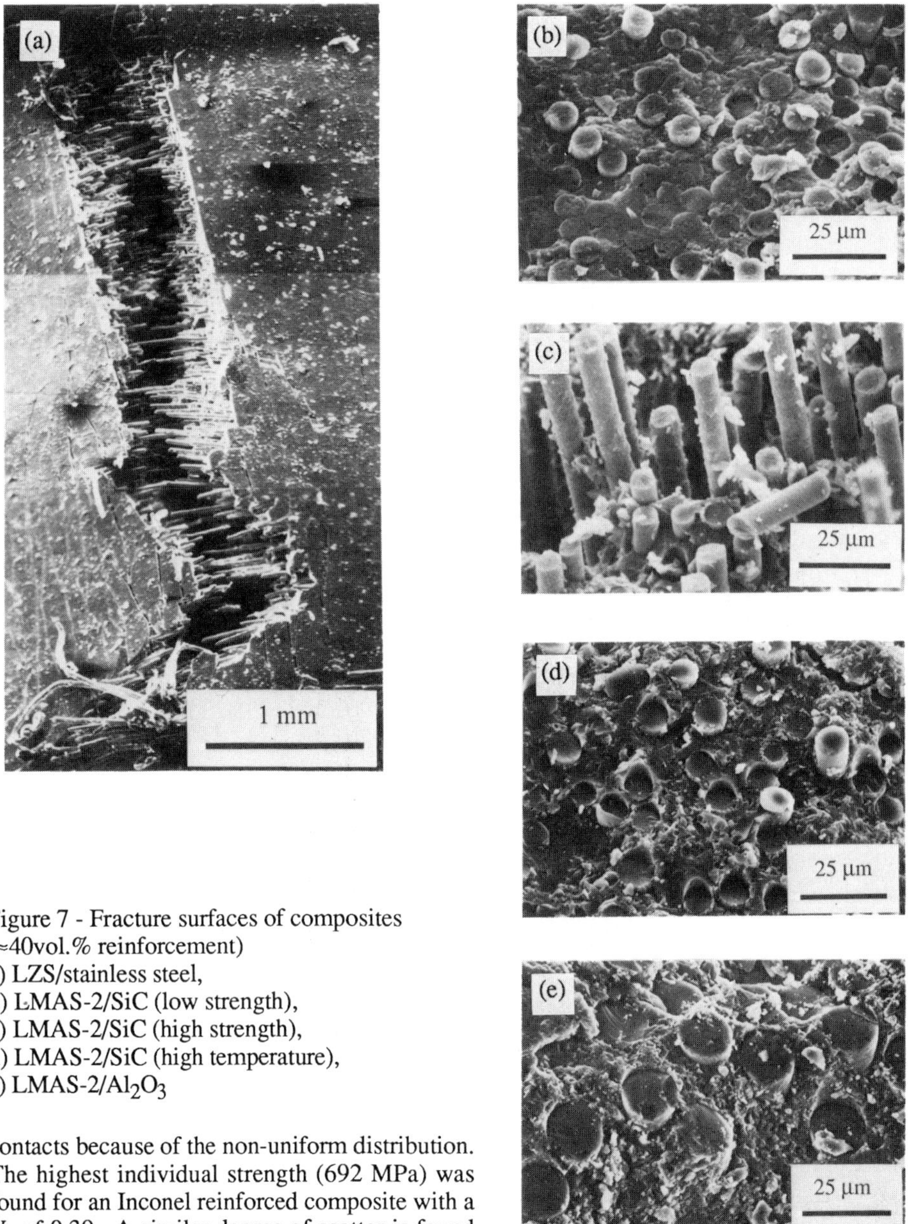

Figure 7 - Fracture surfaces of composites (≈40vol.% reinforcement)
a) LZS/stainless steel,
b) LMAS-2/SiC (low strength),
c) LMAS-2/SiC (high strength),
d) LMAS-2/SiC (high temperature),
e) LMAS-2/Al_2O_3

contacts because of the non-uniform distribution. The highest individual strength (692 MPa) was found for an Inconel reinforced composite with a V_f of 0.39. A similar degree of scatter is found in the work of fracture, WOF, data. The highest individual WOF (46 kJm^{-2}) was found for a 0.4 V_f stainless steel composite. (The mean value for this system was 24.6 kJm^{-2}). Reference to the stress-displacement curves shown in Figure 6 illustrates that the stainless steel reinforced systems can exhibit very large displacements, up to several mm, whilst still supporting a large proportion of the maximum load. If integration is carried out for the whole curve (rather than just to a displacement of four times the maximum load) apparent WOF's approaching 100 kJm^{-2} are found for a number of the stainless steel systems. It is doubtful, however, whether these very high WOF values are of practical value. Overall, the stainless steel reinforcement gave the toughest composites (which may be expected because the thermal expansion mismatch should result in a weaker matrix-filament interface), whilst the Inconel reinforcement (with approximately matched expansion) led to the strongest materials. Extensive fibre rupture and pull-out is associated with the metallic reinforcement, as illustrated in Figure 7(a) for a stainless steel system.

Table 8 - Representative mechanical properties of composites

Matrix	Filament	V_f	Flex. str. (MPa)	WOF+ (kJm^{-2})	E (GPa)	Density (gcm^{-3})
(a) metal filament reinforced						
LZS(a)*	22µm SS	0.29±0.02	390±49	21.3±11.5	56.7±17.1	3.60±0.13
LZS(a)*	22µm SS	0.40±0.01	454±58	31.3±8.6	-	3.93±0.03
LZS(a)	22µm SS	0.29±0.02	352±44	25.0±9.6	71.0±7.3	3.60
LZS(a)	22µm SS	0.40±0.01	421±42	24.6±6.9	75.7±7.6	3.98
LZS(a)	22µm Inc	0.29±0.01	353±101	-	-	-
LZS(a)*	22µm Inc	0.39±0.05	437±147	21.2±8.2	46.5±19.6	4.25±0.21
LZS(a)	22µm Inc	0.39±0.05	533±75	17.9±6.5	88.0±19.3	4.25±0.22
LZS(a)*	12µm Inc	0.41±0.06	530±41	10.4±3.5	68.0±9.5	4.07±0.20
LZS(a)	12µm Inc	0.41±0.06	462±89	13.0±2.0	87.4±14.1	4.07±0.30
LZS(b)	8µm SS	≈0.20	460	6.0	-	-
LZS(b)	22µm SS	≈0.20	265±48	19.7±0.1	-	-
LZS(b)	22µm SS	≈0.40	497±155	59.0±32.0§	-	-
	µwire:-					
pyrex	{Cu}	≈0.15	111±18	0.7±0.2	35.2±11.8	3.20±0.10
pyrex	{Ni-Si-B}	≈0.20	255±42	1.8±0.3	42.5±2.7	2.8
(b) ceramic fibre reinforced						
LMAS-2(a)	SiC	≈0.4	250±105	-	89.2±4.3	-
			{443±28; 208±55}			
LMAS-2(b)	SiC	≈0.4	119±16	-	68.7±2.9	-
LMAS-2(a)	Al_2O_3	≈0.4	190±20	-	96.2±4.4	
MAS-2(a)	SiC	≈0.4	1015±83	29.4±3.1	96.7±21.4	-

SS = stainless steel, * 16mm span, + WOF quoted up to a crosshead-displacement corresponding to 4x that of the maximum load, except for § quoted to zero load.

The degree of scatter associated with the mechanical property data for the LMAS glass-ceramic reinforced by SiC fibres is also relatively high in the case of the composites processed at the lower temperature; see Table 8. In this case, however, it is noteworthy that the results fall into two clearly defined modes of behaviour. Composites failed either during the linear portion of the load-displacement curve (mean strength 208 MPa) or, alternatively, they failed significantly after a well-defined proportional limit had been achieved (mean 443 MPa); see Figure 6. Reference to the fracture surfaces shown in Figure 7 (b) & (c) shows that the lower strength composites failed by fracture through the filaments in the plane of the matrix, whilst the higher strength materials show some fibre pull-out. Equivalent composites fabricated at the higher temperature showed considerably less scatter in the data, but were also much weaker (mean strength only 119 MPa). No pull-out was observed for these materials, as noted in Figure 7(d). The fracture mode of these ceramic fibre composites was normally relatively brittle (although, as described earlier, deviation from linearity is noted in many of the load-displacement curves). The fact that extensive fibre pull-out effects are not observed for these systems is most probably due to the degree of thermal expansion mismatch between fibre and matrix ($\alpha_{matrix} > \alpha_{fibre}$). The composites containing the alumina fibres also showed less scatter in the flexural strength data, but these materials were not exceptionally strong (mean strength of 190 MPa). Fracture of the fibres again occurred within the plane of the matrix; see Figure 7(e). The fracture mode of these composites was also normally brittle. The lack of fibre pull-out effects in this matched thermal expansion system is probably related to the greater degree of chemical bonding which is expected between the alumina fibres and the glass-ceramic matrix.

Work is at an early stage on the matched thermal expansion MAS/SiC composites. Preliminary data do indicate, however, that these materials are significantly stronger than the other composites, exhibiting a mean strength of 1015 MPa. This high strength is coupled with relatively high work of fracture values, of the order of $30kJm^{-2}$. The fracture behaviour is very different from that of the other ceramic fibre systems, samples showing very extensive fibre pull-out coupled with progressive delamination of the matrix.

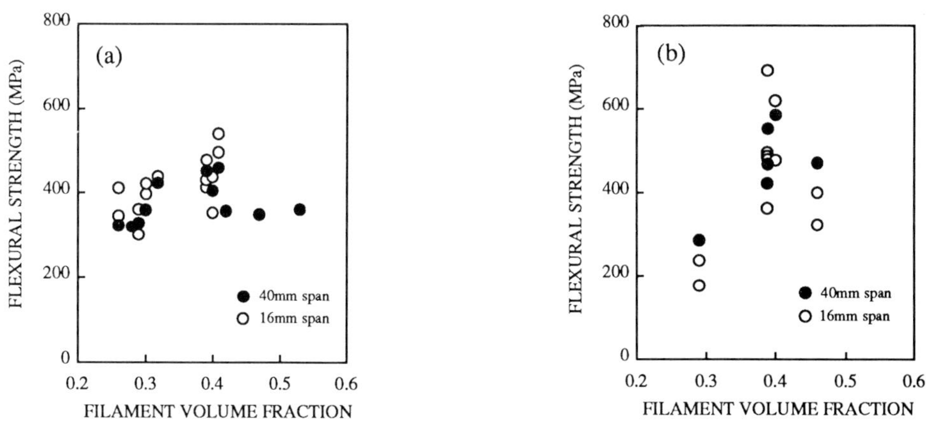

Figure 8 - Flexural strength as a function of volume fraction of metal filament
a) 22μm diameter stainless steel, b) 22 μm diameter Inconel

5. CONCLUSIONS

a) Uni-directionally aligned metal filament reinforced composites exhibiting relatively high strength and toughness have been prepared using a lithium zinc silicate glass-ceramic matrix. Strengths of up to a mean of ≈ 530 MPa have been achieved, coupled with *practical* values for work of fracture of the order of $25kJm^{-2}$.

b) The highest strengths are obtained for the metal filament reinforced systems when the expansions of matrix and filament are approximately matched, whilst higher toughness is found for systems exhibiting a small mismatch in expansion ($\alpha_{fibre} > \alpha_{matrix}$).

c) Uni-directionally aligned metal filament composites have also been prepared directly by the hot-pressing of Cu and Ni-Si-B alloy Taylor-microwires.

d) The thermal expansions of lithium magnesium alumino-silicate and magnesium alumino-silicate glass-ceramics have been approximately matched to that of alumina, whilst that of a magnesium alumino-silicate has been matched to that of SiC. Some prototype composites have been prepared from these systems. The best composites exhibit bending strengths in excess of 1GPa coupled with values for work of fracture of the order of $30kJm^{-2}$.

e) It has been confirmed that useful glass-cermic matrix composites can be prepared by consolidation at temperatures around T_g of the starting glass, ie. it is not necessary to employ high temperatures $>T_{liq}$ to obtain viable materials.

Acknowledgements

The authors are grateful to Messrs. N. J. Mattingley and J. J. R. Davies of AERE, Harwell, for preparing the composites used in this investigation, and for carrying out the density determinations and some of the mechanical testing. We are also grateful to Mrs. E. Prior and Mrs. S. Vail of AWE, Aldermaston, for carrying out the microscopy, and to Messrs. M. Clay and T. Carter, also of AWE, for the X-ray diffraction studies.

References

Aveston J A 1972 in *The Properties of Fibre Composites, NPL Conf. Proc.* (Guildford: IPC Science and Technology Press), pp 63-74
Bekaert Metal Fibre, N V Bekaert S A, B-8550 Zwevegem, Belgium
Brennan J J and Prewo K W 1982 *J. Mater. Sci.* **17** 2371
Dawson D M, Preston R F and Purser A 1988 *Silicates Industriels* **53** 129
Donald I W 1987, *J. Mater. Sci.* **22** 2661
Donald I W and McMillan P W 1976 *J. Mater. Sci.* **11** 949
Donald I W and McMillan P W 1977 *J. Mater. Sci.* **12** 290
Donald I W and Metcalfe B L 1990 to be published in the *Procs. of the 7th CIMTEC World Ceramics Congress*, Italy
Donald I W, Metcalfe B L and Bye A D 1988 *J. Mater. Sci.* Letters **7** 964
Donald I W, Metcalfe B L, Wood D J and Copley J R 1989 *J Mater. Sci.* **24** 3892
Marotta A, Buri A and Branda F 1981 *J. Mater. Sci.* **16** 341
Marotta A, Buri A, Branda F and Saiello S 1982 in *Advances in Ceramics 4*, edited by Simmons J H, Uhlmann D R and Beall G H, Amer Ceram Soc, Columbus, pp 146-152
Metcalfe B L and Donald I W 1989 *paper presented at the 2nd International Conference on Ceramic-Ceramic Composites*, Mons, Belgium.
Metcalfe B L and Donald I W 1990 - this Conference
Phillips D C 1974 *J. Composite Mater.* **8** 130
Thakur R L and Thiagarajan S 1966 *Glass and Ceramic Bull.* **13** 33

Paper presented at Int. Conf. on New Materials and their Applications, Univ. of Warwick, 1990

Continuous fibre reinforced glass and glass ceramic matrix composites by sol-gel processing

V Gunay, P F James, F R Jones and J E Bailey

School of Materials University of Sheffield, Elmfield, Northumberland Road, Sheffield, S10 2TZ

ABSTRACT: Continuous carbon and SiC fibre reinforced composites with glass or glass ceramic matrices were produced by sol-gel processing and by combined sol-gel-powder filler processing followed by hot pressing. The matrices included sodium borosilicate glass, combined silica glass and borosilicate glass, and lithium aluminosilicate or magnesium aluminosilicate glass ceramics. Fibres were infiltrated with borosilicate or aluminosilicate precursor solutions and with borosilicate or silica yielding solutions combined with borosilicate glass filler. The density, porosity, flexural strength and fracture behaviour of the composites were investigated.

1. INTRODUCTION

It has been demonstrated that continuous fibre reinforced glass and glass ceramic matrix composites have excellent strength, stiffness, toughness and dimensional stability and have considerable potential in various engineering applications (Sambell et al (1972); Phillips (1972, 1974); Prewo (1982); Prewo et al (1986)). One established method of preparing glass matrix composites is to infiltrate the fibres with a slurry consisting of fine glass powder mixed with a carrier liquid and organic binder. This is followed by heat treatment to remove the volatiles and hot pressing. To obtain a glass ceramic matrix further heat treatment is required to crystallize the glass after densification is complete.

Recently sol-gel processing routes to composites have been investigated as possible alternatives to the slurry method. In these routes gels are produced from metal alkoxides and/or inorganic salts as precursors, using hydrolysis and polycondensation reactions in aqueous or alcoholic solutions (Roy (1956), Dislich (1971)). Although these methods are commonly referred to as 'sol-gel' routes, in many cases, particularly where alkoxides are used, true sols containing colloidal particles are not involved. Thus the term solution-gel processing would be more accurate, although for convenience 'sol-gel' will be used here. The gels are dried and sintered at relatively low temperatures into solid glasses or crystalline ceramics. Often large shrinkages are involved and cracking during slow drying may be a serious problem in forming large bulk samples. However, bulk samples have also been prepared by hot pressing gels and gel powders (Dislich (1971), Jabra et al (1980)).

A number of investigations of gel processing of glass and glass ceramic matrix composites have been carried out. Fitzer and Gadow (1986) obtained high strength and fracture toughness values of 950 MPa and 10.5 MPa $m^{1/2}$ respectively for hot pressed composites consisting of SiC coated carbon fibres in a gel derived silica glass matrix. Qi and Pantano (1987) reported strength and fracture values of around 350 MPa and 10 MPa $m^{1/2}$ respectively for hot pressed gel derived borosilicate glass matrix composites reinforced with carbon fibre paper. A strength of 83 MPa was obtained by Partlow (1988) for a cordierite glass ceramic matrix composite reinforced with BAS fibre coated with BN. Menessier et al (1989) reported increases in strength and toughness values when the SiC fibres were coated with pyrolitic carbon in lithium aluminosilicate matrix composites. Other relevant investigations include those of Lee and Hench (1985), Hyde (1988) and Hegeler and Bruckner (1989).

Potential advantages of sol-gel processing over the established slurry method (Qi and Pantano (1987)) include; more homogeneous fibre infiltration giving a more uniform microstructure; effective infiltration enabling a reduction of the hot pressing temperature and consequently a limitation of degradation and redistribution of the fibres; and ability to tailor the matrix composition so that the thermal expansion behaviour and fibre-matrix interface chemistry of the composite may be controlled.

In the present paper the fabrication of continuous fibre reinforced glass and glass ceramic matrix composites was investigated using sol-gel processing methods and hot pressing. The fibres used were carbon (Grafil XA-S, unsized, high strength) and, to a limited extent, SiC (Nicalon 202). The matrices were a borosilicate glass, a lithium aluminosilicate (LAS) glass ceramic and a magnesium aluminosilicate (MAS) glass ceramic. The main objective was to study the effects of the processing variables on the properties of the composites, including final density, strength and fracture behaviour. The hot pressing behaviour of both the composite preforms and of gel-derived powders without fibres were studied. Comparison was made of composites with borosilicate glass matrices prepared by both gel processing and by the glass powder slurry infiltration method. Some composites were prepared by a combination of the sol-gel and slurry infiltration methods.

2. EXPERIMENTAL

2.1 Preparation of matrix materials in powder form

The expected compositions of the matrices were as follows (in wt%): for the borosilicate glass 81 SiO_2-13 B_2O_3-4 Na_2O-2 Al_2O_3; for the glass ceramic LAS1 73.0 SiO_2-23.0 Al_2O_3-4.0 Li_2O; and for the glass ceramic MAS2 59.7 SiO_2-22.4 Al_2O_3-17.9 MgO. The borosilicate glass was similar in composition to commercial 'Pyrex' (Corning Code 7740).

The precursors used for the preparation of the borosilicate (BS) glass were $Si(OC_2H_5)_4$ ('TEOS'), boric acid (H_3BO_3), aluminium nitrate and sodium acetate as the sources of SiO_2, B_2O_3, Al_2O_3 and Na_2O respectively. A solution was made in ethanol in the presence of the required amount of H_2O for hydrolysis and HCl acid as catalyst. The detailed procedures of solution preparation are given elsewhere (Gunay et al (1990) for 'Process 2'. After gelation samples were dried at 50°C and calcined at 400°C for up to 20 hours to remove residual water and organic groups. During drying or calcining the gel fragmented either into small pieces or a fine glassy powder.

A silica yielding solution with a controlled gelation time was also prepared containing TEOS, ethanol, water and HCl. This was used in the fibre infiltration experiments described later. The molar H_2O/TEOS ratio was 10 and the molar HCl/TEOS ratio was 0.1 whereas the volume ratio Ethanol to TEOS was one.

For the LAS and MAS compositions TEOS, Al-nitrate, Li-nitrate and Mg-acetate were used as precursors. In the LAS case the required quantities of TEOS, H_2O, HCl (catalyst) and methanol (solvent) were thoroughly mixed at room temperature for one hour. Al-nitrate was then dissolved in the solution and after a further hour Li-nitrate was also dissolved to produce a clear solution. In the MAS case the procedure was similar except that Mg-acetate was added instead of the Li-nitrate. The solutions were allowed to gel at room temperature or at 50°C. Again the gels were dried and calcined as described above to produce fine powders. Gels and powders were also produced using only metal alkoxides as starting compounds but similar results were obtained and the above procedure of using TEOS with metal salts was preferred for ease and convenience.

The densification under hot pressing of the calcined gel powders was investigated at a series of temperatures. Details are given by Gunay et al (1990). The gel powder was sieved to below 53 μm particle size and 3g samples were hot pressed in a nitrogen atmosphere in a graphite die to produce circular discs 22 mm in diameter and 2 to 3 mm in thickness. Samples were heated at 15°C/min to the operating temperature and held for 30 min at a constant pressure (22 MPa). Pressure was applied when the operating temperature was reached and released at the end of the holding time. For comparison, melted glass powder of nominally the same composition as the borosilicate gel powder, sieved to below 53 μm, was also hot pressed under the same conditions.

2.2 Composite processing

Composite fabrication consisted of two steps: (i) the preparation of fibre-gel (or slurry) prepregs, and (ii) thermal treatment and final densification by hot-pressing.

In the first step, prepregs were prepared using a combination of the well known wet-winding procedure followed by vacuum impregnation. In the wet winding process the fibres were drawn through a bath of the partially hydrolyzed solution or a combined sol and powder slurry or a slurry only. Because of the low solid yield in the case of the sol it was necessary to carry out vacuum impregnation repeatedly. After each infiltration the prepregs were allowed to gel, dried at room temperature for a day, at 50°C for a day, and finally thermally heat treated at 300-400°C for 3 hours. This cycle was repeated several times. In the case of the powder slurry or combined slurry, which was a gel precursor solution filled with fine glass powders, only one impregnation was sufficient to prepare the prepregs.

In the second step, individual plies were assembled together in a stack with fibres sligned and placed in a graphite die for hot-pressing. This was performed at various temperatures for 15 to 30 minutes in a nitrogen atmosphere with 10 MPa pressure. After pressing the samples were allowed to cool slowly in the furnace. Further details of the procedure are given by Gunay et al (1990).

Disc shaped composite samples were fabricated with a diameter of 50 mm and thickness of 2 to 4 mm. Test pieces were cut using a diamond wheel and tested in the cut condition. Flexural strength and Young's modulus was measured by three point bending using at least five specimens for each composite. A span of 30 mm was used and the test specimens were on average 2.5 mm thick. Density measurements using a mercury balance and powder x-ray diffraction were carried out on the composites and hot pressed matrix samples.

3. RESULTS AND DISCUSSION

The hot pressing results for the borosilicate gel and melt derived glass powders are shown in Fig 1. The rate of densification of the gel powder was considerably slower than that of the glass powder at all temperatures studied. This can be explained by the earlier occurrence of crystallization in the gel powder compared with the glass powder. Thus x-ray diffraction revealed some crystallization (cristobalite) after hot pressing the gel powder at 750°C whereas none was detected in the 750°C glass sample. Extensive crystallization was observed in both the gel and glass samples hot pressed at 800°C. This suggests that crystallization inhibited the viscous sintering particularly in the case of the gel powders. Moreover, calculated viscosity values from isothermal hot pressing densification data showed that the effective viscosities of the gel were higher than those of the glass (Gunay (1990)). In addition to the effects of crystallization, a difference in chemical compositions between the gel and glass (Gunay (1990)) probably contributed to the differences in viscous flow and densification behaviour.

The hot pressing data for the LAS and MAS gel powders is shown in Figs 2 and 3. In each case a maximum density was achieved at around 1000°C. For LAS1 the maximum density achieved was 93% of the theoretical density for a glass of this composition (2.45 g/cm^3). The slight drop in density above 1000°C may be attributed to the crystallization of the powders during hot pressing. X-ray diffraction showed that some crystallization was present in the 900°C sample and extensive crystallization in the 1100°C sample. The phase detected was β spodumene solid solution. Similar results were obtained for the MAS2 gel powder. The glass theoretical density was 2.58 g/cm^3 in this case. At 1100°C spinel and cordierite phases were observed.

The property results for the borosilicate matrix composites are given in Tables 1 and 2 and for the LAS and MAS matrix composites in Table 3 as a function of the hot pressing temperature. The results comprise the density of the composite, flexural strength (with standard deviation), flexural modulus (E) and measured volume fractions of fibres (V_f), matrix (V_m) and total porosity (V_p). The volume fractions for the C fibre composites were determined from density measurements and the weight change after removal of the fibres by oxidation. The values for the SiC fibre composites were obtained approximately from optical microscopy. The flexural strength is the mean of five samples. The flexural modulus value was calculated from the initial linear part of the load-deflection curve and is the average for five samples. The type of fracture behaviour is indicated as brittle (B) when the load-deflection curve showed a sudden sharp drop at failure, or as tough (T) when the curve showed a gradual decrease in load at the onset of failure.

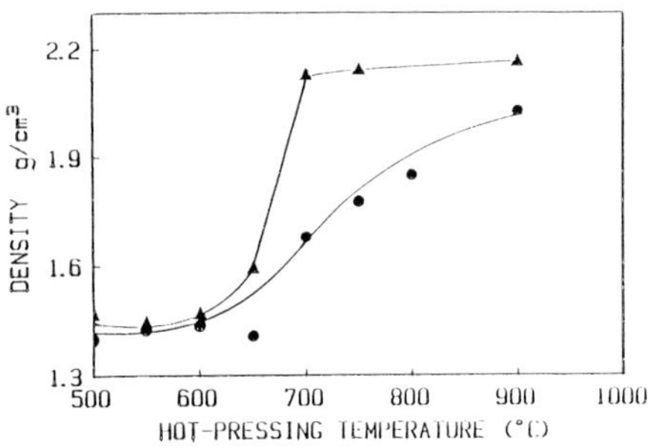

FIG 1 Density against hot pressing temperature; heating rate 15°C/min; holding time 30 min; pressure 22 MPa; borosilicate gel (o) and glass (Δ) powders

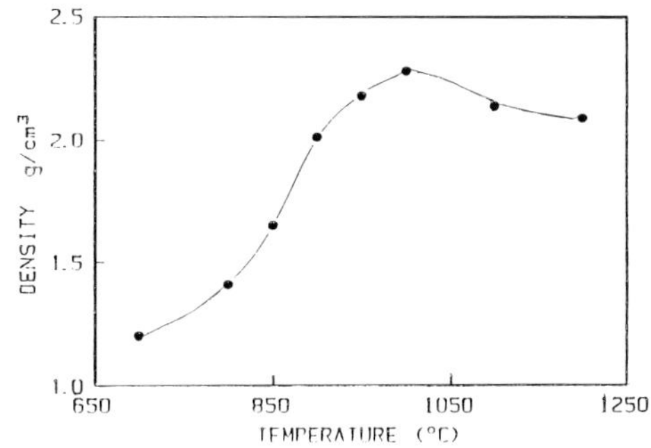

FIG 2 Density against hot pressing temperature for LAS-1 gel-derived powders; for other details see Fig 1

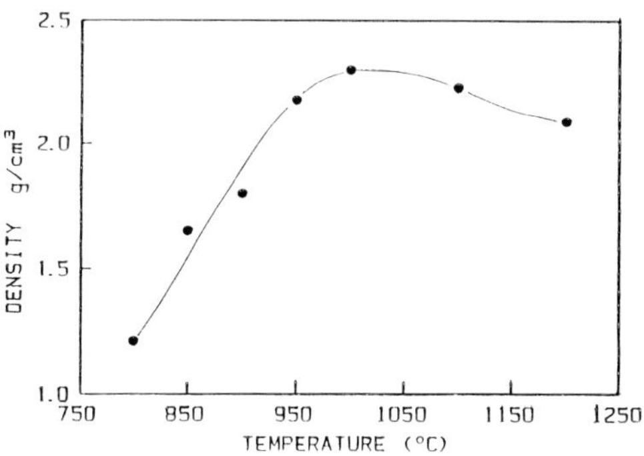

FIG 3 Density against hot pressing temperature for MAS-2 gel derived powders; for other details see Fig 1

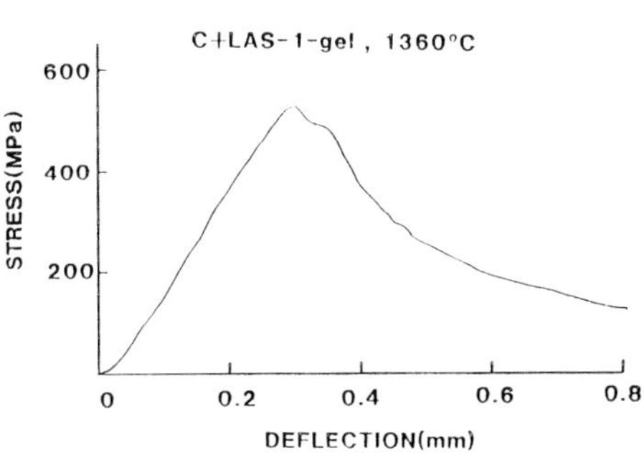

FIG 4 Typical fracture behaviour of carbon reinforced LAS-1 gel derived glass ceramic matrix composite hot pressed at 1360°C

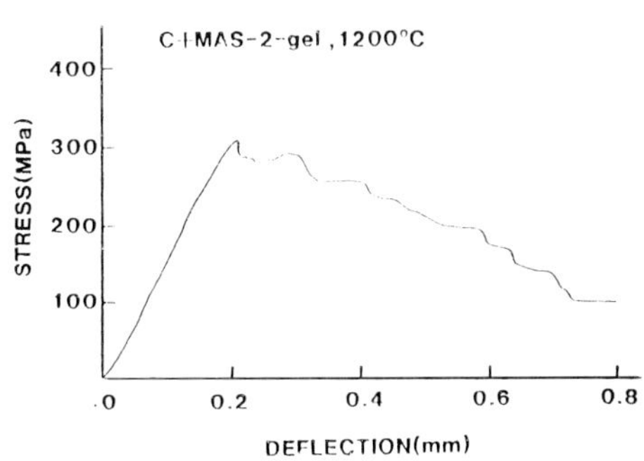

FIG 5 Typical fracture behaviour of carbon reinforced MAS-2 gel derived glass ceramic matrix composite hot pressed at 1200°C

TABLE 1 BS glass matrix composites prepared by sol-gel, slurry, and combined sol-gel/slurry routes

	C + gel derived BS Glass			C + Melted BS Glass		
Hot pressing temperature (°C)	900	1100	1200	1000	1100	1300
Flexural Strength (MPa)	102	234	128	230	263	271
(Standard deviation MPa)	12	60	18	46	15	29
E GPa	46	-	76	29	43	43
Density (g cm^{-3})	1.45	1.53	1.56	1.51	1.67	1.91
V_f (vol %)	37	51	42	18	32	22
V_m (vol %)	35	27	36	53	48	68
V_p (vol %)	28	22	22	29	20	10
Fracture	T	T-B	B	T	T	T-B

	SiC + gel derived BS Glass		$SiC + SiO_2$ gel + BS Glass Filler		
Hot pressing temperature (°C)	900	1100	1100	1200	1270
Flexural Strength (MPa)	179	251	239	454	490
(Standard deviation MPa)	13	10	28	18	31
E (GPa)	28	49	22	36	47
Density (g cm^{-3})	1.60	1.81	1.93	2.05	2.11
V_f (vol %)	50	60	43	35	40
V_m (vol %)	25	20	42	55	50
V_p (vol %)	25	20	15	10	10
Fracture	T	T	T	T	T

TABLE 2 C-fibre composites prepared by combined sol-gel/slurry route

	C + BS gel + BS Glass Filler				$C + SiO_2$ gel + BS Glass Filler		
Hot pressing temperature (°C)	900	1000	1150	1245	1100	1200	1300
Flexural Strength (MPa)	104	248	219	236	242	377	233
(Standard deviation MPa)	4	17	31	15	26	27	16
E (GPa)	25	43	57	84	35	61	73
Density (g cm^{-3})	1.22	1.30	1.37	1.73	1.51	1.75	1.77
V_f (vol %)	28	30	31	38	32	38	38
V_m (vol %)	33	35	37	44	42	48	49
V_p (vol %)	39	35	32	18	26	14	13
Fracture	T	T	B	B	T	T	T-B

TABLE 3 LAS and MAS gel derived glass ceramic matrix composites

	C + MAS 2 gel			C + LAS 1 Gel		SiC + LAS 1 Gel
Hot pressing temperature (°C)	1200	1300	1400	1300	1360	1330
Flexural Strength (MPa)	261	210	252	321	469	402
(Standard deviation MPa)	41	49	24	37	43	64
E (GPa)	58	62	79	93	70	75
Density (g cm^{-3})	1.55	1.69	1.73	1.57	1.67	2.02
V_f (vol %)	46	49	50	38	44	45
V_m (vol %)	28	31	32	36	36	30
V_p (vol %)	26	20	18	26	20	25
Fracture	T	T	T	T	T	T

Several clear general trends were evident for increase in the hot pressing temperature for all the composites. The composite density increased and the volume fraction of porosity decreased. The flexural modulus also increased showing a clear correlation with increase in density and decrease in pore fraction. The flexural strength likewise tended to increase with hot pressing temperature.

The porosities were relatively high in all the composites. For example, the C-fibre/gel derived borosilicate (BS) matrix composite hot pressed at 1200°C had a porosity around 22%. However, the C-fibre/melted BS glass composite hot pressed at 1300°C had a porosity of only 10%, and the C-fibre composite with the SiO_2/BS glass matrix produced by the 'combined' processing route had a porosity of only 13% after hot pressing at 1300°C.

One reason for the high porosities was non-uniform infiltration of the fibres during processing as discussed previously (Gunay et al (1990)). This was a particular problem for the sol-gel derived composites (BS, LAS and MAS) where the gels underwent large shrinkages. Another reason for the porosity was crystallization of the gel derived matrices during hot pressing. As discussed earlier the gel derived BS powders densified more slowly than the melted BS glass powders. This could partly explain the failure of the composites to fully densify since viscous flow of the uncrystallized glass between the fibres would be inhibited in the presence of crystals. The low porosities (10 and 13%) for the composites prepared from BS melted glass powder were probably obtained because the hot pressing temperature (1300°C) was above the liquidus temperature for the pyrex composition (expected to be about 1077°C), so no crystals were present. Crystallization also occurred in the LAS and MAS gel matrices during hot pressing. In this case crystallization of the matrices to glass ceramics was desired but only after densification had been achieved by viscous flow of the glass. Hence crystallization of the glass occurred too early inhibiting full densification.

Typical results for the flexural tests on the carbon fibre-gel derived LAS and MAS glass ceramic matrix composites under three point bending are shown in Figs 4 and 5. The load-deflection curves showed clear evidence for tough fracture behaviour. Similar curves with tough behaviour were observed for all the LAS and MAS composites (Table 3), and for the SiC fibre-gel derived BS matrix composites and SiC fibre-gel derived SiO_2 with BS glass filler, the latter prepared by the combined sol + glass powder filler method (Table 1). Tough behaviour was also observed for the various C fibre-BS matrix composites (Tables 1 and 2) at the lower hot pressing temperatures. However, the BS gel derived composites exhibited pronounced brittle fracture for hot pressing temperatures above 900°C. This was probably the result of chemical interactions between the carbon fibres and the BS precursor solution during impregnation and firing, leading to the formation of a strong interfacial bond between fibres and matrix and consequent brittle failure. Similar effects were reported by Rice et al (1984). By contrast, the non-brittle failure of the glass slurry infiltrated composite after hot pressing at 1100°C may be attributed to weak chemical interactions between the carbon and slurry during impregnation.

The composites prepared from the BS gel + BS melted glass filler (Table 2) also showed more pronounced brittle failure than the composites produced from the SiO_2 gel + BS glass filler. The former hot pressed at 1150°C showed brittle failure whereas the latter hot pressed at 1200°C showed tough failure. In these composites a thin gel derived glass coating was deposited on the fibres during processing. The lower viscosity of the BS coating compared with the SiO_2 coating could suggest a stronger chemical interaction with the fibres for the former composites accounting for the brittle failure observed.

The BS gel derived composites with C or SiC fibres gave mean flexural strengths up to 234 and 251 MPa respectively. These results are comparable to those reported by Qi and Pantano (1987) on C-fibre paper reinforced gel derived 'Pyrex' composites of around 340 MPa, but are low compared with those reported by Phillips (1974) and Prewo (1982) on slurry infiltrated and hot pressed C fibre reinforced 'Pyrex' glass composites. Reasons for the relatively low strengths achieved in the present work include the high porosities in the composites, some fibre misalignment during processing and the chemical interactions between fibres and matrix mentioned above.

In the present work the best mean strength result of 490 MPa was obtained for the SiC fibre reinforced matrix consisting of gel derived SiO_2 with BS glass filler after hot pressing at 1270°C (Table 1). The same system hot pressed at 1200°C gave 454 MPa. These composites also exhibited tough behaviour. The best result of 377 MPa for the C fibre-glass composites was found for the matrix of gel derived SiO_2 with BS glass filler (Table 2) after hot pressing at 1200°C. It may be speculated that the SiO_2 gel coating on the fibres and filler particles is beneficial in reducing chemical interactions between fibres and matrix while helping viscous sintering during hot pressing at 1200°C and above.

The mean strengths of the C fibre reinforced MAS glass ceramic matrix composites Table 3 ranged from 210 to 261 MPa, values which are considerably higher than those reported elsewhere for similar composites prepared by sol-gel processing (see for example Partlow (1984)). Higher strengths were obtained for the LAS matrix composites, the best results being obtained for the C-LAS material hot pressed at 1360°C (469 MPa) and the SiC-LAS material hot pressed at 1330°C (402 MPa). These are encouraging results for the gel processing route but are still lower than results obtained with the glass powder slurry infiltration method (Sambell et al (1972)).

4. SUMMARY AND CONCLUSIONS

The rate of densification under hot pressing of the gel derived borosilicate (BS) powder was slower than that of the melted glass powder of similar composition. This was caused by the faster crystallization rate of the gel powder leading to a reduction in viscous sintering rate. For the LAS and MAS gel powders crystallization also occurred during hot pressing, maximum densification being achieved at around 1000°C.

Densities and flexural strengths of the sol-gel processed composites increased with hot pressing temperature. Tough fracture behaviour was observed for the LAS and MAS glass ceramic matrix composites and for SiC fibre-gel derived BS composites. However, the C fibre-gel derived BS composites showed brittle fracture probably as a result of a stronger interfacial bond between the fibres and matrix. Mean strengths of 490 MPa for the SiC fibre-gel derived SiO_2-BS glass filler composite and 469 MPa for the C fibre-gel derived LAS composite were observed.

The results obtained with gel processing are encouraging. In particular, the use of a sol-gel derived glass phase plus filler is promising as a way of promoting viscous sintering and hence lowering the processing temperature. Another useful approach is to coat the fibres with a sol in order to modify the chemical bond between fibres and matrix and improve the fracture behaviour. Promising preliminary results were obtained by Gunay (1990) by coating fibres with an alumina sol prior to infiltration.

ACKNOWLEDGEMENT

V Gunay wishes to thank the Turkish Government for financial support.

REFERENCES:

- Dislich, H (1971). Angew Chem Int Ed Engl, 10 (6), 363.
- Fitzer, E and Gadow R (1986), in "Tailoring Multiphase and Ceramic Composites", Tressler, R E, Messing, G L, Pantano, C G and Newnham, E R, Eds, pp 572, Plenum Press, NY.
- Gunay, V, James, P F, Jones, F R and Bailey, J E (1990), British Ceramic Proceedings, in press, The Institute of Ceramics.
- Gunay, V (1990), PhD thesis, University of Sheffield, in preparation.
- Hegeler, H and Bruckner, R (1989), J Mater Sci 24 1191.
- Hyde, A R (1988), The GEC Journal of Research, 6, 44.
- Jabra, R, Phalippou, J and Zarzycki, J (1980), J Non-Crystalline Solids, 42, 489.

- Lee, B I and Hench, L L (1986), in "Science of Ceramic Chemical Processing", Hench, L L and Ulrich, D R, Eds, pp 231, John Wiley & Sons, NY.

- Menessier, E, Guette, A, Pailler, R, Naslain, R, Rabardel, L, Hosten, B, Macke, T and Lespade, P, (1989). Proc of 3rd European Conf on Composite Materials, ECCM 3, Bunsell, A R, Lamicq, R and Massiah, A, Eds pp 121, Elsevier Applied Science, London.

- Partlow, D P (1988), Advanced Ceramic Materials, 3(6), 553.

- Phillips, D C (1972), J Mater Sci, 7, 1175.

- Phillips, D C (1974), J Mater Sci, 9, 1847.

- Prewo, K M (1982), J Mater Sci, 17, 3549.

- Prewo, K M, Brennan, J J and Layden, G K (1986), Am Ceram Soc Bull, 65(2), 305, 322.

- Qi, D and Pantano, C G (1987), in "Proc of the 3rd Int Conf on Ultrastructure Processing of Ceramics, Glasses and Composites", Mackenzie, J D and Ulrich, D R, Eds, pp 635, John Wiley & Sons, NY.

- Rice, R W, Spann, J R, Lewis, D and Coblenz, W, (1984), Cer Eng Sci Proc 5 614.

- Roy, R (1956), J Amer Cer Soc, 39(4), 145.

- Sambell, R A J, Briggs, A, Phillips, D C and Bowen, D H (1972), J Mater Sci, 7, 676.

Inst. Phys. Conf. Ser. No 111
Paper presented at Int. Conf. on New Materials and their Applications, Univ. of Warwick, 1990

The role of the interface on the densification of sol-gel processed alumina and mullite fibre composites

M Chen, F R Jones, P F James and J E Bailey

School of Materials, University of Sheffield

ABSTRACT: SiC and C fibre reinforced alumina and mullite composites with good fibre dispersion have been prepared by a single stage infiltration using a filler modified sol-gel process. The density and Vicker's hardness of the composites increased with sintering temperature but the densification was inhibited by the fibres. This led to microcracking of the sintered composites. Hot-pressing significantly improved the densification of the composites, achieving 93% of theoretical density for the carbon fibre/mullite composite. The fibre-matrix interfacial bond strength for the composites was determined using the indentation technique. The interfacial shear strength τ compared well with predictions from the microcrack density. The micromechanics of failure of the composites in bend follow the trend in bond strength.

1. INTRODUCTION

Sol-gel processing of ceramics is an advantageous route to fibre composites, because of the potentially low sintering temperature, matrix microstructural uniformity and ease of fibre dispersion. The main disadvantages of low solid yield and high processing shrinkage have been overcome for alumina and mullite by incorporating fillers [Chen et al, 1987, 1988, 1989]. Recent studies have shown that high density mullite ceramics could be obtained because viscous sintering can occur in the mullite gel [Aksay et al, 1987, Yoldas, 1987, Chen et al, 1989]. In this paper the effect of hot-pressing on the matrix densification, microcrack formation and the interfacial shear strength of SiC and C fibre composites fabricated by a modified sol-gel process.

3. EXPERIMENTAL PROCEDURE

Full details for composite fabrication have been reported elsewhere [Chen et al, 1988, 1989]. Essentially, Boehmite gel powder, containing 0.2% TiO_2 was used as the precursor for the alumina sol. A Ludox colloidal silica sol (AS40 grade) was used as the silica precursor and 0.5 μm (Alcoa) α-alumina particles were used as the filler. The alumina filler was deflocculated and dispersed in aqueous nitric acid solution at a pH of 3.5. The Boehmite powder was then peptized in the above dispersion and the pH adjusted to 3.5 with HNO_3. The ratio of the Boehmite to filler was chosen so that 80% of the oxide alumina resulted from the filler and the other 20% from the Boehmite. The oxide yield of the mixture was controlled at 27 vol%. This is referred to as the alumina sol. To prepare the mullite sol, an appropriate amount of silica in the form of a sol was added and the pH adjusted with 2 M HNO_3 to 2.5. In this way a fluid "mullite" sol could be obtained with a controlled solid yield of 27 vol%.

Unsized, surface treated high strength carbon fibres (Grafil LXA) and Nicalon SiC fibres (ceramic grade) were used as the reinforcements.

The SiC fibres were fired to 600°C for 5 min to remove the sizing. Then the fibres were aligned in a rectangular mould and infiltrated with the alumina sol or mullite sol to produce alumina (SiC/A) and mullite (SiC/M) composites respectively.

In the fabrication of carbon/alumina (C/A) composites the C fibres were first coated with a thin layer of the Boehmite sol and fired to 500°C for 10 min. The fibre preform was then infiltrated with the alumina sol. A similar technique was adopted for carbon fibre reinforced mullite (C/M) composite, except that a thin coating of Pyrex glass was applied rather than the Boehmite coating. For the hot-pressed C/M composite, the preform was directly prepared by the filament winding technique without precoating the fibres. The fibre/sol mixture gelled in about 20 min. The composite preforms were dried at 50°C for 20 hr and heated up slowly in N_2 to 500°C and then at 5 K min^{-1} to 1000°C and at 10 K min^{-1} to the sintering temperature and held for 2 hrs and finally allowed to cool overnight. Composite samples were also uniaxially hot pressed at 1200-1300°C for 1 hr, using a pressure of 20 MPa.

The indentation technique [Marshall, 1984, Mandell, et al, 1986] was used to estimate the interfacial shear strength of the composites, using the M-400 Hardness Tester (LECO) with the load ranging from 10 to 300 g, and an indentation time of 10 sec. Microhardness of the matrices and fibres was also determined. The flexural strengths of the composites were determined under three point bend, using a span of 40 mm and a ratio of span to thickness from 10 to 20.

3. RESULTS AND DISCUSSIONS

3.1 Sintering behaviour

As shown previously [Chen et al 1989], C and SiC fibre reinforced alumina and mullite composites could be readily prepared by a single stage infiltration using a modified sol-gel technique. After sintering to 1300°C, partially densified composites with good fibre dispersion were obtained. The green densities of the composites are given in Table 1. As shown, all the composites have a density which is approximately 70% of the theoretical density (TD) From the known volume fraction of fibres (V_f) at full density, the relative density of the matrices could be calculated. The density of the matrices were observed to be ≈55% TD which is significantly lower than the corresponding monolithic samples, because the drying shrinkage of the matrix gel is inhibited by the fibres. This is confirmed in Fig 1 where it is shown that the pore size of the SiC/A composite matrix dried at 500°C was larger than that of the corresponding monolithic sample.

The densities of the sintered and uniaxially hot-pressed composites are listed in Tables 2 and 3 respectively. From Table 2, it can be seen that for the former, only a small increase in density occurs, reaching 75% TD for the SiC/M composite. Simple calculations indicate that it is still much lower than the corresponding monolith (ie 61 vs 69% for alumina and 65 vs 82% for mullite matrix). This demonstrated that similarly to the shrinkage on drying, matrix sintering was also inhibited by the fibres. Shrinkage occurred isotropically for both monolithic alumina and mullite during sintering. However, for the

Table 1 Green densities of gels and fibre composites dried at 500°C

	A	M	SiC/M	SiC/A	C/M	C/A
V_f'	0	0	0.4	0.4	0.37	0.52
V_f	/	/	0.29	0.29	0.29	0.37
ρ_c	/	/	0.68	0.67	0.69	0.71
ρ_m	0.58	0.59	0.56	0.55	0.58	0.54

V_f', V_f are theoretical and true volume fractions of fibres respectively, ρ_c, ρ_m are relative densities of the composite and matrix or monoliths respectively.
A: Alumina, M: Mullite, SiC: SiC fibres, C: Carbon fibres

Table 2 Relative densities of composite sintered at 1100-1200°C for 2hrs

T°C	1200	1200	1300	1100	1200	1200	1200	1200
	A	M	M	SiC/M	SiC/A	SiC/M	C/A	C/M
V_f'	0	0	0	0.4	0.4	0.4	0.57	0.47
V_f	0	0	0	0.3	0.29	0.3	0.44	0.35
ρ_c	/	/	-	0.73	0.71	0.76	0.76	0.73
ρ_m	0.69	0.82	0.92	0.62	0.61	0.65	0.58	0.60

Table 3 Relative densities of composites hot-pressed (HP) at 1100-1300°C

T°C	1250	1300	1100	1250	1250	1300
	M	M	SiC/M	SiC/M	C/M	C/M
V_f'	0	0	0.4	0.5	0.53	0.53
V_f	/	/	0.3	0.46	0.46	0.49
ρ_c	/	/	0.76	0.92	0.86	0.93
ρ_m	0.87	0.97	0.66	0.85	0.74	0.86

composites, the shrinkage of the matrix in the fibre direction was inhibited. This can be understood by the formation of a good fibre-matrix interfacial bond and the generation of tensile stresses in the matrix with the subsequent formation of microcracks which crack-open during sintering [Chen et al, 1989, 1989]. The sintering shrinkage which contributes to the densification, is therefore occurring in the direction perpendicular to the fibre alignment. A variation in interfacial shear strength, as shown later, can account for the differing densities obtained for C and SiC fibre composites. Increasing the fibre volume fraction enhances the degree of under sintering because a larger number of shrinkage elements are formed.

Hot pressing has been used with the intention of suppressing sintering induced microcracking and thereby enhance the densification of the composites. In Table 2 it can be seen that the mullite matrix sintered more efficiently than the alumina gel, so that lower sintering temperatures could be used [Chen et al, 1989]. This has been attributed to the presence of the amorphous silica which shows viscous deformation [Aksay, 1986, Chen et al, 1989, Yoldas, 1987]. This study has, therefore, concentrated on mullite composites. Hot-pressing under 20 MPa at 1300°C slightly enhanced the densification of the monolithic mullite from 92 to 97% (density of mullite 3.13 g/cm^3) [Chen et al, 1990]. The sintering shrinkage was mainly confined to the pressure (or through-thickness) direction, and found to be 36.4%. Across the width and perpendicular to the fibres, it was 2.6%. This is confirmed by comparing Tables 2 and 3 where it is shown that the density of the mullite composites is significantly enhanced. For example, at 1250°C the SiC/M composites achieved a relative density of 92% which corresponds to a matrix density of 85%. The higher density resulted from the much reduced microcrack opening. The density of microcracks was not affected as seen in Fig 5.

3.2 Microhardness of the matrices

The microhardness of the matrices and monoliths determined from a M-40 Hardness Tester (LECO), using a load of 100g are shown in Table 4. The Vicker's hardness of the pressureless sintered alumina matrix fired to 1200°C was 3-4 GPa which gradually increased to 5 GPa at 1300°C and to 7 GPa at 1400°C. These were slightly lower than those for the corresponding monoliths. For the composites, the Vicker's hardness of the mullite matrix fired to 1200°C was at 5.6 GPa, slightly higher than that of the corresponding alumina matrix. As expected after hot-pressing, the Vicker's hardness increased significantly to 11.8 GPa for the SiC/M at 1250° and C/M at 1300°C. This correlates well with the improvement in microporosity which is indicated by the density measurements and demonstrates the benefits of hot-pressing.

3.3 Fibre/matrix interfacial bond strength measurements

The indentation method [Marshall, 1984 and Mandell, 1986] enables the interfacial bond strength to be calculated from the applied force on a microhardness indentor to push the end of an embedded fibre into the composite. The schematic of the loading arrangement is shown in Fig 5. The interfacial shear stress, τ can be calculated from the following equation.

$$\tau = \frac{F^2}{4\pi^2 . R^3 . U . E_f} \qquad ......(1)$$

Table 4 Microhardness of alumina and mullite monoliths and composite matrices, sintered for 2hrs or hot-pressed for 1hr at different temperatures. Testing load 100g. The Vicker's hardness is given in GPa.

Sinter Temp T°C	PRESSURELESS Alumina	SiC/A	C/A	C/M	HOT PRESSED C/M	SiC/M
1200	4.17	3.28	3.92	5.6	/	/
1250	/	/	/	/	/	11.8
1300	6.07	4.78	4.98	/	11.8	/
V_f	0	0.29	0.42	0.21	0.49	0.46

Table 5 The interfacial bond strengths τ obtained, of composites from the indentation technique (eqn 1) and the estimated average crack spacing, (eqn 2). X_e is the observed spacing.

System	C/A	SiC/A	C/M	C/M	SiC/M	SiC/M
V_f	0.44	0.29	0.47	0.35	0.46	0.3
T°C	1200	1200	1200	1200	1250HP	1200
τ/MPa	<8.8±0.7	38.1±10	110±9	/	126±34	/
X/μm	/	35.4	/	8	/	19
X_e/μm	100	42	/	16	/	20

d_f(SiC)=15μm, d_f(C)=7μm, σ_{mu}(Al_2O_3)=109MPa, σ_{mu}(Mullite)=200 MPa

Table 6 Flexural strengths and moduli of unidirectional fibre reinforced sol-gel derived

T°C	1200	1200	1200	1200	1250HP	1300HP
System	C/A	SiC/A	C/M	SiC/M	SiC/M	C/M
V_f	0.44	0.29	0.21	0.31	0.46	0.49
E_c/GPa	96	66	30	80	79	72
E_c'/GPa	96	52	47	56	83	110
σ_{cu}/MPa	368	330	300	84	54	96
σ_{cu}'/MPa	1760	290	840	310	/	/

E_c σ_{cu} are experimental quantities; $E_c'\sigma_{cu}'$ are law of mixtures estimates assuming no matrix contribution. HP is hot pressing at 20 MPa where F is the load applied to the fibre end of radius R and elastic modulus, E_f. The displacement of the indented fibre relative to the adjacent matrix, U is given by (b-a)cotΦ, where 2b and 2a are the diagonal length of the indentation on the matrix and the fibre, and 2Φ=136° is the angle between the opposite edges of the indentor. Fig 6 shows the typical hardness impression on the fibres in the polished section of differing composites. The calculated values of τ are given in Table 5.

Generally, the interfacial bond is higher for the mullite composites. For the C/A composite, with a 10g load an indentation equivalent to 2b=14μm was made. The true load, F applied on the C fibres is difficult to estimate accurately, because the interface was too weak to allow indentation of the fibre. A maximum value of τ could only be obtained. Taking E_f=225GPa, τ is estimated to be less than 9MPa. For the other composites, 2b and 2a can be readily measured and the true load obtained from the relationship $F=4(HV)\ a^2/1854$ where (HV) is the Vicker's hardness of the fibres, which was measured to be 15.33 GPa and 9.01 GPa for the SiC and C fibres respectively. The maximum fibre displacement was found to be 4μm which means that the debonded length is small in comparison to the specimen thickness of 1cm, confirming the validity of the technique. The values of τ given in Table 5 are in good agreement with those reported by [Jamet, 1986, Chen et al, 1989, 1989, Singh 1988]. The interfacial shear strength of the C/A composite has a value typical of a purely frictional bond. This is an expected result since the reaction of C with alumina is unlikely at the sintering temperature of 1200°C. On the other hand, for SiC/A, the silica rich layer on SiC fibres [Clark et al, 1985] may react with alumina to form mullite at 1200°C, creating a chemical bond at the interface and an increased value of τ. Similarly, for SiC/M composite, the good chemical bond can be understood in terms of the amorphous silica fusing with the silica-rich surface of the SiC fibres. The stronger bond between C and mullite, which is in contrast to C/A composite, probably arises from a reaction between the fibre surface and the viscous amorphous silica to form SiC at the interface during sintering (Singh, 1987, Chen et al, 1989]. In a previous paper [Chen, 1989] we were able to estimate the value of τ from the crack density in a SiC/M composite at 130 MPa. This compares favourably with the value obtained by indentation for a hot pressed sample of 126 MPa, demonstrating the expected result that hot pressing does not modify the interfacial chemistry.

The different interfacial shear strengths give rise to the fracture surfaces shown in Fig 4. The degree of pull-out follows the trend in τ given in Table 5.

3.4 Microstructure of the composites

As shown in Fig 5 the matrices were microcracked perpendicularly to the fibres. These cracks were shown previously to have been formed during sintering and thermally as a result of the mismatch in thermal expansivities of the fibres and matrix [Chen et al, 1989, 1989]. For SiC/M, differentiation between the two is not possible but the crack opening suggested that as with C/M the majority form during sintering allowing the matrix fragments to subsequently shrink independently during continued sintering and cooling. By comparing a with b and c with d in Fig 5, it is demonstrated that hot-pressing constrains the mechanism responsible for crack opening. It is especially noticeable for the

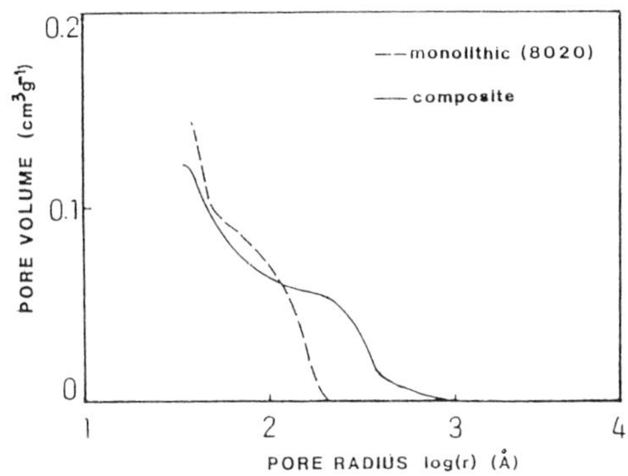

Fig 1. The pore size distribution of the C-fibre reinforced composite (V_f=0.42), in comparison with the monolithic Mullite gel, both dried at 500°C.

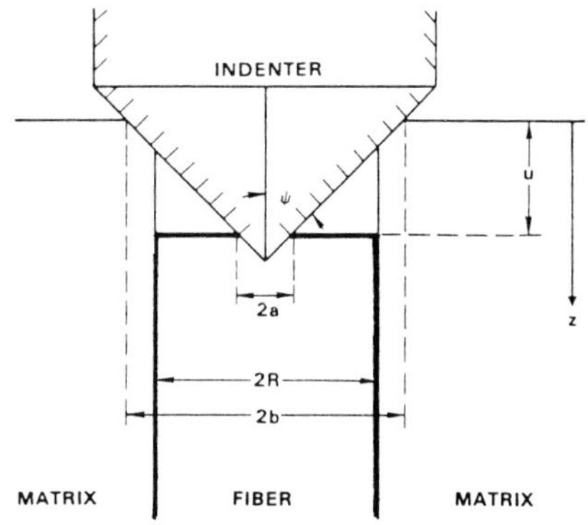

Fig 2. Schematic of the indentation method for measurement of matrix/ fibre bond stress.

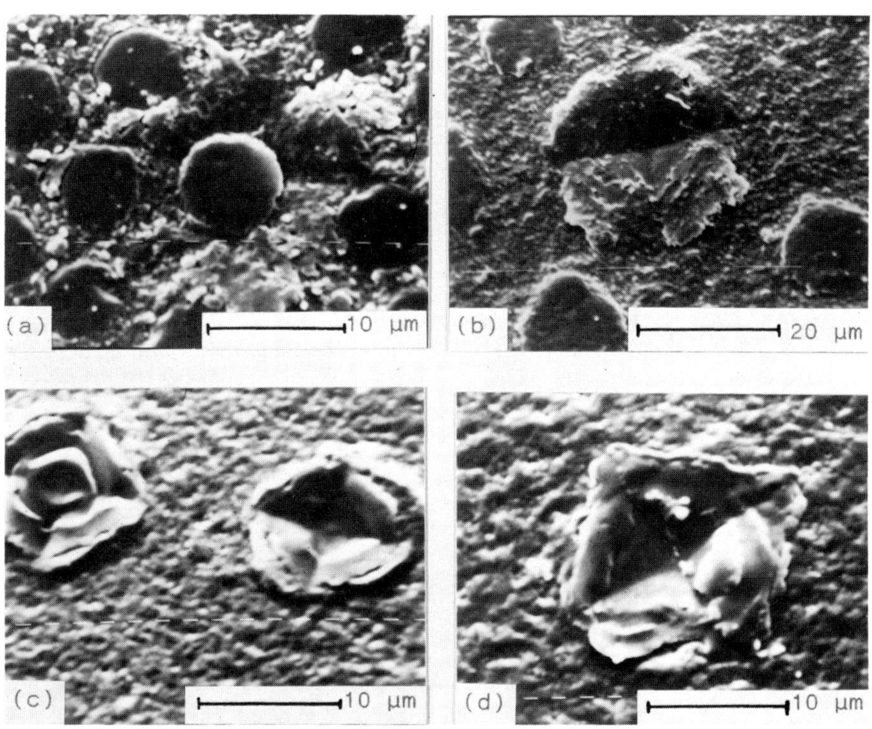

Fig 3. SEM micrograph of fibre in composites after indentation with the pyramid of the M-40 microhardness tester, a) C/A, b) SiC/A, c) C/M, d) SiC/M.

mullite composites in which sintering is more efficient. However, microcracking is not prevented. Furthermore, it is possible that the sintering induced microcracks may have been suppressed under hot-pressing but that thermally induced ones form on cooling. Further study is needed to differentiate between the two mechanisms.

The average crack spacing for the C/A (1200°C), C/M (1200°C) and SiC/M (1200°C) unidirectional composites are 100, 42, 16 and 20 μm respectively (Table 5). The varying crack spacing for each composite is a reflection of the different interfacial bond strengths. It is possible to examine this hypothesis, to a first approximation, by predicting the crack spacing from the measured values of τ. The Cooper and Sillwood equation [Cooper and Sillwood, 1972], relates the minimum crack spacing, x' to the interfacial shear strength τ;

$$x' = V_m \sigma_{mu} R/(2V_f \tau) \qquad ......(2)$$

where σ_{mu} is the ultimate strength of matrix, R is the radius of the fibre. The minimum crack spacing has been shown by Kimber and Keer [Kimber, 1982] to be statistically related to the measured average crack spacing x;

$$x = 1.34x' \qquad ......(3)$$

The average crack spacing will be reduced with increasing τ but is counteracted as σ_{mu} increases.

Taking the value of R as 7.5 μm and 3.5 μm for SiC and C respectively, the average crack spacing x, has been calculated from the value of τ obtained from indentation measurements. The results are given in Table 5. The close agreement between the predicted and experimental crack spacings validates the argument that the microcracking occurs as a direct result of the interfacial reactions which occur during sintering.

3.5 Mechanical properties

The micromechanics of the composites under three point bend loading are illustrated in Fig 6 by the load-deflection curves. The mechanical properties calculated from these tests are summarized in Table 6. The load-deflection curves in Fig 6 describe the differing fracture behaviour. The rule of mixtures has been used to predict the theoretical moduli of these composites with a microcracked matrix which is considered to be non-load bearing. The initial flexural moduli (E_c') agree well with the law of mixtures Young's modulus of the composite (E_c') demonstrating a varying degree of matrix contribution to the composite stiffness. In the case of SiC, where a good interfacial bond with mullite is achieved, the low value for E_c must reflect fibre weakening during fabrication at 1300°C. The formation of Silicon Carbides as a result of carbon fibre interaction with mullite may also have resulted in a reduction in fibre modulus.

The strength predictions from the simple law of mixtures are, as expected, poor. However, the trends are consistent with the nature of the interfacial fibre-matrix bond. For example, for C/A the low interfacial shear strength leads to shear failure and fibre pull-out which gives rise to a tough fracture but the ultimate strength is low. For C/M, the stronger bond coupled to a degradation in fibre strength is responsible for the brittle failure illustrated in Figs 4 and 6.

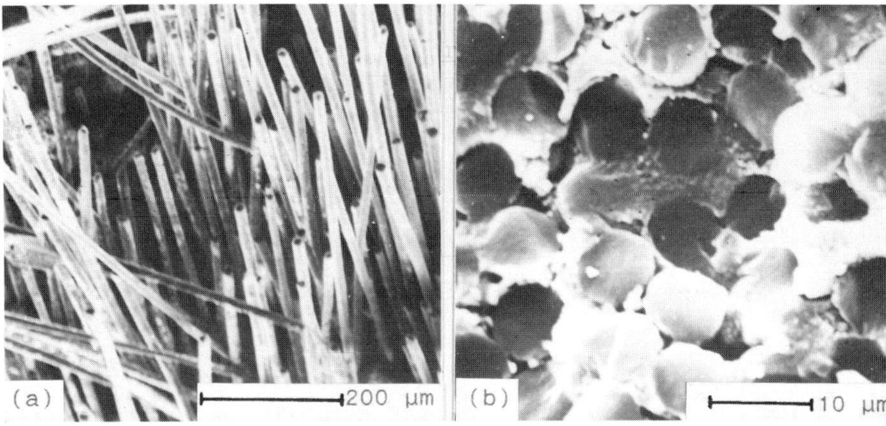

Fig 4. SEM micrograph of composite fracture surfaces, a) SiC/A (1200°C) showing fibre pullout typical of a weak fibre matrix interface, b) C/M (1300°C hot-pressed) showing brittle failure and the absence of fibre pull-out typical of a strong interfacial bond.

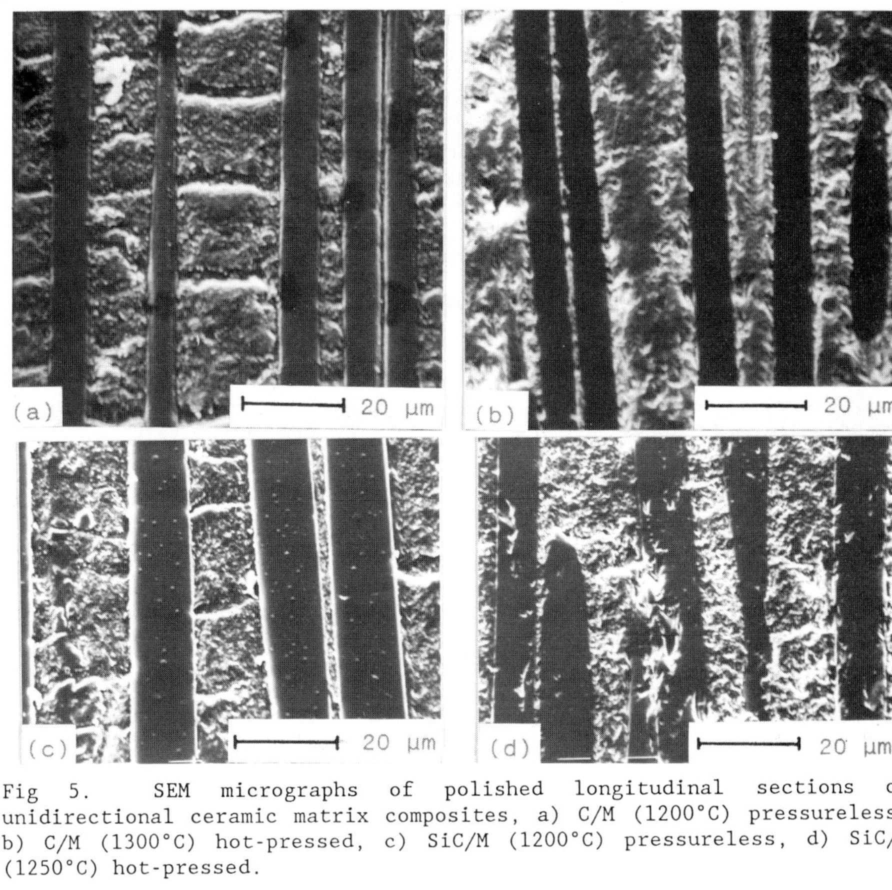

Fig 5. SEM micrographs of polished longitudinal sections of unidirectional ceramic matrix composites, a) C/M (1200°C) pressureless, b) C/M (1300°C) hot-pressed, c) SiC/M (1200°C) pressureless, d) SiC/M (1250°C) hot-pressed.

For SiC/M composites, the strong interfacial bond and the high degree of matrix microcracking ensures that brittle failure occurs. The measured value of composite strength is therefore a function of the statistics of the retained strength of SiC fibres and the presence of strength reducing flaws. The differing strength of the pressureless and hot pressed SiC/M composites at 1200/1250°C may arise from the higher stress intensity associated with the 'fine' microcracks of the latter in comparison to the 'wide' microcracks of the former. This observation also appears to apply to the hot-pressed C/M composites.

In conclusion, the relationship between interface formation and composite performance has been illustrated and the needs for control indicated.

4. CONCLUSIONS

It has been demonstrated that fibre composites can be prepared via a modified sol-gel process with single stage infiltration. Sintering of the matrix within the composites was found to be inhibited by the fibres leading to the formation of microcracks. From the density microhardness and shrinkage measurements, it has been shown that the densification process in the mullite composites is significantly improved by hot-pressing. A density of 92-93% TD was achieved for the SiC/M and C/M composites at 1250°C and 1300°C respectively. This is attributed to viscous sintering in the presence of the amorphous silica in the matrix and a reduction in the crack-opening of the sintering and/or thermally induced microcracks. The interfacial bond strength of the composites have been measured using the intentation technique. The interface in the C/A composite essentially results from friction whereas for mullite, chemical interaction with the C or SiC fibres leads to large interfacial shear strengths, but brittle composites. The density of microcrack formation has been shown to correlate closely with the development of the interfacial fibre-matrix bond.

REFERENCES

Aksay I A, Shin W Y and Sarikaya M, in "Ultrastructure Processing of Ceramics, Glasses and Composites" (J D Mackenzie and D R Ulrich, eds) John Wiley & Son, New York, 1987, pp 393-405.

Chen M, James P F, Jones F R and Bailey J E (1989), in "Developments in the Science and Technology of Composite Materials', ECCM 3, (A R Bunsell, P Lamicq, A Massiah, eds), Elsevier Applied Science, London, 1989, pp 87-92.

Chen M, James P F, Jones F R and Bailey J E (1989), in "Annual Convention 1989, University College of Swansea", The Institute of Ceramics, 5th-7th April 1989.

Chen M, James P F, Jones F R and Bailey J E (1990), unpublished work.

Clark T J, Arons R M and Stamtoff J B (1985), Ceram Eng & Sci Proc 6 576.

Cooper G A and Sillwood J M, J Mat Sci 7 (1972), 325-33.

Coyle T W, Guyot M H and Jamet J F (1986), Ceram Eng & Sci Proc 7 947.

Kimber A C and Keer J G, J Mat Sci Letters (1982), 1, 353.

Mandell J F, et al (1986), in "Composite Materials-Testing and Design", STP 893, ASTM, J M Whitney, ed. pp 87-108.

Marshall D B, (1984), J Am Ceram Soc 67, C-259.

Singh R N and Brun M K, (1987), Ceram Eng & Sci Proc 8, 636.

Yoldas B E, (1987), see Aksay et al (1987) pp 333-45.

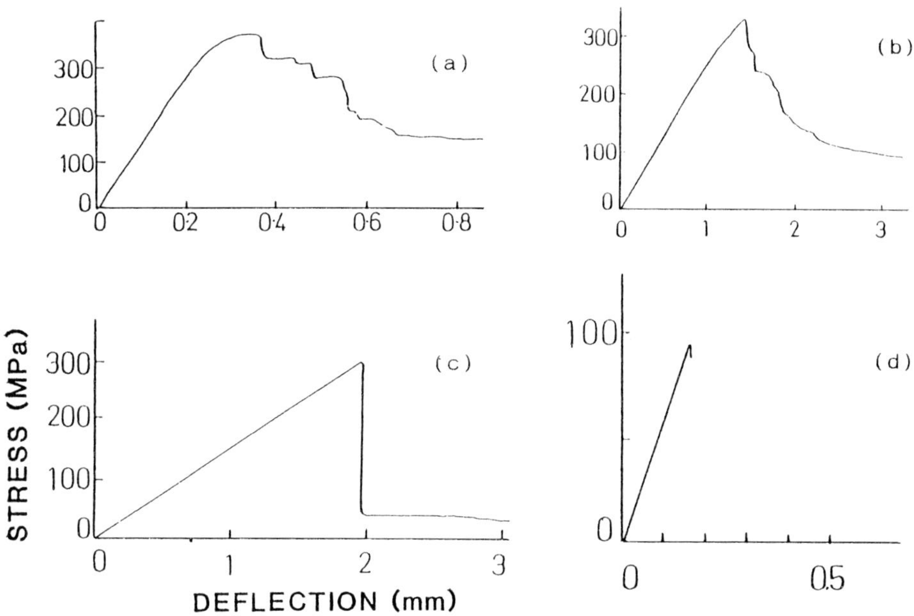

Fig 6. Typical fracture behaviour of the composites sintered at 1200°C, under three point bending loads, a) C/A, (V_f=0.42), b) SiC/A, (V_f=0.29, c) C/M, (V_f=0.21) and d) SiC/M, (V_f=0.29).

Paper presented at Int. Conf. on New Materials and their Applications, Univ. of Warwick, 1990

The preparation of oxidation resistant silicon carbide–silicon carbide composites by chemical vapour infiltration (CVI)

D Bashford

Divisional Manager, Fulmer Research Limited, Stoke Poges, Slough, Berkshire, SL2 4QD

ABSTRACT: There is an increasing need for ceramic matrix composites which can provide useful mechanical properties and have an appreciable life expectancy under harsh thermocyclic and oxidative conditions. Continuous fibre reinforced ceramics need to be prepared such that they possess benign fracture characteristics to give a good balance between strength and fracture toughness. CVI is particularly suitable for preparing SiC-SiC composite shapes provided specific attention is given to: preform construction; fibre to matrix interfaces; infiltration techniques and overall composite microstructure.

1. INTRODUCTION

Ceramic Matrix Composites (CMC's) have a significant role to play in a number of long term applications, notably gas turbine engines, spaceplanes, space propulsion systems and heat exchangers[1]. The intended uses expose these materials to temperatures beyond 1300°C in oxidative environments where thermal cycling and thermal shock will be prevalent. As confidence in CMC's grows, then the expectations on temperature capability and working lives will increase correspondingly. The incentive to develop CMC's stems from the success of carbon-carbon composites as high temperature engineering materials and the need for inherent oxidation resistance. Carbon-carbon composites remain the present choice for components experiencing temperatures of 1600°C - 2200°C and if oxidising environments are encountered then a myriad of protection systems have evolved to extend life expectancy. CMC's with inherent oxidation resistance are in their infancy but sufficient progress has been made with silicon carbide systems to promote development. In the first instance stability to 1600°C is sought for SiC-SiC before venturing to systems including alumina, zirconia, hafnium compounds, high temperature borides and oxides.

All matrices for CMC's possess extremely high melting (sublimation or dissociation) points which hinder processing at realistic temperatures. By using Chemical Vapour Deposition (CVD) or the sister technology of vapour infiltration (CVI) it is possible to form ceramic matrices from gaseous phases at modest temperatures, circa 1000 - 1200°C. CVD/CVI is also a low pressure process (<1 atmosphere) which allows the densification of fibre preforms to a near net shape component. The versatility of CVD/CVI will allow the preparation of composites with controlled composition. This will give longevity at high temperature in avoiding chemical incompatibility and diffusion reactions between fibre and matrix whilst providing desirable fracture characteristics.

Fulmer Research Limited has a long history in CVD and now possesses a CVI facility capable of continuous, uninterrupted operation. Work is presently aimed at i) developing oxidation protection systems for carbon-carbon composites based on SiC infiltrations, diffusion and overlay coatings, and ii) the formation of SiC-SiC composites for use in space bi-propellant thrusters. Such thrusters have in the first instance to sustain repeated thermal cycling to 1425°C under oxidising conditions.

In developing CMC's it is important to bear in mind the intended application; the life expectancy sought and the residual mechanical properties necessary to sustain the design. In many instances the materials will perform a "thermostructural" function, i.e. thermal integrity will be required coupled with modest mechanical properties. Maintaining the integrity will be the key issue as for example: 2000 thermal cycles (life 5 hours) at 1500°C in an oxidising environment is considerably more harsh than five (5) hours isothermal at 1500°C in inert gas.

2. CHARACTERISTICS OF SIC/SIC COMPOSITES

For CMC's to be viable engineering materials they have to possess a degree of toughness and avoid the brittle catastrophic mode of failure shown by monolithic ceramics. With continuous fibre reinforcement, composites can exhibit benign fracture characteristics and an element of "pseudo ductility". The desired fracture phenomenon relies on modest fibre to matrix bonding and the ability to sustain matrix microcracking under load with fibre pullout at ultimate failure. Under increasing load, the mode of failure is one of progressive microcracking of the matrix as it is a low strain to failure material. The ultimate strength of the composite essentially being the in-situ strength of the fibres. From this scenario special attention must be paid to a number of parameters in preparing stable high temperature composites.

i) Select fibres which are thermally stable beyond 1300°C and ideally have a stoichiometric composition.
ii) Use fibres which can be woven or braided into cloths (fabrics) and multidirectional preforms to allow the preparation of net-shape components.
iii) Create an interface between fibre and matrix, usually as a fibre coating, which acts as a modest shear strength layer to ensure fibre/matrix decoupling under load.
iv) Create a matrix phase which is coherent in providing shear and compressive support within the composite. Also, to achieve control of the matrix chemistry, its porosity and microcracking behaviour.

It is simple to describe the ideal composite, but far more difficult in practice due to the chemical instabilities and diffusion mechanisms which are promoted by high temperatures beyond 1300°C. It is now appreciated that the interface layer between fibre and matrix has a crucial role to play in guaranteeing toughness and determining the working life of the composite.

For SiC-SiC composites the choice of commercial fibres is limited essentially to Nicalon (Si-C-O), Tyranno (Si-C-Ti-O) and AVCO (CVD-SiC). The latter possesses good thermal stability but due to its large fibre diameter is intractable and cannot be woven. Nicalon is the most widely used as compared to the more recent Tyranno. From the point of view of thermal stability neither is ideal because they contain additional elements to Si and C. Whilst they are fibres of good tensile strength, both exhibit progressive degradation and loss of strength at temperatures in excess of 1200°C. For use above 1200°C, composites have to rely on residual strength provided by the fibres.

The interface between SiC fibre and SiC matrix needs to be refractory in nature and stable at the intended temperatures of operation. Also it should be chemically compatible with fibre and matrix. The modest shear strength characteristics are best provided by a 'soft' ceramic of which pyrolytic carbon and boron nitride deposited by CVD have become the natural choices. These fibre coatings are effective in thicknesses of less than one micron. The carbon can be deposited at 1000°C, whilst boron nitride requires 1300 to 1900°C which suggests opportunities for fibre degradation.

An additional feature of CMC's prepared by CVI is the residual porosity present which can be 10-15%. This results from sealed off cavities within the fibre tows and between fabric plies where the unfiltrating gases have been unable to deposit and complete the densification. In fibre reinforced plastics such porosity would be deemed unacceptable but in CMC's it can be tolerated. This is primarily due to the modest mechanical properties sought from such composites and the conjected help porosity may give to "fracture toughness" in aiding crack stopping and deflection. Clearly, excessive porosity cannot be tolerated as the composite loses integrity not least in interlaminar shear strength.

3. PROCESSES AVAILABLE FOR SIC/SIC COMPOSITES

There are two principle means by which a SiC matrix can be formed with fabrics or preforms of SiC fibres:

i) CVI/CVD.
ii) Polymer pyrolysis, e.g. converting polycarbosilane to SiC.

Both are low pressure infiltration techniques having their own advantages and disadvantages, which are summarised in the following table.

Process	Advantages	Disadvantages
CVI/CVD	* Even, controlled deposition giving a coherent matrix. * Fibre coatings and matrix built up in single operation. Modest reaction temperatures. * Dopants can be added to matrix.	* Long process times (30-150 hours). * Expensive capital equipment.
Polymer Pyrolysis	* Useful to bind fibre preforms. * Cheaper process than CVI. * Modest process temperatures.	* Repetitive infiltrations and pyrolysis to achieve densification. * Long process times with slow heating rates. * Considerable shrinkage on pyrolysis gives matrix cracking.

Either process will give SiC-SiC composites with theoretical densities around 85-90% and the remainder porosity. Such composites will offer good thermal stability is essentially being made of silicon carbide. In terms of engineering uses where the retention of reasonable mechanical properties is sought for prolonged periods in oxidising environments then the CVI composite is preferable. This mainly results from the even, controlled deposition in a single process giving a coherent matrix. This provides the composite with integrity that is less achievable from repeated polymer pyrolysis densification steps. It is felt that polymer pyrolysis may be useful in an initial processing step to provide a bound, handleable, preform which can be subsequently densified by CVI. Fulmer are experimenting with both direct infiltration into a preform held within a graphite mould and open diffusive densification of a bound preform. Figure 1 gives an example of a SiC-SiC composite prepared at Fulmer.

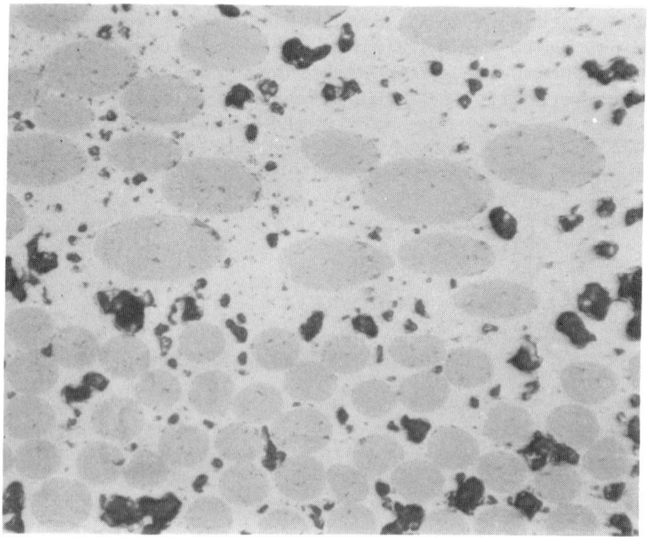

Figure 1. SiC-SiC Composite prepared by CVI.

Chemical Vapour Deposition (CVD) relies on a chemical reaction being made within gaseous phases such that material deposits on a hot surface. With CVI the gaseous mixture is passed through fibre preforms, where the fibres are 12-15 micron in the case of SiC, and very slow deposition conditions have to be used to control the densification. As the SiC, Nicalon and Tyranno fibres currently available begin to degrade at about 1200°C the reaction temperatures need to be lower than this. The principle reactions used in preparing SiC-SiC composites are:

i) $CH_3SiCl_3 + xH_2 \rightarrow SiC + 3HCl + xH_2$

Methyltrichlorosilane (MTS) at 1000-1200°C and 5 Torr.

In this reaction the MTS is dissociated to SiC with hydrogen acting as a carrier gas.

ii) $HSiCl_3 + CH_4 + xH_2 \rightarrow SiC + 3HCl + H_2 + xH_2$

Trichlorosilane at 1000-1100°C and 5 Torr.

With trichlorosilane, a reaction is induced with methane to give SiC and hydrogen again acting as a carrier gas.

iii) For preparing the carboneous fibre coating methane is cracked at 1000°C.

$CH_4 \rightarrow C + 2H_2$

The actual infiltration and densification can be controlled in a number of ways which is determined experimentally for each component.

a) the gas mixture is allowed to diffuse through the fibre preform by placing it in a stream of reactive gas,
b) the gas mixture is forced through the preform. This gives higher deposition efficiencies than a).
c) Pressure pulsing is employed to promote higher rate of deposition.
d) Differential temperatures (gradient) are established within the fibre preform to allow infiltration by back filling[2]. This is only feasible with thicker composite sections of around 10-15 mm.

Figure 2 provides a diagramatic representation of the CVI facility at Fulmer which at present can handle composites 150 mm x 100 mm. Using furnace designs based on either inductive or resistive heating, components several metres in dimension can be infiltrated and the process is by no means limited to carbon and silicon carbide. In the future, uses will be found within high temperature composites for: BN, Al_2O_3, AlN, TiB_2, TiC and HFC, all of which can be handled by CVD/CVI.

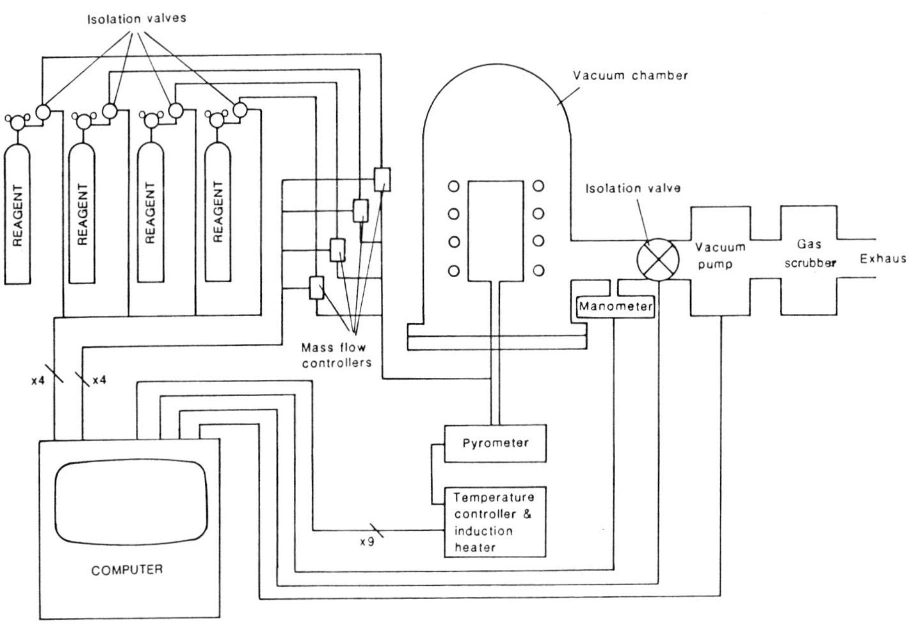

Figure 2. Diagrammatic Representation of CVI Facility at Fulmer Research Limited.

4. APPLICATIONS AND THE IMPLICATIONS FOR THE PROCESSING ROUTE

For ceramic matrix composites particular attention has to be placed on a systems approach to Materials Selection - Design - Manufacturing. The three are inexorably linked in that CVI is effectively a net shape processing route where the full composite is created simultaneously with the component itself. As silicon carbide is a hard ceramic, it is imperative that little or no machining is necessary. Indeed any machining would be viewed as distinctly deleterious in view of the limited toughness of the composite.

To indicate the types of component currently envisaged for SiC-SiC composites the following are highlighted:

20N Bipropellant Thruster for Spacecrat Attitude Control

Such a thruster would be around 115 mm in length with a combustion chamber, throat (4 mm i.d.) and exhaust cone 50 mm in diameter. This represents a fairly small intricate component. The thruster would operate at 1400-1700°C running on an oxidising combustion mixture for short but repetitive operations. The thruster would experience severe thermal shock on start up and impulse times are measured in seconds. At 1400°C a life expectancy of 5000-10000 cycles would be sought with a total life of 20 hours at temperature or 10 years operation. At 1700°C the life expectancy would diminish and a total life around 5 hours would be more realistic.

Thermal Protection Shingles for Hermes Spaceplane

Spaceplanes are reusable and have to withstand repeated re-entry into the earth's atmosphere. The nose-cone, wing leading edges and most other aerodynamic surfaces experience heating on re-entry due to high velocity gas flows(3). These temperatures are in the range 700-1700°C depending on location. For many of the surfaces it is appropriate to use SiC-SiC shingles particularly in areas experiencing temperatures to approx. 1300°C. The SiC-SiC will be exposed to air at these temperatures, which will be dissociated to atomic oxygen and nitrogen. The function of the shingles is to offer thermal protection to the underlying aluminium alloy substructure with a layer of fibrous insulation sandwiched between. It is intended for the spaceplane to operate for thirty (30) flights and the shingles have to be proven for 100 operations (cycles) to ensure confidence and provide a necessary safety margin. The total time at maximum temperature will be in the order of 30 hours.

Exhaust Petals for Gas Turbine Engines

The requirements for gas turbine engines are in many respects more exacting than thruster or thermal protection uses because the working life is intended to be longer between replacements. High reliability is required coupled with the maximum service interval feasible. Typically, life expectancy of several hundred hours would be sought for continued operational use at 800°C with thermal excursions to 1300°C and repetitive operation.

5. CONCLUSIONS

* Silicon carbide fibre reinforced silicon carbide (SiC-SiC) is currently the best inherently oxidation resistant CMC.

* Such composites prepared by CVI do possess limitations, notably on the thermal stability of the fibres which show the onset of degradation at 1300°C.

* Processing by CVI gives composites of 85-90% theoretical density, with a coherent matrix phase and the option of forming an interface in the form of a fibre coating.

* SiC-SiC composites prepared by CVI have already been used for prospective bi-propellant thrusters, spaceplane thermal protection systems and gas-turbine engine components.

6. REFERENCES

1. Bashford, D.P. and Wilson, J. Ceramic Matrix Composites and Carbon-Carbon Composites for High Temperature Space Applications. Fulmer Research Limited Report R1176/D7/February 1990 prepared under ESA Contract No. 7090/87/NL/PP.

2. Stinton, D.P. et al, Advanced Ceramics by Chemical Vapour Deposition Techniques. Ceram. Eng. Sci. Proc., 9[7-8] p705-722 (1988).

3. Lacombe, A., Ceramic Matrix Composites Materials for Aerospace Plane Thermal Structures. ESA WPP-004 (October 1988), Advanced Structural Materials: Design for Space Applications - A Workshop, 23-25 March 1988, ESTEC, Noordwijk, The Netherlands.

Inst. Phys. Conf. Ser. No 111
Paper presented at Int. Conf. on New Materials and their Applications, Univ. of Warwick, 1990

Microstructural traceability of composites

Brian Bousfield - Technical Marketing Manager - Buehler UK Ltd
Chairman - The European Society for Microstructural Traceability

ABSTRACT
New materials are often developed to function in an environment critical to the correct structural condition; they have to withstand structural 'shock' be it physical or temperature. It follows therefore that the preparation of the materials for microstructural analysis is vitally important. The 'New Concept' (Bousfield 1987) and the 'Development of the New Concept' (Bousfield 1989) has illustrated important changes necessary in the preparation of the material surface in order to achieve what is called 'sample integrity'.

INTRODUCTION
The so called 'New Concept' introduces technically derived ways of not just preparing specimens but also a non-commercialised way of expressing or designating the procedure. With these two factors it was possible to make progress in the goal for traceable standards in microstructural preparation applications. A society (TRAC 92) was created (1989) with one of its objectives being European traceability by 1992.

Traceability of materials preparation involves giving consideration to three basic points:-

(a) The preparation procedure (Integrity in the shortest time)

(b) The references i.e. sample material, photomicrographs, image analysis etc.

(d) The custodian

This paper will relate to these three points before applying the principles to actual composite materials.

THE PREPARATION PROCEDURE
The procedure, as illustrated in the introduction, is based on method selection and designation. The method of selection is now a systematic approach based on:

(a) Total depth of damage at starting point
(b) A known series of steps where the least depth of damage remaining after each application is known
(c) The optimum time for each step
(d) The Z axis or stock removal required at each step
(e) The most attractive selection to suit the user without sacrificing integrity

The designation of this information in a non-commercialised manner has required the introduction of standards in referencing different manufacturers grinding and polishing surfaces from uninformative trade names to technically derived numbers.

SAMPLE INTEGRITY

This is the name given to a sample when the preparation has faithfully revealed the true structure i.e. it exhibits integrity. This then is the name given to this stage in the preparation procedure i.e. the sample integrity stage. It consists of a series of multiple single point cutting actions, the abrasives used being progressively smaller. This stage is not to be confused with any polishing steps; integrity is to be achieved through these cutting action and only when sample integrity is achieved shall any polishing, if required, take place (polishing is often not necessary with hard materials).

SURFACE CLASSIFICATION

Surface classification must be available for all commercially produced surfaces and it is advantageous to reference by numbers. Silicon carbide paper as well as the new charged surfaces should also have some relationship.

Since a relationship exists between deformation/structural damage and stock removal we are able to designate surface numbers from material removal.

The concept was to create a series of numbers from 10 to 0 where 10 was the most aggressive removing the most stock yet yielding the greatest damage; the reducing numbers being related to an arithmetical reduction in damage to the theoretical zero. This classification was achieved by relating stock removal per unit time using the same size abrasive.

Datum for Classification.

The concept having established a position of merit in the market place it was to be expected that suppliers of surfaces (for which there are quite a few) would wish to classify their products in order to relate to the 10 to 0 classification. This raised an interesting question of what surface should be used as the datum from which all others are related.

The Buehler Metlap No.10 was proposed and is currently being used within some U.K. organisations until an independent standard can be agreed. (To be fair to all suppliers an independent standard is being pursued through the European Society for Microstructural Traceability and is hoped to be shortly agreed).

FIGURE 1.

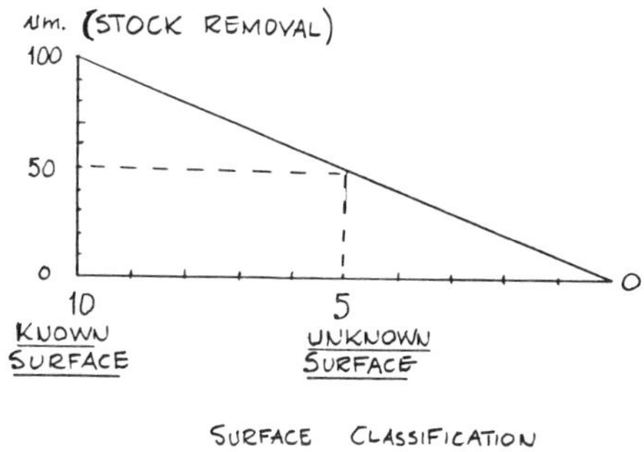

Having established a datum surface such as No.10, the unknown surface can soon be designated as indicated in fig. 1 by using the same size abrasive on both the known and unknown surface for a given time and calibrating the stock removal difference.

It was never originally intended to increase the surface number higher than 10, however, in the event of a new surface being created having a greater stock removal with the same abrasive and identical time, then it must be designated accordingly.

Deformation Classification

What we are ultimately leading up to is a surface classification for all materials whereby the roughness average or surface finish takes second place to the resultant damage remaining in the material after each operation.

PHOTO. 1

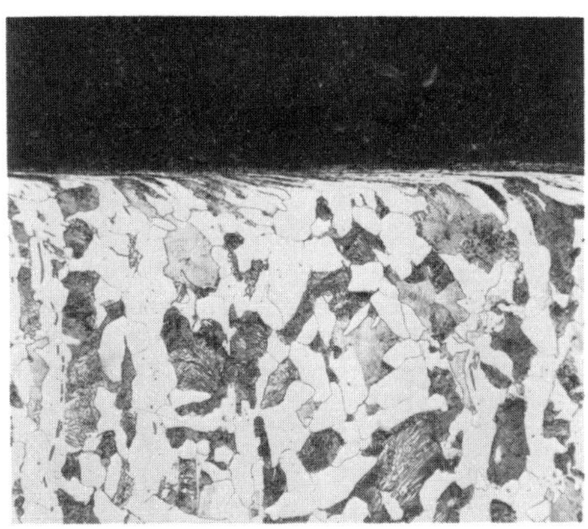

Photo. 1 illustrates the deformed layer of a carbon steel; the object of the exercise is to define a depth of damage and to progress the preparation stages until we ultimately reach a point where the material is faithfully revealed with integrity.

Traditionally, we would remove this damaged layer with the first preparation stage leaving another smaller damaged layer eventually with successively smaller abrasives achieving integrity. If, however, we consider removing just the amount of material which is the limit of integrity available with that particular surface and leave remaining existing damage beyond the scope of that particular abrasive we need to define two factors viz:

(a) the depth of existing deformation

and (b) the limit from which no improvement takes place with each successive surface and abrasive type

Figure 2.

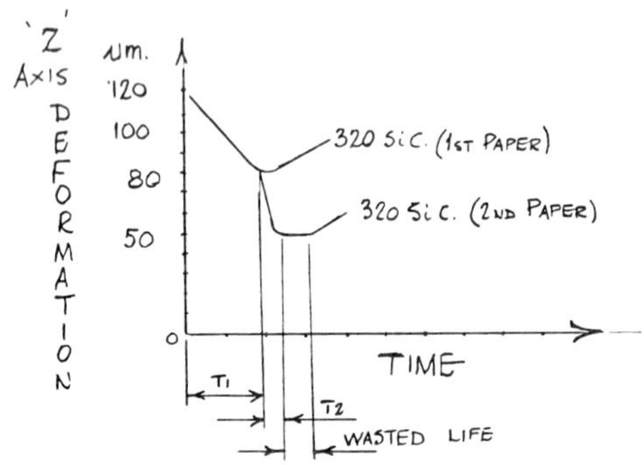

Consider the first stage with a 0.37 steel as being 320 grit SiC and the estimated total damage to be 120 micron (note when estimating the damaged layer from Photo. 1 there is evidence of the top fragmented surface then a definite deformed surface, depicted in the pearlite, as a result of the shear stress but beyond this point an allowance has to be made for residual stress).

This residual stress reveals itself more obviously with softer materials sometimes resulting in twinning; with hard ductile materials add 25% to the deformed layer dimension, with soft materials 50%. If the material had been cut for example on an abrasive cut-off machine then the residual damage can be determined in advance for each material thus avoiding the need to estimate.

From Fig. 2 it can be seen that more than one sheet of silicon carbide paper would be necessary to achieve the optimum condition for that selected grade, it also shows that continued use of the first paper would result in a deterioration of the damaged layer and finally that the second paper would not be fully utilised.

Clearly from the above analogy it can be seen how inefficient the initial selection had been with an initial damage of 120 micron; had the initial damage been 100 micron then 320 SiC grade would have been ideal. If the surface chosen was to be a charged surface (diamond on Metlaps) as opposed to a degrading surface (silicon carbide) then its efficiency is not necessarily time dependent.

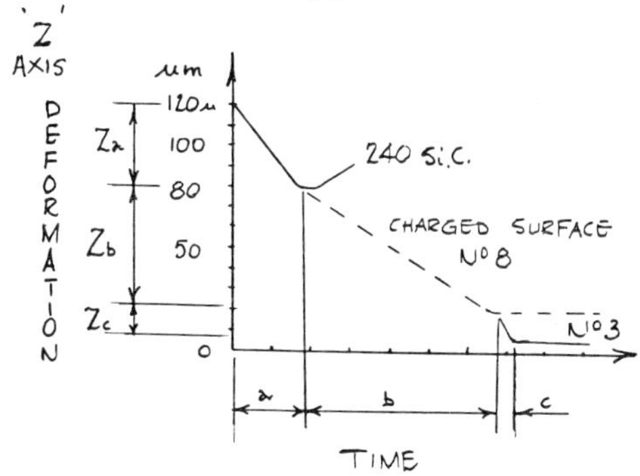

Figure 3.

Fig. 3 shows a combination of SiC and charged surfaces; notice how each of the two chosen 'charged' surfaces reaches a point from which prolonged time does not improve nor does it degrade with time. (Although this is true for this material it must not be assumed to be the case with all materials.) Also notice how the final surface does not give total integrity, it does, however, approach integrity much beyond the scope of degrading abrasives and is often just polished from this condition.

Time

Preparation procedures quote a given time at each step, this was considered to be a variable depending on type/size of machine, number and area of samples, efficiency and type of abrasive, (blocky abrasives not always being the most efficient) type of lubricant, etc. Due to all these variables it was possible to designate a perfectly sound technically derived procedure yet achieving inconsistent results.

We can see from Fig. 3 that this weakness in the concept has been overcome and that with the first step operating for time 'a' the important dimension being the Z axis, i.e. 'Za'. The preparation procedure would then be correct if Za, Zb and Zc were quoted instead of time.

Interpreting the Z Axis

There is a move within Europe to create data and standards for modern engineering materials (VAMAS, BRITISH STANDARDS, NPL, etc) it follows that a standard must designate that which can be audited and will be traceable. No material preparation technique can therefore be data based without the Z axis at each step.

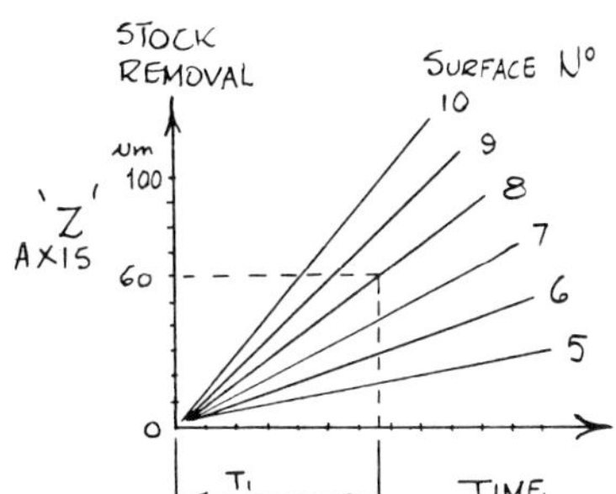

Figure 4.

FIG 5 STOCK REMOVAL CHART
(CHARGED ABRASIVE)

Figure 4 shows how a stock removal chart can be created for specific machines with a given set of conditions. If the Z axis was quoted as 60 micron for surface No. 8 then a time of T1 would be used.

Figure 5.

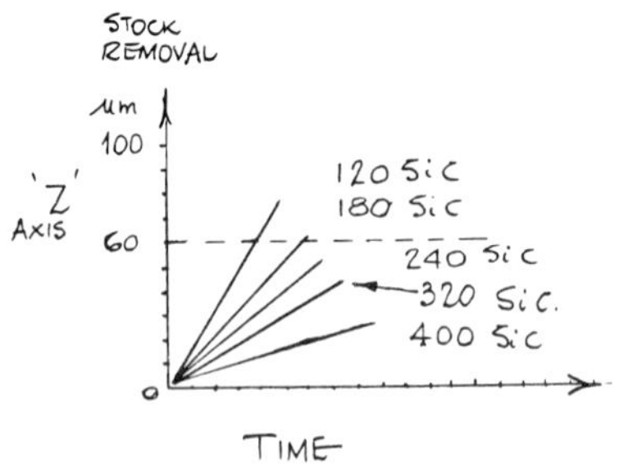

FIG 6 STOCK REMOVAL CHART (DEGRADING ABRASIVE)

The question relating to the life of degrading surfaces can also be overcome by the use of the Z axis as shown in figure 5 where the use of 2 papers would be necessary should 320 grit SiC be chosen.

Another very useful and accurate way to determine the Z axis at each stage is to make a microhardness impresssion in the sample after the planar grinding stage, this allows the reduction in the diagonals to be noted after each step. To convert this figure into a depth measurement simply multiply by 0.202.

$$\text{i.e.} \quad Z = \frac{\tan 22 \times \text{diagonals}}{2} = 0.202$$

POLISHING

The surface classification of polishing cloths as described in the December'88 issue of Metals and Materials (Numbered 0 to 10 in increased nap and decreased surface abrasion) has raised another factor, that of polishing cloth backing.

Classification of the polishing surfaces was made from the results of :-

(i) reflectivity measurements from cloths used without any abrasive

(ii) height measurement of cloth nap

(iii) depth measurement from a ball test

(iv) fibre concentration

Data Bank (Z Axis Curve)

If we consider the preparation possibilities of 0.37 steel and the permutations, there are many, some giving integrity, others not. Those that could faithfully reveal the true structure could be incorrectly used and degrade the surface. 0.37 steel is quite accommodating to less than optimum conditions, modern engineering materials though are quite sensitive to less than optimum conditions and must therefore be carefully prepared to a correct procedure. The question that must be addressed is, "what is the correct procedure and how can we resource the information?"

One essential requirement of the correct procedure is that it must have integrity, that is, it must reveal the true structure. From this point we could define the correct procedure as that which was carried out in the shortest possible time, or that which employed the least number of steps.

One could also see a reason why the best method could be the one that utilised those consumables readily available or those which are the least expensive. There could be occasions where only one method can satisfy the first requirement, that of integrity, in which case there would only be one method.

There could also be a case for having a less than maximum integrity threshold for occasions such as grain size counts or hardness penetration. This and other different parameters could be pre-determined with values rating from, say, 1 to 9 having been selected from a menu structure prior to computer selection. The results could be referenced as data only or as illustrated with a graphic output.

Figure 6 illustrates how the various Z axis curves could be stored relative to their deformation and time ratio. There are two types of curves, the infinite continuous (charged surfaces) and the finite fixed distance (degrading surfaces). The object is to call up the method to the conditions required. Data based preparation methods must when displayed represent a continuous line to be a satisfactory method.

Figure 6.

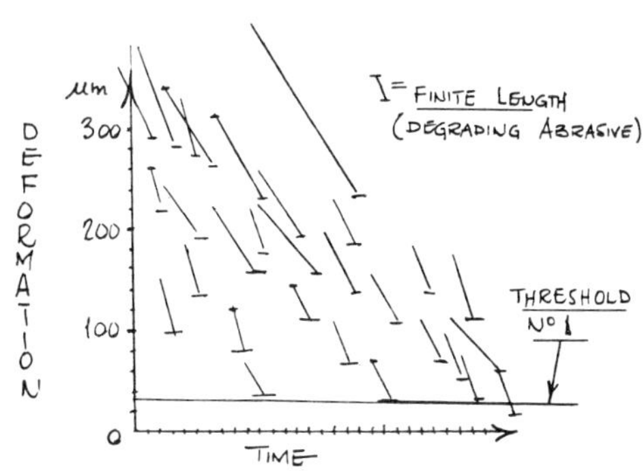

FIG 7 DATA BANK for 0.37 STEEL

Figure 7 illustrates at (a) an incorrect procedure, (b) an acceptable 5 step and (c) an acceptable 2 step. The next question is if only one method is be quoted as a national standard for each material (as should be the case) then which one would be quoted. If "Integrity in the shortest time" is considered to be the criterion, a totally different curve to say the 3 given at figure 7 may be created.

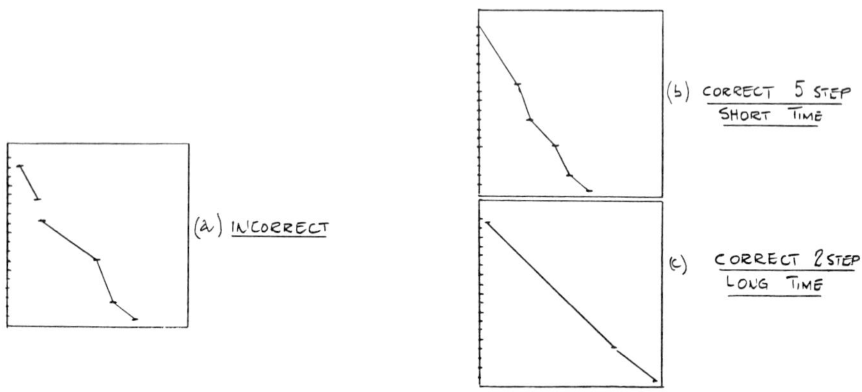

Figure 7. DATA BASE GRAPHIC PRESENTATION

THE REFERENCES

Having defined the preparation route it is important with the faithfully revealed specimen to establish as many references as possible such as a photomicrograph, image analysis, reflectivity measurement etc. Sample material representative of the specimen must also be available. Where possible these references should be traceable to some international standard, as yet the only reference body is the European Society for Microstructural Traceability (TRAC 92). The application of standards in microstructural auditing is to use the referenced sample material in conjunction with the production material at set intervals. The prepared reference material must conform to the referenced photomicrograph and any other referenced document in order to confirm the validity of the prepared production material.

The use of this particular traceable standard does not have to apply to that one production material only, it could, if desired, be used for other materials. When specifying subcontractors this standard should be specified in order to have a uniform traceable standard.

EXAMPLES OF ADVANCED ENGINEERING MATERIALS PREPARATION

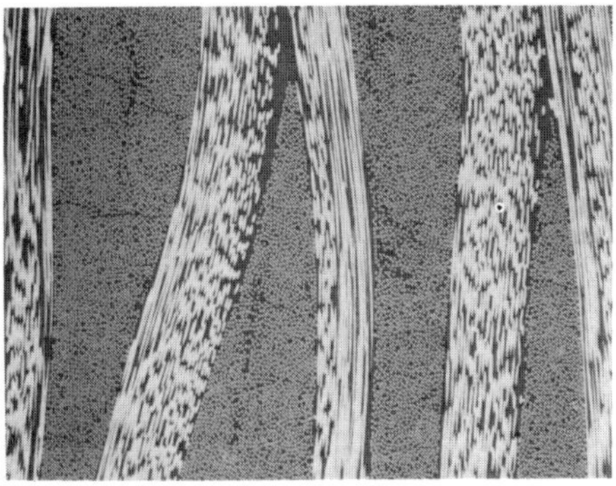

NO. 1 CARBON FIBRE P.M.C.

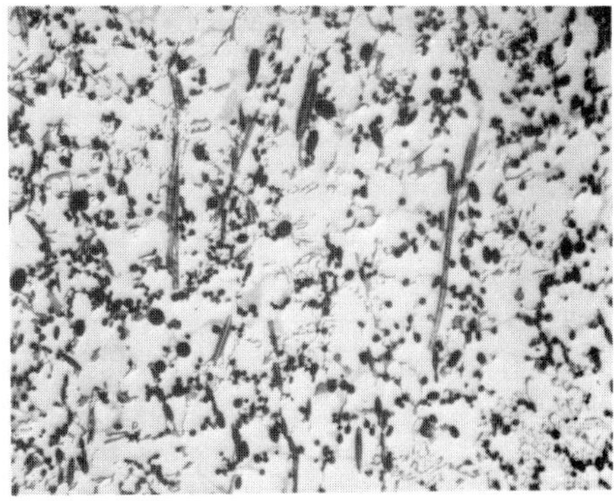

NO. 2 ALUMINA FIBRE M.M.C.

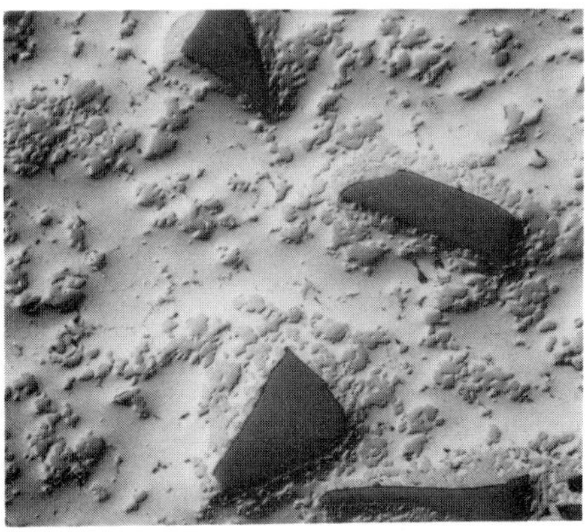

NO. 3 SiC PARTICLES M.M.C.

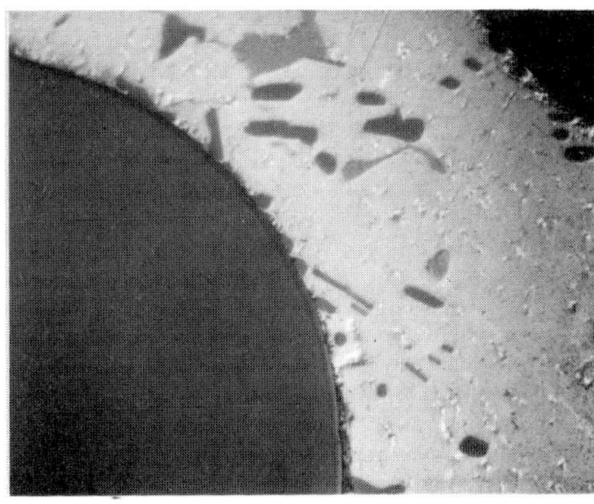

NO. 4 BORSIC FIBRES M.M.C.

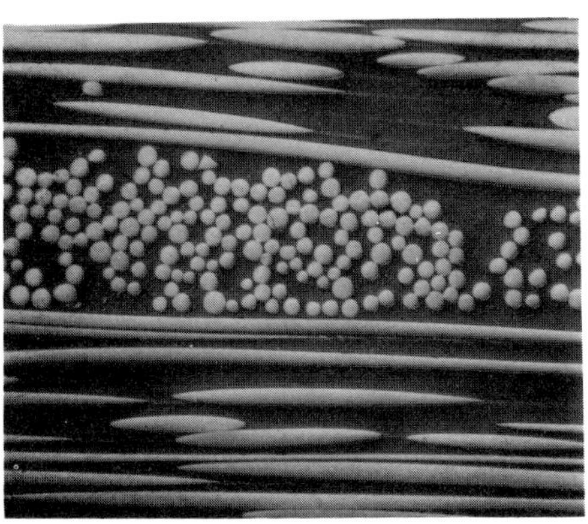

NO. 5 NIKALON FIBRES G.M.C.

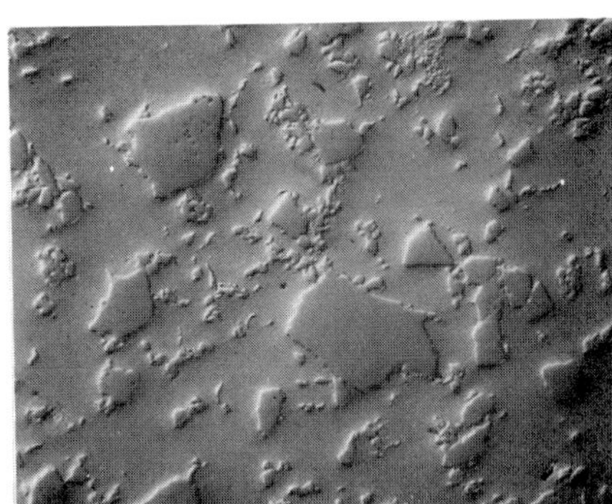

NO. 6 TITANIUM CARBIDE M.M.C.

METHODS FOR ADVANCED ENGINEERING MATERIALS

MATERIAL	Carbon Fibre Reinforced poly ether ether ketone (MICROGRAPH 20x OBJECTIVE) 1
SECTIONING TECHNIQUE	
MOUNTING TECHNIQUE	

	SURFACE	ABRASIVE SIZE AND TYPE	LOAD PER SAMPLE	WHEEL SPEED	HEAD ROTATION	Zμ
PLANAR GRINDING STAGE	No: 10	30 micron diamond	5	250	Contra	TILL PLANE

	INTEGRITY SURFACE	ABRASIVE SIZE AND TYPE	LOAD PER SAMPLE			
SAMPLE	No: 4	9 micron diamond	5	30	Contra	28
INTEGRITY	No: 1	3 micron diamond	5	100	Comp	6
STAGE	No: 1	0.05 Alumina	2-1/2	100	Contra	5 MIN

	POLISHING SURFACE	ABRASIVE SIZE AND TYPE	LOAD PER SAMPLE	WHEEL SPEED	HEAD ROTATION	
POLISHING	No: 8	0.05 Alumina	2-1/2	100	Contra	
STAGE						

MATERIAL	Alumina Fibres in LM 13 (MICROGRAPH 40x OBJECTIVE) 2
SECTIONING TECHNIQUE	
MOUNTING TECHNIQUE	

	SURFACE	ABRASIVE SIZE AND TYPE	LOAD PER SAMPLE	WHEEL SPEED	HEAD ROTATION	Zμ
PLANAR GRINDING STAGE	No: 10	30 micron diamond	5	250	Contra	T. P.

	INTEGRITY SURFACE	ABRASIVE SIZE AND TYPE	LOAD PER SAMPLE	WHEEL SPEED	HEAD ROTATION	
SAMPLE	No: 4	9 micron diamond	5	30	Contra	15
INTEGRITY	No: 1	1 micron diamond	5	100	Comp	3
STAGE	No: 1	Colloidal Silica	2-1/2	100	Contra	2 MIN

	POLISHING SURFACE	ABRASIVE SIZE AND TYPE	LOAD PER SAMPLE	WHEEL SPEED	HEAD ROTATION	
POLISHING	No: 8	Colloidal Silica	2-1/2	100	Contra	
STAGE						

METHODS FOR ADVANCED ENGINEERING MATERIALS

MATERIAL	Aluminium / Silicon Carbide MMC (MICROGRAPH 20x OBJECTIVE) 3
SECTIONING TECHNIQUE	
MOUNTING TECHNIQUE	

PLANAR GRINDING STAGE	SURFACE	ABRASIVE SIZE AND TYPE	LOAD PER SAMPLE	WHEEL SPEED	HEAD ROTATION	Z μ
	No: 10	30 micron diamond	5	100	Contra	T.P.

SAMPLE INTEGRITY STAGE	INTEGRITY SURFACE	ABRASIVE SIZE AND TYPE	LOAD PER SAMPLE	WHEEL SPEED	HEAD ROTATION	
	No: 4	9 micron diamond	5	30	Contra	22
	No: 1	1 micron diamond + Col. Silica High pH	5	100	Contra	2

POLISHING STAGE	POLISHING SURFACE	ABRASIVE SIZE AND TYPE	LOAD PER SAMPLE	WHEEL SPEED	HEAD ROATION	
	No: 8	Colloidal Silica	2-1/2	100	Contra	

MATERIAL	Borsic Fibres in Aluminium (MICROGRAPH 100x OBJECTIVE) 4
SECTIONING TECHNIQUE	
MOUNTING TECHNIQUE	

PLANAR GRINDING STAGE	SURFACE	ABRASIVE SIZE AND TYPE	LOAD PER SAMPLE	WHEEL SPEED	HEAD ROTATION	Z μ
	Resin Bond	320 Grit Diamond	5	125	Comp	T.P.

SAMPLE INTEGRITY STAGE	INTEGRITY SURFACE	ABRASIVE SIZE AND TYPE	LOAD PER SAMPLE	WHEEL SPEED	HEAD ROTATION	
	No: 10	6 micron diamond	5	125	Comp	20
	Silk	3 micron diamond	7-1/2	250	Comp	5
	Silk	1/4 micron Low pH	7-1/2	100	Contra	20 MIN

POLISHING STAGE	POLISHING SURFACE	ABRASIVE SIZE AND TYPE	LOAD PER SAMPLE	WHEEL SPEED	HEAD ROTATION	
	No: 8	Colloidal Silica	2-1/2	100	Contra	

METHODS FOR ADVANCED ENGINEERING MATERIALS

MATERIAL	Nikalon Fibres (Silicon Carbide) in Pyrex Matrix (MICROGRAPH 20x OBJECTIVE) 5
SECTIONING TECHNIQUE	
MOUNTING TECHNIQUE	Longitudinal Fibres aligned parallel to direction of Abrasive Force

PLANAR GRINDING STAGE	SURFACE	ABRASIVE SIZE AND TYPE	LOAD PER SAMPLE	WHEEL SPEED	HEAD ROTATION	2μ
	Resin Bond	320 Grit fixed diamond	5	125	COMP	T.P.

SAMPLE INTEGRITY STAGE	INTEGRITY SURFACE	ABRASIVE SIZE AND TYPE	LOAD PER SAMPLE	WHEEL SPEED	HEAD ROTATION	
	No: 4	6 micron diamond	5	30	CONTRA	18
	No: 2	6 micron diamond	5	120	CONTRA	4
	No: 1	Coll. Silica High pH + 1 micron diamond	5	120	CONTRA	2

POLISHING STAGE	POLISHING SURFACE	ABRASIVE SIZE AND TYPE				

MATERIAL	Titanium Carbide in Titanium (MICROGRAPH 40x OBJECTIVE) 6
SECTIONING TECHNIQUE	
MOUNTING TECHNIQUE	

PLANAR GRINDING STAGE	SURFACE	ABRASIVE SIZE AND TYPE	LOAD PER SAMPLE	WHEEL SPEED	HEAD ROTATION	2μ
	No: 10	30 micron diamond suspension	5	125	Contra	T.P.

SAMPLE INTEGRITY STAGE	INTEGRITY SURFACE	ABRASIVE SIZE AND TYPE	LOAD PER SAMPLE	WHEEL SPEED	HEAD ROTATION	
	No: 4	9 micron diamond suspension	5	30	Contra	19
	No: 1	1 micron diamond + Col. Silica Low pH	10	200	Contra	3

POLISHING STAGE	POLISHING SURFACE	ABRASIVE SIZE AND TYPE	LOAD PER SAMPLE	WHEEL SPEED	HEAD ROATION	
	No: 8	Colloidal Silica 0.06 micron	2-1/2	125	Contra	

THE CUSTODIAN

As mentioned above the only custodian of standards currently is TRAC 92, they will catalogue and supply these standards. With multi-national companies, for which there are many, there could be a case for inter-company custodians, this system however can only be a short term measure since the last thing we want is to have a subcontractor preparing identical materials to different standards.

With the introduction of 'data-based abrasive surface curves' the question of a standard procedure will be resolved; without the references and custodian body it will however be an empty standard.

CONCLUSIONS

The auditing of materials preparation to traceable standards is no longer beyond our reach. We have the key to open this door, the extent to which we use it must be increased. Industry could take it on board immediately; academia could establish procedures for data basing Z axis curves. Many materials are very sensitive to the techniques of specimen preparation, without traceable standards the results at best can only be subjective the analysis however is not just subjective but worthless.

REFERENCES

(1) B. Bousfield - The New Concept 1987

(2) B. Bousfield - Metals & Materials 1988, 4,758-761.

(3) B. Bousfield - Microstructural Traceability of Composites and Ceramics (Composites 1989)

(4) B. Bousfield - Metals & Materials 1990, 6,146-148.

Paper presented at Int. Conf. on New Materials and their Applications, Univ. of Warwick, 1990

Applications of low and high T_c thin film devices to metrology

J C Gallop

National Physical Laboratory, Teddington, Middx. TW11 0LW, UK

ABSTRACT: Thin films of high temperature superconductors already exhibit very promising properties compared with other forms of these complex perovskites. Early applications of devices made with the HTS materials may be expected to come in the field of precise measurements, in view of the importance of conventional thin-film superconducting devices to this field.

1. INTRODUCTION

A review of applications of low and high temperature superconducting thin films to metrology is timely for two reasons. First the discovery of superconductivity in new perovskite cuprate materials at temperatures up to 125K (Bednorz and Muller (1986), Wu et al. (1987), Grant et al. (1988)) has revolutionised the prospect of applications (and SQUIDs and Josephson effects have already been demonstrated at the temperature of liquid nitrogen). Second, remarkable advances have been made in recent years in the sensitivities achieved using SQUIDs, and the sophistication of thin film fabrication of Josephson junction circuits. This has lead to new applications and to an upsurge of interest in measurement applications of superconductivity.

The first section of this paper introduces the fundamental properties of superconducting weak links, the Josephson effects and SQUIDs. The second section explains how a SQUID responds to external magnetic fields and basic magnetometer and gradiometer configurations are described, including some magnetic measurement applications, covering a wide range of physical parameters. A fourth section is devoted to electrical metrology using Josephson junctions and SQUIDs. Finally the new ceramic superconductors with much higher transition temperatures promise to change dramatically the range and take up of SQUID applications. The results so far reported are summarised and some speculation about future directions is included in the final section.

1.1 Macroscopic Quantum Behaviour and Flux Quantisation.

Superconductivity and cryogenics have revolutionised measurements made with the highest precision over the past twenty years, mainly due to the discovery of the Josephson effects and the development of SQUIDs, both of which rely fundamentally on macroscopic quantised behaviour of superconducting systems. We begin by summarising some key results which explain the behaviour of macroscopic devices in terms of the quantum mechanics of a single particle.

The true complexity of superconductivity was only realised some 45 years after its initial discovery when it became clear that the 'pairing' of electrons in a superconductor is responsible for its remarkable properties. There is not space to discuss this further here, so we will take the highly correlated state of the pairs as the justification for describing the whole collection of electrons by a single 'macroscopic' wave function Ψ (also known as the superconducting order parameter), a complex quantity possessing both an amplitude and a phase:

$$\Psi = \Psi o(r,t)\exp(i\phi) \quad ...(1)$$

The spatial variation of the phase of the macroscopic wave function in the superconductor may be changed by applying electric, magnetic or gravitational fields to a superconducting metal. For this review we consider only magnetic fields. It is a basic result of quantum mechanics that the Schrodinger equation for a charged particle in a magnetic field described by a vector potential A(r) is

$$(1/2m)(-i\hbar\nabla-2eA(r))^2\Psi = E\Psi$$

where q is the charge of the particles which for all superconductors so far discovered q=2e. It is straightforward to show that the effect of the vector potential is to multiply the zero field wave function Ψ_o by an additional factor exp(-2ieA(r).dl/h) in going from one point to another a short distance dl away. Now wave functions have to be single valued and, as a result, the total phase change around any path which remains within the superconductor must change by $2n\pi$, where n is an integer. Thus

$$(2e/\hbar)\int A.dl = 2n\pi \quad ...(2)$$

For a piece of superconductor with a hole through it magnetic fields are expelled from the metal itself. However, for contours which surround the hole, n can take any integer value whatever. As a result the normal component of the magnetic flux linking the hole must be quantised, in units of h/2e. (h is Planck's constant and e is the electronic charge)

$$\Phi = \int B.dS = nh/2e = n\Phi_o \quad ...(3)$$

The magnetic flux quantum Φ_o is very small (2.10^{-15}Wb) and quantisation is achieved by screening supercurrents which flow on the metal surface. Any value of external flux Φ_x may be applied to the ring (of self-inductance L) but quantisation is maintained by a circulating current is which flows around the ring so that

$$\Phi_i = \Phi_x + Li_s = n\Phi_o \quad ...(4)$$

1.4 The Josephson Effects

That the flux quantum is so small and furthermore is an exact ratio of two fundamental constants has an obvious appeal to those interested in precise measurement. However in bulk superconductors quantisation is an 'all or nothing' effect. Flux quanta cannot be manipulated singly or in small numbers. To make major progress with cryogenic precision metrology it proved necessary to find a way to smoothly vary the phase of the wave function with an applied magnetic fields. Josephson provided a recipe for this when he proposed a thin film superconducting tunnel junction with an insulating barrier thin enough to allow electron pairs to tunnel between two superconducting electrodes. In this situation the phase difference of the superconducting wave function across the barrier adjusts itself according to the simple relationship:

$$i = i_2\sin(\phi_1-\phi_2) = i_c\sin(\Delta\phi) \quad ...(5)$$

The critical current i_c of the junction sets a limit to its supercurrent carrying capabilities. More surprising, if a current greater than i_c is driven through it a direct voltage V appears across the junction but at the same time an alternating supercurrent flows through it at frequency f

$$f = 2eV/h \qquad ...(6)$$

This behaviour may be explained by a simple argument based on time dependent perturbation theory but alternatively a more physical picture arises from noting that when a finite voltage exists across a weak-link the phase difference of the wave functions on either side begins to increase, each change by 2q corresponding to one flux quantum passing through the link in a direction perpendicular to the current flow (Prance et al. 1986). Thus there are two Josephson junction operating regimes. For low currents the representation we use is in terms of coherent transfer of electron pairs from one electrode to the other. At current levels exceeding i_c the picture changes to one of coherent transfer of flux quanta through the junction, in the direction at right angles to the passage of electron pairs.

2. THE SQUID: A JOSEPHSON JUNCTION IN A SUPERCONDUCTING RING

Imagine a combination of the two quantum mechanical superconducting components we have already discussed. The result is known as a SQUID (Superconducting QUantum Interference Device), and is capable of unprecedented measurement precision for a large range of physical parameters, some of which will be outlined below. We now have a superconducting ring which tends to quantise the magnetic flux linking it, but the weak link interrupting it allows us to control the flow of pair charges or flux quanta. This is intrinsically a quantum mechanical electrical circuit, characterised by the conjugate nature of flux and charge operators.

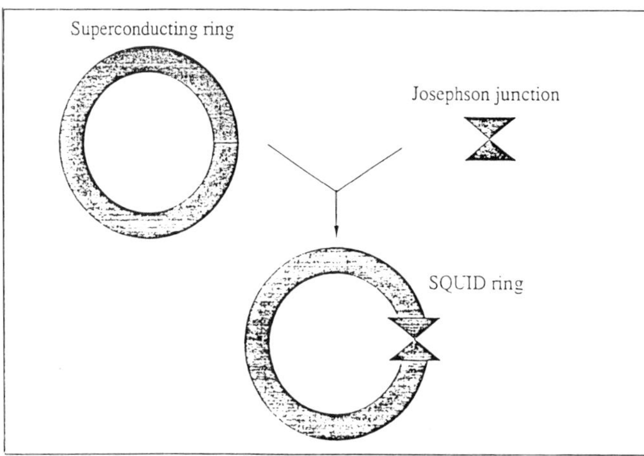

Fig.1 A SQUID represents the combination of two superconducting quantum circuit elements, a flux quantising ring and a Josephson junction.

In one version of a practical system a SQUID ring containing a single weak-link is coupled to a parallel resonant L-C circuit, by magnetically coupling the inductor to the superconducting ring. Flux modulation $\Phi_x(t)$ is applied by driving the L-C circuit at or near its resonant frequency (typically 20MHz). For low modulation amplitudes the SQUID ring response is reversible but as the amplitude is increased hysteretic behaviour begins. The energy dissipation associated with the execution of hysteresis loops draws stored energy from the resonator causing the voltage amplitude of oscillation to fall. A small change in the time-averaged value of the applied magnetic flux $<\Phi_x>$ can bring about a change from non-hysteretic to hysteretic behaviour the energy drawn from the circuit lowers the amplitude of Φ_x below the level required to execute more hysteresis loops. It will take a number of cycles of the oscillation (~Q, the quality factor of the L-C circuit) to restore the amplitude to its original value, whereupon more hysteresis loops will be executed. This 'relaxation oscillation' process typifies the resonant-dissipative mode of operation of the single junction (or r.f.) SQUID. There is not space in this review to present a more complete analysis of this mode. It may be succinctly summarised by stating that the radio frequency impedance of the device is a periodic function of the static magnetic flux which is applied to it with a period which is, not unexpectedly, Φ_0.

2.2 D.c SQUIDs

A d.c SQUID consists of a ring of superconductor, interrupted by two Josephson junctions, usually with similar properties. The introduction of the second junction allows a time independent voltage to be established across the two parallel-connected junctions by a direct bias current. It has been convincingly demonstrated that more sensitive devices may be made, with rather simpler electronics than their r.f counterparts. The basic mode of operation of the d.c SQUID may be explained qualitatively by treating the device as possessing three distinct components which are inseparably connected: an oscillator, a parametric amplifier and a microwave detector. The oscillator consists of the two junctions biased at finite voltage so that an oscillating supercurrent at the Josephson frequency is driven around the superconducting loop. The second element, an inductive parametric amplifier, arises from the fact that the amplitude, phase and frequency of the circulating supercurrent depend periodically on the flux Φ_x applied to the ring. Finally the d.c response of the junctions to the amplitude of the circulating supercurrent performs the detection function.

For both r.f and d.c devices which are available commercially it is possible to subdivide one period, corresponding to one flux quantum, to about 1 in 10^5. High quality, small capacitance Josephson junctions, together with systems which are very well screened from electromagnetic interference have enabled a number of research laboratories to set up SQUIDs, again of both r.f and d.c types, which are even more sensitive. Expressing their detection limit in terms of an input energy sensitivity, the best SQUIDs can sense a change of 10^{-34}J/Hz, a value which is compatible with predictions of the Heisenberg uncertainty principle, $\Delta E.\Delta t > \hbar/2$. This implies that the best SQUIDs should be able to detect single photons at audio frequencies. The available signal bandwidth at present extends from d.c up to at least 200 MHz.

2.3 Practical SQUID Devices

The first commercially available SQUIDs used point-contact or thin film microbridge junctions since these were the most easily fabricated types at the time. In the past ten years, driven by the goal of producing a Josephson computer, many groups around the world have committed a great deal of effort to developing reliable multi-layer thin-film deposition techniques for preparing reproducible, small capacitance, high current density tunnel junctions, the classic form of weak-link proposed by Josephson. Modern commercially available SQUIDs use robust junctions made from thin films of niobium (deposited by sputtering or electron beam evaporation), with refractory tunnel barriers (typical thickness 1-2 nm) made from Al_2O_3 or MgO. These thin barriers are strong enough to remain undamaged by evaporation of the niobium counter electrode. Film patterning is carried out using the photolithographic techniques developed for semiconductor applications. There are many variants available, including direct pattern writing with an electron beam and removal of thin films in vacuo using reactive ion etching. Bulk superconducting pieces surround the thin film junctions and planar coupling coils to provide flux compression and magnetic shielding functions.

3. MAGNETIC MEASUREMENTS WITH SQUIDS

A SQUID ring senses magnetic flux changes directly, although measurement of flux density B is more generally required. Instead of using the SQUID ring as a direct sensor an intermediate coil arrangement known as a flux transformer is generally employed. An entirely superconducting d.c transformer makes use of the zero resistance property of superconductors to collect flux density from a large sensing coil, concentrating it into the much smaller area of the SQUID ring. This action arises from flux conservation in the two coils. A change in B_x causes a change in is which in turn applied a changing flux Φ_x to the SQUID ring. Maximum sensitivity to B_x is found when primary and secondary inductances are equal. A low-noise SQUID in a well shielded environment is easily capable of sensing changes ΔB as small as 1fT. This is very much smaller than a typical fluctuation in the earth's magnetic field (1nT, even far from man-made interference). In a laboratory the magnetic field fluctuations may be three orders of magnitude higher, so very great attention must be given to magnetic shielding, normally carried out by enclosing the field sensitive region inside a strongly superconducting shield. Although initially flux transformers were generally made from bulk superconducting wire, with the ends spot-welded to give a superconducting join, many designs for planar thin film transformers are now in use.

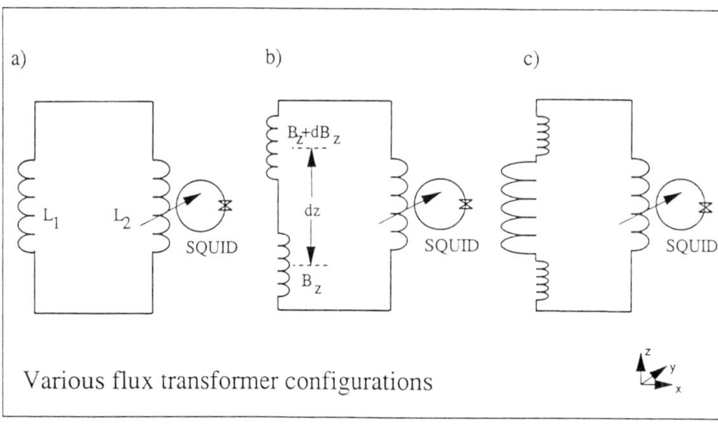

Fig.2 Schematic diagram of various types of superconducting flux transformer. The three types are described in the text.

The energy sensitivity of a SQUID depends on minimising both the inductance of the superconducting ring and the capacitance of the Josephson junctions. Thus E, the Nyquist noise energy in unit bandwidth, referred to the SQUID input, at an operating temperature T is

$$E = 16kT(LC)^{1/2} \text{ J/Hz} \quad ...(7)$$

Thus physically small SQUID rings (~10μm diameter) with low inductance exhibit the best noise performance while a large flux transformer primary coil is required to maximise the sensitivity to flux density changes. Close magnetic coupling has been very effectively achieved between these very different sized circuits by the so-called 'square washer' thin-film SQUID design. This consists of a multi-turn planar spiral secondary coil (of the same inductance but more turns with smaller area than the primary) overlays a thin film washer with a large outer dimension equal to the secondary coil, but with a much smaller inner hole which forms the SQUID ring itself. A thin slot cuts the washer with the two Josephson junctions of the d.c SQUID connected in series across this slot to complete the SQUID ring. (A schematic diagram of this arrangement is shown in Fig.3). The magnetic coupling between secondary and SQUID can be as high as 80% with this arrangement.

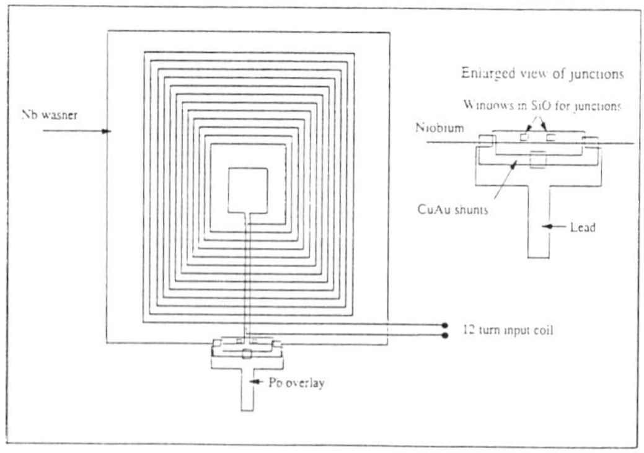

Fig.3 Planar d.c SQUID with 'washer' configuration for tight inductive coupling.

3.1 Field Gradient Measurements

In a majority of SQUID applications the primary of the flux transformer does not consist of a single simple coil. Instead, more usually, some combination like that shown in Fig.2b is employed, in which there are two primary coils with n turns each, identical except that they are wound in opposite senses. A spatially-uniform external field change will produce equal but opposite flux changes in the two coils, summing to zero and giving no change in the flux applied to the SQUID ring (obviously this is only exactly true for perfectly matched coils). If the applied field change has a non-zero spatial gradient in the direction of the separation of the coils the cancellation will not be exact and an output signal will be produced. For the arrangement of Fig.2b the coil system senses changes in dB_z/dz. The flux change at the SQUID is

$$d\Phi = MnA(dB_z/dz)dz/(L_1+L_2) \quad ...(8)$$

where A is the coil area. The field gradient tensor has nine components and off-diagonal terms may be measured by arranging the axes of the primary coils to be orthogonal to their separation.

A SQUID coupled to a gradient-sensitive primary is able to operate in magnetically noisy environments. If the background field B is changing randomly with time, but in a spatially uniform manner over the volume occupied by the coils no change in output of the SQUID will occur. However a nearby localised magnetic source (a fluctuating dipole for example) will in general produce different field changes at each of the primary coils, resulting in a net SQUID signal. Thus nearby signals can be detected in the presence of much larger, more distant fluctuations. By this means SQUIDs have been used to measure the very small magnetic field changes associated with brain activity (~0.1pT), in rather noisy magnetic environments where the field fluctuations may be three or more orders of magnitude larger.

Thin film techniques have been used to make planar gradiometers which are of increasing importance in real SQUID applications, particularly in the field of biomagnetism. The precision of photolithographic techniques means that the two components of a first-order gradiometer coil can be made accurately equal and parallel. The planar geometry introduces the restriction that only off-diagonal components of the magnetic field gradient tensor can e be measured and also there is the requirement that an insulating cross-over of the superconducting films. This represents no problem for conventional superconductors, where the SQUID thin-film fabrication process already involves many different levels of processing. Using sophisticated thin-film fabrication techniques (Knuutila et al.) arrays of SQUID sensors and their associated gradiometer coils allow simultaneous data capture at as many as 37 spatially separated points around the skull, so that real-time imaging of the current flows associated with neural activity deep in the brain may be made. Even more complex primary coils may be used to sense higher order spatial derivatives of the magnetic field changes (see Fig.2c). For example two identical first derivative coils connected back-to-back sense changes in d^2B/dz^2.

A SQUID provides the essential part of the most sensitive displacement transducer now in operation. This is at the heart of a cryogenic cooled gravity wave detector. A high purity thin niobium diaphragm is placed between two thin film superconducting spiral coils, connected in series opposition across the input coil of a SQUID. The complete flux transformer winding carries a persistent supercurrent. A small movement of the diaphragm towards one of the spiral coils and away from the other changes the self inductance of each, causing the net circulating supercurrent to change in order to maintain flux quantisation in the circuit. The current change produces a flux change at the SQUID ring which may be detected. Mechanical resonances of the diaphragm have very high Q values and the frequency of one of them may be electronically tuned to the resonant frequency of the massive bar which acts as the antenna for gravity wave detection. Displacements of the ends of the bar as small as 10^{-19}m are detectable.

4. ELECTRICAL MEASUREMENTS USING SQUIDS AND THE JOSEPHSON EFFECTS

A SQUID may be simply converted from a flux detector to a device for measuring voltage or current. If a coil is inserted in the ring any current flowing in it applies a magnetic flux which the SQUID will detect. The sensitivity to direct voltages is extremely high ($<10^{-20}$V), far smaller than the random noise voltage levels associated with most measurement situations unless both the source impedance and temperature are very low. The current sensitivity is also good ($\sim10^{-13}$A), leading to accurate current ratios and current stabilisers from room temperature current sources. Other electrical quantities, such as a.c amplitude and r.f attenuation may also be accurately measured with SQUIDs. In addition a broad-band ultra-low noise amplifier has been built using thin film tunnel junction d.c SQUID (Hilbert and Clarke). A 20 turn planar input coil is coupled to a d.c SQUID of the general form shown in Fig.3. It includes two thin-film Nb-NbOx-PbIn tunnel junctions (2μm square), shunted with 3Ω resistors to provide some damping. The amplifier has a gain of about 20dB at frequencies up to 100MHz, with a noise figure of only 1.0K, in agreement with the design predictions. This is already the best reported performance for such an amplifier and modified design parameters should lead to further improvements. Demonstrated reliability and ease of use should lead to widespread applications.

4.1 A Quantum Voltage Standard.

The alternating supercurrent in a voltage-biased junction is an oscillating flow of electric charge which will emit electromagnetic radiation, whose frequency is voltage tuneable from zero frequency up to perhaps 1THz using conventional superconductors. Such Josephson oscillators are interesting but little used yet, due to the rather low power output from a single junction (<1nW). If instead a weak link in the finite voltage state is irradiated with microwaves of frequency f the passage of flux quanta can be synchronised to the applied electromagnetic field, pulling the I-V characteristic into a series of steps, equally spaced in voltage by $\Delta V=hf/2e$. This effect forms the basis of the voltage standard which has been used at a number of national metrology institutes since the early 1970's. The step spacings of two Josephson junctions made of different superconductors have been compared and shown to be equal to 2 parts in 10^{16} and the same group has measured the gravitational red-shift in the chemical potential of electron pairs as they change height by only 7cm in the earth's gravitational field (Jain et al.).

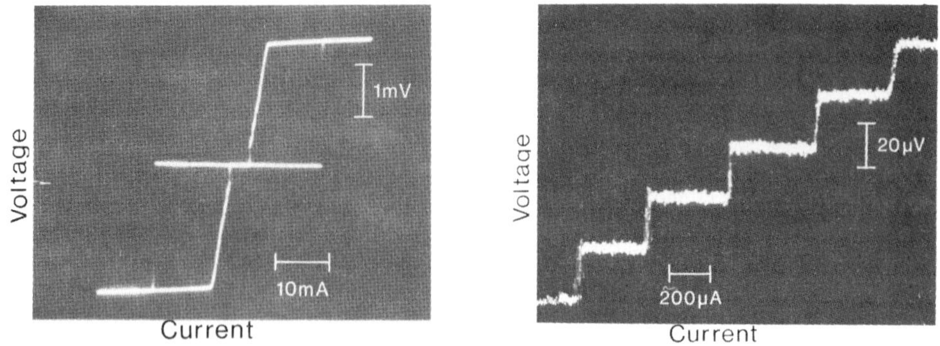

Fig.4 Current-voltage characteristics of Josephson tunnel junction a) without and b) with microwave irradiation. Note the scale change between a) and b).

As the voltage provided by a Josephson junction is only a few millivolts the need to relate this to a standard-cell e.m.f. (~1V) with an accuracy of a few parts in 10^8 led to a number of important developments in precision cryogenic metrology, such as current stabilisers, resistance and transformer ratio devices. The bulky nature of the voltage standard system is at least partly accounted for by the complicated potentiometry associated with this large difference between junction and standard cell voltage. A recent development (see for example Niemeyer et al. 1985) shows that a very much more convenient system is possible. Sophisticated thin film device fabrication methods have allowed junction arrays to be made with carefully matched properties and their electrical characteristics can be designed so that microwave induced steps in the characteristic extend to the current axis. With an array of 1440 junctions, each biased on a voltage step induced by 90 GHz microwave radiation a total voltage of greater than 1V appears across the array without the need for individual current bias leads to each junction. Using such an array a compact solid-state microwave source and a junction array in a small liquid helium cryostat providing a convenient and transportable voltage standard at the 1 in 10^{12} level of accuracy, which would not need highly skilled research staff to run it.

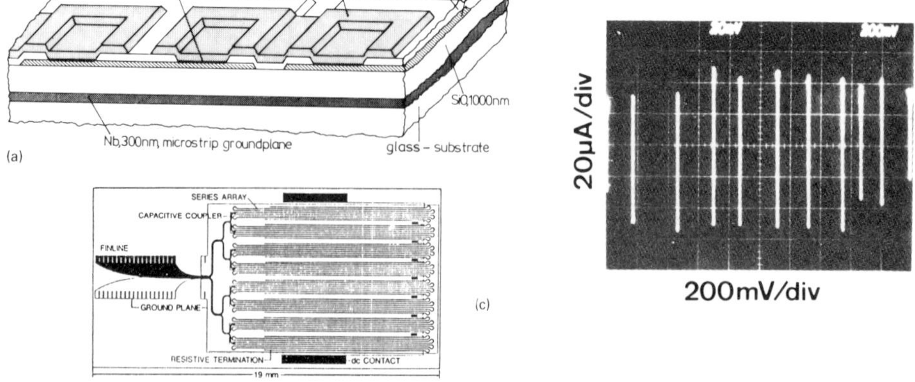

Fig.5 a) & b) detail of and c) I-V characteristic of 1440 junction array (Niemeyer et al.)

4.2 Josephson High Frequency Mixers

The extreme non-linearity of the current-phase relationship in a Josephson junction as embodied in equn.5 provides the basis for low noise high frequency mixer operation. Very high harmonic generation is achieved and mixing between incident frequencies which differ by a factor of 1000 has been demonstrated. An estimate of the upper frequency limit to which Josephson junctions will operate is provided by the observation that when the energy hf of a photon of frequency f exceeds the energy gap 2Δ of the superconductor from which the junction is made the electron pairs responsible for the superconductivity can be dissociated. For conventional superconductors the upper frequency limit for mixing is about 2THz. A low noise mixer can also be used as the basis of a sensitive heterodyne detector and millimetre wave receivers using Josephson devices have been extensively investigated. More recently these have generally been superseded by another form of thin-film superconducting device, the 'quasi-particle mixer'. This has the same structure as a Josephson tunnel junction but the non-linearity used arises from the sharp changes in the I-V characteristic of the junction at or near a bias voltage corresponding to the energy gap: $V=\Delta/e$ (see Fig.4). The Josephson supercurrent is sometimes suppressed by the application of a static magnetic field. If the high frequency currents induced in the junction by the signal are to have their maximum amplitude the impedance shunting the junction must be large. This requires that the capacitance of the junction must be small, which in turn implies the need for very small area junctions with high critical current density j_c. Photolithography on thin-film devices has allowed junction dimensions to be reduced to around 1μm, where the shunt capacitance will be only of order 1fF. Series arrays are usually used in radio-astronomy receivers to improve impedance matching to antenna structures and to increase the dynamic range of the receiver. Antenna coupling structures are often made from patterned superconducting thin films integrated on the same substrate as the junctions.

5. THIN FILM SQUIDS USING HIGH TEMPERATURE SUPERCONDUCTORS

Three years after the discovery of the first liquid nitrogen cooled superconductor progress towards real applications of the new materials has been excellent (although not rapid enough to satisfy some of the expectations raised by the initial excitement following Bednorz and Muller's discovery). The highest transition temperature (T_c) stands at 125K for the compound thallium barium calcium cuprate ($Tl_2Ba_2Ca_2Cu_3O_{10+y}$) and the number of separate families of ternary or quaternary cuprates exhibiting high temperature superconductivity has increased to six. Some of these are detailed, along with some of their physical parameters, in Table 1. Critical current densities at 77 K, in thin film samples at least, are at the level required for many power applications (>10^6 A/cm^2), although a number of serious problems, such as the inability of the materials to pin magnetic flux lines in high magnetic fields, still require to be solved. Theoretical understanding of the new materials still lags well behind experiment.

Table 1

Composition	Transition Temp. T_c (K)	Penetration depth λ (nm)	Coherence Length ζ (nm)	Upper critical field B_{c2} (T)
$(LaSr)_2CuO_4$	38		3.7(ab) 0.7(c)	125(ab) 24(c)
$YBa_2Cu_3O_7$	93	140	3.1(ab) 0.51(c)	210(ab) 35(c)
BiSrCaCuO	110	180	4(ab) 0.21(c)	400(ab) 19(c)
$Tl_2Ba_2Ca_2Cu_3O_{10}$	125	210		

5.1 Low Frequency Metrological Applications

The first simple three-layer Josephson junctions, consisting of two HTS thin films separated by a uniform insulating layer only 1nm thick, have only recently been fabricated (Laibowitz et al.) and at present these do not exhibit ideal or even reproducible properties. This remains a stumbling block to the large-scale exploitation of HTS electronics, such as the Josephson computer. However for situations where only one or two Josephson junctions are required it has proved possible to produce d.c SQUIDS using fabricated 'edge junctions' in which a deposited thin film of $YBa_2Cu_3O_7$ (YBCO) is covered with a protective layer of BaF. A raked edge is produced on this two-layer film by ion beam etching before it is exposed to an oxifluoridation plasma which provides the thin insulating layer. Finally an overlay of YBCO is deposited over the edge and is patterned to give two protruding fingers (about 5 μm wide) which complete a superconducting ring (20x20μm). This achievement of fabricated three-level Josephson junctions must be seen as a considerable breakthrough.

In HTS thin films the grain boundaries which are such a disadvantage for high current carrying applications prove to be a blessing for prototype measurement applications in that they provide Josephson junctions which in the current absence of reliable fabricated junctions may be used in SQUID and other active superconducting electronic measuring devices. The junctions thus formed, although at present little is known about the actual microscopic structure which gives rise to them, they are really quite respectable junctions. For example we have recently shown that in thin films of YBCO with critical current densities at 77K of some $100A/cm^2$ the junctions between the 10um sized grains have a capacitance as low as 10^{-14}F, which means that millimetre wave supercurrent can flow through the junction without being effectively shunted by a displacement current (Gallop et al.). Low capacitance is important for low noise d.c SQUIDs (equn. 7) as well as for high frequency mixing. At NPL we have used a photolithography 'lift-off' technique to pattern d.c SQUIDs in polycrystalline thin films of BSCCO (see Fig.6). HTS SQUIDs with very reasonable performance operating at 77K have been made using these grain boundary junctions. The best reported device of this type, made by IBM, has shown an energy sensitivity of $6x10^{-30}$J/Hz (Koch et al.), better than that of commercially available liquid helium cooled conventional SQUIDs so we may expect applications of such devices in the near future.

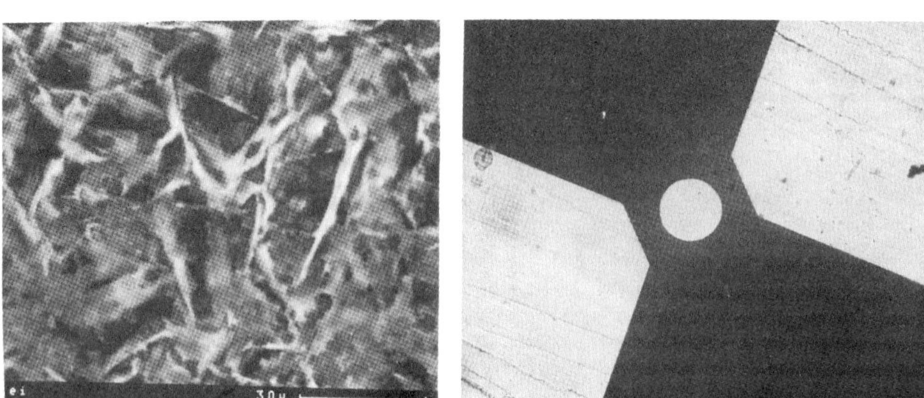

Fig.6 Photomicrographs of BSCCO film showing a) grains and b) d.c SQUID structure.

As we have mentioned above, practical SQUID devices require flux transformers, the properties of which contribute to overall system performance. Early reports on polycrystalline HTS thin films suggested that these materials possessed an additional source of low frequency noise which would seriously limit the performance of HTS SQUIDs. On cooling through the transition temperature large numbers of magnetic flux quanta are trapped through thin polycrystalline films, concentrated at the weak potential minima at grain boundaries. At the relatively high temperature of operation (say 77K) thermal fluctuations are sufficiently large to cause thermal activation of flux quanta from one metastable minimum to an adjacent one and this 'flux creep' in the flux transformer film will produce a changing magnetic flux coupled to the SQUID ring. This noise contribution whose amplitude has an approximately 1/f frequency dependence was found to be very high in early thin films. Recently the advent of epitaxial thin-films with a much reduced density of grain boundaries have shown a dramatic reduction in flux creep noise (Ferrari et al.) and combined with the demonstration that similar high quality films of YBCO may be deposited in a three layer structure, with an insulating barrier between, (Kingston et al.) suggests that the major problems associated with thin film flux transformers for HTS SQUIDs have been solved.

6. SUPERCONDUCTING THIN-FILM MICROWAVE FREQUENCY STANDARDS

Although the prospect of a higher operating temperature, associated with the higher transition temperatures of the new materials is of dominant significance in many electrical engineering applications, this is only one factor in metrological applications of HTS. Of perhaps equal importance is the higher energy gap $2\Delta(T)$ which has been shown to scale approximately with T_c for the new materials. A consequence of the larger gap is that the frequency response of Josephson mixers and detectors should be extendible into the infra-red, perhaps to 20 THz compared with the 1.5 THz limit so far achieved with conventional superconductor junctions. Similarly, since the SQUID has been shown to be an unusual form of parametric up-converting amplifier, its gain is proportional to the pump frequency, which in turn is ultimately limited by the maximum frequency at which supercurrents can flow. To make use of the expected increased frequency limit carefully fabricated and near ideal low capacitance, high current density tunnel junctions will be required and at present this has not been achieved.

Another important area of possible high frequency applications not yet discussed is that of high Q passive microwave devices. At NPL we have been using cavity perturbation techniques to evaluate the real and imaginary parts of the surface impedance of samples of the new superconductors as a function of temperature. Provided that the superconductor is the dominant source of loss in the microwave circuit the Q value attained is inversely proportional to the surface resistance R_s. Most of the specimens we have tested show low R_s values when compared with copper at liquid helium temperatures, though not as low as conventional superconductors such as niobium. To be practically useful HTS materials must be significantly less lossy at their operating frequency $\omega/2\pi$ than copper at 77 K. A simple expression for R_s in terms of temperature, frequency and energy gap describes conventional superconductors adequately:

$$R_s(T) \; \alpha \; (\omega^2/T)\exp(-2\Delta/kT) \qquad ...(9)$$

At present ohmic losses in polycrystalline thin films or bulk samples are orders of magnitude higher than equn. 9 predicts and this is believed to arise from loss mechanisms associated with grain boundaries (which we have seen are used to good effect in planar SQUIDs). Measurements at NPL and elsewhere have already demonstrated that single crystals of HTS material approach much more closely the theoretical behaviour of equn. 9 and rather recently single crystal epitaxial thin films deposited by laser ablation on high quality dielectric substrates such as MgO have shown losses which are close to ideal down to at least 50 K, performing better than copper up to at least 150 GHz. High quality HTS films now seem good enough for real passive microwave applications and a number of groups are patterning these films to produce resonator and filter structures.

Two such thin film resonators, a microstrip configuration and a coplanar transmission line are shown in Fig.7.

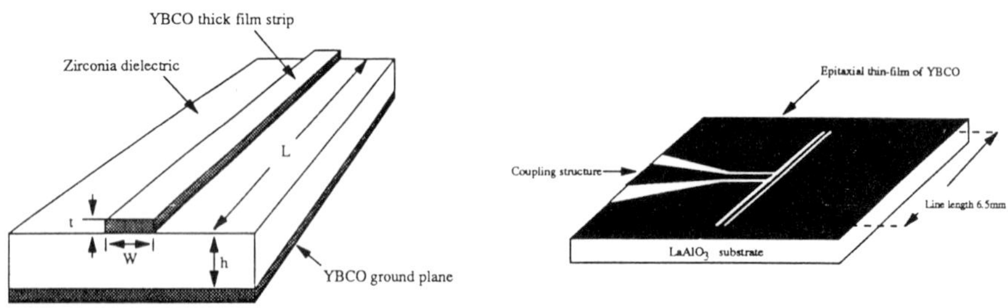

Schematic of HTS microstrip resonator

High Q coplanar transmission line resonator

Fig.7 Thick film microstrip resonator on zirconia substrate (ICI) and coplanar transmission line with Q of 4000 at 10GHz and 77K (Valenzuela et al.)

7. CONCLUSION

Two recent developments in high temperature superconductor thin film technology suggest that applications of the new materials may be expected in a relatively short time: the quality of in-situ films, particularly their microwave properties, has improved dramatically and the first fabricated Josephson junctions have been produced. As with low temperature superconductors, early applications using these two breakthroughs may be expected in the field of precision measurement. Two rather different areas of exploitation may emerge initially. Low frequency applications of SQUIDs as precision transducers at around 77 K seem promising, for use in areas where liquid helium is unavailable or inappropriate. Noise thermometry and displacement transducers may represent two applications of this type. Also as high quality HTS thin films become more widely available for high frequency use, both novel passive and active devices may be expected to emerge, some no doubt operated at liquid helium temperatures, in order to exploit the highest available Q values and frequency response.

REFERENCES

M.J.Ferrari, M.Johnson, F.C.Wellstood, J.Clarke, A.Inam, X.D.Wu, L.Nazar & T.Venkatesan, Nature, 341, 723 (1989)

V.Foglietti, R.H.Koch, W.J.Gallagher, B.Oh, B.Bumble & W.Y.Lee, Appl. Phys. Lett. 54 2259 (1989)

J.C.Gallop, A.J.Marsh, W.J.Radcliffe & C.D.Langham, Phys. Scripta, to be published (1990)

C.Hilbert & J.Clarke, J.Low Temp. Phys. 61, 263 (1985)

A.K.Jain, J.E.Lukens & J.S.Tsai, Phys. Rev. Lett. 58, 1165 (1987)

J.J.Kington, F.C.Wellstood, P.Lerch, A.H.Miklich & J.Clarke, Appl. Phys. Lett. 56 189 (1990)

R.H.Koch, W.J.Gallagher, B.Bumble & W.Y.Lee, Appl. Phys. Lett. 54 951 (1989)

J.Knuutila, Proc. European Workshop on Biomagnetism (1989) to be published.

R.B.Laibowitz, R.H.Koch, A.Gupta, G.Koren, W.J.Gallagher, V.Foglietti, B.Oh & J.M.Viggiano, Appl. Phys. Lett. 56, 686, (1990)

H.Prance, T.P.Spiller, J.E.Mutton,R.J.Prance, T.D.Clark & R.Nest, Phys. Lett. 115A 125 (1986)

W.J.Radcliffe, J.C.Gallop, C.D.Langham, M.Gee & M.Stewart, Electronics Letters 24 1085 (1988)

A.A.Valenzuela, B.Dalmaans & B.Roas, Electronics Lett. (1989)

Inst. Phys. Conf. Ser. No 111
Paper presented at Int. Conf. on New Materials and their Applications, Univ. of Warwick, 1990

Passive microwave device applications of high temperature superconductors

M J Lancaster

Birmingham University, School of Electronic and Electrical Engineering, Birmingham B15 2TT.

ABSTRACT: This paper reviews the state-of-art of the applications of superconductors in the area of passive microwave devices paying particular attention to the use of high temperature superconductors. The text concentrates mainly on those applications where the device can be constructed using planar circuit techniques.

1. INTRODUCTION

Table 1 gives a list of passive microwave device applications of superconductors. The list excludes all those devices which use any tunnelling effects within the superconductor. It is the intention of this paper to review the state-of-art of these applications of high-temperature superconductors in the area of passive microwave devices. However, the areas of application are large and to limit the length of the paper microwave planar circuits have been taken as a theme. Planar circuit resonators, filters and delay lines are discussed. Because the majority of the applications come about due to the reduced surface resistance of the superconductor over conventional conductors Section 2 of the paper will discuss surface resistance in some detail.

2. SURFACE RESISTANCE

At microwave frequencies current always travels along the surface of a conductor. In the case of a conventional conductor the characteristic depth to which it penetrates is given by the classical skin depth ($\delta = (\omega\mu\sigma/2)^{-\frac{1}{2}}$). For a high temperature superconductor at microwave frequencies this characteristic depth is the magnetic field penetration depth and is independent of frequency. A quantity which characterises the conductor at high frequencies in therefore not a bulk impedance but a surface impedance. It is defined by the ratio of the surface tangential electric field to the surface tangential magnetic field. The real part of the surface impedance is the surface resistance which is the resistance of a square of any size of surface with units of ohms. The surface impedance of a conventional conductor is given by

$$Z_s = (1 + j)\left(\frac{\omega\mu}{2\sigma}\right)^{\frac{1}{2}} \qquad \text{........(1)}$$

In a superconductor at a temperature above absolute zero not all of the electrons have condensed into the superconducting state. It is the remaining normal

Resonators
- Frequency references (Gallop 1989)
- Oscillators (Everard 1989)
- Characterisation
- Cavity resonators (Martens et al 1988, Bohn et al 1989).
- Helical resonators (Everard et al 1989)
- Planar resonators e.g. microstrip (see text).
- Tunable resonators (see text).
- Coaxial resonators

Delay lines
Dispersive delay lines (Ralston 1985).
Filters (see text).

Antenna
- Dipole (Khamas et al 1988)
- Loop (Wu et al 1989)
- Array

Matching networks
Combiners
Coaxial transmission lines (Allen et al 1964)
Duplexers
Screening
Switches

Table 1 Passive Microwave Applications of Superconductors

electrons that contribute to a residual resistance at all frequencies other than d.c. The surface resistance due to these normal electrons is only significant when the reactance of the superconducting electron pairs increases (with frequency) so that it no longer shorts out the conduction path. A phenomenological model which appears to work for the high temperature superconductors at microwave frequencies is a model developed by London in 1934 called the two fluid model. The surface impedance calculated using this model at microwave frequencies is

$$Z_s = \frac{\omega^2 \mu_o^2 \sigma \lambda^3}{2} + j \mu_o \omega \lambda \qquad \text{........(2)}$$

where λ is the magnetic field penetration depth and σ is the normal state conductivity. Equation (2) assumes that the frequency is much less than the gap frequency and also that the penetration depth is less than the normal skin depth. This is true for high temperature superconductors at microwave frequencies.

Equations (1) and (2) form the basis for a method of assessment of superconducting material. Figure 1 shows the variation of surface resistance with frequency for conventional copper conductors and superconducting Niobium. Notice that the frequency variation is proportional to $\omega^{\frac{1}{2}}$ and ω^2 respectively. This means that at some frequencies the surface resistance of the two materials will be equal. This frequency is known as the crossover frequency and is a measure of how good the superconductor is compared with copper. The surface resistance of a few high temperature superconducting materials is also included in Figure 1. A comprehensive review of different surface resistance measurements around the world is given by Muller (1989). At present the maximum crossover frequency of bulk or thick film material is about 20GHz and of thin film material is about 200GHz (Klein et al, 1989) but these values may be expected to rise with improved materials. It is mainly the reduced surface resistance over conventional conductors which makes superconductors attractive to applications. Hence most of the applications will be at frequencies below the crossover frequencies.

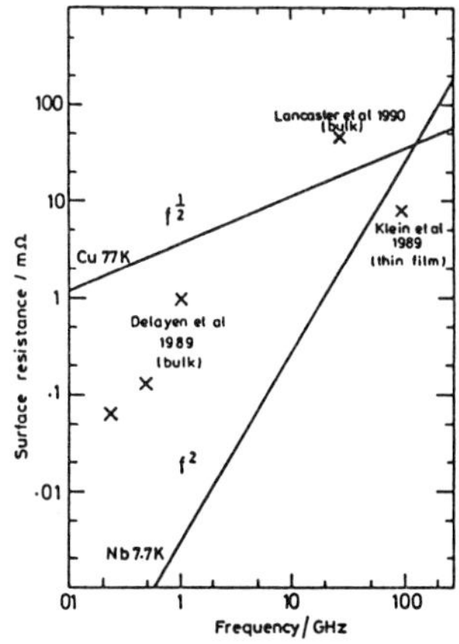

Figure 1

Measured surface resistance of YBCO at 77K

3. PLANAR TRANSMISSION LINES

Planar TEM mode transmission line circuits, which include microstrip and stripline, form a central role in major microwave applications. As well as performing interconnections between components they form important components and signal processing devices themselves.

Figure 2 shows a number of different geometries for these circuits; some geometries are more suitable for superconducting circuits than others as discussed below.

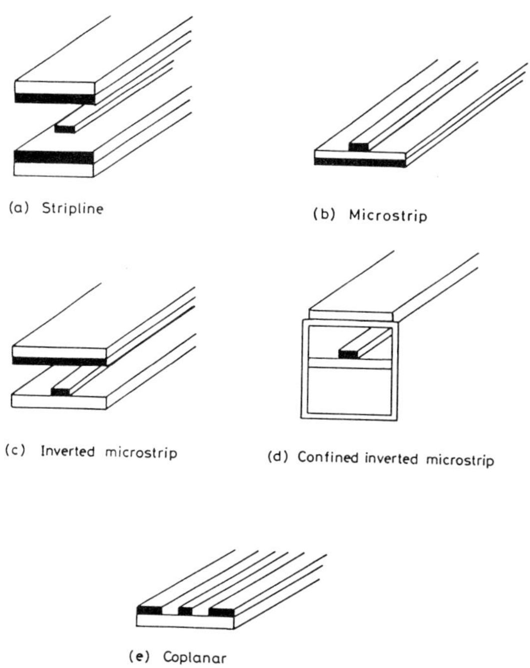

Figure 2 Planar circuits

There are three types of loss associated with planar circuits. The ohmic loss due to the finite conductivity of the conductor, the dielectric loss due to the conductivity associated with the support dielectic and the radiation loss associated with the radiation of electromagnetic energy from the structure. Under most practical circumstances the conductor losses dominate, and this is the reason for the use of a superconductor in these planar circuits. Numerical values for these losses are discussed in Section 4 on resonators. It is clear that once a good superconductor is used, if the total loss in the transmission line is to be reduced still further, then the dielectric and radiation losses must also be reduced. The stripline geometries of Figure 2(a) reduces the radiation loss to a negligible amount as does the confined inverted microstrip of Figure 2(d). The inverted microstrips of Figures 2(c) & (d) allow the majority of the field to be in air, thus reducing the dielectric loss substantially. In addition with the present state of technology it is difficult to produce a multiple layer structure superconductor/dielectric/superconductor by deposition of the three layers so the coplanar line of Figure 2(e) seems attractive because of its simple construction. Of course the three layer structure can be made by deposition of superconductors on both sides of the dielectric crystal. However, thin dielectric layers for certain applications discussed below are difficult to fabricate by this method. Another technological problem which applies to both thin films and thick films is substrate compatibility with the film. Good epitaxially grown thin films are available on strontium titanate. However, this substrate is wholly unsuitable for microwave applications because of its very large dielectric loss tangent and dielectric constant. Other lower loss substrates have more recently been introduced; they include $LaGaO_3$ and $LaAlO_3$ (Sandstorm 1988, Simon 1988). A similar problem exists with thick films where a good substrate would be alumina, widely used for conventional microwave circuits. Superconducting circuits have been shown to work on alumiña (Abell, 1990). However, these films need to be thick and only the upper layers are superconducting, because an amorphous layer of non-superconducting material forms between the substrate and the superconductor due to reaction with the alumina. One structure of transmission line which helps to overcome these problems is the inverted microsrip of Figures 2(c,d) where the field is mainly between the dielectric supports, producing current flow on the upper surface of the superconductor rather than between the dielectric and superconductor as in conventional microstrip.

4. RESONATORS

Resonators form the basis for a number of microwave device applications such as oscillators and filters. They are also of use in the characterisation of superconducting and dielectric materials (Takemoto 1989). Two types of resonator geometry are commonly used; the linear resonator and the ring resonator shown in Figs. 3(a) and (b) respectively. The linear resonator has the advantage that analysis of the coupling of the microwave signal to the circuit is available in closed form (Edwards 1981) whereas with the ring resonator the coupling gap size has to be determined

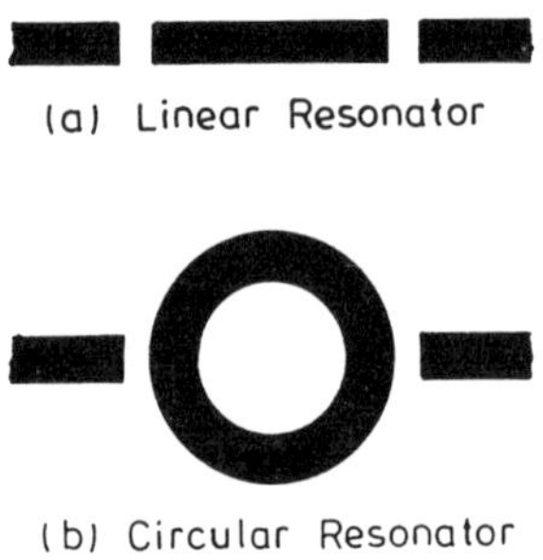

Figure 3

experimentally. An advantage of the ring resonator is that the radiation loss is lower (Edwards 1981).

When a wide microstrip line of the type in Fig. 2(a) is formed into a linear resonator the Q values associated with the various types of loss are

$$Q_c = \left(\frac{\omega \sigma_c \mu}{2} \right)^{\frac{1}{2}} d$$

$$Q_d = \frac{\omega \epsilon}{\sigma_d} = \frac{1}{\tan(\delta)}$$

$$Q_r = Z_o \frac{\pi^2 c^3 \epsilon}{\omega^2 d^2} \qquad (d \ll \lambda)$$

where Q_c, Q_d and Q_r are the conductor Q, the dielectric Q and the radiation Q respectively. σ_c and σ_d are conductivities of the conductor and dielectric and $\tan(\delta)$ is the loss tangent for the dielectric material. Figure 4 shows a comparison of these results graphically. In the calculation of the radiation Q the approximation of two parallel wires with opposing currents has been taken in order to simplify the calculations. This represents a worst case situation because as the microstrip becomes wider the radiation will be reduced.

A number of authors have demonstrated resonators using planar circuit techniques using both low and high temperature superconductors. DiNardo, Smith and Arams (1971) demonstrated a microstrip resonator using thin film lead on an alumina substrate showing Q values as high as 5×10^5 at 14.3GHz. More recently resonators have been constructed out of high temperature superconductors on a number of different types of planar device. Table 2 summarises some of the results. It can be seen that there are a wide variety of results reflecting the different geometrical structures and substrates. The results of Valunzuela (1989) were limited by the radiation loss of the circuit, which produces a substantially lower Q.

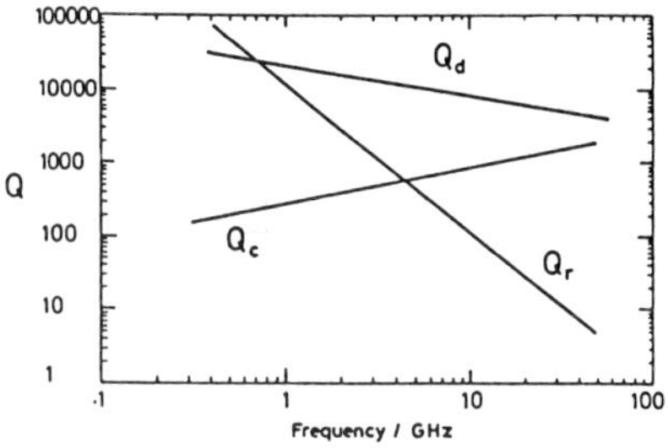

Figure 4 Comparison of Q due to conductor, radiation and dielectric losses.

5. KINETIC INDUCTANCE EFFECTS

From the reactive part of the surface impedance for a superconductor given in Equation (2) a surface inductance can be written as

$$L_s = \mu_o \lambda$$

For a strip of conductor width w, the inductance per unit length is therefore given by

$$L_k = \frac{\mu_o \lambda}{w} \qquad \text{........(3)}$$

This inductance is called the kinetic inductance. This is because the energy storage mechanism associated with the inductance is the kinetic energy of the super electrons. The use of kinetic inductance has been discussed by a number of authors (Mason and Gould 1969, Pond et al 1987, Hohenwarter et al 1989) for use in the slowing of the electromagnetic waves on transmission lines. This can easily be understood by considering a wide microstrip line of the type of Fig. 2(a), where the width of the line is w and the depth of dielectric is d (w >> d). The capacitance and inductance

Author	Superconductor/dielectric/ circuit	Frequency GHz	Q	T
DiNardo et al, 1971	lead/lucalox/microstrip	14.3	2×10^5	4.2K
McAvoy et al, 1989	lead/sapphire/stripline	7.8	6×10^5	
	YBCO/sapphire/stripline	2.8	2×10^4	
DiIorio et al, 1988	YBCO/Nb/microstrip	1	$R_s = 1\text{m}\Omega$	4.2K
Takemoto et al, 1989	YBCO/Gold/MgO/microstrip	6.6	83	40K
Valenzuela et al, 1989	YBCO/MgO/Coplanar	8.8	1300	77K
	YBCO/MgO/Coplanar	9.1	3300	4.2K

Table 2 Examples of Planar Circuit Resonators

per unit length of such a transmission line can easily be shown to be

$$C = \frac{\epsilon w}{d} \qquad \text{........(4)}$$

$$L_m = \frac{\mu d}{w} \qquad \text{........(5)}$$

The phase velocity on the TEM transmission line is

$$v_p = \frac{1}{\sqrt{(LC)}} = \frac{1}{\sqrt{((2L_k + L_m)C)}} \qquad \text{........(6)}$$

Substituting in the values for L_k, L_m and C gives

$$v_p = \frac{c}{\left(\frac{2\lambda}{d} + 1\right)^{\frac{1}{2}}} \qquad \text{........(7)}$$

where c is the velocity of the wave on the transmission line for large d. There is a slowing of the wave for values of dielectric thickness less than the penetration depth. Such thicknesses of dielectric are rather impractical. However, Equation 7 can be modified (Swihart 1961) if the superconducting film thickness is taken into account.

$$V_p = \frac{c}{\left[1 + \frac{2\lambda}{d}\coth\left(\frac{t}{\lambda}\right)\right]^{\frac{1}{2}}} \quad \text{........(8)}$$

For thin dielectrics and thin films it is possible to get a showing of the wave on the transmission line. Of course with thin conductors the attenuation rises substantially. This is one of the reasons why superconductors are proposed for this application. It should also be noted that the phase velocity is dependent upon λ, and λ can be altered by temperature, magnetic field or a d.c. bias current, enabling either variable delay lines or variable frequency resonators to be made. (Anlage 1989, Blunt 1989). Table 3 shows a summary of some experimental results.

Author	Structure	d/nm	t/nm	Slowing/%
Mason et al	Indium/tantalum oxide	30	40	2.4
Hohenwarter et al	Nb/Nb_2O_5	20	15,35	.7
Pond et al	NbN/Si:H	25.0	15	.7

Table 3

Calculations by Pond (1987) indicate that for a niobium nitride microstrip line with a hydrogenate silicon dielectric layer a slowing of about 0.1% can be obtained for layers of dielectric and niobium nitride of thickness 10nm. The calculated loss due to the surface resistance of the superconductor at 2GHz in this case is about 10^{-4} Np/m although for this particular dielectric the dielectric losses are much higher (1 Np/m).

6. FILTERS

Filters are distinguished from resonators in that they are designed to operate in a constant impedance system usually of 50Ω. Thus, in the passband they will have a minimum loss, whereas resonators generally play off this loss against a higher Q. It is not the intention to describe how planar filters are designed in this paper - there are a number of excellent books (Gupta, 1981) and some very compre-

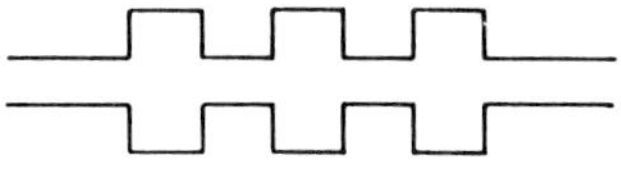

(a) Low pass filters

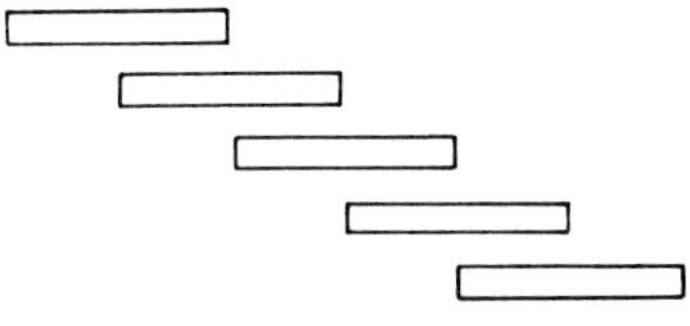

(b) Band pass filters

Figure 5 Planer circuit filters

hensive filter design computer programs. Figure 5 shows examples of two filters. The filtering action is obtained either by the cascading of different impedance lengths of transmission line (Figure 5(a)) or by the use of frequency selective coupling between transmission line segments (Figure 5(b)). There are many different forms of microstrip filter.

The advantage of using superconducting elements for filters is that they can provide low loss and high Q frequency responses. Low loss filters are particularly important at the front end of sensitive receivers to improve the noise and interference characteristics. For example, a low loss filter made out of superconducting elements has been designed to operate with deep space probes where an extremely sensitive system with low noise temperature is required (Bantista et al, 1985). These filters built by Bantista were designed to operate at a center frequency of 8.45GHz with an equal ripple bandwidth of 150MHz. Filters were made out of NbTi, lead and copper for comparison purposes. The copper filter had a 1dB insertion loss at room temperature which reduced to 0.5dB at 4.7K while the superconducting filters had a loss of 0.1 ± 0.06dB. The theoretical loss for the superconducting filter was 10^{-4} dB. This was not achieved due to losses within the SMA connectors and contacts.

A filter made out of high temperature superconductors has been demonstrated by Guo et al (1989). The filter had a center frequency of 1.6GHz, a bandwidth of 333MHz and a Tchebyscheff response with a 1dB ripple within the passband. It was constructed in stripline with aluminium ground planes and three YBCO filter elements. Results show an average insertion loss within the passband of 0.54dB for the superconducting filter and 1.8dB for an identical filter with brass elements measured at 77K.

7. DELAY LINES AND DISPERSIVE FILTERS

Figure 6(a) shows a spiral stripline delay line of the type studied by Lynch et al (1983). Delay lines of the Meander line structure of Figure 6(b) had been demonstrated earlier by Gandolo et al (1968) and Passow et al (1972). The structure of the spiral delay line consisted of a niobium ground plane, a sapphire (or silicon) dielectric polished crystal of thickness of the order 100μm. This was followed by a spiral circuit in niobium and then the upper dielectric layer and finally the second ground plane of niobium. The reason why these delay lines need to made out of superconductor is solely that the losses are reduced when the lines are miniaturised. A spiral delay line was demonstrated in 1983 with a delay of 27ns, an impedance of 52Ω and -47db crosstalk between adjacent spiral loops (Lynch 1983).

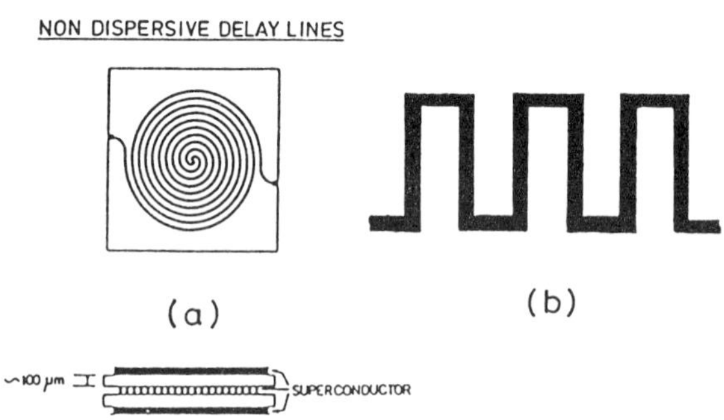

Figure 6 Delay lines

Delay lines are of great importance in signal processing applications, not only because of their inherent use, but also because once an electrical delay is obtained transversal filters of both dispersive and non-dispersive types can be constructed. It should be noted that Surface Acoustic Wave (SAW) filter technology has developed because of the inherent long delay mechanism associated with an acoustic wave travelling over the surface of a crystal (Morgan 1985). Signal processing elements constructed out of superconducting delay lines are seen as large bandwidth and center-frequency devices whereas SAW devices go up to bandwidths of about 1GHz and have longer delays. A comparison of SAW devices and superconducting devices is given by Stern (1985).

One of the major achievements of the superconducting delay line technology has been the demonstration of a dispersive filter by DiIorio et al (1989). Such a filter is formed by two delay lines (in a spiral) coupled at intervals along their length by backward wave couplers. A section of the delay line is shown in Figure 7(a). To explain its operation assume an impulse on the input which travels down the delay line. When it encounters the first backward wave coupler energy is transferred from the input delay line to the output delay line across the coupling gap. The coupled wave is frequency selected by the coupler transferring only frequencies around c/4l. The coupled wave travels backwards along the second delay line to the output. Thus for a long delay line the impulse is a chirp signal as shown in Figure 7(b). Such dispersive delay lines are useful in pulse compression radar systems, compressive receivers and spectrum analysers (Ralston 1985).

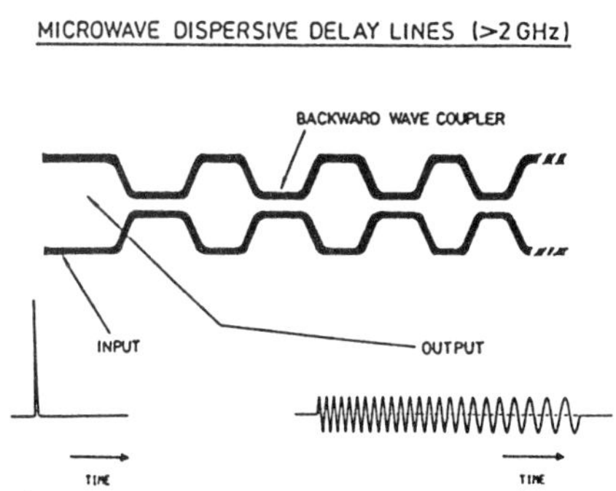

Figure 7 Despersive delay lines

If superconducting dispersive filters are to be used in systems their characteristics, in both amplitude and phase, need to controlled precisely and this has recently been achieved by DiIorio et al (1989) after some 10 years of effort. DiIorio (1989) demonstrates a Hamming weighted dispersive filter with a bandwidth of 2.6GHz, a center frequency of 4GHz and a dispersion or delay of 37.5ns. The filter has excellent agreement with the theoretical design, for example the phase deviation from quadratic was measured to be 3.1° with a design value of 3.5°. The filter is demonstrated in a pulse compression system giving a largest sidelobe level of -32dB. Filters with 6GHz bandwidth, 42ns delay and a phase error of 13° have also been demonstrated. Although the performance specification of SAW devices is generally better (40dB sidelobes) they have a much lower bandwidth.

To date there has been no good demonstration of delay lines made out of the new high temperature superconductors. This is not surprising as the most obvious device to make initially is a resonator which characterises the material much better than a delay line. One attempt at forming a simple delay line out of YBCO is by Hornak et al (1989). The line was 31cm long using $LaGaO_3$ as a substrate. The maximum delay was not reported but should have been of the order 3ns. An increase in delay of about 250ps was observed as the temperature approached T_c. No account was taken of coupling between the lines in the spiral in the design, so that a large crosstalk was observed.

CONCLUSION

This paper has discussed only a section of passive microwave device applications and it can be seen that there is considerable worldwide activity. However, the demonstration, investigation and optimisation of microwave devices using high temperature superconductors is still at a very early stage. Much of the effort world wide at present has gone into the characterisation of the new superconductors. As the superconductors improve, more applications at higher frequencies will become viable.

Whether the applications discussed in the paper will be widely used in systems depends largely on the cost:benefit ratio of the devices. Certainly for low temperature superconductors the ratio was too high and they have only been used in very specific systems. With the use of liquid nitrogen as a coolant the cost of systems should be greatly reduced and as we have seen there is, and is going to be, further study into demonstrating the benefits of superconducting microwave devices.

REFERENCES

Abell, 1990. Private Communication.

Allen, R.S. and Nahman, N.S., 'Analysis and Performance of Superconductive Coaxial Transmission Lines', Proc. IEEE, V. 52, pp. 1147-1154, October 1964.

Anlage, S.M., Shortland, H.J. and Beasley, M.R., 'A Current Controlled Variable Delay Superconducting Transmission Line', IEEE Trans. Mag. V. 25(2) pp. 1388-1391, March 1989.

Bantista, J.J. and Petty, S.M., 'Superconducting NbTi and Pb(Cu) Bandpass Filters', IEEE Trans. Magnetic, V.21(2), pp. 640-643, March 1985.

Blunt, F.J., Edgecombe, C.J. and Campbell, A.M., 'Superconducting Transmission Lines of Controllable Properties', IEE Colloquium on 'The Microwave Applciations of High Temperature Superconductors', No. 1989/116, October 1989.

Bohn, C.L. et al, 'RF Porperties of High-T_c Superconductors', IEEE Trans. Magnetics, V. 25(2), pp. 2406-2409, March 1989.

Delayen, J.R. and Bohn, C.L. "Temperature, frequency and r.f. field dependence of the surface resistance of polycrystaline $YBa_2Cu_3O_{7-y}$', Physical Review B. V.40(7), pp. 5151.

DiIorio, M.S., Withers, R.S. and Anderson, A.C., 'Wideband Superconductive Chirp Filter', IEEE Trans. on Microwave Theory and Techniques, V. 37(4) pp. 706-710, April 1989.

DiNardo, A.J. et al, 'Superconducting Microstrip High Q Microwave Resonators", J. Appl. Phys. V. 42., pp. 186-189, Jan. 1971.

Edwards, T.C., 'Foundations for Microstrip Circuit Design, New York, Wiley, 1981.

Everard, J.K.A. et al, 'High Q Helical Resonator For Oscillators and Filters in Mobile Coomunication Systems', Electronics Letters, V. 25(24), pp. 1648-1650, Nov. 1989.

Gallop, J. 'Applications of Low and High T_c Thin Film Devices in Metrology' This conference.

Gandollo, D.A., Boornard, A. and Morris, L.C., 'Superconductive Microwave Meander Lines', J. Appl. Phys., V. 39, pp. 2657-2660, May 1968.

Guo, G., Maclean, T.S.M., Carey, J.R., Lancaster, M.J., and Alford, N., 'Microwave Stripline Filter for 1.6GHz with Superconducting Elements', IEE Colloquium on 'The Microwave Applications of High Temperature Superconductors', No. 1989/116, October 1989.

Gupta, K.C., Garg, R., Chadha, R., 'Computer Aided Design of Microwave Circuits' Artech. House Inc., Massachusetts, 1981.

Hohenwarter, G.K.G. et al, 'Design of Variable Phase Velocity Kinetic Inductance Delay Lines and Their Measured Characteristics When Fabricated by a Simple Nb Based Process', IEEE Trans. Mag., V. 25(2), March 1989, p. 1100-1103.

Khamas, S.K., Mehler, M.J., Maclean, T.S.M., Gough, C.E., Alford, N. and Harmer, M.A., 'High-T_c Supercondcuting Short Dipole Antenna', Electronics Letters, 24, pp. 460-461, 1988.

Klein, N. et al, 'Millimeter Wave Surface Resistance of Epitaxially Grown $YBa_2Cu_3O_{7-x}$ Thin Films', Appl. Phys. Lett., V. 54(8), pp. 757-759, Feb. 1989.

London, H., Nature, V. 133, pp. 497, 1934.

London, H., Proc. Roy. Soc., V. A176, pp. 512, 1940.

Lynch, J.T., Withers R.S., Anderson, A.C., Wright, P.V. and Reible, S.A., 'Multigigahertz Bandwidth Linear Frequency Modulted Filters Using a Supercondcuting Stripline', Appl. Phys. Lett. V. 43(3), pp. 319-321, 1983.

Martens, J.S. et al, 'Microwave Surface Resistance of $YBa_2Cu_3O_{6.9}$ Superconducting Films', Appl. Phys. Lett., V. 52(21), pp. 1822-1824, May 1988.

Mason, P.V. and Gould, R.W., 'Slow Wave Structures Utilising Superconducting Thin Film Transmission Lines', J. Appl. Phys. V. 40(5), April 1969, pp. 2039-2051.

McAvoy, B.R. et al, 'Superconducting Resonator Performance', IEEE Trans. Mag., V. 25(2), 1989, pp. 1104.

Morgan, D.P., 'Surface Wave Devices for Signal Processing', Studies in Electrical and Electronic Engineering, 19 Elsevier, Amsterdam, 1985.

Muller, G. "Microwave properties of High-T_c superconductors." Presented at 4th Workshop on R.F. superconductivity at KEK Tsukuba-Shi, August 1989.

Passow, C.A., Newhouse, V.L. and Gunshar, R.L., 'An Invesitgation of Some Superconductive Microwave Devices, 1972 Applied Superconductivity Conf. Annappolis Md., May 1972.

Pond, J.M., Claassen, J.H. and Carter, W.L., 'Kinetic Inductance Microstrip Delay Lines', IEEE Trans. Mag., V.23(7), March 1987, pp. 903-906.

Pond, J.M., Claassen, J.H. and Carter, W.L., 'Measurements and Modelling of Kinetic Inductance Microstrip Delay Lines', IEEE Trans. Microwave Theory and Techniques, V. 35(12), pp.1256-1262, December 1987.

Ralston, R.W., 'Signal Processing; Opportunities for Superconductive Circuits', IEEE Trans. on Magnetics, V. 21(2), pp. 181-185, March 1985.

Sandstorm, R.L. et al, Appl. Phys. Lett., V. 53, pp. 1874, 1988.

Simon, R.W. et al, Appl. Phys. Lett., V. 53, pp. 2677, 1988.

Stern, E., 'Comparison of Device Technologies for Signal Processing', Proc. Conf., 'Technology for High Speed Signal Processing', Trondheim, Norway, August 1985, pp. 79-95.

Swihart, J.C., 'Field Solution for a Thin-film Superconducting Strip Transmission Line', J. Appl. Phys. V. 32, pp. 461-469, March 1961.

Takemoto, J.H. et al, 'Microwave Ring Resonator Technique for Measuring Microwave Attenuation in High T_c Superconducting Thin Films', IEEE Trans. MTT, V.37(10), pp. 1650.

Valenzuela, A.A. and Russer, P., 'High Q Coplanar Transmission Line Resonator of $YBa_2Cu_3O_{7-x}$ on MgO', Appl. Phys. Lett. V. 55(10), pp. 1029-1031.

Wu, Z., Mehler, M.J., Maclean, T.S.M., Lancaster, M.J., Gough, C.E. and Alford, N., 'High T_c Superconducting Small Loop Antenna', British Electromagnetic Measurements Confenrence', National Physical Laboratory, U.K., 7-9 November 1989.

Inst. Phys. Conf. Ser. No 111
Paper presented at Int. Conf. on New Materials and their Applications, Univ. of Warwick, 1990

Fabrication and applications of whole-wafer $Nb/Al_2O_3/Nb$ tunnel junctions

M.G. Blamire

Department of Materials Science and Metallurgy, University of Cambridge, Pembroke Street, Cambridge CB2 3QZ, U.K.

Abstract: Progressive development of the precision and reliability of processing technology has allowed the fabrication of complex device structures for a wide range of different applications. This review paper discusses these applications in the context of the materials science and physics of the devices themselves. It will be shown that by extending the current technology, and improving the already precise deposition techniques, a number of novel devices may be developed.

1. INTRODUCTION

Before the early 1980s, tunnel junctions were generally fabricated from soft alloys such as lead and bismuth, and manufacture of more complex circuits than small series arrays was the exception. The IBM computer project was at its height in the late 1970s, and the materials problems which resulted in its slow progress were symptomatic of the methods of fabrication then in general use.

The use of Nb as a replacement for soft alloys gave some improvement in terms of better resistance to thermal cycling and corrosion, but progress was hampered by the complex and technically unreliable Nb oxides which were then used as the tunnel barrier (Halbritter 1985, Darlinski and Halbritter 1987). It was the use of Nb with artificial AlO_x tunnel barriers which really gave the considerable improvement in junction quality and reliability and has enabled the construction of sophisticated circuits of various types (Kroger et al. 1981).

This paper outlines the technology of whole-wafer fabrication, and reviews some of the important applications which have stemmed from its development. Recently an number of more complex structures have been produced using this technique which may lead to the production of new types of superconducting device; these are described in section 6.

The most important aspect of whole-wafer processing is the formation of the tunnel barrier under well-defined conditions on a clean metal surface. This is achieved by depositing the complete tunnel junction structure of base electrode, barrier, and top electrode without breaking vacuum. In the Nb/AlO_x/Nb system, this is usually done by sequentially depositing Nb, Al, and Nb, with the completed Al layer exposed to a low partial oxygen pressure for a certain length of time in order to form the barrier. The tunnel barrier formed by this process thus extends over the complete substrate or wafer, and this structure then requires subsequent processing into separate devices. A number of successful routes have been developed for this processing, and a typical method is described in the next section. All depend on a material-selective isolation process which can isolate areas of the barrier without necessarily removing the lower electrode links (Kroger et al. 1981, Gurvitch et al. 1983, Morohashi et al. 1986).

Although the Nb/AlO_x system is the most widely used, alternatives exist, of which NbN/MgO has received considerable attention (Morohashi and Hasuo 1986, Talvacchio and Braginski 1987, Shoji et al. 1987). In general these alternatives have proved less easy to develop to high quality and it is materials considerations which again limit their progress. We shall discuss this further in section 2.

In general, whole-wafer devices have been applied very successfully to digital circuitry and squids where device parameters such as junction area and barrier conductance are not pushed to extreme limits (Hasuo and Imamura 1989). Attempts to apply them in other areas such as sensor development have generally been less successful, but as we shall see in section 5, considerable advances have recently been made in this area, largely as a result of the better understanding of the materials involved.

2. WHOLE-WAFER DEPOSITION

This section describes the deposition and processing routes in use at Cambridge University (Blamire et al. 1987, 1988), and discusses important differences between these processes and those in use elsewhere.

The whole-wafer deposition is carried out in a diffusion pumped ultra-high vacuum (UHV) system, fitted with dual d.c. magnetron targets for liquid nitrogen getter sputter deposition. The devices are deposited on a variable-width tantalum strip heater, but heat is usually applied only during the Nb base-layer deposition. The system reaches a base pressure of $1\text{-}4\times10^{-7}$Pa. Typical outgassing rates are 10^{-7}Pa/s with the substrate heater at temperature, falling to 3×10^{-8}Pa/s after it has cooled. A UHV system is by no means essential, and indeed we are unusual in using this type of system. However, since we are primarily interested in the materials properties of devices, a very low impurity level is important in allowing the effect of various film growth parameters to be investigated. In addition the preparation of epitaxial films and multilayers necessitates UHV equipment. Although r.f. sputtering and electron beam evaporation has been used as the deposition source, the overwhelming majority of device deposition systems are based on d.c. sputtering.

The standard whole-wafer structure consists of a Nb/Al/AlO_x/Nb multilayer, usually deposited on R-plane single-crystal sapphire substrates. For our system, typical Nb deposition parameters are 6nm/min at a target-substrate distance of 110mm and at a pressure of 0.9Pa, but lower grade vacuum systems require considerably higher deposition rates in order to maintain an adequate critical temperature. On top of this Al, commonly 10nm thick, is deposited, but in clean systems such as ours, where the Al is grown on a smooth single crystal Nb surface, as little as 3-5nm is sufficient. The Nb capping layer is always grown at a low temperature to minimise interdiffusion effects.

A typical oxide barrier is 1-2nm thick (Huggins and Gurvitch 1985). Thermally grown Al_2O_3 has proved to be an extremely reliable barrier material and, if properly formed, has sufficiently few defects for devices to obey theoretical tunnelling predictions down to very low temperatures (Fig. 1). In contrast, directly deposited barriers, both AlO_x (Klapwijk 1987), and MgO (Talvacchio and Braginski 1987) have produced inferior results. Further work is likely to emerge in this area since high temperature superconductivity has stimulated interest in deposited barriers and the development of epitaxial barrier structures (Morris et al. 1990).

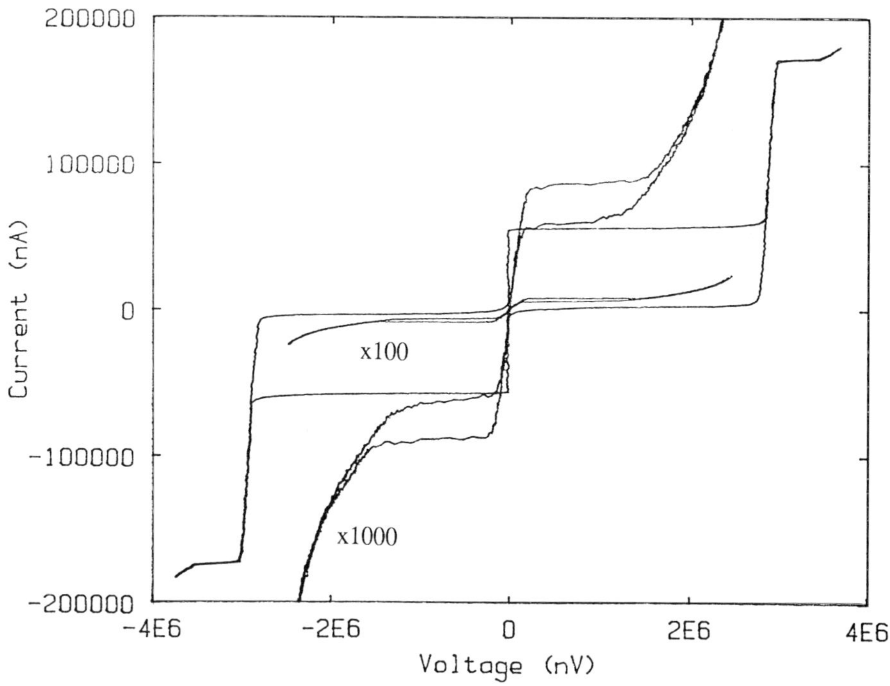

Fig. 1. Voltage current characteristic of a high quality whole-wafer tunnel junction. Measured at 2.2K with 18mT applied parallel to the tunnel barrier. Critical current density $10^6 Am^{-2}$, quality parameter (V_m) 1600mV.

3. WHOLE-WAFER PROCESSING

The whole-wafer method relies on a processing route capable of isolating the region of the structure which is to act as the tunnel junction. The earliest method involved the anodisation of the upper Nb layer between the junctions in order to provide this isolation (Kroger et al. 1981). However, most processes now use the methods developed by Gurvitch et al. (1983) in which the selectivity of CF_4 plasma etching for Nb over Al is exploited. This allows the removal of Nb from around a resist "mesa" which defines the area of the junction, the process terminating at the Al barrier layer.

The isolation between the Nb metallisation layer deposited as the final stage of processing and the lower Nb layer of the original whole-wafer sandwich structure can be provided by several methods; an anodic oxide film grown on the aluminium surface as in the original embodiment of the process (Huggins and Gurvitch 1985, Blamire et al. 1988), by the deposition of a dielectric layer using the same mask (Nakagawa et al. 1986, Morohashi et al. 1985), or an additional resist layer as the lift-off structure (Lumley et al. 1985). All techniques have been used reliably to produce circuits containing thousands of junctions.

4. MATERIALS PROBLEMS IN DEVICE FABRICATION

In the same way that the change of materials from soft to refractory metals resulted in a dramatic improvement in the reliability of devices, so the understanding of the potential failure mechanisms has enabled considerable advances in device processing reliability. The tunnel barrier is the critical element of the structure; although a low Nb T_c will degrade performance, it is a relatively easy parameter to monitor and correct. The barrier, on the other hand, by virtue of its small thickness is difficult to investigate except by the electrical properties of the devices themselves, yet for the same reason is easily damaged during subsequent processing. At an early stage it was realised that failure could be a result of either the initial deposition or of the processing steps, and that it was important to be able to distinguish the two causes, so as to apply the appropriate corrective action. In fact it has turned out that failures are not as clear cut as this; some "processing failures" are the result of problems latent in the original structure.

4.1 Deposition Problems

It has become clear from work produced by a number of groups that the junction characteristics of the completed devices are strongly dependent on both the thickness and morphology of the Al layer which is to form the barrier: it has to be sufficiently thin to prevent suppression of the Nb gap (Huggins and Gurvitch 1985), yet sufficiently thick to avoid oxygen penetrating to the Nb (Morohashi et al. 1985, Imamura and Hasuo 1989b). We have also shown that microstructure of the Al is controlled by the condition of the Nb surface (Blamire et al. 1988). Studies of this sort have required high precision methods of structural analysis. Whilst conventional techniques such as edge-on TEM have been used to probe the structure of the barrier region (Chang et al. 1987, Tanabe et al. 1986), one of the most important techniques in analysing the barrier region has turned out to be anodisation spectroscopy, in which the voltage response during the growth of an anodic film is monitored and can be related to the thickness of the layers of different materials (Huggins and Gurvitch 1985, Blamire et al. 1988, Imamura and Hasuo 1989b). This technique has a resolution capable of resolving individual atomic planes and so it can additionally be used to monitor the epitaxial relationship at interfaces (Blamire et al. 1989b).

4.2 Process-Induced Failures

In processing the devices, apart from obvious causes of failure such as breakdown of the dielectric isolation layer, the main problems stem largely from the deleterious release of any intrinsic stress in the multilayer structure. Much research time has been spent analysing the effects of stress in the Nb layers on the properties of finished devices (Imamura and Hasuo 1989a), and great trouble is now taken to provide conditions in which deposition-induced stress is minimised. However, the concentration of any residual stress which is present by the mesa formation process still leads to an increased failure rate for sub-micron devices (Nakagawa et al. 1986). For this reason alternative approaches have been developed which release the stored stress progressively, and to some extent isolate damaged regions. Our SAWW process (Blamire et al. 1989a) shown in Fig. 2 has already yielded the smallest whole-wafer devices produced anywhere (Kirk et al. 1989), and seems capable of extension to yet smaller areas.

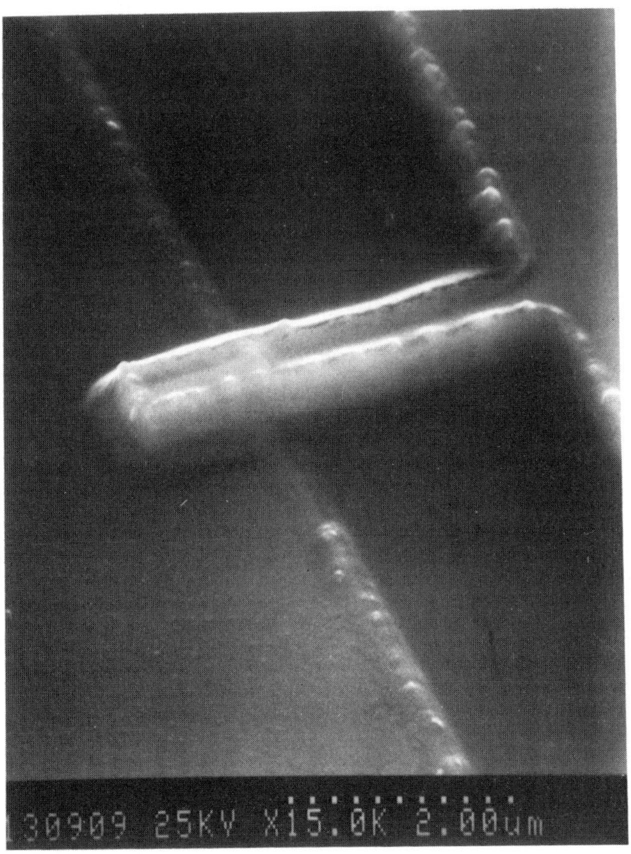

Fig. 2. 0.45μm^2 whole-wafer tunnel junction fabricated using the SAWW processing route. The device is formed by the overlap of the composite Nb/Al counterelectrode strip over the original trilayer structure. The counterelectrode acts as its own etch mask.

5. APPLICATIONS OF WHOLE-WAFER TUNNEL JUNCTIONS

Applications of tunnel junctions range from the large-scale integration involved in digital applications to single junction devices such as r.f. squids. Starting from the simplest we will review these various uses and discuss possible new applications in each area.

5.1 Squids and Sensors

Most of these applications are based on one or two junctions fed by appropriate conventional electronics. Squids operate by sensing flux changes in a loop which is broken by one (r.f. squid) or two (d.c. squid) jospephson junctions (see for example Wellstood et al. 1984). To a large extent the junction is relatively less important in squids than in other devices, since there is no evidence that the quality of the the devices is of particular importance; it is the coupling together of the various inductances in the drive circuits which presents more practical difficulties. Simple squid systems, both d.c. and r.f., have been in commercial production for a number of years and large scale applications such as medical imaging are actively being developed.

A variety of potential sensor applications have been proposed, generally based on the principle of directly counting excitations produced by an incident photon or particle. Very high quality tunnel junctions are required for these applications, since any alternative conduction paths within the devices (such as microshorts) will provide a generally noisy background current on which the signal will be superimposed. A typical application is as a spectrometer for soft X-rays (Gare et al 1989); each incident photon generates a number of excitations (quasiparticles) within the superconducting electrode, and by measuring the charge transferred across the barrier, an estimate of the photon energy is obtained.

5.2 Millimetre Wave Mixers

The quantum theory of mixing demonstrates that under ideal conditions the noise performance of a mixer based on tunnel junctions will greatly exceed that of semiconductor devices (Tucker and Feldman 1985). For this reason a number of research groups are actively engaged in the development of mixers for frequencies ranging from 100 to 1000GHz. Again these devices use only a small number of tunnel junctions, but the twin constraints of a very high conductance barrier and small area junctions have until recently restricted the application of whole-wafer junctions to the lowest frequencies; in general the best practical mixers are currently fabricated from soft alloys (Blundell and Gundlach 1987, Richards and Hu 1989). However, improved processing techniques such as those described earlier are now producing junctions with the correct characteristics for high frequency mixing (Lichtenberger et al. 1989), and the robustness and reliability of whole-wafer devices is allowing the development of space-bourne telescope applications.

5.3 Large-Scale Integration

With the exception of the development of the NBS voltage standards, after the collapse of the IBM project integrated applications of tunnel junctions were more or less dormant. However, over the past 7 years, a number of Japanese electronics companies have been developing the whole-wafer technology required for the development of large scale digital applications.

As their impressive list of publications shows, companies such as Fujistsu have invested heavily in a coherent long-term programme of materials development (see for example Hasuo and Imamura 1989). It is not a coincidence that they, rather than American companies such as IBM and Westinghouse, are now in the position of demonstrating prototype circuits. These devices offer impressive performance; Fujitsu's microprocessor operates at 100 times the design speed of the gallium arsenide circuit on which the design is based, and ETL and NEC have both recently announced the completion and successful testing of a complete though small scale computer system containing a microprocessor and 1K ram together with control logic.

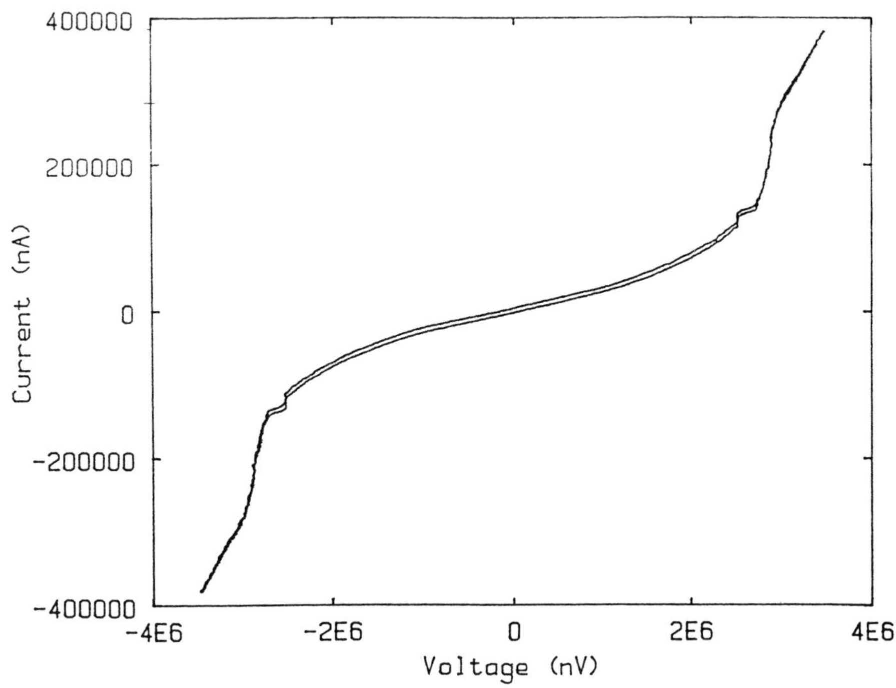

Fig. 3. Characteristics of Nb/AlOx/Al/AlOx/Nb double barrier structure measured at 4.2K. The sub-gap step structure occurs as a result of enhancement of the critical temperature of the middle Al layer as a result of quasiparticle tunnel extraction.

6 FUTURE DEVELOPMENT OF WHOLE-WAFER PROCESSING

The whole-wafer process for the fabrication of tunnel junctions is now a relatively mature technology. Work is still required to enable the construction of high frequency mixers and to improve the quality of sensor devices, but there is unlikely to be any interesting physics emerging from such developments. In contrast the fabrication of more complex structures via whole-wafer technology is an exciting and expanding area.

The idea of putting two tunnel junctions in close proximity is not new (Gray and Willemsen 1978, Mitsen 1979, Faris et al. 1983), and a number of authors have exploited this technique to investigate various aspects of non-equilibrium superconductivity (Gray 1978, Chi and Clarke 1979 and see Schon 1982). Although large gap enhancements under tunnel injection have been predicted (Eliashberg and Ivlev 1985), experimental results have shown enhancements of less than 5% of the equilibrium value at any temperature. These previous experiments have used a three-terminal SIS'IS arrangement fabricated by successive evaporation of soft alloys. This has the twin disadvantage of potential surface contamination during remasking, and the relatively poor quality layers between the barriers.

Using a UHV whole-wafer technique we have fabricated a number of novel structures, including a stacked series array of 4 AlO_x tunnel barriers with Nb interlayers (Blamire et al. 1989c), a set of $Nb/AlO_x/Al/AlO_x/Nb$ devices with the central Al layer in the latter devices as thin as 5nm (Fig. 3), and demonstration three terminal devices (Fig. 4). We have recently shown that tunnel injection at an appropriate voltage bias can result in an enhancement of the Al T_c over the equilibrium value for the same film of over 100%, over an order of magnitude larger than values previously reported. We thus have an electronic method of doubling the transition temperature of a very thin film.

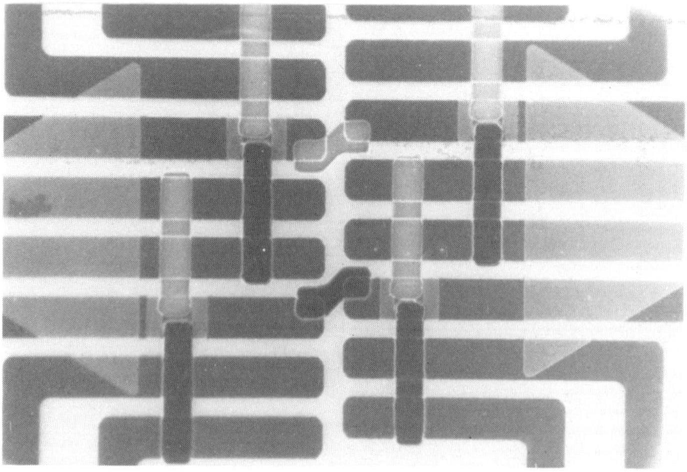

Fig. 4. Micrograph of 4 12x12μm^2 three terminal whole-wafer tunnel devices. The base connection is through the vertical leads, and the two other connections emerge on the horizontal leads.

Although our programme is still at an initial stage, it is clear that complex barrier structures manufactured using UHV techniques will enable more pronounced effects to be observed. In the same way that MBE growth of semiconductor heterostructures has produced and quantum well devices and various quantum magnetic effects, so the eventual aim of this programme is to achieve larger mean-free paths by epitaxial growth of the different layers (Talvacchio and Braginski 1987), in the reasonable expectation that exciting new effects will continue to emerge.

ACKNOWLEDGEMENTS

Most of our work described in this paper was done under the direction of Dr. J.E. Evetts, and performed in collaboration with the other members of the superconductivity and thin films group in the Department. It is supported by SERC, ROE, ESA (Contract No. 7898/88/NL/PB), U.S. Naval Research Laboratory, and the Royal Society.

REFERENCES

Blamire M G, Evetts J E and Hasko D G 1989a *IEEE Trans. Magn.* ***MAG-25*** *1123*

Blamire M G, Huang K H, Somekh R E, Kirk E C G, Morris G W and Evetts J E 1989b *Appl. Phys. Lett.* ***55*** *732*

Blamire M G, Somekh and Evetts J E 1989c *IEEE Trans. Magn.* ***25*** *1135*

Blamire M G, Somekh R E, Barber Z H, Morris G W and Evetts J E 1988 *J. Appl. Phys. 64 6396*

Blamire M G, Somekh R E, Lumley J M, Morris G W, Barber Z H and Evetts J E 1987 *J. Phys. D: Applied Phys.* ***20*** *1159*

Blundell R and Gundlach K H 1987 *Int. J. Infrared & Millimetre Waves 8 1573*

Chang C C, Gurvitch M, Hwang D M and Blonder C W 1987 J. Appl. Phys. **61** 5089

Chi C C and Clarke J 1979 *Phys. Rev.* ***B 19*** *4495*

Darlinski A and Halbritter J. 1987 *J. Vac. Sci. Tech. A5 1235*

Eliashberg G M and Ivlev B I 1985 *Non-equilibrium Superconductivity* (North Holland: D N Langenberg and A.I. Larkin).

Faris S M, Raider S I, Gallagher W J and Drake R E 1983 *IEEE Trans. Magn. MAG-19 1293*

Gare P, Engelhardt, Peacock A, Twerenbold D, Lumley J and Somekh R E 1989 *IEEE Trans. Magn.* ***MAG-25*** *1351*

Gray K E and Willemsen H W 1978 *J. Low Temp. Phys.* ***31*** *911*

Gray KE 1978 *Appl. Phys. Lett.* ***32*** *392*

Gurvitch M, Washington M A and Huggins H A 1983 *Appl. Phys. Lett.* ***45*** *472*

Halbritter J 1985 *IEEE Trans. Magn.* ***MAG-21*** *858*

Hasuo S and Imamura T 1989 *Proc. IEEE* ***77*** *1177*

Huggins H A and Gurvitch M 1985 *J. Appl. Phys.* ***57*** *2103*

Imamura T, Hasuo S 1989a *IEEE Trans. Magn.* ***MAG****-25 1119*

Imamura T, Hasuo S 1989b *IEEE Trans. Magn.* ***MAG****-25 1131*

Klapwijk T M 1987 *Proc. 1986 MRS Fall Meeting*

Kirk E C G, Hasko D G, Blamire M G, Evetts J E and Ahmed H 1989 *Proc. Microcircuit Engineering 89 (in press)*

Kroger H, Smith L N and Jillie D W 1981 *Appl. Phys. Lett.* ***39*** *280*

Lichtenberger A W, McClay C P, Mattauch R J and M J Feldman 1989 *IEEE Trans. Magn.* ***MAG-25*** *1247*

Lumley J M, Somekh R E, Evetts J E and James J H 1985 *IEEE Trans. Magn.* ***MAG-21*** *539*

Mitsen K V 1979 *Sov. Phys. Solid State* ***21*** *2007*
Morohashi S and Hasuo S 1986 *J. Appl. Phys.* ***60*** *3774*
Morohashi S, Hasuo S and Yamaoka T 1986 *Appl. Phys. Lett.* ***48*** *254*
Morohashi S, Shinoki F, Shoji A, Aoyagi M and Hayakawa H 1985 *Appl. Phys. Lett.* ***46*** *1179*
Morris G W, Somekh R E, Barber Z H, Williams E J, Schmahl W W and Evetts J E 1990 *Proc. 1989 MRS Fall Meeting*
Nakagawa H, Nakaya K, Kurosawa I, Takada S and Hayakawa H 1986 *Jpn. J. Appl. Phys.* ***25*** *L70*
Richards P L and Hu Q 1989 *Proc IEEE* ***77*** *1233*
Schon G 1982 *Physica* ***109&110 B*** *1677*
Shoji A, Aoyagi M, Kosaka S and Shinoki F 1987 *IEEE Trans. Magn.* ***MAG-23*** *1464*
Talvacchio J and Braginski A I 1987 *IEEE Trans. Magn.* ***MAG-23*** *859*
Tanabe K, Asano H and Michikami O 1986 Jpn. J. Appl. Phys. **25** 183
Tucker J and Feldman M J 1985 *Rev. Mod. Phys.* ***57*** *1055*
Wellstood F, Heiden C, and Clarke *J. Rev. Sci. Instr.* ***55*** *952*

Inst. Phys. Conf. Ser. No 111
Paper presented at Int. Conf. on New Materials and their Applications, Univ. of Warwick, 1990

Growth of as deposited thin superconducting films of YBCO using plasma assisted laser ablation

D.S. Misra, A.A.C.S. Lourenco and S.B. Palmer
Department Of Physics, University Of Warwick, Coventry, CV4 7AL

ABSTRACT: Thin films of $Y_1Ba_2Cu_3O_{7-\delta}$ are grown using plasma assisted laser ablation. Nd:YAG as well as excimer laser are used for the deposition. A comparison between various superconducting properties of insitu grown YBCO films by Nd:YAG and excimer laser is presented. Influence of the presence of oxygen plasma on the growth of YBCO films is described. Preliminary results of various measurements on a YBCO/PrBCO/YBCO heterostructure are also presented.

1 INTRODUCTION

As deposited thin superconducting films of $Y_1Ba_2Cu_3O_{7-\delta}$ (referred to as YBCO hereafter) have been grown successfully by several groups using a variety of techniques (Leskela et al 1989, Humphreys et al 1990). Laser ablation is one such technique using which YBCO films with transition temperatures (T_c) of 85-90 K and critical current density (J_c) of $10^5 - 10^6$ A/cm^2 have been grown epitaxially on $SrTiO_3$ (100) substrates by different groups (Wu et al 1987, Koren et al 1988, Witanachchi et al 1988, Koren et al 1989, Ying et al 1989, Chan et al 1989, Singh et al 1989, Bendre et al 1990). However most groups working with laser ablation have concentrated on using excimer lasers which are relatively expensive. Use of other lasers for the growth of as deposited thin superconducting films of YBCO has been somewhat limited. A need for comparison between various properties of in-situ grown thin superconducting films of YBCO by excimer laser and other types of lasers clearly exists. We show below that as deposited thin superconducting films of YBCO can be grown using Nd:YAG laser. A comparison between various properties of the films deposited with Nd:YAG and excimer lasers using a plasma assisted laser ablation technique developed in our laboratory is also presented.

2 EXPERIMENTAL

Deposition of thin films of YBCO is carried out in a system shown in Figure 1. A cross shaped cell of pyrex glass with length 40cm and a diameter of 8cm is used for the ablation. Single crystals of $SrTiO_3$ (100 orientation), polished to optical finish, are used as substrates and held onto a stainless steel block which can be heated to 600° C. Substrates are held onto the block with the silver conductive paint. The substrate temperature (T_s) is measured by connecting a tiny thermocouple onto a dummy substrate. A sintered pellet of YBCO is used as a target. Laser beams are focussed onto the target using a quartz lens. A Nd:YAG laser (λ=1.06μm, pulsewidth=35 ns) is used in the energy range 200-400 mJ and the fluence on the target is 15-20 J/cm^2. The XeCl excimer laser(λ=308 nm, pulsewidth=15 ns), on the other hand, generates a fluence of 1-3 J/cm^2 on the target. A fine cappillary tube is used to carry the oxygen near the substrate. A high tension electrode is present in the system and is used to generate a plasma of O_2 during the deposition. The voltage(dc) during discharge is $\simeq$ 300V and the current is $\simeq$12 mA. The target is manually rotated from outside in such a way that the laser beams scan a 25 mm^2 area on the target. T_s is kept $\simeq$ 550-570 C for all the depositions. No change in T_s is observed even after irradiating the substrates with the plasma of O_2 for one hour. Cooling of the substrate by the laser plume has been noted by Koren et al (1988) and we believe that 570° C is an upper estimate of T_s in the present case. The partial pressure of O_2 is $\simeq$ 180 mtorr during the growth. The deposition chamber is filled to $\simeq$500 torr of O_2 subsequent to the depositins, unless specified otherwise, and the films are cooled at a rate of $\simeq$25° C/min.

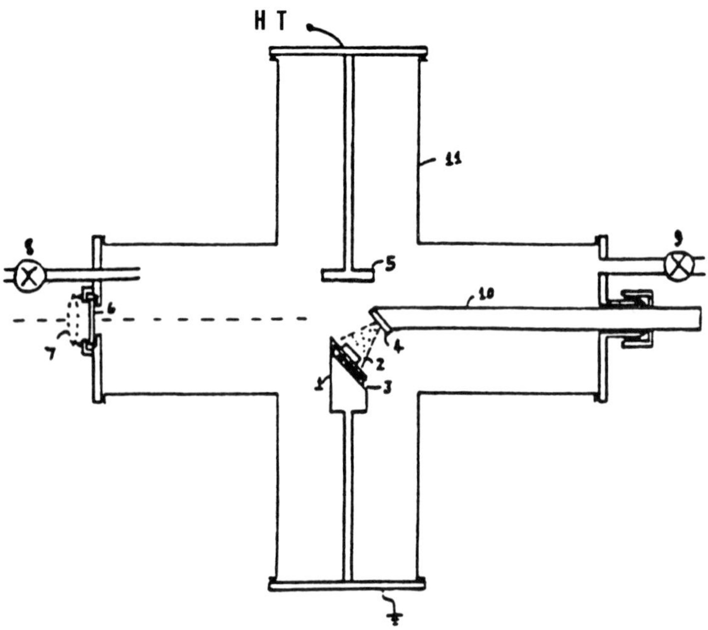

1. Substrate Holder 2.Substrate 3.Heater
4. Target 5.Anode(High Tension Electrode) 6.Quartz Window
7.Biconvex Quartz Lens 8. Needle Valve 9. Valve to Pump
10. Stainless Target Holder Rod 11. Pyrex Glass Cross

Figure 1: Plasma assisted laser deposition system.

Films are used for various measurements in their as deposited form without any post annealing. Gold is used as contact material onto the top of the films. A conventinal four probe technique is used for measuring dc resistivity and J_c(T). A criteria of 1.0

μV is used for J_c measurement. No patterning is done on our films (Humphreys et al 1989) and we believe that the J_c values reported in Table 1 are the lower estimates in our films.

3 RESULTS AND DISCUSSION

As deposited films are shining, smooth and black in appearance. We also, however, observe the presence of microparticles on the top surface of the films under scanning electron microscope in agreement with Koren et al (1989). Scanning electron micrographs showing the top surface of the films prepared using Nd:YAG and excimer laser, respectively, are shown in Figure 2(b) and (a).Microparticles can clearly be seen. Interestingly, the density of the microparticles is much smaller for the film prepared by excimer laser. It may be related to the wavelength of laser, as also noted by Koren et al (1989).

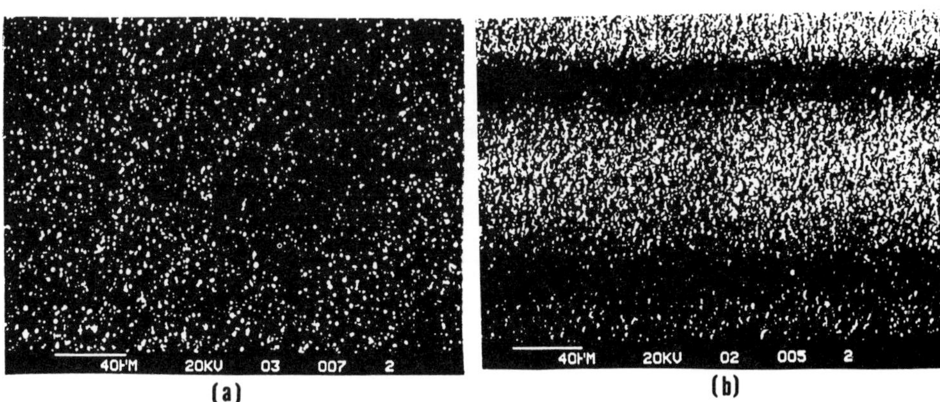

Figure 2: Electron micrograph of the top surface of the film (a) deposited using XeCl excimer and (b) Nd:YAG laser.

X-ray diffraction (XRD) has been done to compare the epitaxial growth of the films deposited using the Nd:YAG and the excimer laser . An XRD plot of 2θ vs intensity is shown in Figure 3. Strong and sharp peaks arising due to reflections from (00L) planes of YBCO can be seen in the films deposited using Nd:YAG as well as excimer laser. Only a small amount of poly crystalline phase (arising due to reflection from (110) and (103) planes of YBCO) are present. This implies that good quality epitaxial films of YBCO can be deposited using both Nd:YAG and excimer lasers. Resistance as a function of temperature (R-T) of the films deposited using Nd:YAG and excimer laser is shown in Figure 4. Although the films deposited using Nd:YAG and excimer laser show similar resistance at room temperature , their resistivity (ρ) differ. A typical film deposited with Nd:YAG laser shows $\rho(300\ \text{K}) \simeq 800$ $\mu\Omega$ .cm wheras the one deposited with an excimer laser shows $\rho(300K) \simeq 200\mu\Omega$.cm.

The films deposited with Nd:YAG as well as excimer lasers show metallic characteristics above the onset of superconductivity. The resistance of the films decreases as a function of temperature as shown in Figure 4. The superconducting transition begins at $\simeq$ 92 K. Zero resistance is observed at $\simeq$ 82 K in the films deposited with Nd:YAG and excimer lasers. Preliminary results described above show that as deposited superconducting thin films of YBCO can be grown epitaxially onto the $SrTiO_3$ (100) substrates using Nd:YAG and excimer lasers with $T_s \simeq 550^o$ C. We believe that the epitaxial growth at 550° C is possible due to presence of oxygen plasma. The influence of O_2 plasma on the growth of YBCO films is demonstrated below. We also show that the post deposition filling of the chamber with high partial pressure of O_2 and the subsequent slow cooling of the films may not be necessary. Heterostructures of YBCO/PrBCO/YBCO have also been grown in our laboratory and results of various measurements on the heterostructure are also presented.

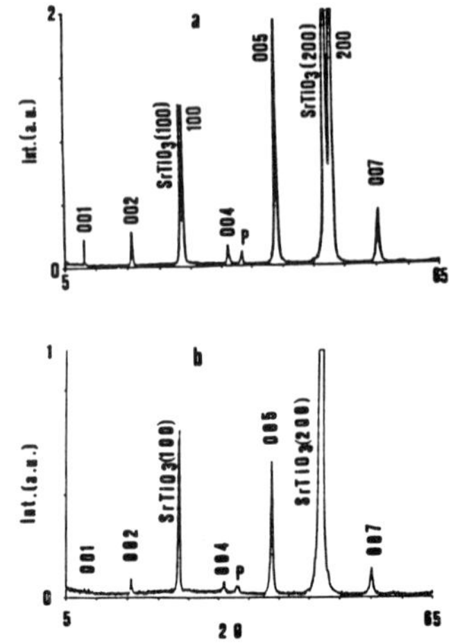

Figure 3: XRD plot of the YBCO film (a) deposited using XeCl excimer and (b) Nd:YAG laser.

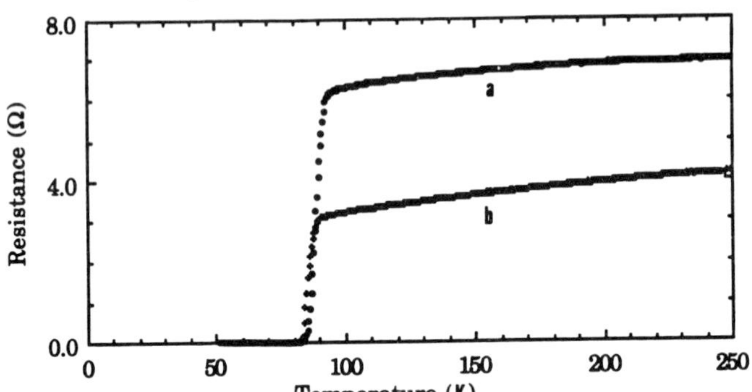

Figure 4: Resistance vs temperature curves of the YBCO films (a) deposited using XeCl excimer and (b) Nd:YAG laser.

S1, S2, and S3 are the films grown in different conditions. S1 was deposited in $\simeq$ 180 mtorr partial pressure of flowing molecular oxygen; the high tension electrode was not used at any stage so there was no plasma of O_2. The deposition chamber was filled to a high partial pressure of O_2 subsequent to deposition ($\simeq$ 500 torr) and the film was cooled quickly by switching the heater off (cooling rate $\simeq 25^o$/min). S2 was deposited in the presence of a O_2 plasma and was cooled at the same rate as S1 with the low pressure O_2 plasma retained. S3 was deposited in the presence of O_2 plasma and the deposition chamber was filled with 500 torr of O_2, the plasma was shut off prior to cooling the film at the same rate as S1 and S2.

Various superconducting properties of S1, S2, and S3 are presented in Table 1. XRD plots of 2θ vs intensity for the films S1, S2, and S3 are shown in Figure 5.

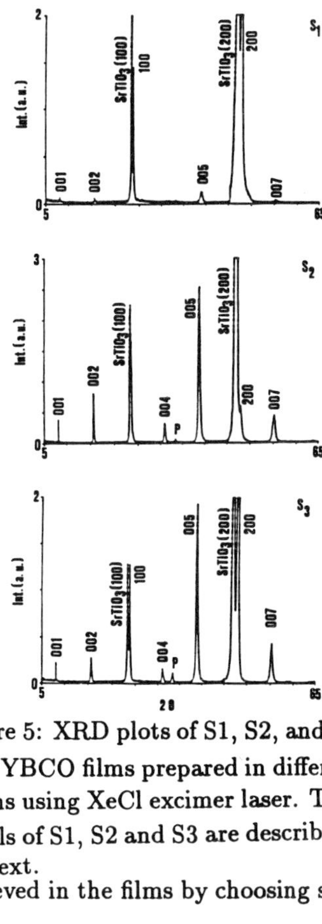

Figure 5: XRD plots of S1, S2, and S3 ; three YBCO films prepared in different conditions using XeCl excimer laser. The details of S1, S2 and S3 are described in the text.

S1 deposited without plasma shows strong peaks arising due to reflections from (100) and (200) planes of YBCO. The peaks arising due to reflections from (00L) planes are weak in intensity. S2 deposited and cooled in a plasma in contrast shows strong peaks arising due to reflections from (00L) planes and the peak due to reflections from (100) and (200) planes are reduced significantly. S3 on the other hand shows strong peaks due to reflections from (00L) planes as well as (100) and (200) planes. S2 and S3 are different only in one sense that subsequent to deposition of S2, the cell is not filled to a high pressure of O_2 and the film is cooled in plasma, a fact which appears to have some influence on the nature of growth of YBCO films. Small amount of polycrystalline phase (denoted by P in Figure 5) is also present in S2 and S3. S1 seems to have grown with its a axis perpendicular to the surface of $SrTiO_3$. S2 has a dominantly c axis growth and S3 appears to have a mixture of the two. The nature of growth of S1, S2 and S3 seems to tie up very well with their thickness as shown in Table 1. S1 is $0.1\mu m$ thick, S2 is $0.27\mu m$ and S3 is $0.21\mu m$. It also appears that a desired growth can be achieved in the films by choosing suitable deposition conditions.

ρ-T results on S1, S2, and S3 are shown in Figure 6. S1 is nearly metallic in the temperature range 90 K$\leq$ T $\leq$ 300 K. A broad superconducting onset begins in S1 $\simeq$ 90 K and the zero resistance is reached at $\simeq$ 60 K. S2 and S3 in contrast are very metallic down to $\simeq$ 90 K, where a superconducting trasition begins. Zero resistance temperature (T_c(Zero)) for S2 is $\simeq$ 85 K and for S3 is $\simeq$ 87 K. The width of transition, ΔT, for S2 and S3 is $\simeq$ 4K and 2K respectively. A comparison of various superconducting properties of S1, S2 and S3 is also shown in Table 1. Results of ρ-T measurements again emphasize the importance of the presence of the O_2 plasma during growth. S1, which we believe to be deficient in oxygen , deposited without plasma has a T_c(Zero)$\simeq$ 60 K which is about 25 K lower than that of S2 and S3. S2 and S3 which are cooled in different environments of O_2 on the other hand do not show strikingly different superconducting properties. Their resistivity in the normal state (at 250 K) differ by a factor of 2. S2 has a J_c greater than 10^4 A/cm^2 at 70 K while S3 shows a similar value at 82 K. Small difference in superconducting properties

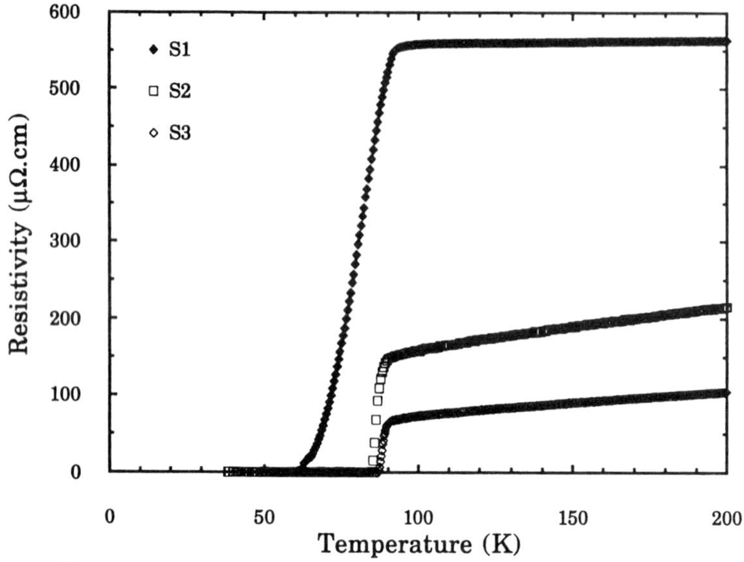

Figure 6: Resistivity vs temperature plots of S1, S2 and S3.

of S2 and S3 may be related to the fact that S2 was grown on a used substrate while S3 was grown on a fresh substrate. It may be mentioned that various properties of S2 and S3 are comparable to those reported by other groups in their as deposited films grown by laser ablation and other techniques. Deposition of heterostructures of

Table 1

Sample	thickness	ρ(250K)	J_c(A/cm^2)	T_c(Zero)
S1	0.1μm	563.4$\mu\Omega$.cm	-	60 K
S2	0.27μm	229.1$\mu\Omega$.cm	10^4 at 70 K	85 K
S3	0.21μm	115.8$\mu\Omega$.cm	10^4 at 82 K	87 K
YBCO(B)	0.2μm	235.6$\mu\Omega$.cm	-	79.5 K
YBCO(T)	0.2μm	192.4$\mu\Omega$.cm	-	80.5 K

YBCO/PrBCO/YBCO is done in the same cell in one straight step without breaking the vacuum at any stage. Two sintered pellets of $Y_1Ba_2Cu_3O_{7-\delta}$ and $Pr_1Ba_2Cu_3O_7$

are used as targets and the laser beam is switched from one to other by manipulating the target holder from outside. A $SrTiO_3$ substrate with predeposited gold contact was used for growing the heterostructure. A portion of the gold contact on the $SrTiO_3$ is covered during deposition to facilitate the contact with the bottom layer of YBCO. The thickness of top and bottom layer of YBCO are 0.2μm and that of PrBCO layer in between is 150 A^o. Deposition of all three layers is done in presence of a O_2 plasma at $\simeq$ 180 mtorr and $T_s \simeq 570^o$ C. The device is soaked in $\simeq$ 500 torr of O_2 subsequent to the deposition and then cooled in the manner as S1, S2 and S3. Gold contacts on the top of the heterostructire are deposited in a separate coating unit. Results of ρ-T measurements on the top and bottom YBCO layer of the heterostructure, XRD and the resistance vs temperature of the junction are reported together with preliminary results of the current-voltage (I-V) measurements done on the junction.

XRD result on the heterostructures show similar epitaxial growth as S3. Strong and sharp peaks arising due to reflections from (00L) planes and (100) and (200) planes of YBCO are seen (see Figure 7). Results of ρ-T measurements done on the top and bottom layer of the heterostructure YBCO(T) and YBCO(B) are presented in Figure 8. Resistivity of the YBCO(T) and YBCO(B) are 192.4 $\mu\Omega$.cm and 235.6 $\mu\Omega$.cm

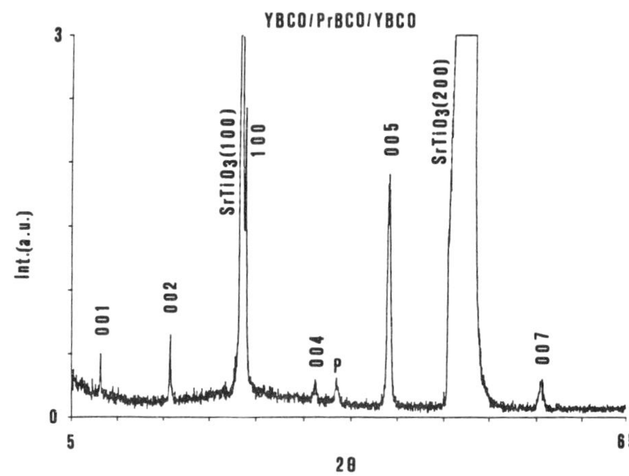

Figure 7: XRD plot of the heterostructure YBCO/PrBCO/YBCO showing the sharp and storng peaks arising due to reflections from (00L), (100) and (200) planes of YBCO along with the peaks due to $SrTiO_3$.

respectively (see Table 1). Top and bottom layers of YBCO show similar resistivity as S2 and S3. As is evident from Figure 8 the YBCO(T) and YBCO(B) layers are not as metallic as S2 and S3 perhaps due to being in close contact with the insulating PrBCO layer. T_c(Zero) for YBCO(T) and YBCO(B) are 80.5 and 79.5 K respectively while the ΔT values are 6.0 K and 7.0 K , indicating that the superconducting properties of the top and bottom layers of the heterostructure are similar.

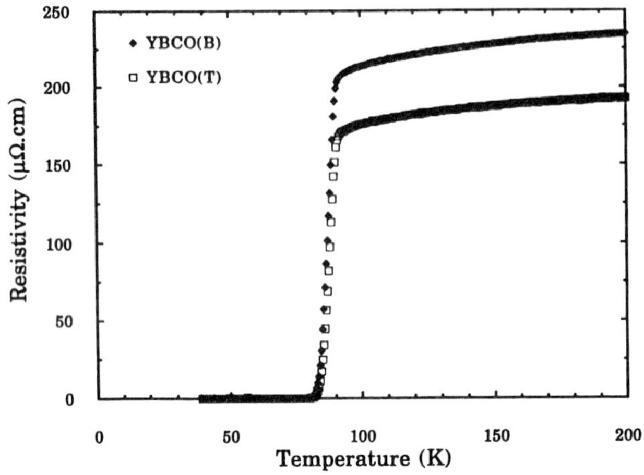

Figure 8: Resistivity vs temperature curves of the top and bottom layers of YBCO of the heterostructure (YBCO(T) and YBCO(B) respectively).

The resistance of the heterostructure (measured with current flowing from the YBCO(B) through the PrBCO layer to the YBCO(T) layer) is plotted as a function of temperature in Figure 9. The heterostructure shows a resistance R_H of $\simeq$ 16.0 Ω at room temperature which is greater by about a factor of 3-4 as compared to the

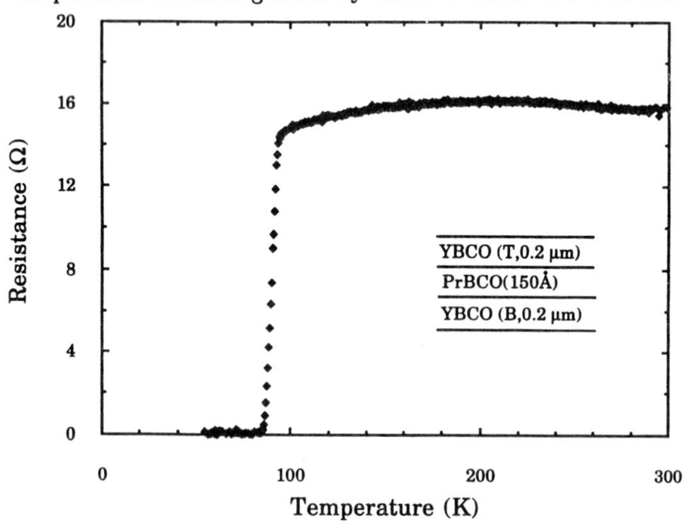

Figure 9: Resistance vs temperature plot of the junction of the heterostructure.

resistance of YBCO(T) or YBCO(B). It should be noted that two other heterostructure reported in the literature both of very different geometries also have a resistance of $\simeq 16\Omega$ at room temperature (Rogers et al 1989, Hiraka et al 1990). $R_H(T)$ shows a slightly semiconductive behaviour down to 200 K followed by a nearly metallic behaviour down to 90 K where the superconducting onset begins. The zero resistance is reached at $\simeq$81 K. We do not observe a metallic behaviour followed by a semiconductive one as seen by Rogers et al (1989). It may be related to the fact that the thickness of PrBCO layer in our case is much smaller than theirs. A background resistance observed by Hiraka et al (1990) is not seen in the present case. I-V measurements on the heterostructure were carried out at various temperature in a constant current mode. I-V curves at 60 K shows S-N-S like nonhystertic characteristic similar to as observed by Rogers et al (1989). A constant voltage step similar to phase slip centers as reported on conventional superconductors is seen. However the device area is large in our case (a few mm^2) and we are planning to do I-V measurements on devices of various area and different PrBCO layer thickness before we conclude that a true Josephson tunneling is being seen in this novel class of materials.

ACKNOWLEDGEMENT:

Helpful discussion with Dr. Don Mc Paul are gratefully acknowledged. One of us (DSM) feels indebted to Dr. Kruppa for his kind and invaluable assistance during the preparation of this manuscript. Financial support from S.E.R.C for the laser and superconducivity research is acknowledged.

REFERENCES

Bendre S T, Koinkar V N, Vispute R D, Vishwanathan R, Dhote A M, Chaudhari S M, Kanetkar S M and Ogale S B (1990), Solid state comm.,73,345

Chan Siu-Wai, Chase E W, Wilkens B J and Hart D L (1989), Appl. Phy. Lett., 54, 2032

Hirata K, Yamamoto K, Iijima K, Takada J, Terashima T, Bando Y,and Mazaki H (1990), Appl. Phy. Lett., 56, 683

Humphreys R G, Satchell J S, Chew N G, Edwards J A, Goodyear S W, Blenkinsop S E, Dosser O D and Cullis A G (1990), Supercond. Sci. Technol., 1, 1

Humphreys R G, Satchell J S, Chew N G and Edwards J A (1989), Appl. Phy. Lett.,54,75

Koren G, Polturak E, Fisher B, Cohen D and Kimel G (1988), Appl. Phy. Lett. 53, 2330

Koren G, Gupta A, Baseman R J, Lutwyche M I and Laibowitz R B (1989), Appl. Phy. Lett., 55, 2450

Leskela M, Truman J K, Mueller C H and Holloway P H (1989), J. Vac. Sci. Technol., A7, 3147

Rogers C T, Inam A, Hegde M S, Dutta B, Wu X D and Venkatesan T (1989) Appl. Phy. Lett., 55, 2032

Singh R K, Narayan J, Singh A K and Krishnaswamy J (1989), Appl. Phy. Lett., 54, 2271

Witanachchi S, Kwok H S, Wang X W and Shaw D T (1988), Appl. Phy. Lett., 53, 234

Wu X D, Dijkkamp D, Ogale S B, Inam A, Chase E W, Miceli P F, Chang C C Tarascon J M and Venkatesan T (1987), Appl. Phy. Lett., 51, 861

Ying Q Y, Kim H S, Shaw D T and Kwok H S (1989), Appl. Phy. Lett., 55, 1041

Paper presented at Int. Conf. on New Materials and their Applications, Univ. of Warwick, 1990

Chemical vapour deposition (CVD) of high T_c thin films

M. L. Hitchman*, D. D. Gilliland*, D. J. Cole-Hamilton# and S. C. Thompson#
*Department of Pure and Applied Chemistry, University of Strathclyde, Glasgow G1 1XL
#Department of Chemistry, University of St. Andrews, St. Andrews, KY16 9ST

ABSTRACT: This paper reviews the CVD of high T_c films. Some aspects of precursor chemistry are discussed and considerations of deposition processes suggest some strategies for controlling layer growth and improving layer quality.

1. INTRODUCTION

The discovery in late 1986 by Bednorz and Mueller of a ceramic mixed oxide of La, Ba and Cu (LBCO) which became superconducting at 35K, some 12K higher than any previously available material, followed a few months later by that of Wu et al (1987) that by substituting Y for La one could obtain a ceramic oxide (YBCO) which was superconducting at 98K, 21K above the boiling point of liquid nitrogen, has lead to unprecedented levels of research activity in the area of superconductivity. This activity, whilst now more consolidated, continues unabated, and although the ceramic materials are still in the embryonic stage and there is much work to be done before they emerge from the laboratory to make a significant technological impact it is generally agreed that one of the most promising areas for the application of the new superconductors is in electronic circuitry. Circuits with no resistance and hence no time constant and reduced noise will undoubtedly lead to enhanced electronic performance (Gosling, 1988), and for electronic circuits thin films of the new high T_c materials will be required. As a result there has been considerable interest in the use of techniques such as sputtering, laser ablation, electron beam and molecular beam evaporation as well as wet chemical processes. Often with these processes, though, there is a problem associated with low chemical potential of oxygen during deposition which results in an oxygen deficient layer and consequent loss of superconducting properties. The use of CVD for preparing thin superconducting layers has the advantage that deposition could be carried out at a high chemical potential of oxygen. Furthermore, CVD is a chemical process which can be controlled to give the oriented or even epitaxial grain growth that seems to be essential for high T_c characteristics. Also, once thin films of high temperature superconductors with properties suitable for routine microelectronic circuits can be prepared, the deposition onto many samples at the same time, the high degree of both intra- and inter-sample uniformity, and the good conformality of layers on patterned structures all suggest that CVD will be a potentially powerful technique for preparing thin film superconductors.

2. PUBLICATIONS ON CVD

As a result of this potential, since mid-1988 when the first publication describing the CVD of high T_c films was published by Berry et al there have been some 40 further publications reporting on the formation of YBCO with a few decribing the formation of the Bi, Sr, Ca, Cu, O (BSCCO) and the Tl, Ba, Ca, Cu, O (TBCCO) families. In all of these papers detailed information is usually given on the experimental conditions used and Table 1 summarises some of this information.

Table 1 Summary of Publications of CVD of High T_c Superconductors

Author	YBCO Precursors Y	Ba	Cu	Td/ °C	Pd/ torr	Oxidant	Substrate	jd/ nm min^{-1}	Tc/ K	T_0/ K	Jc/ 10^5Acm^{-2}	Analysis
Berry (1988a)	DPM	DPM	ACAC	400	760	O_2+H_2O	MgO	8	80	36	-	SEM XRD
Yamane (1988a)	DPM	DPM	DPM	900	10	O_2	YSZ	13	89	80	-	SEM XRD
Abe (1988)	DPM	DPM	DPM	600-780	76	O_2	MgO/$SrTiO_3$	2-167	85	65	-	SEM XRD EPMA
Nakamori (1988)	DPM	DPM	DPM	400-600	76	O_2	MgO	40-100	76	60	-	SEM EPMA
Zhao (1988)	DPM	HFOD	ACAC	700	5	H_2O	MgO	10-30	90	66	-	SEM XRD EDAX
Yamane (1988b)	DPM	DPM	DPM	900	5	O_2	$SrTiO_3$	33	-	84	0.2	SEM XRD
Panson (1988)	HFOD	HFOD	ACAC	500	760	O_2+H_2O	$SrTiO_3$/Al_2O_3	77	90	70	-	SEM EDAX XRD
Kurosawa (1988)	DPM	DPM	DPM	900	10	O_2	YSZ/Al_2O_3	16-33	-	-	-	SEM XRD
Yamane (1988c)	DPM	DPM	DPM	800	10	O_2	YSZ	16	83	43	-	XRD
Yamane (1988d)	DPM	DPM	DPM	800	10	O_2	YSZ	13-16	-	-	-	SEM XRD
Shinohara (1988)	DPM	HFA	HFA	600	10	O_2	$SrTiO_3$	6	83	65	-	ICP XRD
Zhang (1989a)	DPM	DPM	DPM	650	50	O_2	YSZ	170	93	84	-	SEM EDAX XRD
Oda (1989)	HFA	HFA	ACAC	500-700	10	O_2	Al_2O_3/Si	4-15	-	-	-	AES RHEED EDAX
Watanabe (1989)	DPM	DPM	DPM	850	5	O_2	$SrTiO_3$	-	>90	89	>4	SEM
Dickinson (1989)	DPM	DPM	DPM	600-900	-	O_2	$SrTiO_3$/Al_2O_3	-	90	68	-	SEM EDAX XRD
Noh (1988)	DPM	DPM	DPM	500	76	O_2	$SrTiO_3$	-	90	88	-	SEM EDAX XRD
Tsuruoka (1989a)	DPM	DPM	DPM	600-800	40	N_2O/O_2	$SrTiO_3$/MgO/Si	-	88	79	-	XRD
Tsuruoka (1989b)	DPM	DPM	DPM	800	40	O_2	$SrTiO_3$	15	-	88	-	XRD
Yamaguchi (1989)	DPM	DPM	DPM	750	4	O_2	$SrTiO_3$/ Metal	250-500	92	89	0.013	SEM XRD
Radpour (1989)	DPM	DPM	DPM	780	5	O_2	BaF_2/YSZ	-	-	80	-	-
Inoue (1989)	DPM	DPM	DPM	660-740	0.2	O_2	MgO	16	89	57	-	ESCA XRD EDAX
Kanehori (1989)	DPM	DPM	DPM	700	10-1.5	O_2	$SrTiO_3$	5-13	86	83	-	SEM XRD
Yamane (1989a)	DPM	DPM	DPM	600-900	10	O_2	$SrTiO_3$	12-15	-	91.5	4	SEM XRD
Schmaderer (1989)	DPM	DPM	ACAC/HFA	900	4	O_2	$SrTiO_3$/YSZ	16	-	95	10	SEM XRD

Author	YBCO Precursors Y	Ba	Cu	Td/ °C	Pd/ torr	Oxidant	Substrate	jd/ nm min^{-1}	Tc/ K	T_0/ K	Jc/ 10^5Acm^{-2}	Analysis
Muto (1989)	-	-	-	-	-	-	-	-	-	92	18	EPMA SEM
Kobayshi (1989)	DPM	DPM	DPM	-	-	-	$SrTiO_3/MgO$	-	~85	80	-	XRD
Noge (1989)	DPM	DPM	DPM	850	-	-	MgO	-	-	86	0.1-1	XRD
Yamane (1989b)	DPM	DPM	DPM	800-900	10	O_2	$SrTiO_3/YSZ/MgO/Al_2O_3$	16-70	-	93	0.3	XRD SEM
Peng (1989)	DPM	DPM	ACAC	<300 Plasma	2	O_2	Various	6-16	110	77	-	XRD
Bai (1989)	DPM	DPM	DPM	300-500 Plasma	2	O_2	YSZ	40-60	92	79	-	XRD SEM

Author	BSCCO Precursors Bi	Sr	Ca	Cu	Td/ °C	Pd/ torr	Oxidant	Substrate	jd nm min^{-1}	Tc/ K	T_0/ K	Jc/ 10^5Acm^{-2}	Analysis
Yamane (1988e)	TE	DPM	DPM	DPM	910	1	O_2	MgO	133	-	78	-	SEM XRD
Berry (1988b)	TP	DPM	DPM	ACAC	500	760	O_2	MgO	8-10	83	73	-	SEM XRD
Natori (1989)	TP	DPM	DPM	DPM	750-810	1	O_2	MgO	16	110	80	-	SEM XRD ICP EPMA
Yamane (1989c)	TE	DPM	DPM	DPM	770-850	1	O_2	MgO	16-50	110	77	-	SEM XRD
Zhang (1989b)	TP	DPM	DPM	ACAC	550	2	O_2+H_2O	MgO	35-50	110	75	-	SEM EDAX XRD

Author	TBCCO Precursors Tl	Ca	Ba	Cu	Td/ °C	Pd/ torr	Oxidant	Substrate	jd/ nm min^{-1}	Tc/ K	T_0/ K	Jc/ 10^5Acm^{-2}	Analysis
Zhang (1989c)	DPM	DPM	DPM	DPM	600	50	O_2	Al_2O_3	83	100	94	-	SEM XRD EPMA
Richeson (1989)	CP	DPM	HFOD	ACAC	600	760	O_2+H_2O	YSZ	-	120	100	-	SEM XRD

Notes: Precursor Ligands: DPM dipivaloylmethane; ACAC acetylacetone; HFOD heptafluorooctanedione; HFA hexafluoroacetylacetone; TE triethoxy; TP triphenyl; CP cyclopentadiene. Substrates: YSZ yttria stabilised zirconia.

The three most important parameters for superconductivity are temperature, current density and magnetic field and the critical values of each marking the transition to superconductivity. The factors affecting these transitions will, of course, depend on layer characteristics and these will in turn depend on growth conditions in general and particularly on the chemistry of the deposition process. The effect of reaction parameters on layer properties is, though, at this stage by no means fully understood and clearly it is important that a better understanding of the chemistry of layer growth is obtained, particularly as improvements in layer characteristics are sought. Of course, in some instances recipes may be found to allow layer specifications to be met. But, while recipes for controlling processes can be useful, it is still not very satisfactory if there is no real scientific understanding underlying such recipes. In this paper we discuss the sort of basic understandings of physicochemical characteristics of reactions and deposition processes which are necessary to give increased insights into the factors controlling deposition. Before doing that, though, we first consider some of the features of precursors needed for oxide growth.

3. PRECURSORS FOR CVD

The deposition of metal oxides from the vapour phase has been known for some years, but the development of suitable volatile precursors for each particular materials system is a challenging area of resarch. Basically two approaches may be followed, either involving the introduction of the oxygen from an external source (usually O_2 or H_2O) or involving the decomposition of a precursor or several precursors which already contain M–O bonds. An example of the former is the growth of SnO_2 using $SnCl_4$ and water (Advani 1979). The growth of ZrO_2 films from $Zr(O^tBu)_4$ exemplifies the latter (Bradley and Faktor 1959). Both processes have advantages and disadvantages. In many ways direct decomposition of a single precursor has attractions since only one material is required, the growth system can be very simple, and problems with pre-reaction between the metal compound and O_2 or H_2O, which can lead to deposition in areas other than the substrate, can be avoided.

In the growth of mixed metal oxides, the system is necessarily complex. The problems of pre-reaction may be more severe, particularly if one precursor reacts more readily with O_2 or H_2O than the others, so that selective deposition and serious problems with stoichiometry control will result. Ideally, therefore, the growth of mixed metal oxides from volatile oxygen containing precursors in the absence of an external oxygen source is a highly desirable and challenging target.

What is required for the successful growth of thin films of high T_c superconductors are volatile oxygen containing compounds of the metals which will break down at high temperatures to give metal oxide layers. Many of the precursors that have been used for direct metal oxide deposition are metal alkoxides (Bradley 1989). It is believed that the mechanism of their decomposition is as shown below for a tertiary alkoxide, involving transfer of a hydrogen atom from the carbon atom β to the oxygen to give hydroxy species followed by loss of water and/or alcohol.

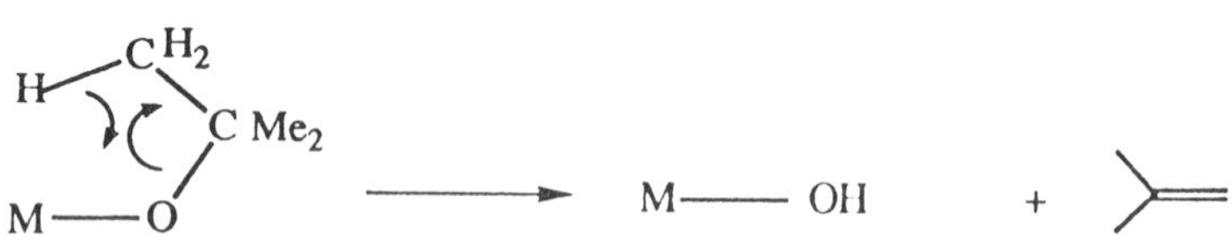

However, a general problem arises because the alkoxide oxygen can donate a lone pair of electrons to a different metal atom to form a bridge. This oligomerises the alkoxide and greatly reduces its volatility. Taking the specific case of group IV alkoxides, bridging does occur to give 6–coordination. The driving force here is the tendency of the metal to attain as high a coordination number as possible. Such bridging can be prevented by the use of bulky groups (e.g. ${}^{t}Bu$). Thus $[Zr(O^{t}Bu)_4]$ is monomeric with the four butoxide groups effectively shielding the metal from further coordination by other $[Zr(O^{t}Bu)_4]$ units (Bradley et al 1978).

Unfortunately, for elements in oxidation state 2 or 3 the problems are even more severe. Bridging can occur to give extended arrays as there are only two or three negative alkoxide groups. This means that extremely bulky alkoxides would be required to prevent oligomer formation and the bulk of these alkoxides will itself reduce the volatility even of monomers (Bradley et al 1954). Our preliminary results with barium, using very bulky alkoxides, suggest that it may be possible to reduce the oligomerisation. However, sublimation appears to be accompanied by decomposition and the sublimate is subsequently involatile (Thompson 1990).

Another problem which has been reported for yttrium alkoxides is that, although they can be synthesised as simple $[M(OR)_3]_m$, they sometimes give oxocentred high nuclearity clusters with bridging and terminal alkoxides, which have very low volatility (Bradley et al 1988; Poncelet et al 1989). The source of the oxygen atom has not been fully elucidated, but it probably comes from an alkoxide or alcohol molecule via a mechanism similar to that shown above rather than from external water or oxygen. Similar problems have been reported for barium alkoxides (Hubert–Pfalzgraf et al 1989). Fortunately copper has been reported to have a volatile alkoxide. $[Cu(O^{t}Bu)]_4$, although tetrameric, is sufficiently volatile to be used as a copper source and, under different conditions, can deposit Cu or Cu_2O or CuO (Jeffries and Girolami 1989).

Given that there are problems with the alkoxides of most high T_c superconductor precursors, a different strategy has been adopted. This involves the use of β–diketonate ligands, which may be considered as being special forms of alkoxides containing a β–γ unsaturated ketone substituent (Figure 1a). Coordination of this ligand to the metal via both oxygen atoms blocks some of the sites which might be used for bridging. In reality the structure is delocalised as shown in Figure 1b. A further attraction of these kind of compounds is that they are insensitive to air and moisture, which makes them easy to handle and use. Some possibilities for R and R' are also shown in Figure 1.

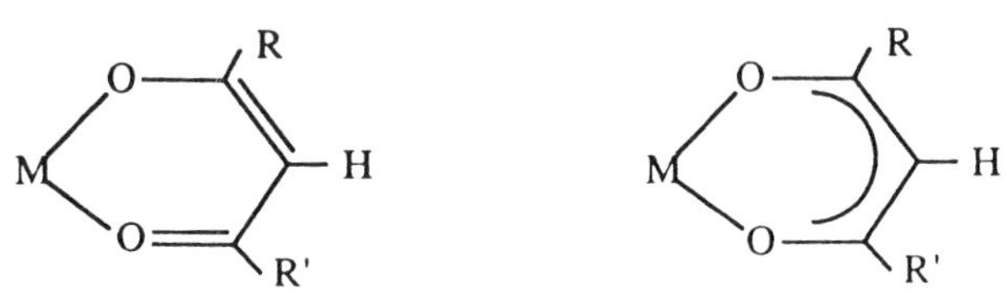

a) b)

Figure 1 Structure of β–diketonate ligands a) One resonance hybrid showing relationship to an alkoxide b) Delocalised structure

R = R' = CH_3	ACAC	R = CH_3, R' = CF_3	TFAA
R = R' = $C(CH_3)_3$	DPM	R = $C(CH_3)_3$, R' = $CF_2CF_2CF_3$	HFOD
R = R' = CF_3	HFA		

For copper, the simple acetylacetonato (2,4–pentanedionato, ACAC) ligand is adequate; indeed $[Cu(ACAC)_2]$ has been the precursor of choice in some instances (cf Table 1).

This complex is simple and cheap to prepare, sublimes at 170°C in vacuo and is stable to air and moisture (Berry et al 1988a). It thus has many of the properties of an ideal precursor, but, as with all ACAC type compounds, requires an extrinsic source of oxygen (H_2O or O_2) for the deposition of oxide films. Although the compound does contain hydrogen atoms on the carbon β- to the oxygen, transfer of this hydrogen to the oxygen with consequent C–O bond rupture cannot readily occur. This is because the product would be a high energy allene type molecule:

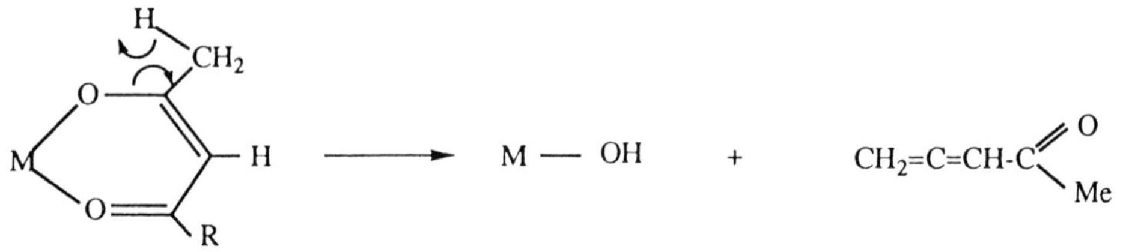

For Y, Ba, Sr, Ca, etc. simple ACACs are not volatile, because van der Waals forces between molecules produce high lattice energies (Seivers and Sadlowski 1978). Intermolecular hydrogen bonding may also occur and, at least for Ba, it is possible that oligomerisation could occur by expanding the coordination geometry from tetrahedral (a) to octahedral (b):

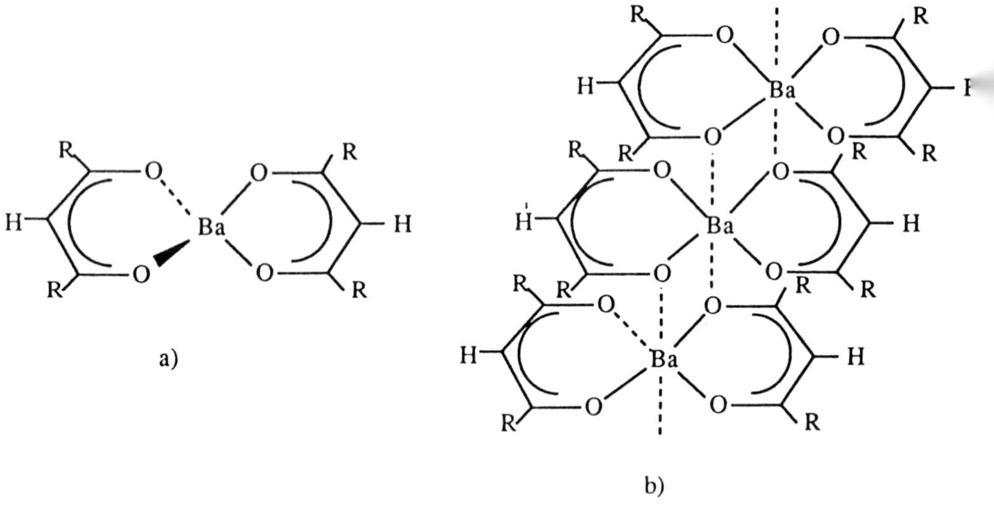

Once again, these problems can be overcome by bulky ligands shielding the metal ion from the intermolecular forces which render it nonvolatile. Thus dipivaloylmethane (DPM) complexes in which the methyl groups of ACAC are replaced by tertiary butyl groups are often sufficiently volatile for use in CVD growth. $Y(DPM)_3$ has been widely used as a CVD precursor for production of thin films of YBCO (cf Table 1). Sublimation temperatures for DPM's and related precursors are shown in Table 2. This table reveals that even the DPM complex of barium is not particularly volatile, so a still more successful approach has been to employ fluorine containing substituents; e.g.

hexafluoroacetylacetone (HFA), trifluoroacetylacetone (TFA) or heptafluorooctanedione (HFOD). The structure of these ligands has already been given above.

Table 2 Sublimation Temperatures for Various β-diketonate Precursors[a]

Precursor	Sublimation temperature/^{o}C	Reference
$[Cu(ACAC)_2]$	140-170	Berry 1988a Gonzalez-O 1988
$[Cu(DPM)_2]$	140-155	Nakamori 1988
$[Y(DPM)_2]$	120-160	Berry 1988a Gonzalez-O 1988 Nakamori 1988
$[Ba(HFA)_2]$	164	Gonzalez-O 1988
$[Ba(HFA)_2]^b$	240-255	Purdy 1989
$[Ba(DPM)_2]^c$	230-240	Berry 1988a
$[Ca(HFA)_2]$	80-120	Purdy 1989
$[Ca(HFA)_2]^b$	180-185	Purdy 1989
$[Sr(HFA)_2]^b$	220-225	Purdy 1989

a "in vacuo" unless otherwise stated; b at 1 atmosphere; c at 10^{-5} torr

Not only is fluorine a much larger atom than hydrogen (thus providing steric blocking) it also removes intermolecular hydrogen bonding and produces much lower van der Waals interactions, the result of fluorine's high electronegativity (Sievers and Sadlowski 1978). As Table 2 shows, this strategy is highly successful and sublimation temperatures close to 200°C are observed (Purdy et al 1989). The volatility of the HFA complexes of group II metals increases on ascending the group. This trend is as expected because of the higher molecular weights but also because metal atoms of lower ionic radius will be better shielded from intermolecular forces (Sievers and Sadlowski 1978).

Since chelation can cut down intermolecular interactions it may be possible to design bidentate alkoxides. These would combine chelate control of oligomerisation with the possibility of direct oxide formation in the absence of extrinsic oxygen sources. To this end we have studied the reaction of barium with 2-methoxyethanol. An alkoxo species is formed but its low volatility suggests either that the ligand bridges two barium atoms rather than chelating or that the barium alkoxide can oligomerise by expanding its coordination number to six (octahedral) via bridging oxygen atoms of alkoxide ligands (Thompson 1990).

4. CHEMISTRY OF DEPOSITION

Although, as Table 1 indicates, there is now a significant number of publications on the CVD of high T_c materials, much of what has been published is of an empirical nature. Certainly, very little consideration has been given to the chemistry of deposition. In the previous section we have discussed the value of understanding the chemistry underlying the design of precursors and such an understanding is no less important for the actual deposition step. Since there are no studies as yet on the chemistry of deposition of high T_c materials we can perhaps illustrate the value of chemical insights by reference to another technologically important material, doped polysilicon. This can be deposited by thermal pyrolysis of a mixture of SiH_4 and a dopant gas at temperatures in the range 550–650°C and pressures in the range 15–130 Pa (Hitchman and Jensen, 1990). In the case where the dopant is phosphorous the overall chemical reaction can be represented by

$$SiH_4(g) + xPH_3/N_2(g) \longrightarrow SiP_x(s) + xN_2(g) + (1.5x + 2)H_2(g)$$

Flow rate ratios of PH_3/SiH_4 (= γ) from 10^{-3} to 10^{-2} are commonly used. The actual chemistry of deposition is, however, much more complex than the simple overall reaction would suggest. Consideration of the known gaseous and surface chemistry of silicon in the presence of phosphine has suggested an overall reaction scheme as shown in Figure 2 (Hitchman et al 1987). The strong adsorption of phosphine on silicon leads to a blocking of surface sites and a dramatic fall in growth rate for a very small rise in γ. Furthermore, the blocking of the growth from SiH_4 adsorption allows there to be a more significant contribution to growth by SiH_2 and Si_2H_6 adsorption. However, these species have a much higher heterogeneous reaction probability than that of SiH_4 and so in a conventional LPCVD reactor with wafers oriented perpendicular to the gas flow and parallel to one another significant depletion of SiH_2 and Si_2H_6 occurs during their diffusion into the inter-wafer spacing. An understanding of the chemistry of this process suggests the use of a cage-like structure around the wafers in order to remove the highly active homogeneous species and to restore a major, albeit diminished, contribution from SiH_4. This procedure is now standard for the CVD of a number of microelectronic materials such as SiO_2 and SIPOS where similar mechanisms to that shown in Figure 2 are believed to operate (Hitchman and Jensen 1990).

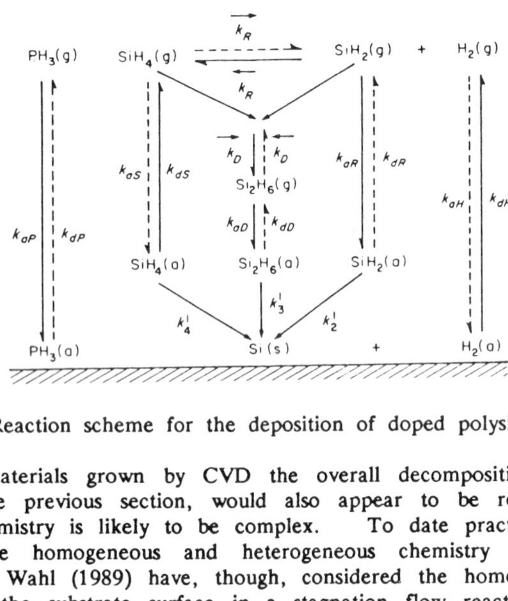

Figure 2 Reaction scheme for the deposition of doped polysilicon

For high T_C materials grown by CVD the overall decomposition of a precursor, as illustrated in the previous section, would also appear to be relatively simple, but in practice the chemistry is likely to be complex. To date practically nothing has been reported on the homogeneous and heterogeneous chemistry of deposition kinetics. Schmaderer and Wahl (1989) have, though, considered the homogeneous mass transport of reactants to the substrate surface in a stagnation flow reactor and have concluded from a comparison of the observed growth rates with their calculated values that their deposition is probably under mass transport control rather than any form of kinetic control. The slight variation of the growth rate in the temperature range 600–900°C observed by Yamane et al (1989a) certainly indicates mass transport control. On the other hand, the variation of growth rate from 4–15 nm min^{-1} over a temperature range 500–700°C reported by Oda et al (1989) could suggest kinetic control. We have shown in a different context (Hitchman 1981a) that there is usually a wide transition region between pure kinetic and pure mass transport control and the only way to examine this in more detail is to make studies over a wide temperature range in a reactor with well defined mass transport. The stagnation reactor used by Schmaderer and Wahl is such a reactor, so the possbility for a more comprehensive study is there. Interestingly Noh et al (1989) have used a rotating disk reactor which has previously been shown (Hitchman 1981a and b) to be a very powerful system for studying the basics of deposition processes. Apart from noting that layers with uniformly distributed grains were obtained, the potential of rotating disk systems does not seem to have been realised. One very important aspect of any deposition study is, as we have already indicated, the dependence of growth rate on temperature. Although some of the papers have examined this very few details are given. Oda et al (1989) and Tsuruoka et al (1989a) are exceptions and do plot growth rate as a function of temperature. What is more interesting about Tsuruoka's results, though, is the significant fall in growth rate when N_2O is used as an oxygen source rather than oxygen. Our work (Hitchman et al 1981c) on SIPOS prepared from SiH_4 and N_2O has strongly suggested the inhibition of growth by N_2O adsorbed on the growing SiO_x surface and thus a similar effect could be operating here. This could be of some significance when one considers the factors affecting the crystallinity of layers, as discussed in the next section.

The conclusion here on the chemistry of deposition is that more basic studies are needed to both obtain a better fundamental understanding of layer growth and to optimise conditions for improved layer quality.

5. LAYER PROPERTIES

In the previous section the importance of understanding the effect of reaction parameters on growth rates and hence layer thickness and uniformity has been indicated. For high T_c materials, though, the critical values of temperature, current, and magnetic field which mark the transition to superconductivity appear to be more dependent on layer structure and morphology than on layer thickness. The crystal structure required for superconductivity is that of a defect compound (Edwards et al 1987) with, for example, a deficiency of oxygen in the perovskite structure of YBCO, resulting in orthorhombic superconducting phases when in $YBa_2Cu_3O_x$ x has a value in the range 6.5 to 7 and the loss of superconductivity for the tetragonal form when $x < 6.5$. Clearly loss of oxygen has a strong influence on crystal structure and this can be readily seen in XRD patterns of layers (e.g. Yamane et al 1989a). Furthermore, the oxygen content critically affects the superconducting properties. Any deposition must therefore ensure oxides are formed in the right crystalline structure and that the oxygen is retained in the lattice. As a result in the publications of CVD of high T_c materials a range of post-deposition treatments to ensure adequate oxygen incorporation into the layers is reported. Often this involves a post deposition annealing (see below) but almost without exception involves cooling in an atmosphere usually of oxygen but also in air (Shinohara et al 1988). Cooling rates vary from $\sim 2^{o}C\ min^{-1}$ (e.g. Nakamori et al 1988) to $\sim 20^{o}C\ min^{-1}$ (e.g. Kurosawa et al 1988). However, since it is by no means clear what affects the oxygen incorporation there is scope for considering other ways of achieving this. Yamane et al 1988c, 1989a) have found that the same post-deposition oxygen treatment results in different c axis length and observed T_c values for layers grown at different temperatures, suggesting structure differences affect the oxygen diffusion process. The use of oxygen containing precursors so one is less dependent upon solid state diffusion could be another advantage of novel precursor chemistry.

Oxygen content is not the only important factor in determining superconducting characteristics. Another factor is film orientation or even epitaxial grain growth (Dimos et al 1988; Schmaderer and Wahl 1990). Defect oxides are highly anisotropic both physically and electronically. The coherence length for Cooper pairs in the a-b plane for YBCO is 10–20 nm, but it is only 0.5–0.7 nm along the c axis. It has been found that while high current densities can be achieved in the a-b plane, conduction in the direction of the c axis is often much less (Caplin 1988). This problem can be reduced by producing films with a preferred orientation of the c axis perpendicular to the direction of electron flow. This preferred orientation may be achieved by the deposition on a substrate which is lattice matched with the a-b axis (strontium titanate – Yamane et al 1988b) or by deposition onto a substrate at high temperatures ($> 900^{o}C$) which tends to promote oriented growth (Kurosawa et al 1988); the reason for this oriented growth is by no means certain but may be related to the lower free energy for the formation of the crystal phase perpendicular to the c axis (Kurosawa et al 1988). Thus control of grain growth and orientation is important in obtaining desirable superconducting properties, particularly high values of J_c. As just indicated higher deposition temperatures can facilitate oriented growth and this can also, of course, promote grain growth in general with the formation of large single crystal domains with few voids among the domains (Yamaguchi et al 1989). For single crystal films, J_c values of 10^6 A cm^{-2} have been reported (Yamaguchi et al 1989) and this could be associated with the reduced possibility of misalignment of grain boundaries which reduces J_c (Verhoeven and Gibson, 1988). However, whilst it is thus important to promote grain growth in general and grain growth of the correct orientation in particular, the use of high temperatures either for deposition or post-deposition annealing can have undesirable effects arising from interdiffusion between the deposited layer and the substrate (Derks et al 1988).

Kanehori et al (1989) have found that grain growth is also promoted by deposition at low pressures. Typically for deposition below atmospheric pressure the pressures used are in excess of 10 torr (see Table 1), but using 1.5 torr Kanehori et al were able to obtain enhanced grain growth. At the lower pressure of 1.5 torr the deposition rate was 37.5% of that at 10 torr and this will mean there will be a greater opportunity for surface diffusion to promote nucleation. This effect is seen quite clearly in the case of polysilicon deposition to which we referred in the last section. It is found in that case that there is a critical temperature ~ 580°C above which polycrystalline layers are grown whilst below this temperature layers are largely amorphous. If one considers the crystalline growth rate (j_{cg}) of amorphous silicon films (Zellama et al 1979) and compares it with the polysilicon deposition rate (j_d) then it is found that above ~ 585°C, $j_{cg} > j_d$ and below it $j_d > j_{cg}$. Therefore for deposition temperatures (T_d) ~ 580°C deposition occurs faster than crystallisation can occur and films remain largely amorphous, but for $T_d >$ ~ 580°C the crystallisation rate exceeds the deposition rate and so films become increasingly more crystalline. However, it is not just a matter of growth rate. Kinsbron et al (1983) have also reported on the basis of TEM observations that there is a critical deposition temperature below which deposited films are amorphous, but they have shown too that deposition time as well as deposition temperature is important. TEM micrographs show that for T_d = 640°C a fine columnar crystalline structure is produced throughout the film. This is not unexpected since at this temperature $j_{cg} \sim 2j_d$. For T_d = 591°C, j_{cg} is still $> j_d$ and so again one might expect a high degree of crystallinity throughout the film, but it is only material near the interface which appears crystalline. This is material which is put down in the initial stages of deposition. More recently deposited material near the surface is not crystallised. Therefore in addition to j_{cg} one must consider nucleation time before crystallisation occurs. Zellama et al (1979) have, from conductance measurements on partially crystallised amorphous films, derived nucleation or induction times for crystallisation as a function of temperature. For example, at 590°C the induction time is estimated to be ~ 110 min and this is why for layers grown at that temperature it is only the material which was deposited in the first stages of deposition that show crystallisation. Similar effects could apply to high T_c materials. The situation can be complicated by layers being left in a hot reactor for varying lengths of time after deposition is completed and as a result unintentional annealing occurring. Unless such effects are taken into account varying optimum deposition conditions and annealing conditions will continue to be reported in the literature.

A further consideration is also suggested from the work on polysilicon in that it is found that material grown amorphous and subsequently annealed is of a much higher device quality than material grown in polycrystalline form. Raman, elastic light scattering and TEM observations all show that annealed amorphous grown material is of very good structural perfection with little internal strain, in contrast with polycrystalline layers which have very large volumes of imperfect and poorly crystallised material (Harbeke et al 1984). In particular, Raman linewidths of annealed layers as a function of deposition temperature show that the material deposited at low temperatures have linewidths close to the value for bulk single crystal silicon whilst material deposited at high temperatures have much broader linewidths. This indicates that while there is little disturbance of the crystal lattice for the annealed low temperature group for the high temperature group the material after annealing remains in a highly disturbed state with some poor and some good crystallisation. Poorly crystallised material will be highly undesirable for superconducting applications because of high internal stress (e.g. dislocations and stacking faults) which will almost certainly produce a reduction in critical current density. Again the need to carefully control growth is indicated.

In this context, it is also interesting to note the effect of N_2O on growth (Tsuruoka et al 1989) mentioned in the previous section. Not only is the growth rate reduced compared with that in oxygen, but also crystallisation occurs at a lower temperature. This will obviously be as a result of the reduced rate of growth allowing time for surface diffusion to occur. The use of other gaseous oxidants (e.g. NO) would be interesting to explore.

Finally, the effect of a dopant on film morphology and crystal orientation is worth considering. For phosphorus doped polysilicon it is found that (Harbeke et al 1983) annealed layers deposited at low temperatures exhibit slightly more strain than undoped layers, but in the case of doped layers the strain does not show as much dependence on deposition temperature. Phosphorus doping seems to prevent the poor crystallisation observed for undoped layers deposited at higher temperatures and this is probably related to the presence of the phosphorus impurity enhancing silicon self-diffusion (Fairfield and Masters 1967). A similar effect may be operating in the Pb doping of BSCCO films where highly oriented layers with significantly improved surface morphology are obtained (Zhang et al 1989d).

6. CONCLUSIONS

The potential for commercial exploitation of thin film high T_c materials is considerable. CVD will undoubtedly play an important role in those developments. Real success will only be achieved though if the technological developments are accompanied by studies of the chemistry of deposition. This review has indicated how a basic understanding could lead to improved layer quality and superconducting characteristics.

7. ACKNOWLEDGEMENT

The support of The Associated Octel Co. Ltd. is gratefully acknowledged.

8. REFERENCES

Abe H, Tsuruoka T, and Nakamori T 1988 *Jap. J. of Appl. Phys.* **27** 1473

Advani G N, Jordan A G, Lapis C H P and Longin R L 1979 *Thin Solid Films* **62** 361

Bai G R, Tao W, Wang L M, Xie L M, Zhang X K and Huang J 1989 *Appl. Phys. Lett.* **55** 194

Bednorz J G and Mueller K A 1986 *Z. Phys. B.* **64** 189

Berry A D, Gaskill D K, Holm R T, Cukauskas E J, Kaplan R and Henry R L 1988 *Appl. Phys. Lett.* **52** 1743

Berry A D, Holm R T, Cukauskas E J, Fatemi M, Gaskill D K, Kaplan R and Fox W B 1988 *J. of Crystal Growth* **92** 344

Bradley D C 1989 *Chem. Rev.* **89** 1317

Bradley D C, Chudzyuska H, Fugo D M, Hursthouse M B and Mazid M A 1988 *J. Chem. Soc. Chem. Commun.* 1258

Bradley D C and Faktor M M 1959 *Pure Appl Chem* **9** 5425

Bradley D C, Mehotra R C and Grant P D 1978 *Metal Alkoxides* (London, Academic) and references therein

Bradley D C, Saad M A and Wardlaw W J 1954 *J. Chem. Soc.* 3488

Caplin D 1988 *Nature* **335** 204

Derks W P T, van Hal H A M and Langereis C 1988 *Physica* **C156** 62

Dickinson P H, Geballe T H, Sanjurjo A, Hildenbrand D, Craig G, Zisk M, Collman J, Banning S A and Sievers R E 1989 *J. Appl. Phys.* **66** 444

Dimos D, Chaudhari P, Mannhart J and LeGoues F K 1988 *Phys. Rev. Letts.* **61** 13

Edwards P P, Harrison M R and Jones R 1987 *Chem. Brit.* **23** 962

Fairfield J M and Masters B J 1967 *J. Appl. Phys.* **38** 3148

Gonzalez–O C, Schachner H, Tippmann H and Trojer F J 1988 *Physica C* **153–155** 1042

Gosling W 1989 *Superconductivity: The Opportunities, Implications and Applications* (London: IBC Technical Services Ltd) pp 25–33

Harbeke G, Krausbauer L, Steigmeier E F, Widmer A E, Kappert H F and Neugebauer G 1983 *RCA Rev.* **44** 187

Harbeke G, Krausbauer L, Steigmeier E F, Widmer A E, Kappert H F and Neugebauer G 1984 *J. Electrochem. Soc.* **131** 675

Hitchman M L, 1981a *Prog. Crystal Growth and Charact.* **4** 249

Hitchman M L and Curtis B J 1981b *Progr. Crystal Growth and Charact.* **4** 283

Hitchman M L and Kane J 1981c *J. Cryst. Growth* **55** 485

Hitchman M L, Ahmed W, Shamlian S and Trainor M 1987 *Chemtronics* **2** 147

Hitchman M L and Jensen K F 1990 *Chemical Vapour Deposition* (New York: Academic) Ch. 3

Hubert–Pfalzgraf L G, Massiani M C, Papiernik R and Poncelet O 1989 *J. de Physique* **C5** 981

Inoue M, Takase E, Takai Y and Hayakawa H 1989 *Jap. J. of Appl. Phys.* **28** L1575

Jeffries P M and Girolami G S 1989 *Chem. of Materials 1 8*

Kanehori K, Sughii N, Fukazawa T and Miyauchi K 1989 *Thin Solid Films* **182** 265

Kinsbron E, Sternheim M and Knoell R 1983 *Appl. Phys. Letts.* **42** 835

Kobayashi N, Iwasaki H, Kawabe H, Watanabe K, Yamane H, Kurosawa H, Masumoto H, Hirai T and Muto Y 1989 *Physica C* **159** 295

Kurosawa H, Yamane H, Masumoto H and Hirai T 1988 *J. of Mat. Sci. Letts.* **7** 1133

Muto Y, Watanabe K, Kobayashi N, Kawabe H, Yamane H, Kurosawa H, and Hirai T 1989 *Physica C* **162–164** 105

Nakamori T, Abe H, Kanamori T and Shibata S 1988 *Jap. J. of Appl. Phys.* **27** L1265

Natori K, Yoshizawa S, Yoshino J and Kukimoto H 1989 *Jap. J. Appl. Phys.* **28** L1578

Noge S, Yamashita T, Wang Z, Matsui T, Kurasawa H, Yamane H and Hirai T 1989 *Jap. J. of Appl. Phys.* **28** L1581

Noh D W, Gallois B, Chern C S, Caracciolo R, Kear B H, Zawadzki P A, Tompa G S and Norris PE 1989 *J. Appl. Phys.* **66** 5099

Oda S, Zama H, Ohtsuka T, Sugiyama K and Hattori T 1989 *Jap. J. of Appl. Phys.* **28** L427

Panson A J, Charles R G, Schmidt D N, Szedon J R, Machiko G J and Braginski A I 1988 *Appl. Phys. Lett.* **53** 1756

Peng D K, Meng G Y, Cao C B, Wang C L, Fang Q, Wu YH and Zhang Y H 1989 *J. de Physique* **C5** 149

Poncelet O, Sartain W J, Hubert–Pfalzgraf L G, Folting K and Caulton K G 1989 *Inorg. Chem.* **28** 263

Purdy A P, Berry A D, Holm R T, Fatemi M and Gaskill D K 1989 *Inorg. Chem.* **28** 2799

Radpour F, Singh R, Sinha S, Tulpule A M, Chou P, Thakur R P S, Rahmati M, Hsu N J and Kumar A. 1989 *Appl. Phys. Lett.* **54** 2479
Richeson D S, Tonge L M, Zhao J, Zhang J, Marcy H O, Marks T J, Wessels B W and Kannewurf C R 1989 *Appl. Phys. Lett.* **54** 2154
Schmaderer F and Wahl G 1989 *J. de Physique* **C5** 119
Shinohara K, Munakata F and Yamanaka M 1988 *Jap. J. of Appl. Phys.* **27** L1683
Sievers R E and Sadlowski J E 1978 *Science* **201** 217
Thompson S C and Cole-Hamilton D J 1990 Unpublished results
Tsuruoka T, Kawasaki R, and Abe H 1989a *Jap. J. of Appl. Phys.* **28** L1800
Tsuruoka T, Takahashi H, Kawasaki R, and Kanamori T 1989b *Appl. Phys. Lett.* **54** 1808
Verhoeven J D and Gibson E D 1988 *Appl. Phys. Letts.* **52** 1190
Watanabe K, Yamane H, Kurosawa H, Hirai T, Kobayashi N, Iwasaki H, Noto K and Muto Y 1989 *Appl. Phys. Lett.* **54** 575
Wu M K, Ashburn J R, Torng C J, Hor P H, Meng R L, Gao L, Huang Z J, Wang Y Q and Chu C W 1987 *Phys. Rev. Letts.* **58** 908
Yamaguchi T, Aoki S, Sadakata N, Kohno O, and Osanai H 1989 *Appl. Phys. Lett.* **55** 1581
Yamane H, Kurosawa H, Iwasaki H, Masumoto H, Hirai T. Kobayashi N and Muto Y 1988a *Jap. J. of Appl. Phys.* **27** L1275
Yamane H, Masumoto H, Hirai T, Iwasaki H, Watanabe K, Kobayashi N, Muto Y and Kurosawa H 1988b *Appl. Phys. Lett.* **53** 1548
Yamane H, Kurosawa H, Hirai T, Iwasaki H, Kobayashi H and Muto Y 1988c *Nippon-Seramikkusu-Kyokai-Gakujutsu-Ronbunshi* **96** 799
Yamane H, Kurosawa H and Hirai T 1988d *Chemistry Letters* 939
Yamane H, Kurosawa H, Hirai T, Iwasaki H, Kobayashi N and Muto Y 1988e *Jap. J. of Appl. Phys.* **27** L1495
Yamane H, Kurosawa H, Hirai T, Watanabe K, Iwasaki H, Kobayashi N and Muto Y 1989a *J. of Crystal Growth* **98** 860
Yamane H, Kurosawa H and Hirai T 1989b *J. de Physique* **C5** 131
Yamane H, Kurosawa H, Iwasaki H, Hirai T, Kobayashi N and Muto Y 1989c *Jap. J. of Appl. Phys.* **28** L8277
Zellama K, Germain P, Squelard S and Bourgoin J C 1979 *J. Appl. Phys.* **50** 6995
Zhao J, Dahmen K-H, Marcy H O, Tonge L M, Marks T J, Wessels B W and Kannewurf C R 1988 *Appl. Phys. Letts.* **53** 1750
Zhang K, Kwak B S, Boyd E P, Wright A C and Erbil A 1989a *Appl. Phys. Lett,* **54** 380
Zhang J, Zhao J, Marcy H O, Tonge L M, Wessels B W, Marks T J and Kannewurf C R 1989b *Appl. Phys. Lett.* **54** 1116
Zhang K, Boyd E P, Kwak B S, Wright A C and Erbil A 1989c *Appl. Phys. Lett.* **55** 1258
Zhang J M, Marcy H O, Tonge L M, Wessels B W, Marks T J and Kannewerf C R 1989d *Appl. Phys. Lett* **55** 1906

Inst. Phys. Conf. Ser. No 111
Paper presented at Int. Conf. on New Materials and their Applications, Univ. of Warwick, 1990

Growth and characterization of thin films of thallium-based superconducting oxides

H.C.Lai, K.D.Vernon-Parry and C.R.M.Grovenor

Department of Materials, Parks Road, Oxford OX1 3PH, U.K.

ABSTRACT: Superconducting Tl-Ca-Ba-Cu-O thin films less than 1μm thick have been prepared on Y-stabilised $ZrO_2(100)$ and MgO(100) substrates using thermal evaporation. No superconducting films have been prepared on sapphire substrates. Using a two-step annealing process, we have been able to prepare reasonably uniform films of the $Tl_2Ca_2Ba_2Cu_2O_x$ (2122) superconducting phase, and have successfully reduced the density of impurity phases by controlling the stoichiometry and morphology of the precursor Ca+Ba+Cu films.

1. INTRODUCTION

The recent discovery of the high temperature superconducting TlCaBaCuO system by Sheng et al (1988) and Hazen et al (1988) has resulted in the development of a family of superconducting phases, several with transition temperatures (T_c) over 100K. Several groups have reported the preparation of Tl-based superconducting thin films on various substrates by pulsed laser evaporation (Liou et al (1989)), sequential electron beam evaporation (Ginley et al (1988)) and sputtering (Kang et al (1988)). These authors report that the best superconducting properties are obtained only if great care is taken in designing the annealing process the films experience after deposition.

Recently, papers by Richeson et al (1989) and Shah et al (1990) have reported that using a two-step annealing technique can improve the uniformity of the superconducting phase and enlarge the grain size. These processes involve prereacting BaCaCu films to a uniform precursor before a thalliation step to form the superconducting phase. If BaF_2 is used as a precursor to the BaCaCu film it is also important to carry out the preannealing step in an oxygen atmosphere saturated with water vapour, (see Richeson et al (1989) and Chan et al (1988) for example). It is commonly observed that impurity phases rich in Ca and Cu are retained in most films even if their superconducting properties are excellent.

This paper presents details on a reproducible method for preparing uniform 2122 superconducting thin films by a two-step annealing process. We have also investigated the effect of various substrates on the morphology, microstructure and superconducting properties of the films.

2. EXPERIMENTAL

Precursor films were prepared on MgO (100), Y-stabilized ZrO_2 (100), and sapphire (1102) substrates by sequential thermal evaporation of BaF_2, CaF_2, and Cu to give metallic ratios of Ba:Ca:Cu of 2:1:2, and prereaction by the annealing treatments described below. Thalliation was carried out in the presence of both bulk 2122 superconductor and pure Tl_2O_3 as the sources of thallium. Precursor films were positioned 1mm above the bulk superconductor in a silver-covered Al_2O_3 crucible.

Grain morphologies and phase purities were analyzed in a scanning electron microscope (SEM) equipped with energy dispersive X-ray analysis (EDS), and by X-ray diffraction. Resistance/temperature plots were made using conventional four point contact techniques with a measuring current of 100μA.

Thin film samples were prepared by the following methods:
(a) As-evaporated Cu-Ba-Ca films were thalliated at 850°C for 30 min in air.
(b) As-evaporated Cu-Ba-Ca films were thalliated at 865°C for 20 min in air.
(c) As-evaporated Cu-Ba-Ca films were preannealed at 850°C for 1 hour in air. Subsequent thalliation was carried out at 865°C for 10 min.
(d) As-evaporated Cu-Ba films were preannealed at 845°C for 3 hours in the presence of water vapour-saturated oxygen. Subsequently, CaF_2 was evaporated and thalliation carried out in air at 845°C for 15 min.
(e) As-evaporated Cu-Ba-Ca films were preannealed at 865°C for 10 min in air. Subsequent thalliation was carried out at 865°C for 10 min.

3. RESULTS AND DISCUSSION:

(1) One-step annealing processes

Figure 1 shows the microstructure of the 2122 phase prepared on MgO by method (a). The connectivity of these films is very poor, although the phase purity is excellent.

Figure 2 shows the microstructure of the 2122 phase prepared on Y-ZrO_2 by method (b). The film is composed of large poorly-connected grains arranged in patches with straight sides parallel to the [100] direction of the substrate. EDS analysis shows that Ca_2CuO_3 particles with spheroidal morphology decorate the surface of the 2122 phase. A conglomeration of unreacted CaO and Tl_2O_3 particles can be observed between the 2122 grains. In our experience, it is very difficult to get rid of these impurities by a traditional one-step annealing process.

(2) Two-step processes

Figure 3 shows a SEM micrograph of the 2122 phase prepared on MgO by method(c). This film is much denser than those described above, with isolated voids through to the MgO substrate. The XRD pattern of the same film, Figure 4, shows that while textured 2122 is the majority phase, traces of unreacted BaF_2, CuO, and impurities of $BaCuO_2$, Ca_2CuO_3, and $CaCu_2O_3$ are present.

Figure 5 shows a SEM micrograph of the 2122 phase made on MgO by method(d). It can be seen that effective preannealing results in significant melting and concomitant grain growth. Bright particles of excess Tl_2O_3 may also be seen. A uniform $BaCuO_2$ film was formed in the preannealing process, and we believe that this is important in preparing a uniform morphology of 2122 film.

Figure 6 shows XRD data at various stages of forming the 2122 phase. Figure 6(a) shows the precursor film after annealing to form $BaCuO_2$ plus a few traces of other precursor phases. Figure 6(b) shows the film after the final anneal to form 2122 with some excess Tl_2O_3.

Neither one-step (method(b)) nor two-step (method(c)) processes have been successfully used to prepare superconducting films on sapphire substrates. XRD traces taken from these samples indicate the formation of aluminide phases by reaction of the films with the substrate, but no trace of superconducting phases.

Figure 7 shows the resistance versus temperature plots for samples prepared by one-step and two-step processes on MgO and Y-ZrO_2 substrates. Films prepared by a single-step annealing process show a drop in resistance, but rarely reach zero resistance above 25K, curve (c). In comparison, films with prepared by two-step annealing exhibit a steep metallic-like slope in the normal region and a superconducting onset temperature of about 108 K. Different two-step processes have a noticeable effect on the zero resistance temperature. The transition width in a sample prepared by method (d) was about 5 degrees. However, a film prepared by method (e), where the BaCaCu precursor film is not completely reacted before thalliation (curve (b)) shows two transitions, one at 108K (2122 phase), and a second at 90K (1122 phase). XRD confirms the presence of these two phases in this film.

4. CONCLUSIONS

(1) The morphology and microstructures of superconducting thin films in the Tl-Ca-Ba-Cu-O system are controlled by the heat treatment process and the choice of substrates. The best heat treatment process that we have discovered so far is when an evaporated Cu-Ba film is preannealed at 845°C for 3 hours in the presence of water vapour-saturated oxygen, subsequent CaF_2 evaporation, and final thalliation in air at 845°C for 15 min. The best films that we have prepared are on MgO substrates, on which the morphology of the films is more uniform than on $Y\text{-}ZrO_2$. No superconducting films have been prepared on sapphire substrates.

(2) The formation of a uniform $BaCuO_2$ film is a most important step in preparing uniform 2122 superconducting thin films.

5. ACKNOWLEDGEMENTS

The authors are indebted to Mr.O.F. Schilling and Dr L Romano for analytical advice on X-ray diffraction and scanning electron microscopy. Professor Sir Peter Hirsch, FRS, is gratefully acknowledged for the provision of laboratory facilities.

REFERENCES

Chan S W, Bagley B G, Greene L H, Giroud M, Feldmann W L, Jenkin K R and Wilkins B J 1988 *Appl. Phys. Lett.* **53** 1443
Ginley D S, Kwak J F, Hellmer R P, Baughman R J, Venturini E L, Mitchell M A and Morosin B 1988 *Physica C* **156** 592
Hazen R M, Finger L W, Angel R J, Prewitt C T, Ross N L, Hadidiacos C G, Heany P J, Veblin D R, Sheng Z Z, EI Ali A and Hermann A M 1988 *Phy. Rev. Lett.* **60** 1657
Kang J H, Kampwirth R T and Gray K E 1988 *Phys. Lett. A* **131** 208
Liou S H, Aylesworth K D, Lanno N J, Johs B, Thompson D, Meyer D, Woolam J A and Barry C 1989 *Appl. Phys. Lett.* **54** 760
Richeson D S, Tonge L M, Wang X K, Marcy H O, Marks T J, Ketterson J B, Cheng R P H and Kannewurf C R 1989 *Appl. Phys. Lett.* **55** 2778
Shah S I, Herron N, Fincher C R and Holstein W L 1990 *Appl. Phys. Lett.* **56** 782
Sheng Z Z and Hermann A M, 1988 *Nature* **332** 138

Fig. 1. An optical micrograph of a 2122 thin film prepared on MgO by method(a), showing a poor connectivity and non-uniform distribution of the 2122 phase.

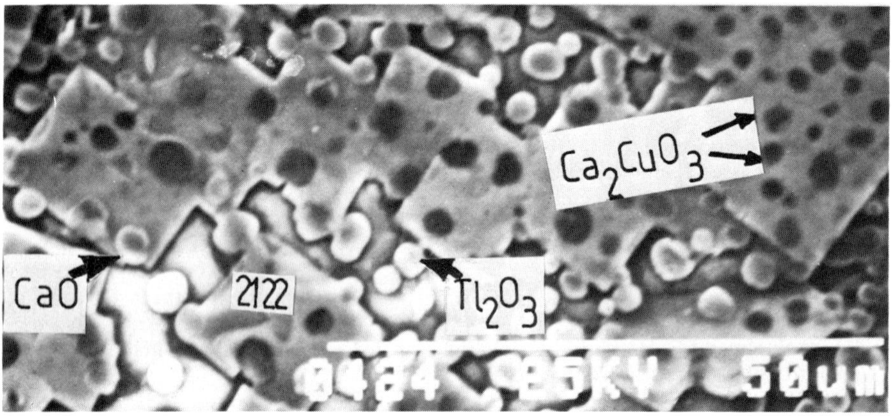

Fig. 2. An SEM micrograph of a 2122 thin film prepared on ZrO_2 by method(b), showing poorly connected 2122 grains separated by large amounts of incompletely reacted impurity phases.

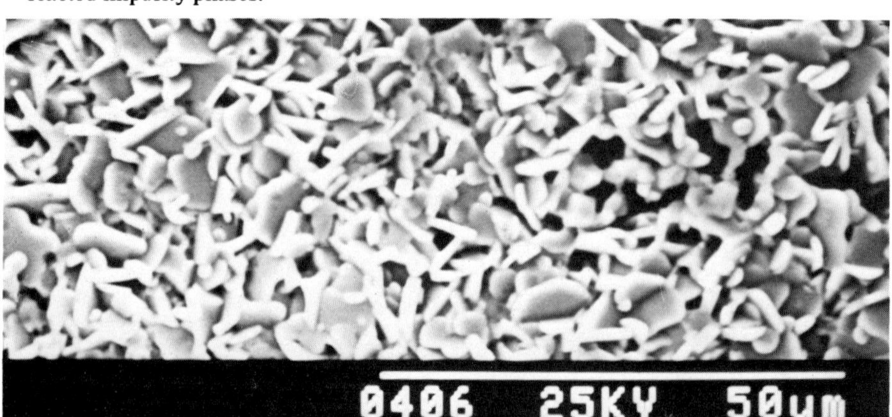

Fig. 3. An SEM micrograph of a 2122 thin film prepared on MgO by method(c), showing a fully dense and fine grained film with a small amount of impurity phases.

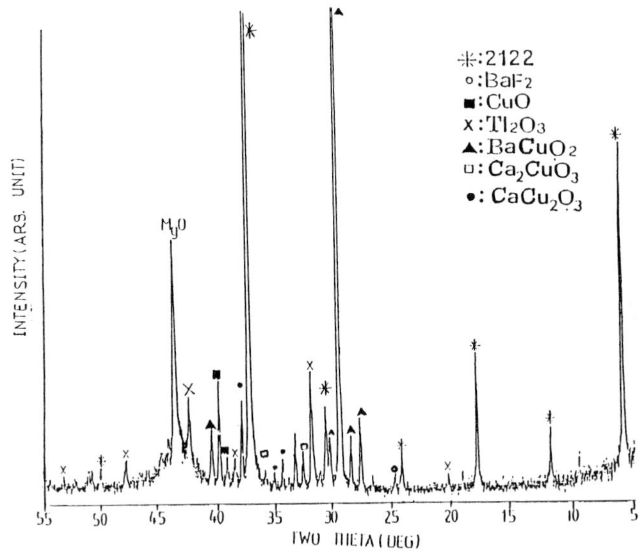

Fig. 4. An X-ray diffraction pattern of the 2122 thin film shown in Fig. 3, showing the presence of impurity phases.

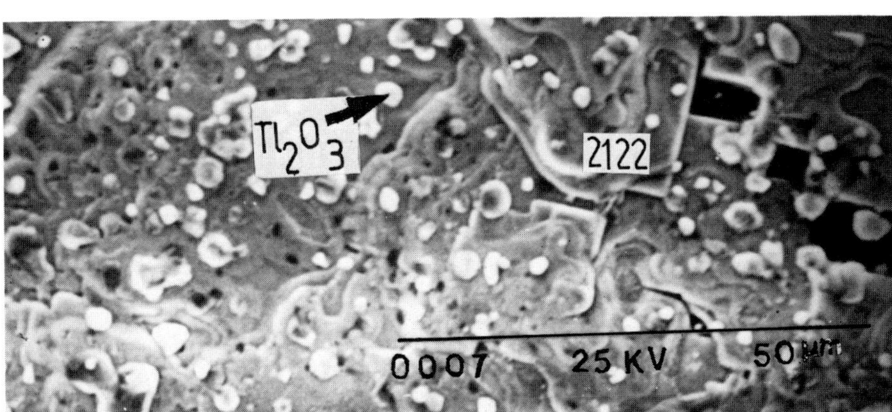

Fig. 5. An SEM micrograph of a 2122 thin film prepared on MgO by method (d), showing a uniformly melted 2122 thin film with a few Tl_2O_3 particles on the surface.

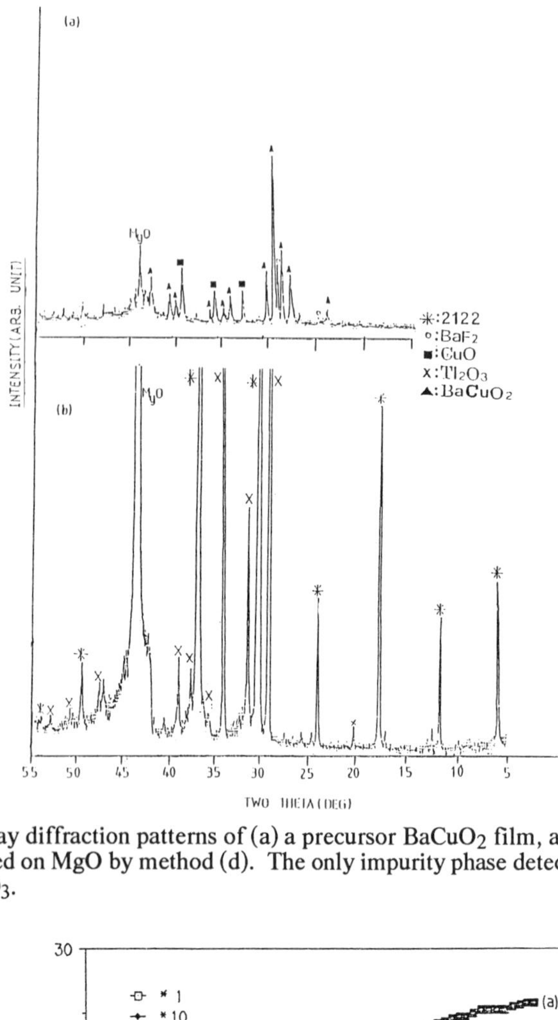

Fig. 6. X-ray diffraction patterns of (a) a precursor $BaCuO_2$ film, and (b) a 2122 thin film prepared on MgO by method (d). The only impurity phase detected by the XRD is excess Tl_2O_3.

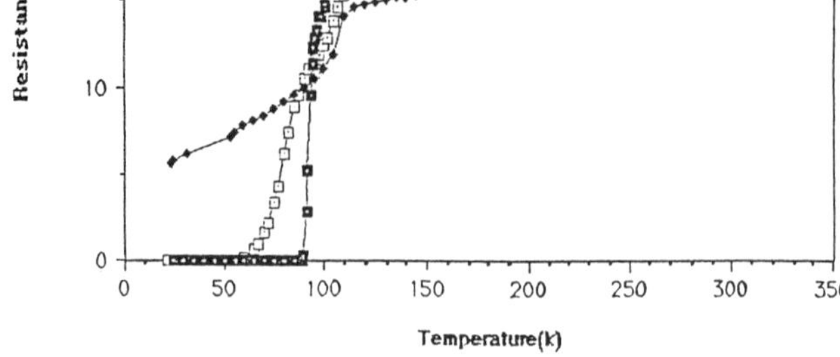

Fig. 7. Resistance/temperature plots for samples prepared by various annealing processes on different substrates; curve (a) on MgO by method (d), curve (b) on MgO by method (e), and curve (c) on ZrO_2 by method (b)

Paper presented at Int. Conf. on New Materials and their Applications, Univ. of Warwick, 1990

Degradation and ageing in $Tl_2Ba_2Ca_1Cu_2O_x$ high temperature superconductors

L T Romano, K D Vernon-Parry, J S Lees*, J Rogers†, G Fletcher†, T Beales†, C R M Grovenor and R T G Tilley*

Department of Materials, University of Oxford, Parks Road, Oxford, OX1 3PH
*School of Engineering, Division of Materials, University of Wales, PO Box 917, Cardiff, CF2 1XH
†Cooksons Technology Center, Sandy Lane Yarnton, Oxford OX5 1PF

ABSTRACT: The degradation processes in 2212 thallium-based superconductors have been investigated using a combination of X-ray diffraction and electron microscopy techniques. The formation of amorphous and subsequently crystalline parasitic phases has been followed, and correlated with the degradation of the superconducting properties.

1. INTRODUCTION

One of the areas of interest for materials scientists working on high temperature superconducting ceramics has been the stability of the superconducting phases when in contact with moisture and acidic or alkaline environments. Degradation of $YBa_2Cu_3O_x$ (YBCO) has been studied by several groups (Barns and Laudise (1987) and Bansal and Sandkuhl (1988)), and it is generally agreed that decomposition of the superconductor is observed, especially under conditions of elevated temperature and high humidity. Hoyle (1988) showed that the presence of barium compounds such as $BaCO_3$ in the sample in a moist environment results in the rapid formation of alkaline conditions on the sample surface and accelerated degradation.

Since the discovery of superconductivity at elevated temperatures in Tl-containing compounds (Sheng and Hermann (1988) and Hazen *et al* (1988)) rather little discussion has been presented in the literature on their stability in moist environments. This is surprising since it has proved difficult to prepare these alloys in phase-pure form because of the extreme volatility of thallium (Narain and Ruckenstein (1989)), and the presence of residual precursor phases that might be expected to accelerate degradation processes in the same way as observed for YBCO. Gopalakrishnan *et al* (1988) and Shih and Qiu (1989) have, however, reported some indications of degradation of the thallium compounds in water.

This paper presents the results of a preliminary study of the ageing effects which occur in predominately single phase $Tl_2Ba_2Ca_1Cu_2O_x$ samples, concentrating on investigation of the role of trace precursor phases on the degradation process. Some data on the degradation properties of both bulk and thin film material has been obtained.

2. SAMPLE PREPARATION AND CHARACTERISATION

Samples of $Tl_2Ba_2CaCu_2O_x$ were prepared by a variety of bulk processing and thin film deposition techniques, and their phase purity was studied with X-ray diffraction (XRD), scanning electron microscopy (SEM), electron probe microanalysis (EPMA) and transmission electron microscopy (TEM) techniques. Studies were were made to establish fine scale elemental distribution and investigate whether any amorphous or microcrystalline phases were present which are difficult to detect by XRD.

Samples used for bulk degradation studies were prepared by the following methods:
i) CaO and CuO powders in a 2:3 ratio were heated at 850-880 °C for 18-68 hours to form Ca_2CuO_3 and CuO. The resulting mixture was ground with Tl_2O_3 and BaO_2 (sufficient to give a 2:2:2:3 ratio), pelletized and fired at 820 °C for 10 hours in either crimped gold foil or covered alumuina crucibles.

ii) $BaCuO_2$, Tl_2O_3 and CaO powders (in quantities to produce a 2:2:1:2 ratio) were ground together, pelletized and fired at 865 °C for 15 minutes in a sealed gold tube in an attempt to produce phase-pure 2212 (after Narain and Ruckenstein (1989)). Similar starting powders were also sintered in alumina crucibles covered by silver sheets.

Figure 1a shows a section of the XRD trace of a sample made by method (i), indicating that it contains only 2212 and a trace of $BaCO_3$. There is no evidence in the XRD of the Ca-containing phases which must be present in this sample to maintain the overall calcium to barium ratio of 1 in the precursor powders. The transition temperature to zero resistance of this sample was about 110K, as shown by curve 1 in Figure 2. The XRD trace in Figure 3a shows that the samples made by method (ii) consist primarily of 2212, with $BaCuO_2$ as the main

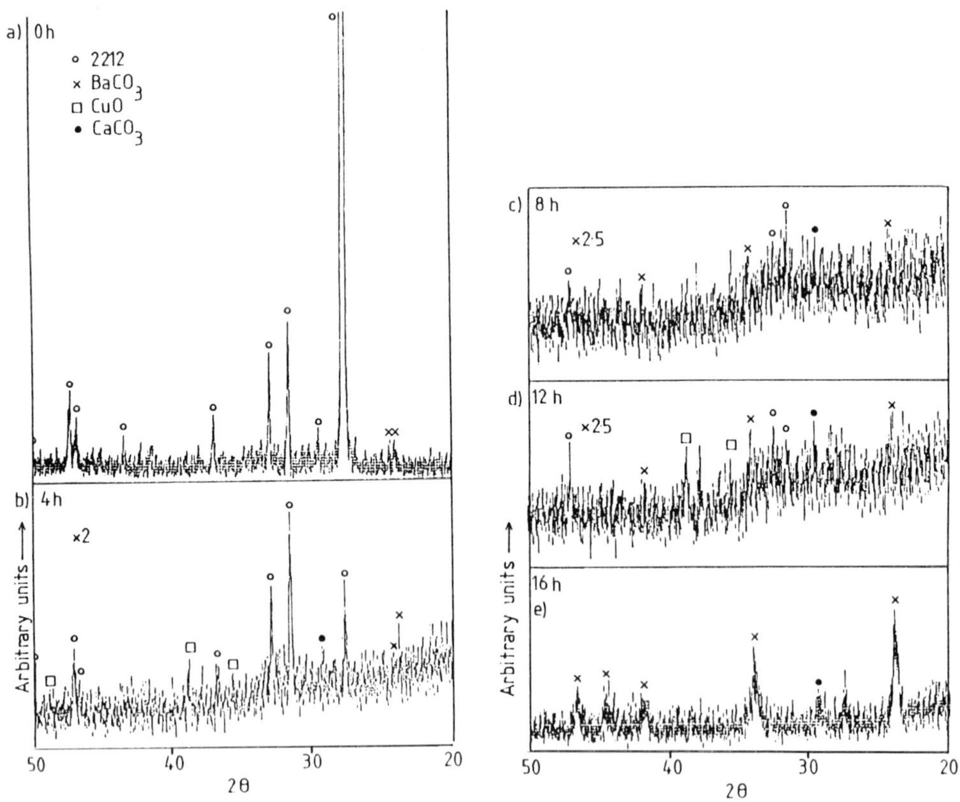

Fig. 1 XRD traces from a sample made by method (i) after 0, 4, 8, 12 and 16 hours of accelerated ageing at 40°C.

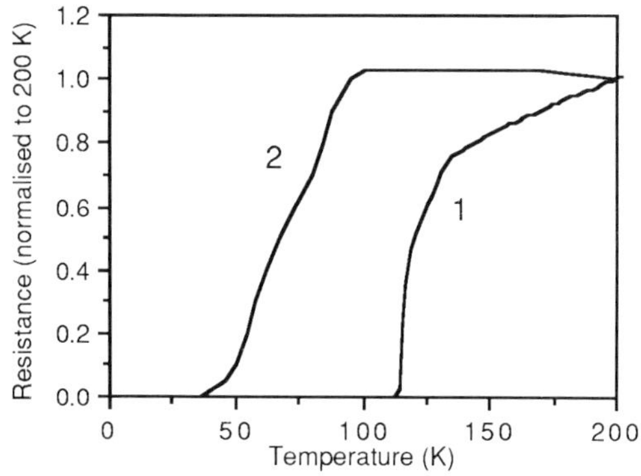

Fig.2. Resistance v temperature curves for samples prepared by method (i). Curve 1 as-sintered and curve 2 after 2 months' storage in a dessicator.

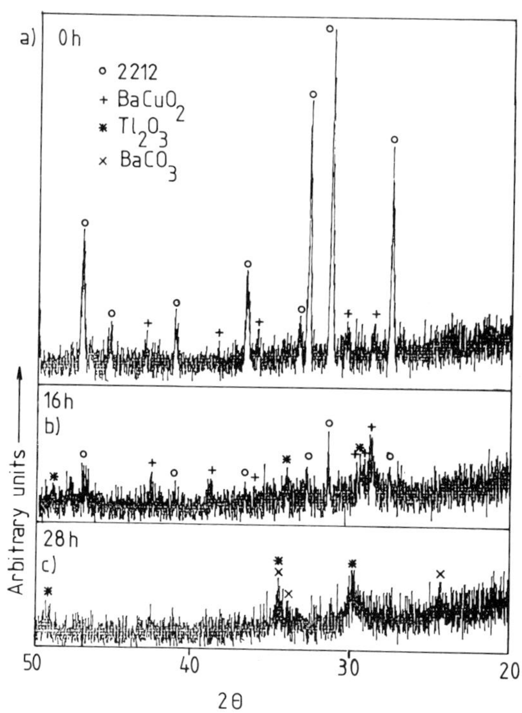

Fig. 3. XRD traces from a sample prepared by method (ii) after 0, 16 and 28 hours accelerated ageing.

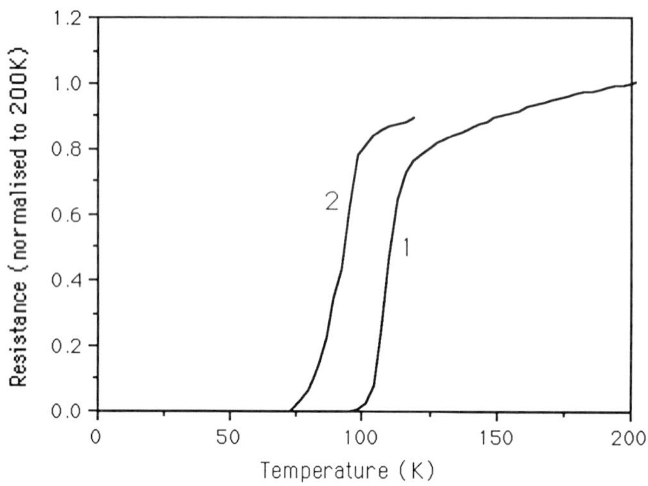

Fig. 4. Resistance v temperature curves for samples prepared by method (ii). Curve 1 is as-sintered and curve 2 after 6 weeks' storage in an evacuated dessicator.

detectable impurity. The zero resistance temperature of these samples is around 93 K, curve 1 in Figure 4. Samples of the same starting composition fired in silver covered alumina crucibles contained significant amounts of free $BaCO_3$.

When a sample prepared by method (i) was analysed in the electron microprobe and TEM, small polycrystalline regions were found containing high levels of (Tl + Ca) or (Tl + Ca + Cu), although in general the elemental distribution was very uniform as would be expected in a sample which is primarily 2212. A few areas rich in Ba were found, consistent with the presence of the free $BaCO_3$ identified by XRD. These phases must accommodate the excess Ca when 2212 is formed in material of overall composition 2223, and are probably related to the Tl_xCa_yO and $Tl_xCa_yCu_zO$ phases discussed, but not precisely identified, by Narain and Ruckenstein (1989). Significant amounts of the amorphous phases identified by TEM in our samples, and we suspect that similar phases are present in all Tl superconductors prepared by a bulk powder sintering process.

Thin films of 2212 were prepared by the thermal evaporation of BaF_2, CaF_2 and Cu and subsequent thalliation at 865°C for 10 - 20 minutes in an alumina crucible sealed with a silver cap. These films contain significant amounts of impurity phases, including $BaCuO_2$ and Ca_2CuO_3 grains, as is described in more detail elsewhere. The transition temperatures to zero resistance of these films are between 80 and 110K.

3. DEGRADATION TREATMENTS AND RESULTS

Samples prepared by methods (i) and (ii) have been studied either after extended storage in a dessicator or exposure to an ambient saturated with water vapour at 40°C for shorter periods, (accelerated ageing). Changes in the pH of water in which 2212 samples have been immersed have also been monitored to give an indication of the rate at which alkaline solutions may be formed on a specimen surface during the phenomenon of atmospheric degradation.

It was found that the degradation behaviour of the samples can be divided into two distinct groups. Those samples which contained free $BaCO_3$ when they were freshly made degraded much faster than samples in which no $BaCO_3$ was present initially.

3.1 2212 + traces of $BaCO_3$

The accelerated ageing process on 2212 samples initially containing free $BaCO_3$ produced the following results. The XRD traces in Figure 1 show a gradual decrease in the height of the peaks characteristic of crystalline 2212, until after 6 hours no distinct crystalline peaks were observed along with the formation of a whitish layer visible on the sample surface. The loss of crystallinity in the XRD pattern indicates the white layer to be amorphous. After 12 hours ageing, peaks characteristic of crystalline $BaCO_3$ became visible and grew in intensity as the ageing process was continued, and in addition peaks corresponding to the presence of $CaCO_3$ and CuO were sometimes present.

TEM/EDX analysis of material obtained by scraping the surface layers off a sample aged for 6 hours showed a large number of amorphous regions with a variety of compositions. Some of these amorphous regions were observed to contain only thallium, and were probably an amorphous thallium oxide. Nearby regions were consequently depleted of thallium, and were observed to be amorphous or microcrystalline, and rich in one or more of the other elements. In some of these samples, the process of warming them from below 100K in contact with air resulted in the formation of large Ba-rich plates on the specimen surface as shown in the SEM micrograph in Figure 5, with a subsequent degradation of the superconducting properties.

Fig. 5. SEM micrograph showing Ba-rich plates formed during the warming of a sample to room temperature.

3.2 2212 + traces of $BaCuO_2$

The samples made by method (ii) in a Au tube showed similar behaviour, although the degradation process was slower. Figure 3 shows the XRD results for this sample after 0, 16 and 28 hours. In these samples the 2212 XRD peaks had almost disappeared after 16 hours of ageing. Evidence for the presence of polycrystalline $BaCO_3$ did not appear until 28 hours.

3.3 Long term degradation

Samples of 2212 + traces of $BaCO_3$ stored in a dessicator for periods in excess of two months also showed rapid degradation of the superconducting phase. XRD analysis indicated the formation of large amounts of crystalline $BaCO_3$, and a white deposit was visible on the sample surfaces. The superconducting transition width in these samples is much broader than seen in the freshly prepared samples, although the onset of the drop in resistance remains around 100K, curve 2 in Figure 2 which is from a sample aged for 2 months. Figure 6 shows an SEM micrograph of a sample aged for 8 months. A thick layer of $BaCO_3$ (at least 0.5 μm) has formed on the sample surface in this time.

Fig. 6. SEM micrograph showing the surface of a sample made by method (i) after 16 hours accelerated ageing. The bright areas covering almost all of the surface are $BaCO_3$.

The superconducting properties of bulk samples of 2212 containing no free $BaCO_3$ also degrade during long term storage. An example of this effect is shown Figure 4 curve 2, for a sample stored for 6 weeks in an evacuated dessicator. The transition temperature is degraded but to a lesser extent than observed for samples made by method (i) (compare to Figure 2). The thin film samples which contained no $BaCO_3$ showed no degradation in superconducting properties after one month in a evacuated dessicator then two months storage on a laboratory shelf. This must be considered a very stringent test of long term stability since the films are only 0.5μm thick.

4. pH RESULTS.

Preliminary tests were carried out to give only an indication of the relative rates at which an alkaline environment was formed around each bulk sample. Approximately 0.5 g of material was added to 50 ml of distilled water for each test. Pure $BaCO_3$ was also tested as a standard and found to change the pH from 7 to 9.2 in 60 seconds, more rapidly than any of the 2212 samples tested. The samples which contained free $BaCO_3$ were decomposed by water more quickly than samples containing only 2212 and $BaCuO_2$. The samples made by method (ii) which were fired in an alumina crucible and contained $BaCO_3$ caused the pH to increase from 7 to 7.5 over 90 seconds, and to 9.7 after 4 minutes. The solution containing the sample made by method (i) showed no change in pH after 5 minutes, but it increased to 7.6 after a further 25 minutes.

The 2212 + $BaCuO_2$ sample caused no change in the pH of the solution after 2 hours, but after 3 hours it had risen to 7.5. Even when this sample was crushed, the pH of the solution rose to only 8 after 2 minutes and then showed no further increase. This indicates that free $BaCO_3$ results in rapid formation of an alkaline environment, but that $BaCuO_2$ is sufficiently stable to resist solution and the pH only increases rather slowly.

5. DISCUSSION AND CONCLUSIONS

This work has indicated that 2212 samples containing even small quantities of free $BaCO_3$ degrade much faster in humid environments than samples containing other impurity phases. In particular, samples in which the major impurity phase is $BaCuO_2$ are relatively stable over quite long periods even in direct contact with water. In addition, we have observed that a film less than 0.5 μm thick containing no $BaCO_3$ stored for over three months showed no evidence of degradation of its superconducting properties. This observation is very strong evidence that in the absence of $BaCO_3$ even thin film samples of 2212 can show long term stability. We suggest that the presence of free $BaCO_3$ leads to the formation of local highly alkaline environments, which stimulate the decomposition of the 2212 majority phase.

The phase degradation process has been studied by detailed TEM microanalysis, showing that the 2212 grains gradually decompose into a wide range of microcrystalline and amorphous phases in samples tested in humid environments, and only at a later stage in the degradation process do larger crystals of $BaCO_3$ grow from this mixture of phases. By the time that $BaCO_3$ can be detected in any quantity on the sample surfaces, the superconducting properties of the entire sample are also severely degraded.

ACKNOWLEDGEMENTS

The authors gratefully acknowledge Professor Sir Peter Hirsch, FRS, for the provision of laboratory facilities, and LTR and K D V-P thank the SERC for support during the course of this work.

REFERENCES

Bansal N P and Sandkuhl A L 1988 *Appl. Phys. Lett.* **52** 323
Barns R L and Laudise R A 1987 *Appl.Phys.Lett.* **51** 1373
Gopalakrishnan I K, Sastrey P V P S S, Gangadharen K, Phakak G M, Yakhmi J V and Iyer R M 1988 *Appl.Phys.Lett.* **53** 414
Hazen R M, Finger L W, Angel R J, Prewitt C T, Ross N L, Hadidiacos C G, Heaney P J Veblen D R, Sheng Z Z and Herman A M 1988 *Phys.Rev.Lett.* **60** 1657
Hoyle R 1988 *Part II thesis* Oxford University
Narain S and Ruckenstein E 1989 *Supercond.Sci.Technol.* **2** 236
Sheng Z Z and Herman A M 1988 *Nature* **332** 55 and 138
Shih I and Qiu C X 1989 *Appl. Phys. Lett.* **53** 523

Paper presented at Int. Conf. on New Materials and their Applications, Univ. of Warwick, 1990

Metal complexes as materials for chemical sensors

John D. Wright
University Chemical Laboratory, University of Kent, Canterbury, Kent CT2 7NH

ABSTRACT: The wide range of chemical sensor applications of metal complexes is illustrated with examples from cation- and anion- selective electrodes, surface acoustic wave device selective coatings, redox mediators for biosensors and semiconducting metal complexes for gas sensors. Some of the complexities of optimising chemical constitution, physical form and operating conditions to produce practical chemical sensors are illustrated for the case of semiconducting metal complex films as gas sensors.

1. INTRODUCTION

The complexes formed by metal ions with organic ligands have for many years attracted the attention of mainstream inorganic chemists. The combinations of metal ions, with their varying sizes, oxidation states, redox properties and preferred geometries, and ligands, with coordinating atom type, electronic energy levels and preferred geometry all capable of fine tuning using the skills of synthetic organic chemists, have provided an exceptionally rich field of exploration. The opportunities for tailoring molecular properties are enormous, and have played an important role in conventional analytical chemistry. Thus, for example, selective complexing agents are used in complexometric titrations and in colorimetric analysis, while metal complexes are useful as redox indicators with finely-tunable end-point potentials.

In recent years the advent of chemical sensors for a wide range of molecules and ions has been producing a quiet revolution in analytical chemistry. Chemical sensors are transducers (generally small) which convert specific chemical concentrations into optical or electronic signals proportional to those concentrations. With current optics and microelectronics technology, such sensors provide the means to measure even very low concentrations reliably, rapidly and selectively using small portable instruments and/or microsensors capable of, for example, measuring environmental quality with good time and space resolution. Conventional analytical chemistry generally cannot provide this combination of advantages. Chemical sensors require two features for their successful functioning: they must incorporate chemically-selective interactions capable of binding the target species to the sensor, and this binding must produce a change in the physical properties of the sensor material which is capable of being detected by electronic or optical means. It is precisely the breadth of these challenges which makes the application of metal complexes as chemical sensor materials so appropriate, as their range of possible chemical and physical variations provides a formidable armoury with which to meet the challenges.

It is therefore not surprising that metal complexes are increasingly being incorporated into chemical sensors in a variety of roles. This article will first summarise the current scope of this field with reference to representative examples in the recent literature, before expanding on some details of one particular class of metal complex sensor materials extensively developed in our laboratories - the organometallic semiconductor nitrogen dioxide sensors. Thus, the versatility of metal complexes as materials for chemical sensing will be demonstrated, and some of the complexities of optimising their chemical constitution, physical form and operating conditions to produce a practical chemical sensing instrument will be explored.

2. SOME APPLICATIONS OF METAL COMPLEXES TO CHEMICAL SENSING

2.1 Selective complexing of metal ions in cation-selective electrodes

Cation-selective electrodes (figure 1) commonly consist of a polyvinyl chloride (PVC) membrane which separates an internal reference solution containing a standard concentration of the ion to be detected from the test solution. Contact is made with the internal reference solution using a suitable reference electrode such as silver/silver chloride, reversible to the counter-ion (in this case chloride), and the cell is completed with a standard reference electrode such as the calomel electrode. In order that the potentials between the two surfaces of the membrane and the reference and test solutions can reflect the specific cation concentrations present, the membrane must be made to interact selectively with the cations in question. Furthermore, the species which achieves this selective interaction must also be capable of incorporation into the PVC membrane. The macrocyclic polyethers such as crown ethers (Pedersen, 1967) (e.g. 1), cryptands (Lehn, 1978) (e.g. 2) and spherands (Cram and Trueblood, 1981) (e.g. 3) fulfil these requirements elegantly. The conformations of the free ligands

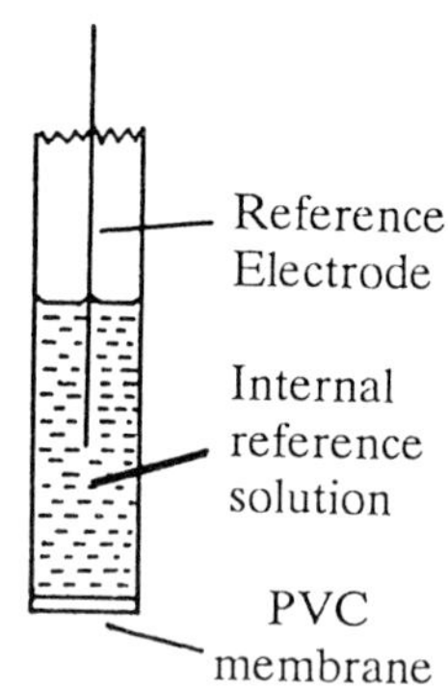

Figure 1.

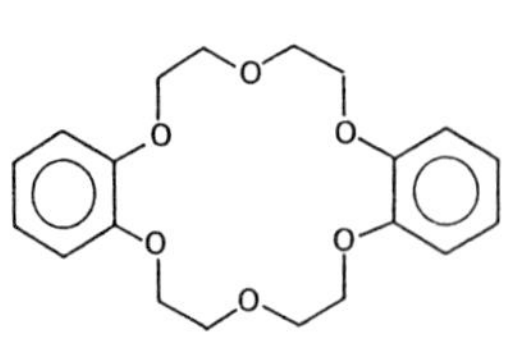

(1)

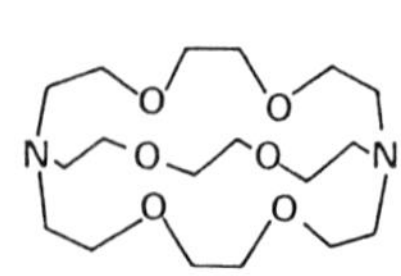

(2)

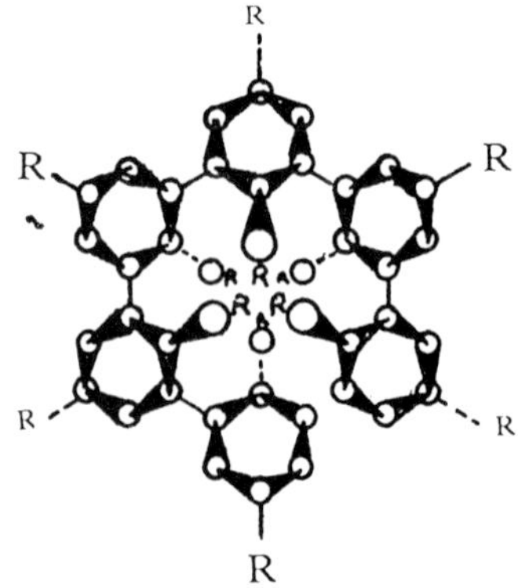

(3)

depend on the polarity of the local environment, and in hydrophilic media the hydrophilic oxygen atoms lie on the exterior of the molecule while the lipophilic organic regions form the interior. However, in lipophilic media the molecular conformation is inverted forming a lipophilic exterior with a hydrophilic electronegative host cavity. The size of this cavity can be tailored by modifying the ligand. Thus the cavity in dibenzo-18-crown 6 (1) is exactly the right size to complex the potassium ion, whereas Na^+ is too small and Cs^+ too large, and so potassium is selectively encouraged to enter the crown-ether-doped PVC membrane. The sensitivity of such a membrane electrode towards potassium ions is approximately 100x higher than that towards sodium ions. In practice, detailed modelling of the selectivity of a particular PVC/ligand combination is not possible. For example, complexing in a membrane is different from complexing in solution because ionic diffusion to the incorporated ligand molecules has to occur, and the enthalpy and entropy changes associated with complexing within a PVC membrane will be different from those in an aqueous medium. Although these problems present considerable challenges, a great deal of synthetic chemistry has been done in this area, and a wide range of ligands exhibiting selective complexing in aqueous solution to different alkali metal ions, for example, are known. These can be used to produce highly cation-selective electrodes, even though the details of the complete functioning of these electrodes are still not fully understood (Edmonds, 1988).

2.2 Selective coordination of anions by metal complexes in anion-selective electrodes

Membrane electrodes selective to anions have in the past generally been formed from polymer membranes incorporating lipophilic quaternary ammonium or phosphonium salts, in which the cations are hydrophobic and strongly prefer to remain in the polymer membrane. The selectivity towards anions in these membranes is achieved largely by exchange of a more hydrophilic anion in the membrane with a less hydrophilic one in the test solution. This gives a selectivity order which is always the same, i.e. $ClO_4^- > SCN^- > I^- > Br^- > Cl^-$, known as the Hofmeister sequence, which is the inverse of the order of free energies of hydration of the ions. To obtain a different selectivity pattern, metal complexes which selectively coordinate specific ligands can be used in the membrane. Thus, a vitamin B12 derivative, in which the cobalt(III) preferentially coordinates nitrite and thiocyanate, provides an electrode selective to these anions. Recently (Meyerhoff, Pranitis, Yim, Chaniotakis and Park, 1989) a series of membrane additives based on synthetic porphyrin samples have been prepared, and a PVC membrane incorporating chloro 5,10,15,20-tetra((triphenyl)phenyl)-porphyrinato manganese (III) (4) shows a potentiometric response to SCN^- more than 10^5x larger than that towards Cl^-. This is believed to arise from the bulk and extra conjugation of the four triphenyl-phenyl groups, and has potential application in the detection of high SCN^- concentrations in biological samples, which have been correlated with high cigarette smoking. Similarly, dichloro(5,10,15,20-tetraphenyl) porphyrinato tin(IV) shows high selectivity to salicylate, with potential application to studies of pharmacological effects of aspirin.

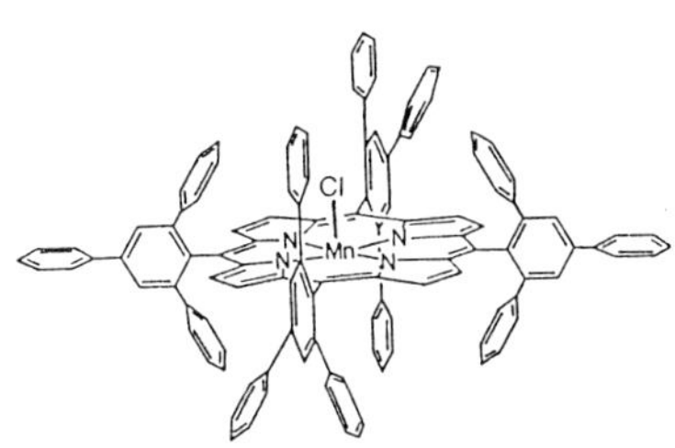

(4)

2.3 Metal complexes as selective coatings for surface acoustic wave sensors

The kinetics of ligand exchange reactions of inorganic complexes have been extensively studied, and one area of particular interest has been such reactions of square planar platinum complexes, where the rate of ligand exchange has been found to be dependent on the nature of the ligand occupying the trans- position in the complex. For olefin complexes with the trans-ligand pyridine, the exchange of one olefin for another occurs via an associative pathway, in which the incoming olefin approaches as the original olefin leaves. For such a mechanism, the reaction kinetics are largely dependent on the nature of the incoming ligand, with no pronounced influence of solvent. This fact can be used to devise a sensor which is specific to vinyl acetate. Molecular orbital calculations for trans-$PtCl_2$(olefin)(pyridine) complexes (5) (Meester, van Dam, Stufkins and Oskam, 1976) indicate that the vinyl acetate complex should be particularly stable. Thus, the ethylene in the $PtCl_2$(ethylene)(pyridine) complex should be readily exchanged for vinyl acetate, with no effect of other solvent vapours on the reaction kinetics. This ligand-exchange process leads to a change in the relative molecular mass, which can be detected using a surface acoustic wave (SAW) device. SAW devices measure the propagation velocity of a surface acoustic wave passing through a thin film of the sensing material. The propagation velocity is dependent on the mass of the film material, amongst many other parameters. A typical SAW device is shown in figure 2, and is in fact a delay line which can be used as a feedback element in an oscillator circuit, so that the oscillator frequency varies according to the surface mass loading.

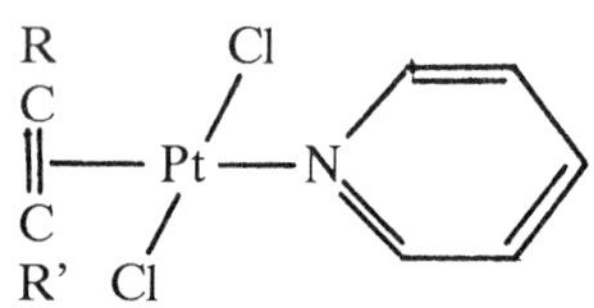

(5)

At 30 MHz resonance frequency, frequency shifts of several hundred Hertz per microgram on areas of the order of 1 cm^2 can be achieved, giving high sensitivity. If such a device is coated with a uniform dispersion of crystals of the platinum complex in a polyisobutylene film, exposure to vinyl acetate vapour produces an initial rate of frequency shift which is proportional to the vinyl acetate concentration (Zellers, 1989). This permits rapid measurement of vinyl acetate concentrations, and the film is readily restored to its initial condition by brief flushing with pure ethylene. No interferences from substances such as methanol, vinyl chloride and ethyl acetate are observed, although the system is temperature-sensitive. Although this system is somewhat complex, the selectivity and the industrial significance of vinyl acetate make this an important application. The high sensitivity of SAW devices can, however, only be used successfully in chemical sensing if highly selective coatings such as this are developed, since the technique detects mass changes only, and mass is a universal property.

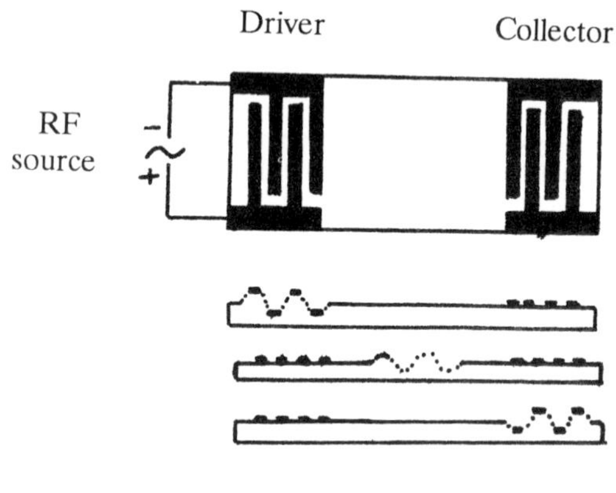

Figure 2

2.4 Organometallic complexes as redox mediators in biosensors

Electrochemical monitoring of the activity of redox enzymes is an important technique in biosensing. Thus, in monitoring blood glucose concentrations in diabetics the glucose-oxidase (GOD) catalysed oxidation of glucose to gluconolactone plus hydrogen peroxide can be used, with electrochemical oxidation of the hydrogen peroxide as a method for measuring the extent of the reaction. The major disadvantage of using this approach is that the glucose assay method depends on a satisfactory supply of oxygen. In whole blood or plasma, the available oxygen may be insufficient, limiting the rate of reaction of the glucose and hence the sensor response. However, if a suitable organometallic complex, typically a ferrocene ($Fecp_2$) (6), is present the electron-transfer processes occur without the participation of oxygen (Cass, Davis, Francis, Hill, Aston, Higgins, Plotkin, Scott and Turner, 1984):

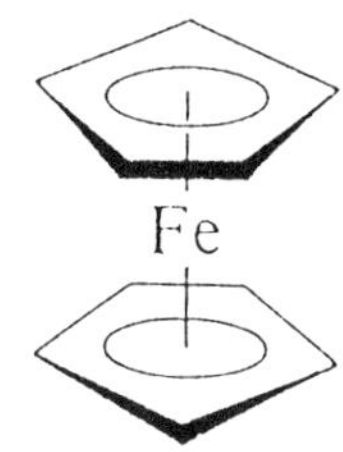

(6)

(At the enzyme:)

$$\text{Glucose} + \text{GOD}_{ox} \rightarrow \text{Gluconolactone} + \text{GOD}_{red}$$

$$\text{GOD}_{red} + 2\text{Fecp}_2^+ \rightarrow \text{GOD}_{ox} + 2\text{Fecp}_2 + 2\text{H}^+$$

(At the electrode:) $$2\text{Fecp}_2 \rightleftharpoons 2\text{Fecp}_2^+ + 2e^-$$

This reaction sequence is nearly independent of oxygen concentrations, and of pH (since the electrode reaction does not involve protons). Ferrocenes are ideally suited as electron transfer mediators since they can be readily substituted in different ways to modify their redox properties, their solubility in a wide range of solvent can also be controlled by appropriate substitution, they are chemically and thermally stable, they can be polymerised or grafted onto polymers or other molecules and they are small enough to interact effectively with the redox-active sections of large enzymes without unfavourable steric factors.

2.5 Metal phthalocyanines as semiconductor gas sensors

Phthalocyanine (7) and its complexes with metals form chemically-stable molecular crystals which do not sublime at temperatures below about 350°C. The electrical conductivity of these crystalline materials is very poor, since the constituent molecules are essentially uncharged and energies of several eV are required to produce charge carriers intrinsically, and since the charge-carrier mobility is small (≈ 1 cm²/v.s) due to the fact that the molecules are separated by van der Waals distances. However, when electron acceptor gases adsorb on the phthalocyanine surface, electron transfer is facilitated, leading to a surface p-doping and large increases in surface semiconductivity. Studies

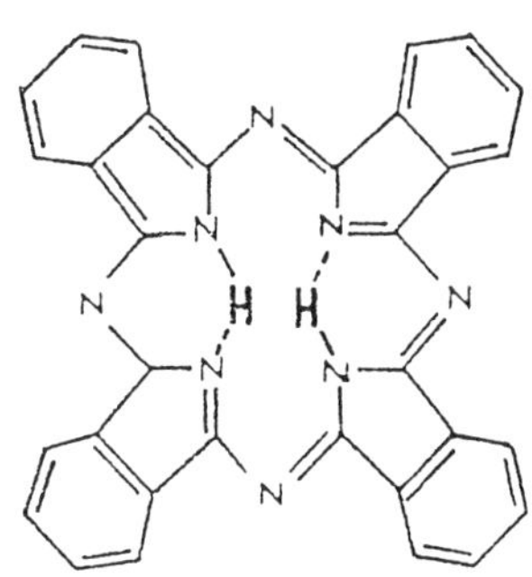

(7)

using single crystals of the materials (van Ewyk, Chadwick and Wright, 1980) have shown that the effect is predominantly a surface phenomenon, and that the log(surface resistivity) falls linearly as log(acceptor gas pressure) increases, with surface resistivity changing by a factor of at least 10^7 before complete surface coverage is reached at a pressure of 10^3 Pa in the case of NO_2. The size of this change permits detection of NO_2 at concentrations down to 1 part per 10^9 of air, and provides the basis for a series of sensors for NO_2, chlorine and other electron acceptor gases (Wright, 1989). The strengths of the interactions between the semiconductor surface and the adsorbed gas, which control the sensitivity, response kinetics and reversibility, can be modified by varying the metal ion, substituting the periphery of the phthalocyanine ring, or using related ligands with smaller or larger conjugated π-electron systems.

3. THE DEVELOPMENT OF NO_2 SENSORS BASED ON PHTHALOCYANINE FILMS

The requirements of selective binding of the analyte leading to an optically or electrically detectable change in the physical properties of the sensor material, which were discussed in the introduction, are necessary but not sufficient conditions for a practical chemical sensor device. A practical sensor should also be reversible, stable (both chemically and physically) and rapid in response (or at least give reproducible response kinetics so that the responses from short sampling pulses are truly representative of the final equilibrium responses), as well as giving reproducible calibration curves which should ideally be the same for all sensors prepared by a particular method. Frequently these criteria can be met by optimisation of both the chemical and physical nature of the sensor material. This process is cyclic, with variations in material parameters leading to improved understanding of sensing mechanisms, which in turn permit further refinement of the relevant material parameters. This will now be illustrated using the example of NO_2 sensors based on phthalocyanine films.

3.1 Comparison of phthalocyanine single crystals and films as NO_2 sensors

Figure 2 compares the responses of a single crystal and a sublimed film (on a rough polycrystalline alumina substrate) of lead phthalocyanine to several cycles of exposure

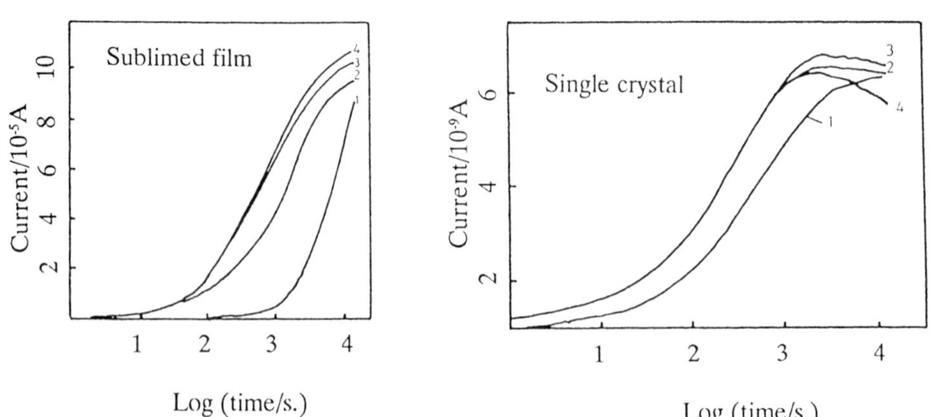

Figure 2.

to 300 ppb of NO_2 in dry air, with reversal in clean dry air between successive cycles, with the sensor at 155 °C in each case. Although the general forms of the responses are similar, the single crystal responds more reproducibly, after the first cycle, whereas the film requires several cycles to achieve reproducible sensitivity and response kinetics. Films are nevertheless more likely to be used in practical devices, because device resistance can be lower (using interdigitated electrodes), crystals of suitable size are not easy to produce and mount on substrates, and they are prone to cracking due to differential thermal expansion relative to the substrate material when operated at the elevated temperatures needed to produce good kinetics and reversibility.

3.2 Effects of film morphology

The crystal structure and morphology of phthalocyanine films significantly influence their gas adsorption properties and the consequent effects on the semiconductivity of the films. The strength of the interaction between the gas and the film surface is a key parameter determining the form of the adsorption isotherm and the electrical consequences of gas adsorption. For a film with a range of different adsorption sites, in contact with air, the adsorption of a strong electron acceptor such as NO_2 will occur most rapidly at the weakest adsorption sites, from which oxygen is most readily displaced. As the NO_2 concentration increases, successively stronger adsorption sites will be populated with increasing difficulty, and the electrostatic attractions between thermally-generated surface p-type charge carriers and NO_2^- on the surface will increase, altering the proportionality constant between surface coverage and conductivity change. Film heterogeneity and degree of crystallinity and crystallite size are dependent on the nature of the substrate, the evaporation rate, the substrate temperature during film deposition, and the thermal history of the film. Most phthalocyanines are polymorphic, and adsorption of oxygen has been shown to depend significantly on the polymorphic form, which changes in many cases on heating or exposing the films to solvent vapours. Even in cases where the polymorphic form is constant for different films, varying the film deposition conditions has been shown to alter the orientation of the crystallites relative to the substrate plane. Since phthalocyanine crystals have anisotropic structures, with stacks of molecules arranged in various ways, varying the crystallite orientation will change both the adsorption sites presented at the film surface and the electrical conductivity of the film as measured in a particular direction. Finally, the above possible variations in film morphology will also influence diffusion down grain boundaries in polycrystalline films, and the chemical reactivity and volatility of the films. These effects have recently been reviewed (Wright, 1989) and a detailed discussion of them all is inappropriate to the main theme of the present paper. However, two effects of film morphology on the kinetics of the conductivity change produced in lead phthalocyanine films on exposure to low concentrations of NO_2 will be highlighted as examples of the importance of these phenomena.

One interpretation of the gradual increase in sensitivity and response kinetics which is observed when a freshly-sublimed lead phthalocyanine film is exposed to NO_2 (figure 2) is that amorphous regions of the film undergo thermally-induced crystallisation. This process, which can be observed by electron microscopy, increases the surface area of the film and the number of active adsorption sites which can lead to enhanced surface charge-carrier generation. This interpretation is supported by the observation that films previously heat treated in air respond more rapidly on initial exposure to NO_2. Also films deposited on substrates maintained at temperatures which optimise the film surface smoothness (Vincett, Popovic and McIntyre, 1981) take longer to reach reproducible response kinetics, as expected for materials having less sites at which crystallisation is initiated.

A second effect of film morphology on the kinetics of response of lead phthalocyanine to NO_2 is that films which have been heated for one hour at 360°C in air show a very much faster (although smaller) conductivity change on exposure to NO_2. This has been interpreted as being a consequence of the substantial increase in crystallite size which is observed in electron microscopy studies of the films (Jones, Bott and Thorpe, 1989a). This reduces the proportion of crystal edges and corners, as well as the range of different crystal faces exposed to NO_2, thus reducing both the proportion of strong adsorption sites and the surface area. Since weaker adsorption sites are populated with NO_2 most rapidly, as explained above, this accounts for both the rapid response and the lower sensitivity.

However, these effects cannot be fully explained in terms of morphological changes. Thus, lead phthalocyanine films which have been cycled as in figure 2 to obtain reproducible responses are found to lose their sensitivity to NO_2 after exposure to ultra high vacuum (10^{-10}mbar), although the original reproducible response can be regained after further exposure cycles to NO_2 (Archer, Chadwick, Miasik, Tamizi and Wright, 1989). Since exposure to vacuum cannot reverse crystallisation processes, some additional phenomenon must be involved. Similarly, the effects observed on heating lead phthalocyanine to 360°C in air are not observed for all phthalocyanines (Jones, Bott and Thorpe, 1989b) whereas similar crystallite-size effects should occur irrespective of the nature of the central metal atom. Such phenomena require consideration of effects of the surface chemical composition of the films.

3.3 Effects of surface chemical composition.

The chemical nature of a phthalocyanine film surface can be changed by chemical reactions occurring during or after film deposition, or by varying the chemical nature of the original material (e.g. by modifying the ligand or using different metal ions). The kinetic effects described above can be related to chemical changes occurring after the film has been deposited. Thus, photoconductivity studies have shown that after exposure of phthalocyanines to low concentrations of NO_2 and reversal of the effect in clean air the photocurrent is still enhanced relative to that for a clean surface, and the action spectrum is also different (van Ewyk, Chadwick and Wright, 1981). This suggests the presence of residual traces of NO_2 bound at the strongest adsorption sites. The strong electron-hole attraction at such occupied sites results in a negligible contribution to the semiconductivity, but this strong surface dipole exerts a repulsion effect on the oxygen species at the surface near the residual NO_2 species. This effect increases the desorption rate of oxygen, resulting in faster response to NO_2. This influence of strongly-bound residual species also explains why even single crystals of phthalocyanine show some increase in response kinetics on initial cycling as in figure 2, and why exposure to UHV necessitates re-initialisation of the films as gas sensors by further exposure cycles. This UHV treatment effectively removes even strongly-bound NO_2. The model also explains why the initialisation process seen in figure 2 is slow. The strong adsorption sites which must be populated by residual NO_2 are also the slowest ones to populate, since they hold oxygen most strongly.

The reason why heating the films to 360°C is only effective in producing faster response for lead phthalocyanine and not for other phthalocyanines can also be explained in terms of a chemical effect. Lead phthalocyanine is non-planar, and this weakens the π-bonding linking the four iso-indole moieties which make up the phthalocyanine molecule. It is therefore particularly prone to oxidative cleavage, which has been demonstrated directly for films exposed to high concentrations of NO_2 (Hobert, Weaver and Wright, 1989). Such reactions are particularly likely at the sites where oxygen is most strongly bound to the surface, and this provides a mechanism whereby these strong adsorption sites are effectively removed from the sensing process. The removal of sites responsible for the slow component of the response reduces the overall sensitivity but the remaining response is fast. This interpretation is supported by the observation of lead oxide species on the surface of the lead phthalocyanine films heated to 360°C (Eccles, Humphrey, Jones and Vickerman, 1987). In planar phthalocyanines, which are less prone to reactions of this type, the strong adsorption sites are not removed by this heat treatment, so the slow components of response remain.

Sublimation conditions can also affect the surface chemical composition of lead phthalocyanine films. XPS studies (Mockert, Graf, Schmeisser, Göpel, Ahmad, Archer, Chadwick and Wright, 1990) show that loss of nitrogen occurs when films are sublimed at temperatures above 400°C. This has been interpreted in terms of thermal fragmentation of the non-planar lead phthalocyanine. Reactions between fragments, leading to the formation of large conjugated oligomers could be the origin of some of the slow-responding strong adsorption sites on the film surface.

Changes in the chemical nature of the film material itself also lead to changes in NO_2-sensing properties. For example, use of the ligand tetraazadibenzo[14]annulene (TADA) (8), which is a weaker π-donor due to its smaller delocalised π-electron system, produces a faster response to NO_2 since surface oxygen species are more rapidly displaced. Similarly, the weaker donor power of the non-planar lead phthalocyanine leads to a faster response to NO_2 than that shown by planar phthalocyanines (e.g. Cupc, AlpcF etc.) (Archer, Chadwick, Miasik, Tamizi and Wright, 1989).

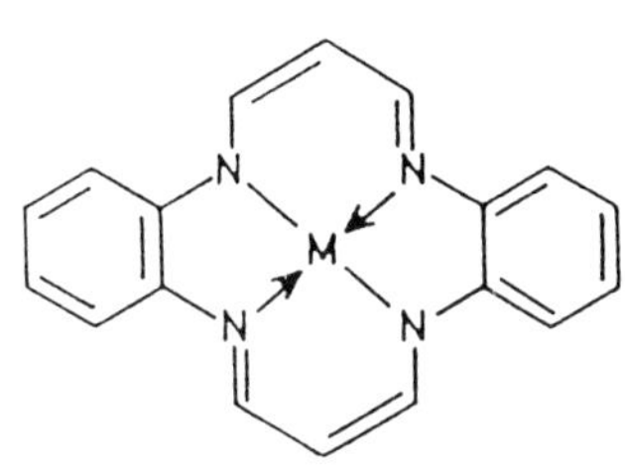

(8)

Finally, impurities in phthalocyanines generally determine both the magnitude and type of semiconductivity for clean films, and these determine both the ultimate lower limit for NO_2 detection and the direction of the conductivity change near this limit. Thus, most phthalocyanines show very low p-type conductivity in the absence of strong electron acceptor gases, and so show measurable conductivity increases even in very low concentrations of NO_2 (e.g. 1 part in 10^9). However, we recently observed that a film of impure fluorogallium phthalocyanine appeared to have n-type properties, so that initial exposure to NO_2 led to a decrease in conductivity via depletive chemisorption. However, if the surface coverage with NO_2 is allowed to increase, for example by using longer exposures, higher gas concentrations or higher temperatures, p-type conductivity ultimately dominates, leading to conductivity increases. These effects can lead to calibration curves whose form depends strongly on the sensor operating temperature and the duration of the sampling pulses, as illustrated in figure 3, which shows responses to 60-second pulses of NO_2, with a decrease in conductivity for 270 ppb of NO_2 at 150°C (a), an increase for 2 ppm at 210°C (b), and a decrease followed by an increase for 2 ppm at 100°C (c). Such

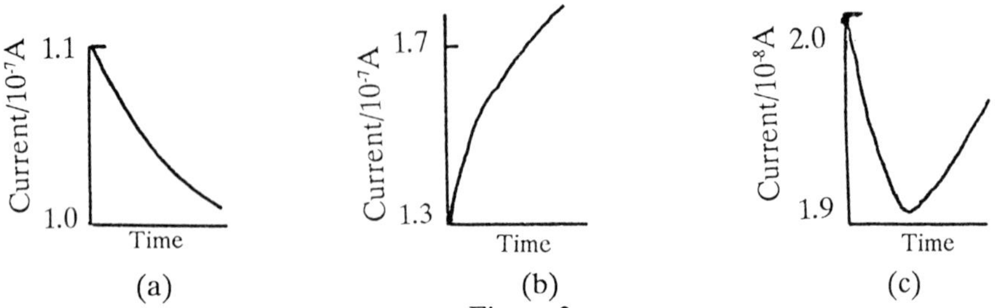

Figure 3

complex behaviour is undesirable in a practical sensor, so careful purification by repeated entrainer sublimation is recommended. Purer films of fluorogallium phthalocyanine do indeed show conductivity increases in NO_2 under all conditions.

3.4 Chemical and physical stability of films

Materials for practical NO_2 sensors should not react under normal measurement conditions nor should they sublime easily at temperatures which would limit the available range of operating temperatures excessively. Thus, fluoroaluminium phthalocyanine, which gradually decolourises over several days in ppm concentrations of NO_2 with a steady decrease in sensitivity, and fluorochromium phthalocyanine, whose films readily change colour from blue to bronze irreversibly in these conditions, are unsuitable materials although the nature of these reactions has not yet been explored. Chemical reactions between lead phthalocyanine and NO_2 at high concentrations (3000ppm) (Hobert, Weaver and Wright, 1989) do not seem to occur in lower concentrations (1 ppm) as XPS studies of films exposed for long periods to these conditions show no changes in surface composition (Mockert, Graf, Schmeisser, Göpel, Ahmad, Archer, Chadwick and Wright, 1990). It is possible, however, that slight reaction under these conditions leads to volatile products which desorb at sensor operating temperatures contributing to a shortening of sensor lifetime. Evaporation of films generally proves to be the main factor limiting sensor lifetime for phthalocyanines (Bott and Jones, 1984). This is a disadvantage particularly for the more volatile TADA materials and for the substituted phthalocyanines which are used in Langmuir-Blodgett films, since the increased solubility achieved by substitution also leads to increased volatility. As the above examples suggest, both reactivity and volatility can be varied, and hence controlled, by suitable modification of metal and/or ligand in the complex.

3.5 Selectivity and interferences.

Effective p-doping of phthalocyanine surfaces can only be achieved by strong electron acceptor vapours. Thus, oxygen has a negligible effect, while H_2S (5 ppm), SO_2 (10 ppm), H_2 (100 ppm), CH_4 (1%) and CO (100 ppm) have no effect (Bott and Jones, 1984). Chlorine produces effects comparable to those of NO_2, but is most unlikely to be an interferent in normal environmental monitoring applications. Since different acceptor/film combinations exhibit different strengths of interaction, chemical variation of the film together with varying operating temperature provides scope for improving selectivity.

Interferences are of two classes: electronically-inactive vapours which adsorb and physically block the surface sites, and electronically-active vapours which inhibit the effects of NO_2 and other electron acceptors. Water vapour is the most serious example of the first class, while ammonia is an example of the second. Water vapour binds tenaciously to phthalocyanine surfaces, probably via hydrogen bonding, and some water is retained even after heating at 150°C in 10^{-10} mbar UHV conditions for 1 hour (Mockert et al. 1990). This leads to a reduction in the amount of NO_2 which a film can adsorb in given conditions, unless these conditions are controlled to minimise attempts by NO_2 to populate sites which are occupied by water molecules, for example by limiting the fractional surface coverage by NO_2. This can be achieved by using short sampling pulses and low operating temperatures (giving slower adsorption) particularly for high NO_2 concentrations where the water interference is worst (Wright, 1989). Interference by electron-donating impurities is more serious, as adsorbed NO_2 at all types of site will be released from the surface by such gases (Kaufhold and Hauffe, 1965), and selective removal of these interferents is not possible since NO_2 will be strongly adsorbed by all suitable materials. However, in environmental monitoring applications such interferences may be less significant as the presence of donor gases alongside NO_2 will mitigate some of its adverse environmental effects.

3.6 Development of an NO_2-sensing instrument based on phthalocyanine films

The above discussions show that phthalocyanine films can best be used as materials for NO_2-sensing when controllable pulse-sampling is used in conditions of reproducible response kinetics. The reproducible kinetics can be achieved by ensuring that the residual adsorbed NO_2 concentration on the film surface is constant at the start of each sampling pulse, so that its effect in enhancing the desorption of surface oxygen species is constant. This can be achieved by monitoring the reversal of the conductivity of the film in clean air after each sampling pulse, triggering the next pulse when a preset background limit has been reached, since the background conductivity reflects the amount of residual adsorbed NO_2. This can be coupled with variation in sampling pulse length and/or sensor temperature to minimise interferences and possible chemical reaction according to conditions, with the whole sequence of operations under microprocessor control (Wilson, Chadwick and Wright, 1989). Thus the knowledge which has been developed of materials features and sensing mechanisms in such films can be incorporated into one intelligent system in which all relevant parameters are continuously optimised.

4. CONCLUSIONS

Metal complexes have already found applications in chemical sensing, using many different sensing principles which take advantage of the diverse properties of metal complexes. In some cases the effects of physical and chemical variation of the metal complex sensing material have been explored in some detail. However, the scope for further developments in this area is extremely wide, and many new applications of metal complexes in chemical sensing may be confidently expected in future.

ACKNOWLEDGMENT

The contribution of the late Dr. T. Alwyn Jones to the development of phthalocyanine sensors and his support and discussions with the author over many years have been an invaluable influence in this field of research.

REFERENCES

Archer, P B M, Chadwick, A V, Miasik, J J, Tamizi, M and Wright, J D 1989 *Sensors and Actuators* **16** 379

Bott, B, and Jones, T A 1984 *Sensors and Actuators* **5** 43

Cass A E G, Davis, G, Francis, G D, Hill, H A O, Aston, W J, Higgins, I J, Plotkin, E V, Scott, L D L, and Turner, A P F 1984 *Analyt. Chem.* **56** 667

Cram, D J and Trueblood, K N 1981 *Top. Curr. Chem.* **98** 43

Eccles, A J, Humphrey, P, Jones, T A and Vickerman, J C 1987 *Proc. SIMS VI Symposium*, Versailles

Edmonds, T E (ed.) 1988 *Chemical Sensors* (London: Blackie) Chapters 2,3 and 9

van Ewyk, R L, Chadwick, A V and Wright, J D 1980 *J. Chem. Soc. Faraday I* **76** 2194

van Ewyk, R L, Chadwick, A V and Wright, J D 1981 *J. Chem. Soc. Faraday I* **77** 73

Hobert, H, Weaver, R C and Wright, J D 1989 *Z. Chem.* **29** 454

Jones, T A, Bott, B. and Thorpe, S C 1989a *personal communication*

Jones, T A, Bott, B and Thorpe, S C 1989b *Sensors and Actuators* **17** 467

Kaufhold, J and Hauffe, K 1965 *Ber. Bunsenges. Phys. Chem.* **69** 168

Lehn, J M 1978 *Acc. Chem. Res.* **11** 49

Meester, M A , van Dam, H, Stufkins, D J and Oskam, A 1976 *Inorg. Chim. Acta* **20** 155

Meyerhoff, M E, Pranitis, D M, Yim, H S, Chaniotakis, N A and Park, S B 1989 in *Chemical Sensors and Microinstrumentation* ed. Murray, R W et al., ACS Symposium Series **403** 26

Mockert, H, Graf, K, Schmeisser, D, Göpel, W, Ahmad, Z A, Archer P B M, Chadwick, A V and Wright J D 1990 *Sensors and Actuators* in press

Pedersen, C J 1967 *J. Amer. Chem. Soc.* **89** 7017

Vincett, P S, Popovic, Z D and McIntyre, L 1981 *Thin Solid Films* **82** 357

Wilson, A, Chadwick, A V and Wright, J D 1989 *U.K Patent App.* 8923924.8

Wright, J D 1989 *Prog. Surf. Sci.* **31** 1

Zellers, E T 1989 in *Chemical Sensors and Microinstrumentation* ed. Murray, R W et al., ACS Symposium Series **403** 176

Paper presented at Int. Conf. on New Materials and their Applications, Univ. of Warwick, 1990

An investigation into the viability of screen printed organic semiconductor compounds as gas sensors

A W J Cranny and J K Atkinson
Department of Electronics & Computer Science
University of Southampton

Abstract

Comparisons of metal based phthalocyanine organic semiconductors deposited both as thin (thermally evaporated) and thick (screen printed) films reveal both forms of the material as having significant potential as gas sensors. The different deposition methods employed in the fabrication of the sensors result in significant differences in the behaviour of the material. Characterisation of both forms of lead phthalocyanine consequently offers some insight into the mechanisms affecting the conductance of this material in the presence of a reactive gas. Whilst both fabrication techniques result in potentially useful sensors, the thick film form of the material is more easily produced at lower cost.

1. Introduction

The use of metal based phthalocyanines for the construction of organic semiconductor gas sensors has been extensively documented (Wright, 1989; Wilson and Collins, 1987; Jeffery et al, 1987). The high sensitivity exhibited by these materials in the form of a change of conductance due to the presence of sub ppm concentrations of oxidising and reducing gases makes their use as sensors very attractive. The principle drawbacks to using metal based phthalocyanines however, have been generally regarded as a lack of stability and a high susceptibility to cross-interference resulting from background gas species other than those actively being detected. The first of these shortcomings can be significantly improved through the use of alternative forms of the material, in particular polymers of phthalocyanine (Jeffery and Burr, 1989).

The high degree of cross-sensitivity exhibited by these materials has been the subject of a research programme aimed at improving the specificity of phthalocyanine gas sensors using signal processing techniques (Atkinson et al, 1989).

The research described here has resulted in the development of an array sensor consisting of several individual phthalocyanine sensors on a common substrate.

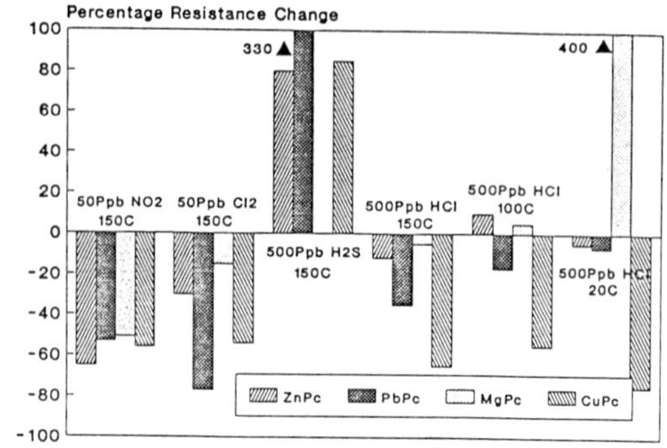

Figure 1 Response of several metal substituted phthalocyanines to various gases and sensor operating temperatures.

Since the responses of phthalocyanines having different central metal atoms can be markedly different for any given gas (figure 1), the choice of a suitable combination of phthalocyanines can be used to obtain unique responses to gases. The output signals of this sensor array can then be used to selectively indicate the concentration of a given analyte using pattern recognition techniques. This approach involves many interdisciplinary activities which include the optimisation of the sensor array construction, the choice and deposition of the phthalocyanine materials and the development of pattern recognition systems. A facility has been developed to allow particular array sensors to be characterised automatically using a computer controlled gas testing rig. This enables the operating temperature of the sensors and the concentration and type of the exposure gas to be closely controlled and monitored. The results obtained from each experimental run are added to a data base from which the laws governing the pattern recognition of the target gases are derived. In this way the pattern recognition system "learns" to recognise the target gases during the sensor characterisation phase.

2. The Sensor Array

Thick-film hybrid circuit techniques (White and Cranny, 1987; Holmes and Loasby, 1976) have been used to realise an array of individually treated gold electrode sites onto which the phthalocyanine sensor materials can be deposited. Each of the array elements thus comprises a platinum resistor, used for temperature control of the individual elements, which is electrically isolated from the electrode pattern above by a layer of insulating material. The structure of the resulting array is shown in figure 2.

The sensor array has been designed to plug into a standard 28 pin integrated circuit socket and is consequently approximately 15 mm x 36 mm. There are five individual gold electrode patterns contained within this area, each with its own platinum resistance heater. In order to obtain as large a degree of thermal isolation between adjacent sensor sites as possible, laser cuts are introduced into the substrate thereby producing an air gap. The air gaps act as a sufficient thermal break to allow adjacent sites to be operated at any temperature up to 200°C without significant cross-channel interference. The individual sensors are actively controlled by an electronic control circuit which senses the values of the platinum resistors and maintains a constant temperature by modulating the heater current flowing in each.

The phthalocyanine layers are deposited onto the sensor array using one of two alternative production methods. One technique is to utilise a conventional low pressure vapour deposition method which produces film thicknesses of typically upto one micron. The area to be deposited is selected by covering the areas which are not to be coated with a PTFE film which acts as a mask. Different masks are then used to deposit each of the chosen phthalocyanines onto the designated electrode patterns of the sensor array.

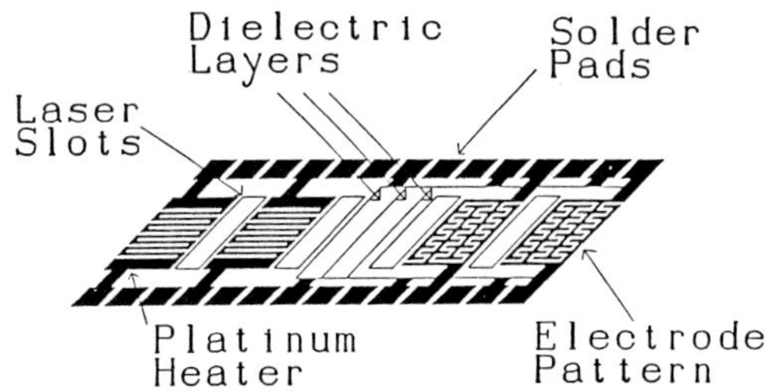

Figure 2 Schematic representation of hybrid sensor array

The alternative method to thermal evaporation is to use thick-film screen printing as a deposition technique. Here the phthalocyanine compounds are first made into screen printable inks, through powdering and the addition of organic solvent carriers, prior to being printed at thicknesses of typically 25 microns. Apart from producing a different morphology, the thick films being more granular and porous in nature, the two production techniques differ in terms of their viability for batch production purposes.

Thick-film techniques are much more conducive to batch processing methods than their thin film counterparts and are consequently much cheaper to realise. The sensors no longer need to be masked since the screen printing process much more precisely defines the deposition area. A major benefit of thick-film printing, however, is that it allows the use of polymer phthalocyanines on the sensor array which is not always possible using thermal evaporation. Indeed it is possible to deposit a wide range of materials onto the array and this method would allow the simultaneous use of inorganic metal oxide sensors on the same array as phthalocyanine sensors.

3. Experimental Results

The effects of both temperature and gas concentration on the response characteristics of thermally evaporated (thin film) and screen printed (thick film) phthalocyanine sensors are reported. Gas response is indicated as a change in the observed dc dark current. Results are presented on NO_2 sensing using lead phthalocyanine.

3.1 Current-Voltage Characteristics

Figure 3 shows the dc dark current-voltage characteristics for thin and thick film lead phthalocyanine (PbPc) sensors deposited on thick film gold electrodes. Electrical properties were measured with devices mounted in a light-tight PTFE chamber in a constant flow of clean air at room temperature. Clearly the characteristics for both forms of phthalocyanine film are shown to be ohmic.

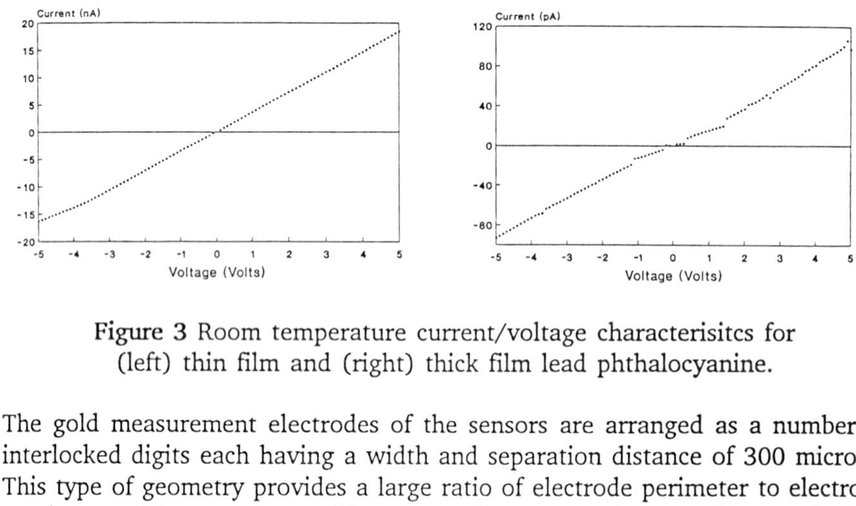

Figure 3 Room temperature current/voltage characterisitcs for (left) thin film and (right) thick film lead phthalocyanine.

The gold measurement electrodes of the sensors are arranged as a number of interlocked digits each having a width and separation distance of 300 microns. This type of geometry provides a large ratio of electrode perimeter to electrode spacing; an important consideration when measuring weakly conducting materials. The measured conductance, σ, of the lead phthalocyanine films can be described by the relation:

$$\sigma = \sigma_0 A/d \quad (1)$$

where A is the area of the measurement electrode interface, d is the distance between electrodes, and σ_0 is the intrinsic conductivity of the phthalocyanine film. Since the cross-sectional area of the film, A, is simply the product of the film thickness, t, and the electrode perimeter, P, then the above equation may be written as:

$$\sigma = \sigma_0 Pt/d \quad (2)$$

The ratio P/d is fixed by the electrode geometry and the values for the measured conductance, σ, for both films can be derived from the slopes of the corresponding dark current-voltage characteristics.

Substitution of measured quantities for the appropriate parameters in the above equation yields appreciably different values for the intrinsic conductivity of PbPc ($2.55 \times 10^{-5}\ \Omega^{-1}m^{-1}$ for the thin film and $5.7 \times 10^{-9}\ \Omega^{-1}m^{-1}$ for the thick film). The lower value exhibited by the thick film can be partly attributed to its greater porosity since the discontinuous film is unlikely to have a high conductivity. However, other factors such as the relative levels of impurities in the films and the possibility of different conduction mechanisms existing, must not be overlooked.

The differences in the intrinsic conductivity manifests itself physically as a large difference in the ambient resistance of sensors formed from these materials deposited on electrodes of the same geometry. Typically, thermally evaporated PbPc films exhibit room temperature resistances of the order 10^8 ohms, while the thicker screen printed films show values in excess of 10^{10} ohms at the thickness employed here. Clearly this is a major disadvantage of the thick film form of PbPc sensor, since a large resistance is difficult to measure accurately and is more prone to electrical noise. However, it is possible to fabricate lower resistance sensors using thick film techniques simply by redesigning the electrode geometry.

3.2 Effect of Temperature on Conductivity

By including the provision for independent temperature control at separate electrode sites on the sensor array (figure 2) it is possible to measure the dc dark current-voltage characteristics as a function of temperature. Such measurements are presented in figure 4 as plots of Log_e conductance versus inverse temperature for samples of thin and thick lead phthalocyanine films.

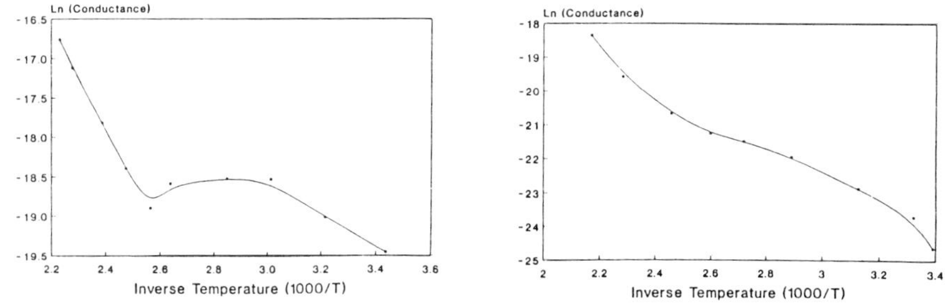

Figure 4 Effect of temperature on (left) thin film and (right) thick film lead phthalocyanine conductivities.

Both film morphologies exhibit a curved Arrhenius plot typical of an extrinsic semiconductor. These characteristics can be attributed to the presence of significant levels of impurities centred about an average energy E_A that is greater than the valence band level, E_V, but smaller than the intrinsic Fermi level. That is, both films exhibit the electrical properties associated with a p-type semiconductor and therefore charge transfer is attained primarily by the movement of holes. The distinctive shape of both curves presented in figure 4 can, therefore, be described in terms of simple solid state theory.

At low temperatures, the slope of the characteristic is proportional to the difference (E_G - E_A) where E_G represents the bandgap of the film. In the case of screen printed thick film lead phthalocyanine films, the conductivity is seen to increase with temperature whereas a slight decrease is observed for thermally evaporated thin films. This would appear to indicate the presence of a greater number of impurities in the thick film material spread over a wider range of energy levels.

As the temperature is increased yet further, a level is eventually reached when the thermal activation of holes from the conduction band to the valence band is possible and the Fermi level reaches its intrinsic value. At temperatures in excess of this value, the conduction process is seen to be intrinsic, increasing linearly with temperature, and the slope of the characteristics is proportional to the bandgap, E_G.

These results have demonstrated that a critical temperature exists, above which lead phthalocyanine based sensors must be operated to ensure intrinsic conduction. The data shows that this critical temperature is similar for both thin and thick film forms of lead phthalocyanine and is approximately 400K.

Bandgap energies have been calculated from the slope of the characteristics in the intrinsic region. For thin film thermally evaporated lead phthalocyanine a value of E_G = 1.15 eV is observed whilst the corresponding figure for screen printed thick film lead phthalocyanine is approximately 0.76 eV. This apparent dissimilarity in values is not easily explained but may be due to physical differences in the structure and/or orientation of the crystals in the films.

3.3 Effect of Gas Exposure

Changes in conductance are observed in organic semiconductors such as PbPc on exposure to certain reactive gases due to changes in the occupancy of adsorption sites at the gas interface, resulting in a modification of the charge distribution.

The density and energy level distribution of potential adsorption sites in a phthalocyanine film depends not only on the level of impurities included in the material, but also on the film morphology. For example, the surface of a phthalocyanine film will present a range of different adsorption sites corresponding to different crystal faces, edges and corners as well as a range of structural defect sites. Since at each of these sites the molecules at the surface will have different local environments, the adsorption interactions will also vary in strength. Such heterogeneity is expected to be greater in thick films produced by screen printing techniques due to their granular, more crystalline structure.

Regardless of the deposition technique used, when phthalocyanine films are first exposed to the air adsorption of atmospheric gases occurs starting at the most reactive sites. The potential of the material to detect a particular gas species, therefore, depends on the ability of that gas to compete in adsorption with the previously adsorbed species from the atmosphere. In the case of phthalocyanines, these background species consist primarily of oxygen and water molecules. Thus, differences between the electron affinity levels of the adsorbed gas and the background species that it displaces result in charge-transfer interactions that are physically represented by measurable changes in conductance.

3.3.1 Response to Nitrogen Dioxide

The transient response characteristics to a 100 Ppb pulse of NO_2 in a background flow of clean air (2 litres/minute) are examined for thin and thick films of PbPc. Results are presented in the form of percentage conductance change information at sensor operating techniques of 25°C and 150°C in both cases. The sample gas was introduced to the sensors at the 5 minute point and removed 5 minutes later.

Figure 5(left) shows that large changes in conductance are observed in thin films of PbPc on exposure to nitrogen dioxide. In contrast, figure 5(right) shows that this effect is not as pronounced when exposure measurements are performed on thick films of the same material. However, in both cases the response and recovery characteristics show a dependence on the operating temperature to varying degrees.

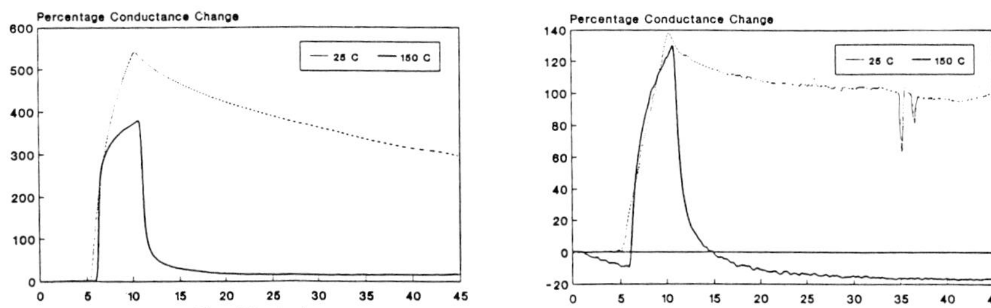

Figure 5 Transient response of (left) thin film and (right) thick film lead phthalocyanine to 100 Ppb NO_2 pulse in clean air.

In the case of thermally evaporated thin films, increasing the operating temperature has the effect of reducing the magnitude of the observed response whilst greatly improving the recovery time from exposure. A trade-off therefore exists when choosing a suitable operating temperature for a viable gas detection device based on thin films of this material. The thick film sensor also exhibits improved recovery times at higher operating temperatures, but unlike its thin film counterpart, does not appear to show any significant reduction in its magnitude of response.

3.3.2 Gas Response as a Function of Temperature

Figure 6 shows the effect of continuous exposure to 100 Ppb NO_2 in dry air as a function of sensor operating temperature for thin film and thick film lead phthalocyanine. Both types of film exhibit faster response times (arbitrarily defined as the time required to reach 90% of the equilibrium resistance value) with increasing temperature. This is to be expected since the rate of diffusion of the reacting gas into the adsorption interface is dependant on temperature. Furthermore, at higher temperatures water molecules are preferentially desorbed with respect to molecules of the reacting gas, effectively increasing the number of potential adsorption sites.

Figure 6(left) also shows that increasing the operating temperature of the thin film PbPc sensor has the effect of reducing the magnitude of the observed change in resistance. Conversely, figure 6(right) indicates that thick film PbPc sensors show an increase in the size of their response with increasing temperature. This increased sensitivity is clearly a major advantage in gas sensing applications where high operating temperatures may be imposed on the sensor by the environment e.g. monitoring NO_x emission levels in car exhaust systems.

Figure 6 Effect of temperature on the response to 100 Ppb NO_2 exposure for (left) thin film and (right) thick film lead phthalocyanine.

3.3.3 Effect of Gas Concentration

Figure 7 shows the responses of thin and thick film PbPc to various concentrations of NO_2 against a background of dry air and at a sensor operating temperature of approximately 150C. Both films exhibit good sensitivity below Ppm levels with increases in response dynamics (magnitude of change and speed of response time) being observed with increasing gas concentration.

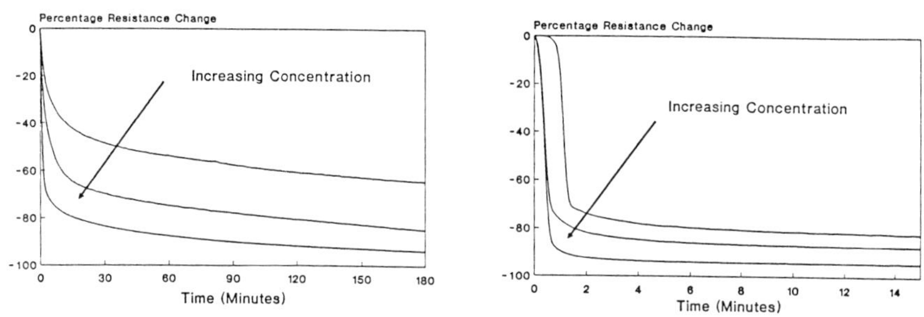

Figure 7 Effect of NO_2 concentration on the response of (left) thin film and (right) thick film lead phthalocyanine sensors operated at 150°C.

4. Summary

The initial results obtained from the characterisation of thick film phthalocyanine would appear to indicate that the material in this form is indeed viable for use in gas sensors. Certainly, the production of thick films of phthalocyanine by screen printing is quicker, easier and much cheaper than the thermal evaporation methods employed with thin films.

Comparison of both types of phthalocyanine film shows them to have similar temperature dependent responses. Both types exhibit an appreciable increase in the speed of response (and recovery) at elevated temperatures although thin films also show a reduction in their magnitude of response with increasing temperature.

Investigation of the temperature dependence of conductivity would appear to indicate that operation above approximately 400 degrees Kelvin would not only improve the performance of thick films but could also avoid problems of repeatability associated with impurity concentrations. Given the nature of the thick film fabrication process it would be difficult to control the levels of impurities and hence operation in the intrinsic region of the conductivity-temperature characterisitc would appear prudent.

Examination of phthalocyanine responses at ambient temperature do however, suggest that the films may have potential in the development of time weighted average sensors. Longer term exposures (several hours) could be inferred from the conductance of a phthalocyanine film contained within a personal monitoring device worn as a lapel badge dosimeter.

At sub ppm levels both film types exhibit increases in conductance in response to increasing concentrations of NO_2. There is, however, a difference in the dependence of the response with respect to concentration. This could be attributed to the difference in porosity of the two films with the thick film being more porous. It has been suggested (Gardner, 1989) that the conduction mechanisms have a strong dependence on the porosity of the material and that the process is diffusion limited. The results obtained so far would appear to support this view although further investigation is currently being undertaken.

References

Atkinson, J.K., Cranny, A.W.J., Jeffery, P.D. and Burr, P.M., "An evaluation of the use of thick-film organic semiconductor sensor arrays in the detection of gases", Proc. Int. Conf. Sensors and Systems, 3, (1989).

Gardner, J.W., "A non-linear diffusion-reaction model of electrical conduction in semiconductor gas sensors", Sensors and Actuators, B1, pp. 166–170, (1989).

Holmes, P.J. and Loasby, R.G., "Handbook of thick film technology", Electrochemical Publications Ltd, (1976).

Jeffery, P.D., Burr, P.M., Cranny, A.W.J. and Atkinson, J.K., "Thick film chemical sensor array allows flexibility in specificity", Proc. Int. Conf. Measurement Exhibition Test and Transducer, 1, pp. 37–57, (1987).

Jeffery, P.D. and Burr, P.M., "Gas sensing properties of polymeric silicon and germanium phthalocyanine films", Sensors and Actuators, 17, pp. 475–480, (1989).

White, N.M. and Cranny, A.W.J., "Design and fabrication of thick film sensors", Hybrid Circuits, 12, pp. 32–35, (1987).

Wilson, A. and Collins, R.A., "Electrical characteristics of planar phthalocyanine thin film gas sensors", Sensors and Actuators, 12, pp. 389–403, (1987).

Wright, J.D., "Gas adsorption on phthalocyanines and its effects on electrical properties", Progress in Surface Science, 31, pp. 1–60, (1989).

Inst. Phys. Conf. Ser. No 111
Paper presented at Int. Conf. on New Materials and their Applications, Univ. of Warwick, 1990

Use of ceramic sensors in the aluminium production process

E Fazio, E Proverbio, R Cigna

Department ICMMPM, University of Rome "La Sapienza", via Eudossiana 18, 00184 Roma - Italy

ABSTRACT: The knowledge in real time of the current efficiency, linked to the CO/CO_2 ratio in the off-take gases, and the alumina concentration in the bath are indeed of paramount importance in order to save energy and to lower production costs in aluminium industry. Starting from the technical literature an yttria-stabilized zirconia probe was selected for the CO/CO_2 ratio determination in the anodic gases. Data obtained from experimental tests in industrial cells are reported. As far as the aluminium concentration in the electrolyte bath is concerned, the use of many metal-metal, metal-ceramic couples is discussed, and results of tests for a W-Cr_2O_3 and a Fe-Cr_2O_3 couple are reported.

1. INTRODUCTION

Optimization of every production process, such as aluminium production, needs computers to control and to analyse process parameters.

Aluminium today is electrolitically obtained from alumina by the Hall-Heroult process at about 1000 °C. The receptacle which contains the produced metallic aluminium also acts as the cathode, while large coke and coal tar pitch blocks, which are consumed during electrolysis, are the anodes.
The main parameters actually under direct computer control are: temperature, cell voltage (V), electrolysis current (I), cell electric resistance (V/I) and anode immersion depth. Major parameters such as alumina concentration in the bath and anode gas composition are monitored only with analytical techniques which though more and more sophisticated, are always time-consuming and are not necessarily in real time.
A great advantage for automation of the electrolytic process would be achieved if these data could be obtained directly as an electric signal.
In considering the electrical and ionic conduction properties of some ceramic, Fray (1987) suggested the development of different sensors to monitor process variables. In fact it is possible due to the ability to

obtain ionic conductors for a wide range of elements, to control oxygen, hydrogen and sodium activities in liquid metal in order to obtain a better product, and to control sodium and lithium activities in the molten salts. Development of such probes is strongly linked to chemical resistance of ceramic materials to attack from molten cryolite, metallic aluminium (tab. 1) or fluorine salts (tab. 2) due to the following reaction:

$$MeO_{n/2} + n/3*Al \longleftrightarrow n/6*Al_2O_3 + Me$$

$$MeO_{n/2} + (n-\delta)/3*AlF_3 \longleftrightarrow MeF_{n-\delta} + (n-\delta)/6*Al_2O_3 + \delta/4*O_2$$

where the value of δ is non-zero when the metal oxide changes valence state as it goes into solution. In such cases the thermodynamics of reaction are influenced by the P_{O2}. When $\delta<0$, the reaction is forced to the right at higher values of the P_{O2}.

Table 1 (from McLeod 1987)
Free energy of reaction for metal-oxide exchange at 1023 °C

Oxide	ΔG(kJ)	Oxide	ΔG(kJ)
PbO_2	-829	SiO_2	-150
WO_3	-733	B_2O_3	-145
SnO_2	-528	TiO_2	-130
WO_2	-482	Na_2O	-124
CuO	-385	Ti_2O_3	-48
Fe_2O_3	-384	ZrO_2	19
NiO	-301	MgO	33
CdO	-295	CaO	72
CoO	-284	La_2O_3	76
Cr_2O_3	-231	Ce_2O_3	110
Cu_2O	-184	Y_2O_3	132

Table 2 (from McLeod 1987)
Free energy of reaction for metal oxide - metal fluoride exchange at 1023 °C

Oxide	ΔG(kJ)	Oxide	ΔG(kJ)
SrO	-190	Bi_2O_3	33
PbO_2**	-156	Fe_2O_3*	59
Y_2O_3	-119	Nb_2O_5	112
SeO_3	-50	V_2O_3	136
PbO	-47	Mn_2O_3	144
Ti_2O_3	-6	PbO_2	148
CoO	-1	VO_2*	174
MnO	24	VO_2	223
Mn_2O_3*	28	WO_3	255

* $\delta=1$
** $\delta=2$

2. CONTROL OF CURRENT EFFICIENCY

For a long time (Pearson 1947, Kestjukev 1963, Saakjan 1963) reaserchers have tried to control the current efficiency (percentage of aluminium produced in relation to the theoretically obtainable amount) by anodic gas analysis using updated versions of Pearson and Waddington's equation (1947)

$$CE = 100 - (\%CO)/2 = 50 + (\%CO_2)/2$$

Gas analysis may be carried out in several ways (Alcorn 1988) employing typical laboratory instruments: orsat gas analyser, infrared, gas chromatography, mass spectrometer. However problems arising from these techniques are not to be neglected.

In this field the use of a ceramic zirconia - based - probe seems to be revolutionary. These probes are today being used in the steel and copper industries, to monitor furnace and IC engines. In these cases probe voltage output is proportional to oxygen partial pressure in the analysed gas (p')

$$V = RT/4F \ln(p''/p')$$

where p" is the oxygen partial pressure in a reference gas (air).
The possibility of using an yttria stabilized zirconia probe to control the CO/CO_2 in off-take gas from the Hall Heroult cells has been widely discussed in a previous paper (Proverbio 1989a).
Carbon monoxide influences probe voltage taking part to the electrochemical reaction:

$$CO' + O^{=} \longleftrightarrow CO_2' + 2e$$

where $O^{=}$ is zirconia reticular oxygen and CO' and CO_2' are the gases near the ceramic surface.
When CO, CO_2 and O_2 are simultaneously present near the ceramic surface (w) the emf obtained can be explained by the following equation:

$$V = f_{O2} \{V^\circ_{O2} + (RT/4F)\ln[P_{O2(air)}/P_{O2(w)}]\} + f_{CO} \{V^\circ_{CO} + (RT/4F)\ln[P^{\frac{1}{2}}_{O2(air)} * P_{CO(w)}/P_{CO2(w)}]\}$$

where f_{O2} and f_{CO} are the fraction of active sites occupied by O_2 and CO respectively, which may be calculated by means of the Langmuir absorption isotherm. Experimental tests carried out with a laboratory equipment using a commercial yttria stabilized zirconia probe (FER) kept in CO/CO_2 stream at variable composition are in good agreement with theory.

More tests have been carried out with the same probe in an industrial aluminium electrolytic cell. The probe was placed into a steel bell above an off-take gas leakage point on the cryolite crust, near almost life-ended anode (fig. 1) Reference air was supplied by a cylinder located at the edge of the cell. Probe temperature and probe output were recorded by a computer in the control room.

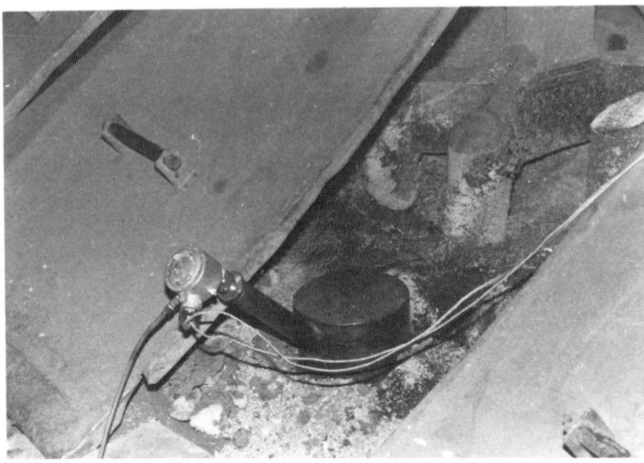

Fig. 1 The zirconia probe during tests

A typical plot of emf vs time is shown in figure 2; in the same picture is also plotted the relative resistance of the cell.

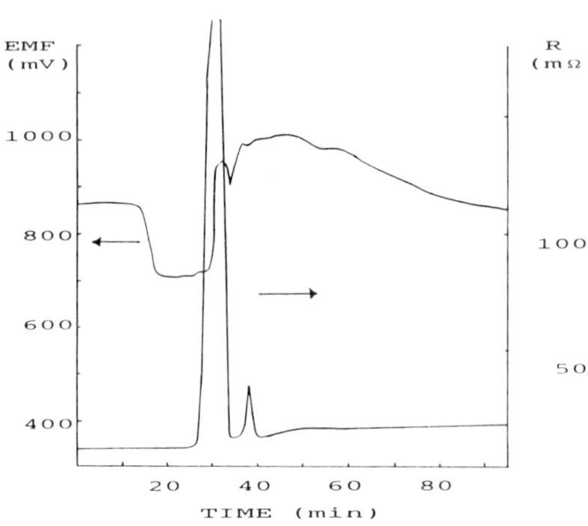

Fig. 2 Results of plant test

In this case, the sudden change of the resistivity is due to the anode effect rising up due to the alumina depletion, which leads to a decrease in cell efficiency. As is shown in the figure 2, robe output voltage change abruptly about 20 minutes before resistivity change and its low value is due to the reaching of the optimum values of the working parameters (CO/CO_2 ratio low, high efficiency, low alumina concentration).

3. CONTROL OF ALUMINA CONCENTRATION

Alumina concentration control plays a major role in correct operation and in alumina fedeeing check. When it becomes approximately as low as 1.5 % by weight,the anode effect take places, carbon anode wettability decreases and, as a consequence, large gas bubbles form the electrode surface causing sudden rising of the cell voltage (up to eight times normal value). Nevertheless major efficiency is reached at low alumina concentration (see previous paragraph).

By disregarding discontinued and time-consuming analytical control techniques, measurement of the activity of aluminium oxide (between activity and concentration there is a narrow correlation) in a fused salt solution can be carried out measuring the emf of the cell

$$Al / Al_2O_3 + \text{fused salts} / \text{Oxygen electrode}$$

using Nernst's equation

$$E = E^\circ - (RT/nF)\ln(a_{Al2O3})$$

A great many reference electrodes have been studied by Grjotheim (1982): gas electrode, oxide electrode and metal electrode. Oxide electrodes have been employed and studied in depth by Rolin (1960, 1962), who detected the possibility of using also galvanic cells whose metal electrode was not aluminium, (aluminium is liquid at the electrolysis temperature and so needs a ceramic container). Also, in a recent patent by Sumitomo Aluminium Company (1981) the use of such kind of galvanic cell i.e. metal-metal couple like TaC-Ta/bath/Al or a metal oxide, ZrC-Zr/bath/Al or a metal oxide has been reported.
The use of a metal-ceramic couple has also been reported by the Authors in a previous paper (Proverbio 1989b). In this case as reference electrode sintered Cr_2O_3, sintered SnO_2

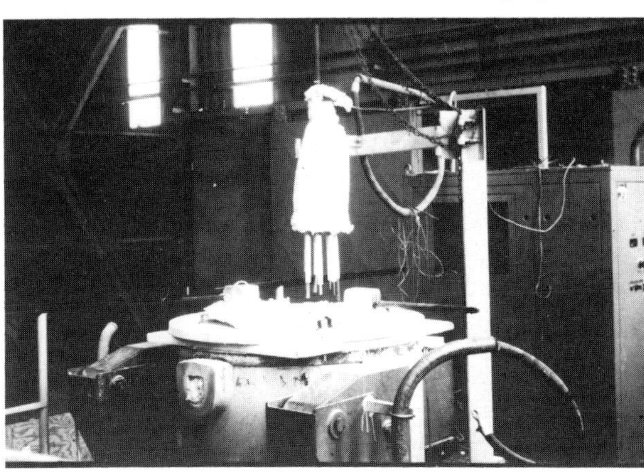

Fig. 3 Test equipment

(with 2% ZnO) and plasma spray Cr_2O_3 coated steel electrode were utilised. Working electrodes were made with Armco iron, tungsten and niobium. Several types of metal-ceramic couples were tested in a 1kA bench scale gas heated electrolytic cell (fig. 3). The bath temperature was kept between 960 and 980 °C. Tests were carried out with and without current flow.

Figure 4 shows emf values obtained with W-Cr_2O_3 and Fe-Cr_2O_3 couples changing alumina concentration without current flow. When putting into the graphite pot a cilindrical carbon anode, and so allowing a current density of about 1 A/cm^2 to flow, couple emf was higly disturbed due to cell geometry which permits a lateral current to flow (metal and ceramic electrodes were put, for geometrical restrictions, in the anular space between carbon electrode and pot wall). In industrial cells there is a minor lateral current flow (current flowing quite exclusively from lower anode surface and liquid aluminium on the bottom of the pot) but tests are still to be carried out.

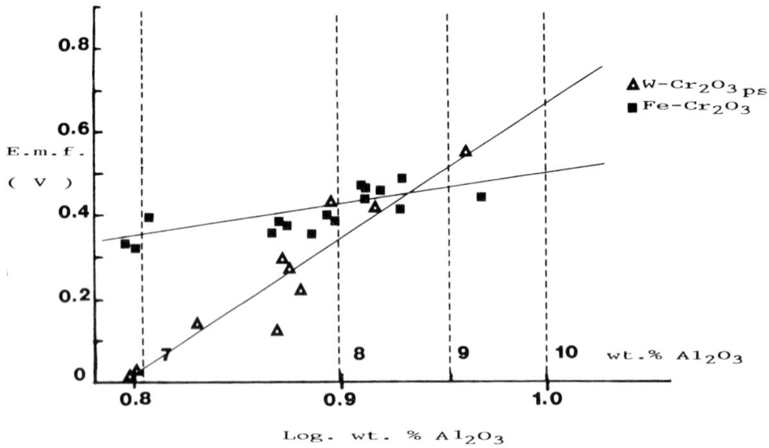

Fig. 4 value of emf for W-Cr_2O_3 and Fe-Cr_2O_3 couples vs alumina concentration

3. CONCLUSIONS

The possible use of ceramic probes developed in laboratory and those probes currently used in different industrial sectors, has been detected in aluminium industry.

Though there are numerous demands to data they are not used so extensively as in the steel industry (Goto 1983).

The employment of a stabilized zirconia probe to control current efficiency seems to be very promising due to facilicies of use and sensibility to anode gas composition changes. However, it should be remembered that other than CO and CO_2 evolving gases, such as CF_4, C_2F_6, SiF_4 and HF, can interfere with adsorbing phenomena reducing free sites fraction on ceramic surface and competing with carbon monoxide in reacting with reticular oxygen.

A solution to alumina concentration monitoring has been detected for a long time but material resistance has always restrained industrial application. A metal/bath/ceramic cell (e.g. W/bath/Cr_2O_3 or Fe/bath/Cr_2O_3) has given good results when tested in a bench scale electrolytic cell but potential instability due to current flow is yet to be resolved.

4. REFERENCES

Alcorn T R, McMinn C J, Tabereaux A T 1988 Light Metals pp 683-695

Fray D J 1987 Mat. Sci. Tec. **3** pp 61-65

Goto K S 1983 Proc. Int. Conf. on Chemical Sensors ed T Seiyama (Amsterdam: Elsevier) pp 333-334

Grjotheim K, Krohn C, Malinovsky M, Matiasovsky K, Thonstad J 1982 Aluminium Electrolysis (Dusseldorf: Aluminium Verlag) pp 194-215

Kestjukev A A 1963 Zvet. Met. **3** p 37

McLeod A D, Lihrmann J M, Haggerty J S, Sadoway D R 1987 Light Metals pp 357-365

Pearson T G, Waddington J I 1947 Faraday Discuss. Soc. **1** p 307

Proverbio E, Cigna R 1989a Euro-Ceramics **3** ed G With, R A Terpstra, R Metsellar (London and New York: Elsevier) pp 636-642

Proverbio E, Carassiti F, Cigna R, Fazio E 1989b Proc. Eur. Conf. on Advanced Materials and Processes, Aachen November 22-24 1989 (unpublished)

Rolin M, Gallay J J 1960 Bull. Soc. Chim. pp 2093-2104

Rolin M, Gallay J J 1962 Elact. Acta **7** pp 153-178

Saakjan P S 1963 Zvet. Met. **12** p 38

Sumitomo Aluminium Co. Ltd Jpn Kokai Tokkyo Koho 1981 Jp patent 81, 158, 889 7 Dec. 1981

Inst. Phys. Conf. Ser. No 111
Paper presented at Int. Conf. on New Materials and their Applications, Univ. of Warwick, 1990

Molecularly-doped polymers for use in electronic imaging

J. Mort
Webster Research Center, Xerox Corporation, Webster, N Y 14580, U. S. A.

ABSTRACT: Digital xerographic imaging plays an increasingly important role in the processing of documents. Polymeric-based photoreceptors are critical electronic components in this imaging process and an important example of materials engineering based on fundamental understanding. These insights have led to the development of molecularly-doped polymers which represent a multi-billion dollar exploitation of polymers as active electronic materials.

1. INTRODUCTION

The original concept of reproducing images by the xerographic process was, and continues to be, a commercial success of historic dimensions with more than one trillion copies being made annually. The original process was based on the analog imaging in which the optical image of the document to be copied was projected on to a page-size, electrostatically charged photoconductive insulator, known as the photoreceptor. In what is essentially electrostatic photography, selective photodischarge in proportion to the light content of the image creates a latent electrostatic image on the photoreceptor surface which replicates the information content of the document. This is then developed by a triboelectrically-charged carbon black or colored pigment-polymer composite, known as toner, and this in turn is fixed or fused to the paper producing a permanent copy (Mort 1989).

2. BASIC OPERATION OF PHOTORECEPTOR

Xerographic photoreceptors combine high resistivity to retain the surface corona charge before light exposure, efficiencies for charge photogeneration approaching unity and adequate transport properties of electronic charge carriers. Speed of carrier motion, determined by the carrier mobility, need only be sufficient that carriers cross the 20-50 micron thick photoreceptor in a time short compared with the progression of the photoreceptor from the light exposure to the development zone. Since this time is of the order of milliseconds, for typical fields used in the xerographic process it is found that mobilities of 10^{-6}cm^2/Vs or higher are sufficient. More important is whether the carriers can move through the photoreceptor thickness without deep traping; otherwise incomplete discharge and reduced development fields occur and the product of the mobility and deep trapping lifetime must be 10^{-6} cm^2/V or larger (Mort 1989). Amorphous materials can meet these requirements together with the need for fabrication in large areas at affordable cost. Although the initial exploitation of xerography depended on the use of amorphous selenium, this material had shortcomings. These included a tendency to crystallize, although alloying with cross-linking elements were found to inhibit this, a lack of sensitivity to red light and cost.

3. LASER PRINTERS

It had been clear from the earliest days of xerography that one could print line by line by writing with light directly on the photoreceptor. What was lacking were the technical means and motivation to do so. The situation began to change with the emergence of affordable helium-neon and solid-state lasers providing the needed sharply focussed points of light, and the gradual diffusion of electronic digital processing into the office which depended on affordable desktop computers (Starkweather 1980). In this context the shortcomings of amorphous selenium were of particular concern because red and infrared photosensitivity were needed to capitalize on these parallel developments.

4. MOLECULARLY-DOPED POLYMERS

Cheaper analogs of inorganic amorphous materials are organic polymers which are also compatible with cost-effective web coating technologies using solutions. Unfortunately polymers are typically quintessential insulators which exhibit little absorption of light except in the ultraviolet and in any case typically do not transport free charge carriers. The issue of charge transport is solved by doping otherwise electrically inactive polymer matrices such as polycarbonates with monomer molecules (Mort 1989). These molecules act as localized sites for free carriers to hop between, thereby providing a sequence of "stepping stones" by which the charge can traverse the polymer film. These dopant molecules are incorporated in concentrations of 10^{20} molecules per cubic centimeter and, forming solid solutions, on the average are separated by only about 10 angstroms. This gives sufficient molecular wave function overlap that the hopping probability between adjacent sites results in carrier mobilities of 10^{-5} cm^2/Vs. If the dopant molecules are donor-like, with a propensity to accept electrons, then hole transport occurs, whereas if the dopant molecules are acceptor-like then electron transport is provided. It is also found that deep trapping lifetimes in these systems are at least minutes so that the carrier range can easily exceed photoreceptor thicknesses. The lack of visible or infrared absorption by the dopant molecules means that photogeneration must occur in a separate layer. These dual-layer photoreceptors, as they are called consist of a thin (about 1 micron) photogenerator layer overcoated for protection by a thicker (twenty to thirty microns) transport layer. A new phenomenon now comes into play, namely, the transfer of charge between two contiguous layers and the somewhat surprising fact is that such injection can occur with almost unit efficiency. Infrared sensitive polymeric-based photoreceptors utilize organic pigments as the source of photogenerated carriers. Many organic pigments manifest field-dependent photogeneration processes, reflecting competitive intra-molecular radiative and non-radiative relaxation processes to photogeneration. Such field dependence, often an issue in copying where the photodischarge can occur over many transits, is less of a problem in laser printing where pixel by pixel writing requires the total exposure to be delivered in fractions of a microsecond. In this case the photogeneration occurs in an essentially constant high electric field.

REFERENCES

Mort, J. 1989 *The Anatomy of Xerography*, and references therein, (Jefferson, NC: McFarland & Company Inc., Publishers)

Starkweather, G. K., 1980 "*High-Speed Laser Printing Systems*", Laser Applications, vol. 4, 125

Inst. Phys. Conf. Ser. No 111
Paper presented at Int. Conf. on New Materials and their Applications, Univ. of Warwick, 1990

Development and characterization of the highly magnetostrictive alloy $Tb_{0.3}Dy_{0.7}Fe_{1.9}$ for use in sensors and actuators

D. C. Jiles

U S Department of Energy
Ames Laboratory
Iowa State University
Ames, Iowa 50011
USA

ABSTRACT: The highly magnetostrictive alloy known as terfenol-D has been developed as a joint project between Ames Laboratory and the Naval Surface Weapons Center. This alloy has a room temperature magnetostriction which can be in excess of 2000 ppm, under a compressive load of 14 MPa, compared with the magnetostrictions in pure metals such as nickel of -35ppm . However, present observations on over one hundred specimens of this material have revealed rather large variations in performance during laboratory tests.

1. INTRODUCTION

It has been known since the 1960's that the heavy rare earth elements such as dysprosium and terbium have very high magnetostriction at low temperatures [1]. Several factors have prevented these materials properties from being utilized in magnetostrictive devices. These factors are principally that the rare earth elements, with the exception of gadolinium, lose their magnetic order at temperatures well below ambient and that the materials have high anisotropy which means that very high fields are needed to change the direction of the moments within a domain. Consequently, the materials have high coercivity.

In the 1970's attention was given to a number of rare earth iron alloys in order to raise the ordering temperature. In iron the Curie temperature is 770°C and so alloys of iron with the rare earth elements in the RE-Fe_2 series of components such as $TbFe_2$, $HoFe_2$ and $DyFe_2$ exhibit magnetic order at ambient temperatures.

Among the group of rare earth-iron alloys, $TbFe_2$ had the highest magnetostriction. Values of $\lambda_{//} - \lambda_{\perp}$ of 2500 x 10^{-6} were measured by Clark in this material [2]. This is shown in Fig. 1. It is seen that although the magnetostriction is very high it also requires a high energy, in the form of a magnetic field strength in excess of 20 kOe to achieve this.

The reason for this high field strength is the high anisotropy of $TbFe_2$. Using the one constant model this has a value of K = -6.3 x 10^6 J/m^3 at ambient temperature, 27°C. A study of the anisotropy constants of the rare earth metals revealed that terbium and dysprosium had anisotropies of different sign. An alloy of these two elements with iron of the form

$$Tb_{1-x}Dy_xFe_2$$

should therefore combine high magnetostriction with low anisotropy. Work by Clark et al. [3] at the Naval Ordinance Laboratory (now the Naval Surface Weapons Center) showed that this was correct and that consequently a highly magnetostrictive alloy was produced which had both the desirable properties of high λ and low anisotropy (Fig. 2). This material which had the composition $Tb_{0.3}Dy_{0.7}Fe_2$ was called Terfenol-D.

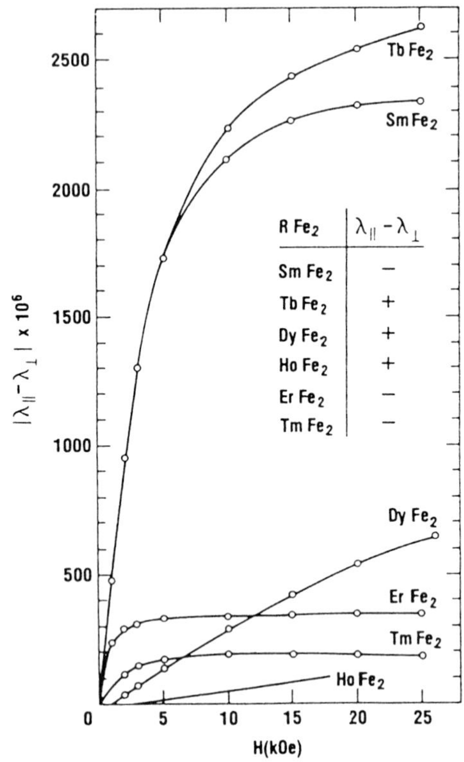

Fig. 1 Magnetostriction of $TbFe_2$ and several other rare earth iron compounds after Clark [2].

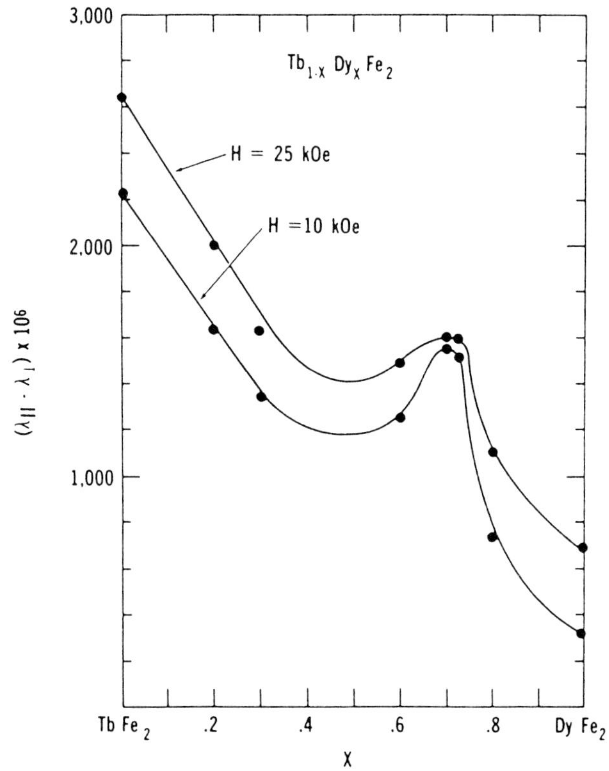

Fig. 2 Magnetostriction of the pseudobinary rare earth alloy $Tb_{0.3}Dy_{0.7}Fe_2$, after Clark [2].

2. DEVELOPMENT OF TERFENOL-D

Further development of the alloy was undertaken at Ames Laboratory, through the Metallurgy and Ceramics division under the direction of Verhoeven and McMasters. One of the disadvantages of this material is the difficulty in growing it along the <111> directions. This problem has not been overcome. At this time the best solution to the problem has been to grow the material along the <112> direction (Fig. 3) which is at an angle of 19.5° from the nearest <111> axis.

Two methods of crystal growth are preferred, the modified Bridgman (MB) method and the free standing zone melt (FSZM). Improvements in the magnetostriction were gradual until in 1987 specimens were produced which had been given a final post process anneal at 950°C with a magnetic field greater than 5kOe applied along the <111> direction perpendicular to the <112> axis of the rod shaped specimens. This lead to dramatic improvements [4] as shown in Fig. 4. This was due to an increased occupancy of domain magnetizations in these directions.

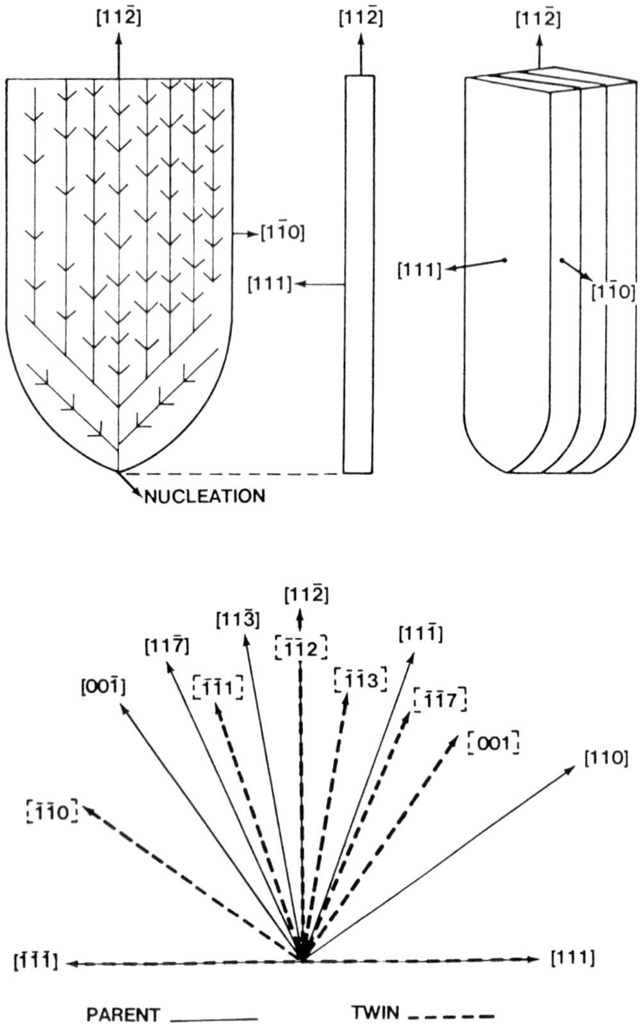

Fig. 3 Growth of twinned crystals of $Tb_{0.27}Dy_{0.73}Fe_2$ along the [11$\bar{2}$] axis, with the [111] axis at right angles to the rod axis, (after Clark, Verhoeven, McMasters and Gibson IEEE Trans. Mag. 22, 973, 1986).

In order to maximize the magnetostrictive response of these materials the cylindrical specimens are subjected to a compressive stress of between 7 MPa and 14 MPa. This causes a further preferential alignment of domain magnetizations normal to the unique axis of the specimen [5]. Since magnetostrictive changes arise from non-180° domain processes, this leads to enhanced magnetostriction.

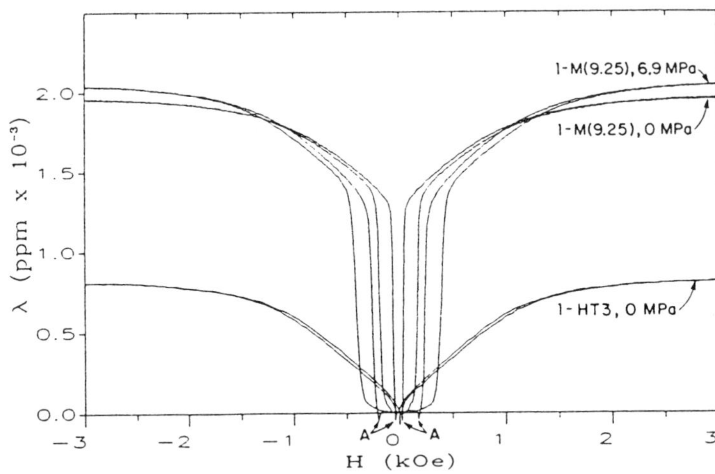

Fig. 4 Improvements in magnetostriction of a specimen of $Tb_{0.3}Dy_{0.7}Fe_{1.96}$ after the magnetothermal post processing anneal (after Verhoeven et al. [4]).

3. CHARACTERIZATION OF MATERIAL

Characterization of the Terfenol produced at Ames Laboratory has been made using a multi parameter magnetic inspection system [6] with an electromagnet containing a gas pressure cell [7] for applying compressive load to the specimens. Results on some of the most interesting specimens, with bulk magnetostrictions in the range 2200×10^{-6} have been reported [8, 9] as shown in Fig. 5. Recently details of the magnetization mechanism have been deduced from the use of Barkhausen and magnetoacoustic emission measurements [10]. Anisotropy calculations have been used together with magnetization and magnetostriction data to determine the distribution of domain magnetizations [11].

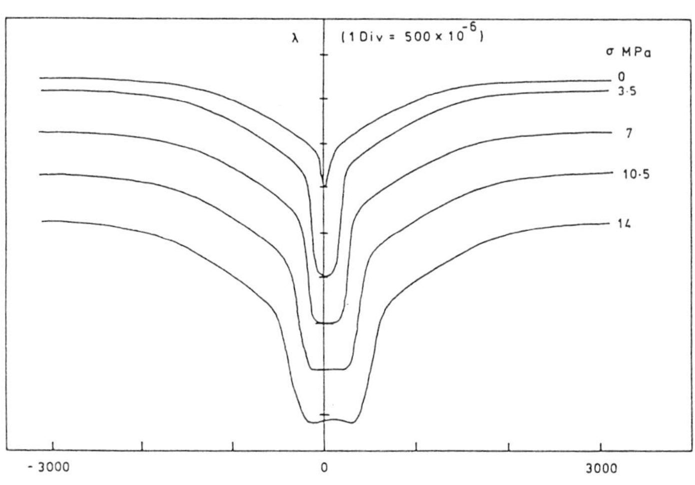

Fig. 5 Magnetostriction of a specimen of $Tb_{0.3}Dy_{0.7}Fe_{1.9}$ as a function of magnetic field under several different compressive stresses (after Jiles et al. [9]).

4. APPLICATIONS

The main applications of Terfenol-D are in low frequency sonar with frequencies typically 1 kHz, positioning devices, valve actuators, antivibration mountings, dynamic load detection and very recently magnetometry. These applications have been discussed by Butler [12, 13]. The most important use is in sonar [14] and several designs of magnetostrictive Terfenol transducers are given in Figs. 6, 7 and 8.

The design and construction of Terfenol based magnetostrictive antivibration mountings has also been successfully carried out by Greenough at the University of Hull.

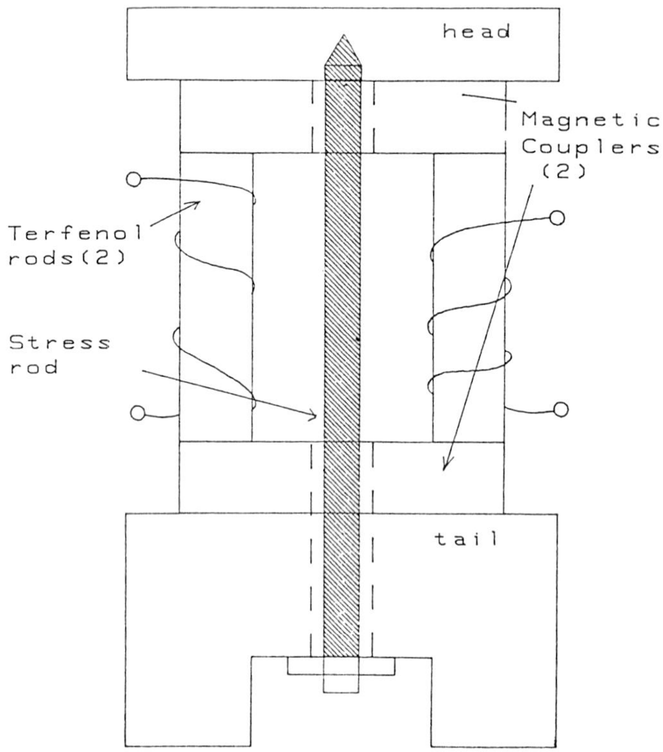

Fig. 6 Design for a "tonpilz" low frequency transducer using Terfenol (after Butler [13]).

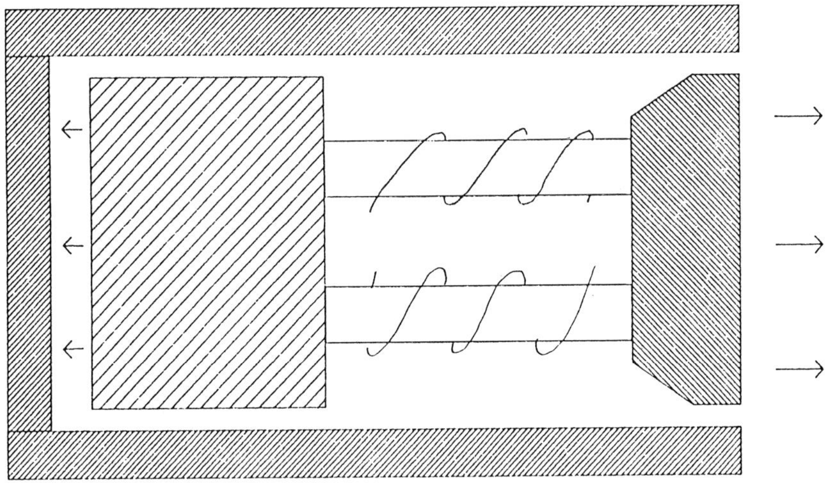

Fig. 7 Design for a piston low frequency sonar transducer using Terfenol (after Butler [13]).

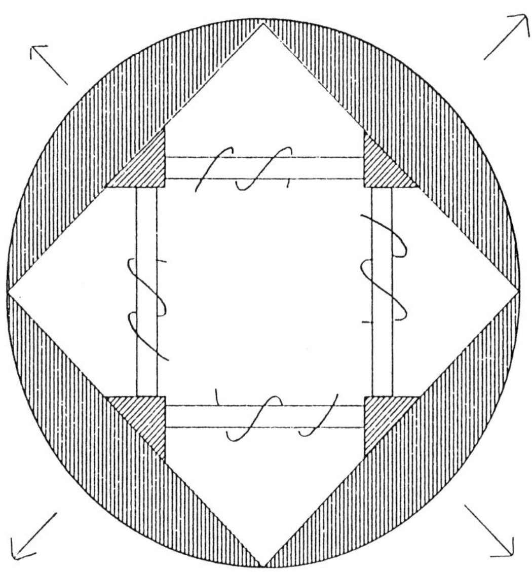

Fig. 8 Design for a "square ring" sonar transducer using Terfenol (after Butler [13]).

5. REFERENCES

1. J. J. Rhyne and S. Legvold, Phys. Rev. 138, A507, 1965.

2. A. E. Clark, in "Ferromagnetic materials" Vol. 1, by E. P. Wohlfarth, North Holland Publishing, 1980.

3. A. E. Clark, R. Abbundi and W. G. Gilmor, IEEE Trans. Mag. 14, 542, 1978.

4. J. D. Verhoeven, J. E. Ostenson, E. D. Gibson and O. D. McMasters, J. Appl. Phys. 66, 772, 1989.

5. A. E. Clark and H. T. Savage, J. Mag. Mag. Mater. 31-34, 849, 1983.

6. D. C. Jiles, J. Appl. Phys. 63, 3946, 1988.

7. J. E. Ostenson, D. C. Jiles and D. K. Finnemore, Rev. Sci. Inst. 60, 278, 1988.

8. D. C. Jiles, J. E. Ostenson and C. V. Owen, J. de Physique 49-C8, 1939, 1988.

9. D. C. Jiles, J. E. Ostenson and C. V. Owen, J. Appl. Phys. 64, 5417, 1988.

10. D. C. Jiles and S. Hariharan, Proceedings of the 34th Conference on Magnetism and Magnetic Materials, Boston, Mass., November 1989. To be published in J. Appl. Phys.

11. D. C. Jiles and J. B. Thoelke, 1990 Intermag Conference, Brighton, England, April 1990.

12. J. Butler, "Use of rare earth materials in low frequency transducers", Proceedings of the Terfenol Workshop, Ames Laboratory, September 19, 1985.

13. J. Butler, "Application manual for the design of Terfenol-D magnetostrictive transducers", Edge Technologies Inc., Technical Briefing, 1988.

14. J. R. Oswin, R. J. Edenborough and K. C. Pitman, "Rare earth magnetostriction and low frequency sonar transducers", Proceedings of the IOP/IEE/IEEE Seminar on Sensors and Amorphous Materials, April 1988.

Measurement of the temperature dependency of the stress-optic coefficient of photoelastic materials

M Ayub, R C Spooncer and B E Jones

Brunel Centre for Manufacturing Metrology
Brunel University, Uxbridge, Middlesex, UB8 3PH, England

ABSTRACT: The accuracy of photoelastic modulated optical fibre sensors depends on the stability of the appropriate materials, which may vary due to time, temperature and applied stress. This paper examines the suitability of three photoelastic materials for the determination of pressure in optical fibre sensors with respect to the temperature dependency of the stress–optic coefficient.

INTRODUCTION

Over the last ten years several optical fibre sensors (OFS) have been developed which use stress induced birefringence to modulate the light coupled between the input and output fibres. Both intrinsic monomode and extrinsic multimode sensors have been reported; the latter may use monochromatic or broadband light (Jones and Spooncer 1983).

Multimode OFS are versatile and can show high sensitivity but sensitivity is a function of the stress–optic coefficient which may vary by a factor of 10^3 between say polyurethane and fused quartz; accuracy is, however limited by the stability of the photoelastic material and suffers from creep, time edge effect, variation of Youngs modulus with stress and temperature dependency of the stress–optic coefficient. Ideally a combination of material properties is needed which will give high sensitivity, i.e. high stress–optic coefficient, coupled with good material stability but often the error–inducing factors seem to increase with the sensitivity particularly in polymeric materials. Unfortunately, data on these factors is sparse particularly on the newer materials such as polycarbonate or polyurethane making the selection of suitable material for a given application somewhat haphazard. The temperature stability of polyurethane rubber and polycarbonate resin under static condition has already been reported by the author (Ayub et. al. 1988) These measurements were carried out on a optical fibre pressure sensor operating over the range 0 – 10^2kPa.

The lower photoelastic sensitivity of fused quartz compared with the polymeric materials makes it unsuitable for use in the 0 – 10^2 kPa test sensor. A second pressure sensor was therefore built operating over the range 0 – 10^5 kPa.

This paper compares the stability of fused quartz with the data previously reported.

PHOTOELASTICITY

Normally isotropic materials may become anisotropic when subjected to mechanical stress and exhibit birefringence in which their refractive index varies with the plane of polarisation of the transmitted light. If the stress in uniaxial and the material is illuminated by plane polarised light, then the light is split into two components polarised at right angles which travel at different speeds; on emergence from the material the two beams differ in phase and can be made to interfere to give an intensity modulation system. The intensity of the transmitted light is given by:

$$I = I_o \, Sin^2 \, (\pi C\sigma t \; /\lambda) \ldots\ldots\ldots (1)$$

where I_o = intensity which would have been transmitted had the 2nd polariser been optically isotropic,
C = stress–optic coefficient of the sample (SOC),
σ = the applied stress,
t = thickness of the sample in the direction of the light beam and
λ = wavelength of light.

The transmitted intensity varies sinusoidally and if the stress increase ($\Delta\sigma$) between an adjacent minimum and maximum is measured, then the SOC is given by:

$$C = \lambda \, / \, 2 \, (\Delta\sigma)t$$

Alternatively this may be expressed in terms of the material fringe value, (f), more commonly used as an engineering unit.

$$f = \lambda \, / \, C = 2\Delta\sigma t \qquad \ldots\ldots\ldots (2)$$

MATERIALS ASSESSED

(1) **Polyurethane Rubber:** Polyurethane rubber is a transparent amber coloured materials with a high stress–optic coefficient, (typical properties are shown in table 1). It is better known under its trade name of Hysol 4485 or Photoflex and is normally purchased in sheet form. Main applications include dynamic or wave propagation studies.

(2) **Polycarbonate resin:** Also known as Makrolon or PSM–1, is transparent, tough, ductile material. It is manufactured by compression moulding and during this process it suffers some permanent initial birefringence, which cannot be removed entirely even by annealing. It is mainly used for two–dimensional model studies.

(3) **Fused Quartz:** Also known as vitreous silica, and is one of the purest materials available commercially. It is used for crucibles and furnaces for high temperature processing, and especially for optical components such as optical fibres, telescope mirrors, lenses and prisms. Some of the physical properties are shown in table 1.

DESIGN AND TESTING OF THE PRESSURE SENSOR

The 0 – 10^5 kPa photoelastic pressure sensor consisted of a cylindrical disc of fused quartz, sandwiched between two backing plates, sealed by o–rings which form symmetrical pressure chambers on either side of the disc. The disc is thus equally loaded on both sides so that a uniaxial compressive stress, free from bending stresses is guaranteed. The material used for the backing plates (high tensile steel) and clamping bolts (type 316 stainless steel) were chosen to give zero differential thermal expansion over the test temperature range of 0 – 50°C. Figure 1 shows a section through the components of the sensor and plate 1 displays the exploded view of the sensor. The optical path is diametric across the disc, which is counterbored to accept collimating GRIN ROD lenses and polarisers (plate 2). Finite element analysis was performed on the disc to determine whether local stress concentrations induced by clamping the disc between the backing plates interfere with the optical path.

RESULTS

The finite element method analysis predicted that, although stress was not uniform throughout the fused quartz disc, the variation along the optical path was less than +/-0.25%. Any inaccuracies induced by clamping loads were therefore ignored.

The sensor was loaded hydraulically and cycled over one complete fringe (0 – 1.4×10^5 kPa), in an environmental chamber at 10°C, 30°C, and 50°C. The material fringe value was calculated from equation (2) but it was found that a differential phase shift was induced in the polymeric polarisers as the temperature varied. At any given temperature this phase shift is constant irrespective of the sensor loading and may be eliminated by averaging the material fringe value measured at $\lambda/4$ and 3 $\lambda/4$ birefringence. Typical results are shown in table 2.

The stability of fused quartz was excellent and showed zero temperature dependency, hysteresis and creep and had negligible locked–up stress. Table 3, compares these data with the previous measured for polyurethane rubber and polycarbonate resin. Hysteresis was determined from loading /unloading cycles, creep was assessed by measuring the time required after the application of a constant load, for the transmitted intensity to fall within 1% of the final value. Locked–up stress was estimated by viewing a back illuminated sample through crossed polarisers.

CONCLUSION

The material stability of fused quartz as the sensing element in a custom–built optical fibre photoelastic pressure sensor has been examined. Temperature dependency, hysteresis and creep were not detectable, locked–up stress was negligible. Its behaviour is far superior to the previously measured for polycarbonate resin and polyurethane rubber but its photoelastic sensitivity is at least twenty times inferior. The superior material stability of fused quartz, therefore cannot easily be exploited in low stress applications.

REFERENCES

Ayub M, Spooncer R C and Jones B E, Proc. Int. Cong. on Optical Science and Engineering, Hamburg, Sept., 1988 (SPIE Vol. 1011, paper 21).

Jones B E and Spooncer R C, IEE Conf. on Optical Fibre Sensors, Conf. Pubn. No. 221, pp 173–177.

Sharples Stress Engineering Ltd., Preston, Lanc., England.

TSL Group PLC, Wallsend, Tyne & Wear, England.

Welwyn Strain Measurement, Basingstoke, Hants., England.

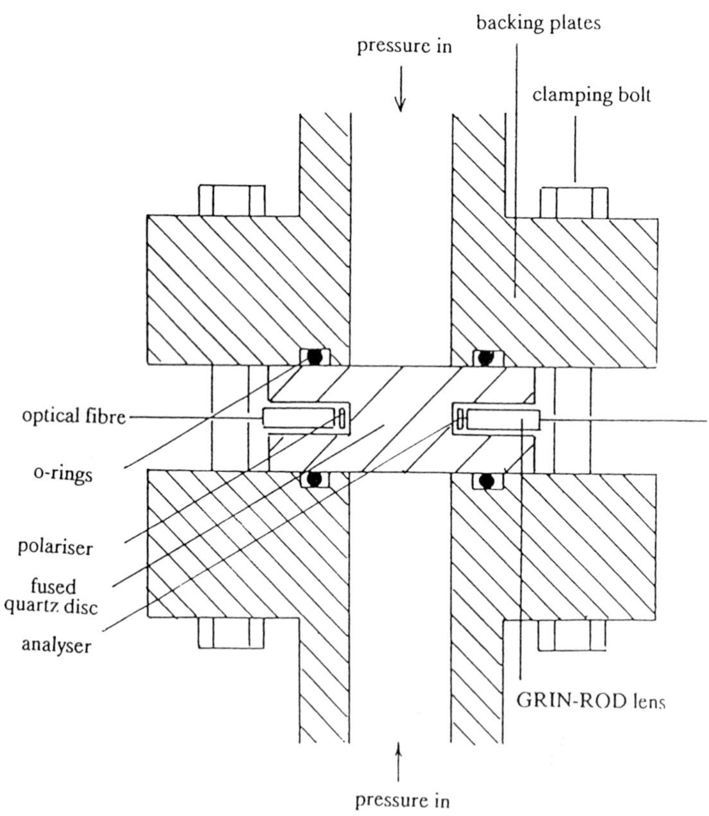

Figure 1. Section through the components of the sensor

MATERIAL	Young's Modulus E (GPa)	Poissons Ratio v -	Material Fringe Value f kPa/m-fr	Manufacturer
Polyurethane Rubber (photoflex)	$0.7\text{-}4.1 \times 10^{-3}$	0.48	0.14-0.17	Sharples Stress Engineering Ltd,
Polycarbonate Resin (PSM-1)	2.5	0.38	7.0	Welwyn Strain Measurement.
Fused Quartz (Vitreous 066)	72.9	0.17	-	TSL Group Plc.

Table 1. Typical properties of photoelastic materials.

MATERIAL	Material Fringe Value (kpa/m-fr) 10°C	30°C	50°C	Temp. Dependency %/°C
Polyurethane	0.20	0.22	0.23	+0.38
Polycarbonate	6.86	6.73	6.58	-0.11
Fused Quartz	180	180	180	none

Table 2. Variation of material fringe value with temperature.

MATERIAL	Max. Hysteresis (%) of span	*Creep (min)	Locked Up Stress
Polyurethane	6.7	16	negligible
Polycarbonate	2.5	0.08	considerable
Fused Quartz	none	none	negligible

* Time required for the value to be within 1% of the actual value.

Table 3. Observed values of hysteresis, creep and locked-up stress.

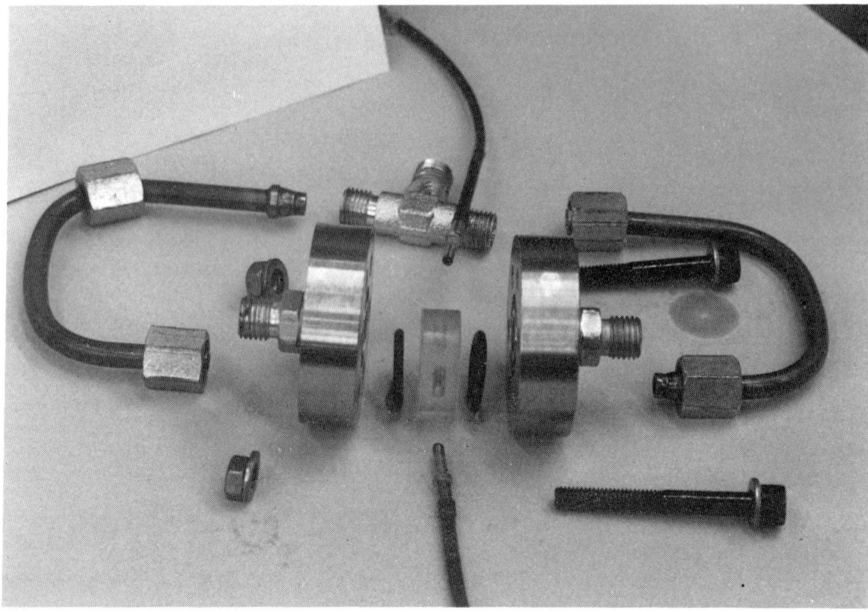

Plate 1. Exploded view of the sensor

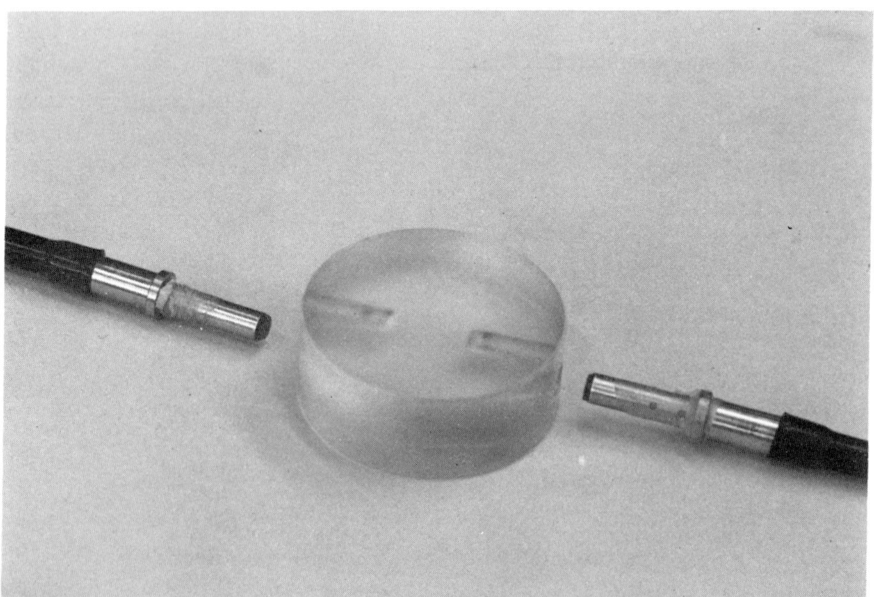

Plate 2. Optical components of the sensor

Inst. Phys. Conf. Ser. No 111
Paper presented at Int. Conf. on New Materials and their Applications, Univ. of Warwick, 1990

All fibre-optic mircromechanical resonator sensor system

Rebwar M A Fatah* and Nigel A D Stokes**

* British Telecom Research Laboratories, Martlesham Heath, Ipswich IP5 7RE, UK.
** Department of Electronic & Electrical Engineering, University College London, WC1E 7JE,UK.

Abstract: The pressure and the temperature sensitivity of micromechanical resonators have been measured; the microresonators are activated and monitored using an all-fibre optic system. Silicon and silicon dioxide devices have been studied and compared.

1. Introduction:

Micromechanical resonator beams can be used as sensor heads in sensor systems. They measure any parameter that interacts with their resonance such as chemical concentrations, viscosity, density, strain, force, mass flow, acceleration, pressure and temperature. The physical parameters that determine the resonant frequency can be found from equations of resonance. These parameters are Young's modulus, density, geometry and end-conditions.

All-optical fibre pressure sensor system based on microfabricated silicon beam structures have sensor heads compatible in size to the cross-sectional area of the fibre. This makes them particularly attractive for in-vivo applications where for instance blood pressure can be directly monitored. Further, the sensing can be achieved remotely since the sensor head is attached to the end of the fibre optic cable. In telecommunications, the microresonant sensor would allow for continuous pressure readings along the cable to be made and also other environmental parameters such as tension and temperature could be monitored if required.

Micromechanical resonators in the form of silicon and silicon dioxide beams are fabricated by anisotropic etching techniques, and coated with either evaporated aluminium or sputtered gold. These devices are driven into resonant vibration using an intensity-modulated laser source and a cw laser at a different wavelength is used to monitor the mechanical vibration. Both laser beams are guided through the same monomode optical-fibre and directed onto the resonator beam surface. The experimental set-up is illustrated in a previous publication (Fatah *et al* 1989).

The resonant frequency of the silicon beam shifts with both an applied differential pressure across the wafer and with a change of the ambient temperature. A comparison has been made between the different materials.

Pressure and temperature sensitivities of microresonators have been reported using bulk optic systems (Uttamchandani *et al* 1987, Thornton *et al* 1988). For devices with more complicated shapes fibre optic systems with two fibres have been used one for activation and the other for detection (Andres *et al* 1985). We report on sensitivities of these devices by guiding both laser beams, activation and detection, through the same optical fibre positioned on the resonator. Furthermore, we show that silicon dioxide resonators sense only temperature and not pressure. Thus for temperature measurements where other sensing methods are not reliable silicon dioxide beams are 'naturally' compensated for any pressure changes in the environment.

2. Consideration of applied pressure:

In the presence of a differential pressure, the tensile stress, σ, varies across a wafer: the stress near the edge and near the centre, for example, are of opposite polarity. If the beam is on a circular wafer area clamped at the edges, and this area is exposed to a differential pressure, ΔP, then two components of the stress can be considered, a radial one, σ_{rad}, and tangential one, σ_{tan}, and they can be expressed as follows (Gieles and Somers 1973):

$$\sigma_{rad} = \frac{3\,\Delta P\,R^2}{8d^2}\left[(3+v)\frac{r^2}{R^2} - (1+v)\right] \qquad 1(a)$$

$$\sigma_{tan} = \frac{3\,\Delta P\,R^2}{8d^2}\left[(1+3v)\frac{r^2}{R^2} - (1+v)\right] \qquad 1(b)$$

where R is the sample radius exposed to pressure, d is the wafer thickness, v is Posson's ratio and r is the distance from the centre of the wafer to the point where the stress is measured. Thus the orientation of the resonator beam on the wafer effects the magnitude of the applied stress.

Neglecting the effect of the etch pit under the beam, the stress at the centre, σ_c, of the sample is:

$$\sigma_c = \frac{3R^2}{8d^2}(1+v)\Delta P \qquad 1(c)$$

If the applied pressure at the back of the wafer is higher than atmospheric, i.e. positive ΔP applied, the front of the wafer assumes a convex shape, Fig 1(b), introducing a tensile stress where the beam is, at the centre of the outer surface. By contrast, if the pressure of back of the wafer is lower than that of atmospheric Fig 1(c), i.e. a negative applied pressure, the front of the wafer then assumes a concave shape introducing a compressive stress at the centre of the outer surface.

Fig 1: The silicon sample with clamped edges is shown under applications of no pressure (a), positive differential pressure (b) and negative differential pressure (c). The fibre optic system and the tube from which pressure applied are not shown. See Fig 2.

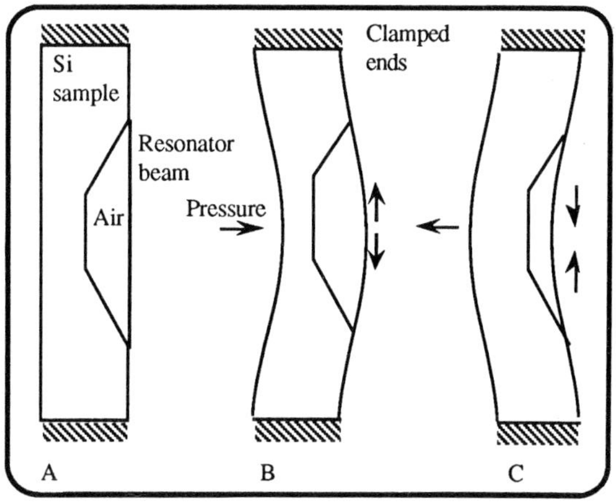

The shift in resonant frequency, f_F, for a beam of rectangular cross-section subjected to an axial force F, can be calculated from (Albert 1982):

$$\frac{f_F}{f_0} = \pm a \frac{L^2 F}{E w t^3} \qquad 2(a)$$

where f_0 is the resonance frequency of the relaxed beam; L, w and t are the length, the width and the thickness of the beam, respectively; E is the Young's Modulus, and a is a constant with value 0.147. The positive and the negative signs are, respectively, applied to the applications of tensile and compressive forces.

If the beam is in the centre of the wafer to which pressure is to be applied, then 1(c) may be substituted in 2(a). Then, putting $F = \sigma w t$ we have

$$\frac{f_F}{f_0} = \pm a \frac{\sigma L^2}{E t^2} \frac{3 R^2}{8 d^2} (1+\nu) \Delta P \qquad 2(b)$$

The deflection of the sample, w_c, is maximum in the centre and can be calculated (Oberg *et al* 1989):

$$w_o = \frac{0.221 \pi R^4}{E d^3} \Delta P \qquad 3$$

3. Pressure sensitivity of resonance frequency:

To show that the beam resonance is pressure sensitive a differential pressure must be applied across the wafer thickness. This can be achieved by application of an air pressure, above or below atmospheric, from the back of the wafer while the front is subjected to atmospheric pressure.

The sample was mounted on a metal ring with a circular cross-section using a commercial epoxy, the metal ring was then connected to a nitrogen cylinder via a tube. The set-up to test the pressure sensitivity of the beam resonance is shown in Fig 2. In preparing the samples for pressure application the resonator could not be positioned accurately at the centre of the wafer and this affected the pressure sensitivity.

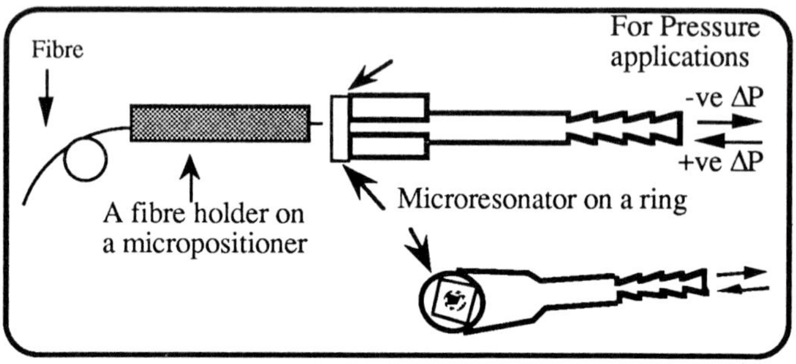

Fig 2: The experimental set-up for pressure sensitivity measurements. The fibre optic system is not shown.

The silicon device (sample A) shown in Fig 3(a) was examined using this set-up. The silicon thickness was 0.5µm and the device was sputter-coated with 65nm of gold. The measured resonance frequency is 225 KHz, the quality-factor is 40 and signal-to-noise ratio 20dB. This frequency shifted to 247KHz when the sample was glued to the pressure ring. The resonance frequency then shifted with application of 4 bar differential pressure across the full wafer thickness (380µm) and over a circular area of 2mm diameter from 247KHz to 250KHz . It was not possible to make an accurate measurement of the position and orientation of the beams on the wafer. The calculated shifted resonant frequency for such a beam positioned at the wafer centre, from 2, is approximately 2.4% per bar. The experimental value of the resonant shift is 0.3% per bar.

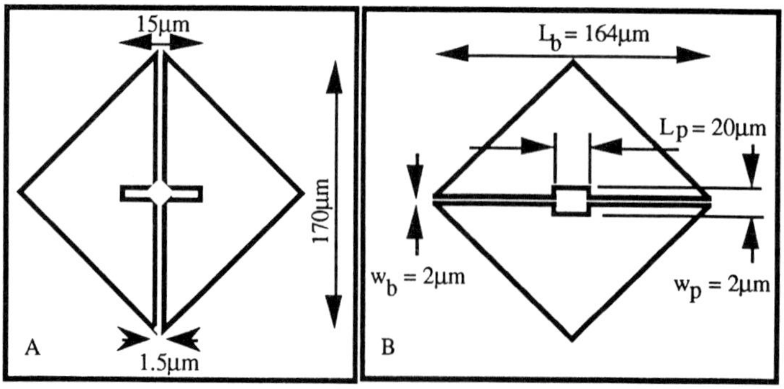

Fig 3: The plan views with dimensions of micro-resonators examined. (Samples A and B)

Considering 2 we note that the pressure sensitivity of a resonator increases with decreasing the wafer thickness. The back of a wafer with on etched beams (sample B), Fig 3(b), were chemically thinned from 380μm to 180μm and then mounted a metal ring with area of 1.8mm in diameter. In this case, the beam thickness is 0.5μm and coated with 200nm of aluminium. The first measured resonance frequency was 179 KHz, the quality-factor 33 and the signal-to-noise-ratio 35dB. The experimental results have shown that the sensitivity of the resonance frequency is 4.1% per bar. Fig 4 shows frequency plotted against applied pressure. The calculated shifted resonant frequency, from 2, is approximately 5.2% per bar.

Fig 4:
The pressure sensitivity of the first resonant frequency of the beam resonator shown in Fig 3(b).

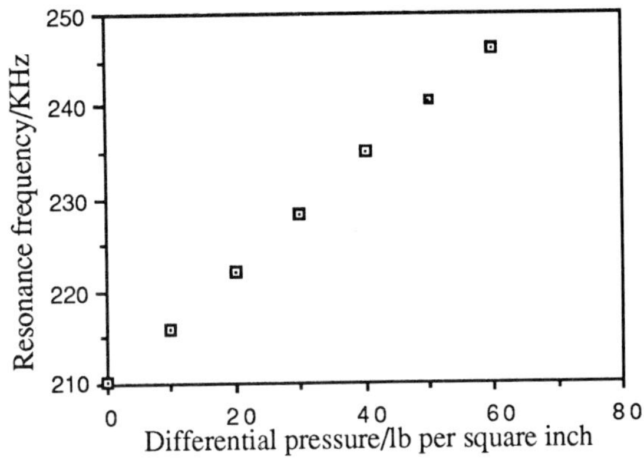

The pressure sensitivity of a beam resonator fabricated from silicon dioxide thermally grown on a silicon wafer was also studied. In order to predict the pressure sensitivity the wafer is considered to be single-material and the interface between the silicon and the silicon dioxide is ignored. The direct application of (2) predicts a greater sensitivity for a silicon dioxide than for an equivalent silicon beam because of the lower value of Young's modulus.

A series of samples of silicon dioxide beams were examined using the set-up described for examining silicon devices. These beams were of similar dimensions to sample B, of thickness 0.5μm and coated with a thin layer of evaporated aluminium or sputtered gold. Differential pressures of up to 4 bar across circular cross-sections of up to 5.0mm in diameter were applied across the full wafer thickness and no shift in the resonant frequency were observed above the resolution bandwidth of 1KHz.

3.1. Discussion:

The results obtained for the pressure sensitivities of resonance frequencies of the silicon beams are of the order predicted. A higher sensitivity is obtained by using a thinned wafer. The position and orientation of the microresonator beams on the wafer is clearly significant. Thus in order to reliably optimise the pressure sensitivity it is desirable to position the beam in the centre of the wafer. The beam stress is also affected by any dislocations or defects which may be introduced during the process of fabrication, and by the surface roughness resulting from the chemical thinning process. Further, the epoxy that to bond the sample to the pressure ring introduced significant stress to the beam. This was indicated by the 6% resonant frequency shift of the beam of sample A on bonding to the metal ring. For the implementation of a practical sensor it will be necessary to consider these factors.

When silicon is thermally oxidised, the silicon dioxide resulting is in a state of compression, hence when a silicon dioxide beam is defined by etching the silicon beneath an oxide strip it deflects either upward or downward. This compression is a combination of intrinsic stress and thermal stress. The intrinsic stress is attributed to the volume expansion resulting from the conversion of silicon to its oxide (Alexandrova *et al* 1988). The thermal stress is induced when the wafer, oxidised at around 1000°C, is cooled to room temperature. When the beam is undercut and allowed to deflect, nearly all the stress is relieved. Fig 5 shows a scanning electron micrograph of a thermally grown silicon dioxide beam on silicon which is deflecting downward.

The deflection of the oxide microbridge can be measured using a scanning electron microscope to be around 5μm . The displacement of the wafer (from 3) is not significant, and we conclude from this that when the differential pressure is applied the bend shape of the beam will change so that the stress remains relieved. Any change in resonance frequency must then result from a change in the residual stress remaining in the bent beam. This residual stress is thought to be about two orders of magnitude lower than the relieved compressive stress (Lin and Pugacz-Muraszkiewiz 1972) and the change in resonance frequency is expected to be almost negligible. This conclusion is supported by the above results, which indicate a frequency shift of less than 0.2% per bar.

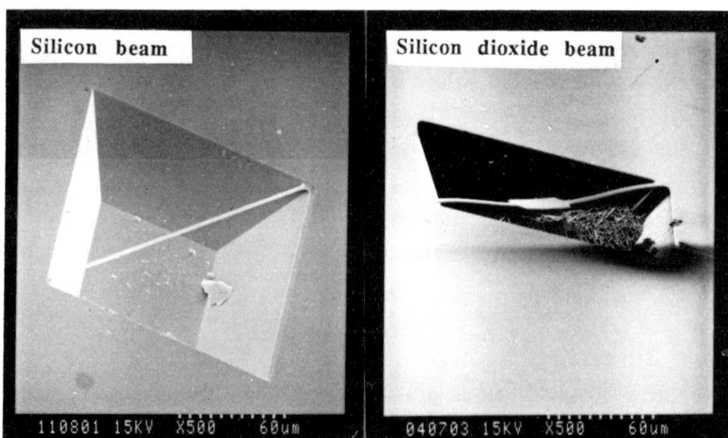

Fig 5: Scanning electron micrographs of a silicon and a silicon dioxide beam grown on silicon. The oxide beam deflected downward by about 5μm.

These results indicate that silicon rather than silicon dioxide will be a more suitable material for pressure-sensitive microresonator beams. However for other sensor applications the pressure-insensitivity of the silicon dioxide microbeams may prove to be a useful property.

4. Temperature sensitivity of resonance frequency:

Temperature sensitivities of resonance frequency were measured by mounting the samples on a Peltier heating block. To attach the wafer a thin layer of heat-sinking compound was used. To monitor the temperature, a thermocouple was fixed to the metal-coated surface of the wafer within 6mm of the resonator. Good thermal contact was ensured by fixing the thermocouple tip with thermally conductive glue.

A silicon device similar to sample A was examined. The dimensions of the beam were as follows: L_b = 175μm, L_p = 21μm, w_b = 3.5μm, w_p = 8μm and silicon thickness 1.0μm. It was evaporation coated with 200nm of aluminium. The experimentally measured resonance frequency was 353KHz with a quality-factor of 62. The resonance frequency is plotted against temperature in Fig 6. It can be seen that the dependence is not linear; however the temperature coefficient of resonance frequency between 5 and 60^{0}C is of the order of -0.17% /^{o}C.

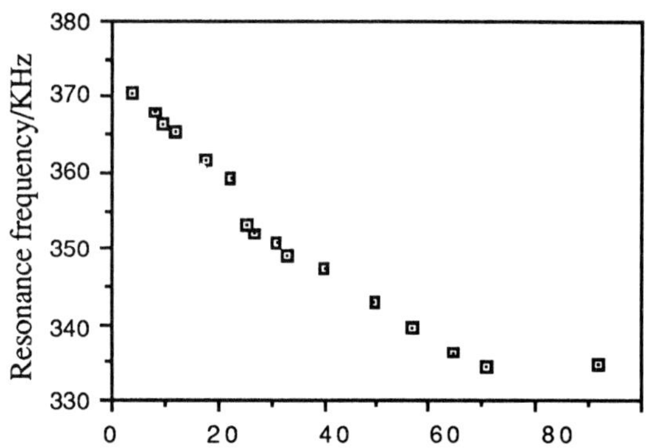

Fig 6:
Dependence of resonance frequency on temperature for a silicon beam.

A silicon dioxide device similar to sample B was also examined. The silicon dioxide thickness was 0.5μm, and the device was evaporation coated with 200nm of aluminium. The experimentally measured resonance frequency was 333KHz with a quality-factor of 40. The resonance frequency is plotted against temperature in Fig 7. The temperature coefficient of resonance frequency between 25 and 60°C is of the order of -0.13% / ^{o}C.

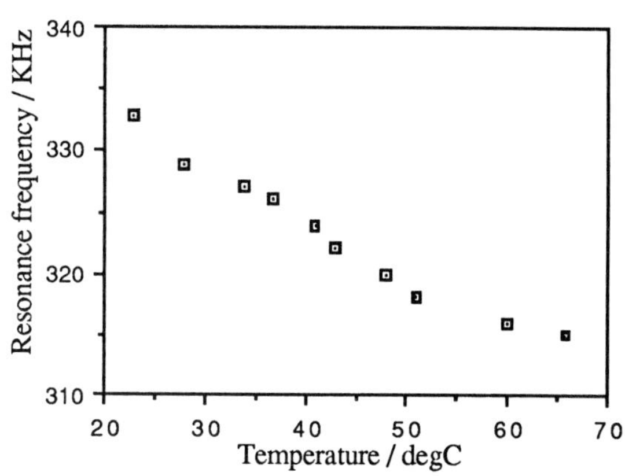

Fig 7:
Dependence of resonance frequency on temperature for a silicon dioxide beam.

4.1. Discussion:

The variation of resonance frequency with temperature is due to a combination of the thermal dependence of material parameters (Young's modulus and density) and the thermal dependence of beam stress. From simple beam theory (Timoshenko *et al* 1974) it is found that the change in material parameters affects the resonance frequency by about +0.01% / °C. The observed negative shift therefore indicates that the dominant factor is the change in beam stress.

At room temperature the silicon beam is under stress which is primarily caused by two effects. The aluminium is under tension thermally induced in the evaporation process and the silicon under tension as it is highly doped with boron as part of the fabrication process. (The tension being caused by the lattice mismatch between silicon atoms and the doped boron.) The tension in both materials, principally in the metal, is reduced as the temperature of the beam is raised since the expansivity of metal is higher than that of silicon. This in turn will reduce the resonance frequency, 2(a), as the experimental result confirms. A reduction in metal thickness is expected to significantly reduce the temperature sensitivity of the beams since this reduces the compression stress induced due to expanding the metal.

The mechanism is more complicated for the oxide beam. Again the observed negative coefficient indicates that the dominant factor is the change in beam stress. At room temperature the evaporated aluminium is under thermally-induced tension, but this keeps the oxide under compression, see Fig 8. An increase in temperature will reduce both the tension and the compression so that the polarity of the temperature coefficient of resonance frequency will depend on the relative thickness of the metal and the oxide. It should thus be possible to compensate for the effect of temperature to a first order by a suitable choice of thicknesses.

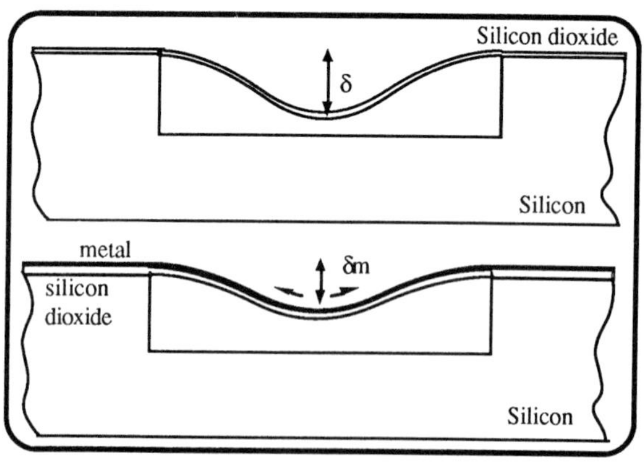

Fig 8:
Stress in a silicon dioxide beam is relieved after undercutting; the metal coating puts the oxide back under compression. $\delta > \delta_m$.

A further method of temperature compensation involves the excitation of a beam and a cantilever using a single multimode fibre. Such a structure is shown in Fig 9. The resonance frequency of the cantilever is independent of the differential pressure across the wafer, and it is therefore straightforward to process the signals received from the two vibrating elements to achieve a temperature-compensated pressure measurement. A multimode optical fibre system similar to that used for the previous work has been assembled. This system was successfully used for activating resonators, including cantilevers such as that characterised in Fig 10. The experimentally measured resonant frequency is 74KHz with a quality-factor of 15.

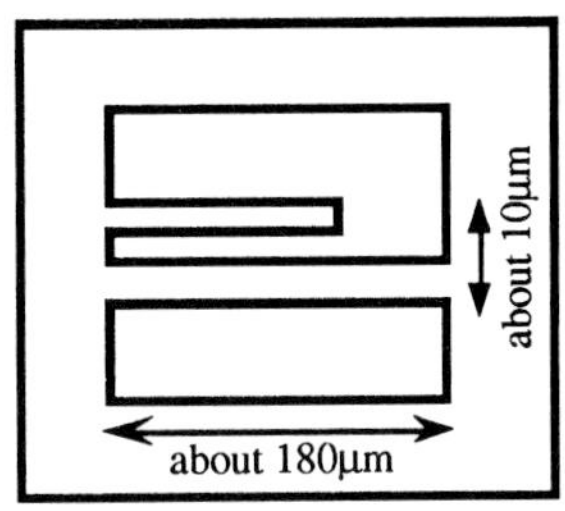

Fig 9:
Microresonator beam pressure sensor with cantilever for temperature compensation.

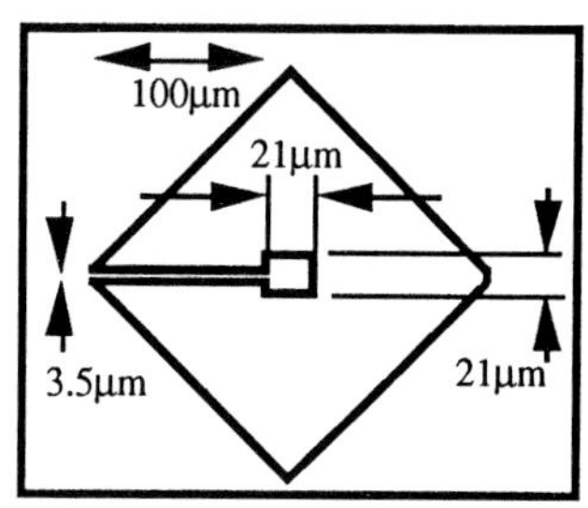

Fig 10:
Silicon cantilever beam resonator which is evaporated with 200nm of aluminium.

5. Conclusions:

The change in resonance frequency of metal-coated silicon beam with applied differential pressure is found to be in agreement with theoretical predictions. The sensitivity may be increased by thinning the wafer on which the device is fabricated. The metal-coated silicon dioxide beam resonance frequency has a very low pressure sensitivity (less than 0.2% per bar). This is thought to be a result of the beam deformation induced during fabrication.

The temperature coefficients of resonance frequency measured are of the same order for the silicon and silicon dioxide beams. The polarity of the temperature coefficient of resonance frequency for a metal-coated oxide beam is thought to depend on the relative thicknesses of the metal and the oxide. A compensation scheme has been proposed.

6. Acknowledgments:

R M A Fatah wishes to thank K H Cameron, A L Matthews, L N Barker and P Cochrane from BTRL for all his comments, discussion and support. He is also grateful for the support of G Parry. N.A.D.Stokes is supported by the SERC. The authors wish to thank S Venkatesh for all her useful discussion and support.

7. References:

Albert, W.C. 1982 Vibrating quartz crystal accelerometer *Proc. 28th Int. Instrum. Symp.* Las Vegas USA 33-44

Alexandrova S, Szekfres A and Christova K 1988 Stress in silicon dioxide films *Phil. Magazine Lett.* 58 33-36

Andres M V, Foulds K W H and Tudor M J 1985 Optical activation of silicon vibrating sensor", IEE Elect Lett, 21 pp 1097-1099.

Fatah R M A , Stokes N A D and Venkatesh S 1989 Activation of silicon michromechanical Resonators in Fibre Optic Systems *Proc. EFCO* Amsterdam pp 418-422.

Gieles, A C M and Somers G H J 1973 Miniature pressure transducers with silicon diaphragm *Philips Tech. Rev.* **33** 14-20

Lin S C H and Pugacz-Muraszkiewiz I 1972 Local stress measurement in thin thermal SiO_2 films on Si substrates

J Appl Phys 43 119-125

Oberg E, Jones F D and Horton H 1989 Machinery's Handbook 23rd Edition (new York 1989) p 297

IEEE Transaction on Electron Devices 26 (12) 1887-979

Thornton K E B, Uttamchandani D and Culshaw B 1988 Novel optically excited resonant pressure sensor *IEE Electr Lett* 24 (10), pp 573-579

Timoshenko, S P Young,D H and Weaver,W 1974 Vibration Problems in Engineering (John Wiley and Sons: 4th ed.)

Uttamchandani D, Thornton K E B and Culshaw B 1987 Optical excited resonant beam pressure sensor", *IEE Electr Lett* 23 (25), pp 1333-1334

Inst. Phys. Conf. Ser. No 111
Paper presented at Int. Conf. on New Materials and their Applications, Univ. of Warwick, 1990

Designing ceramic components—the state of the art

R Morrell

National Physical Laboratory, Teddington, Middlesex, TW11 0LW

ABSTRACT: The successful application of ceramics in engineering requires careful consideration of design of components to minimise the risk of brittle failure. This paper reviews the state of the art in terms of simple design rules for non-critical applications and the techniques for critical design and lifetime estimation. It is concluded that while simple design rules can be useful for low-stress applications, the validity of techniques for critical design is uncertain because of material reproducibility factors.

1. INTRODUCTION

The nature of the atomic bonding in ceramic materials results in a poor ability to deform under localised stress except at very high temperatures. For all intents and purposes, useful ceramics are non-ductile at room temperature. In terms of designing components for engineering service, this poses problems for the engineer (e.g. McLean (1986)). For years he has become accustomed to materials that resist fracture by deformation, allowing redistribution of applied loads, which has enabled him to design with impunity. Brittle materials have been discarded in favour of more-ductile ones because they are more forgiving, such as the move over the last century from cast iron to mild steel for structural work. Dealing with completely non-ductile materials is daunting for all but the specialist.

Consideration of the use of ceramics normally arises because of advantageous properties as shown by test data from test-pieces. Based on such data, the relative performance of materials from different sources can be assessed in relation to the application, and a material chosen. Although this is an essential part of the process, the actual design for mechanical reliability is another matter. Most of the design expertise resides with the manufacturers who try to advise customers as to an appropriate geometry for components, usually based on simple rule-of-thumb experience built up over many years from knowledge of the interaction between their own material fabrication capability, the resulting properties, and the successful utilisation by customers. This understanding is difficult for the customer to appreciate, especially when offered conflicting solutions from different potential suppliers.

The scientific community has studied this problem for many years, and has developed some elegant mathematical techniques for predicting failure probabilities in components subjected to loads in service. Usually, these techniques are based on certain assumptions concerning the nature of brittle failure, the distribution of defects likely to cause failure, the time dependence of fracture and the consistency of material properties. However, there is an increasing body of evidence that the material properties are often not sufficiently reproducible to allow reliance to be placed on such design procedures. A more-pragmatic approach, that of proof-testing, is seen as a better guide to the reliability of individual components. This paper reviews the knowledge base in design of ceramic components, and attempts to identify the areas of concern.

2. SIMPLE DESIGN RULES

2.1 An example

Bearing in mind that non-ductile materials cannot redistribute applied stresses by plastic deformation, the first consideration in design is the overall geometry of the component and the method by which it is supported or attached to other components in the device. It is necessary to avoid localised stress concentrations in the joining process, and once joined, any applied loads, both mechanical and thermal, must be minimised as far as practicable. A classic example is the automobile spark plug which contains a ceramic insulator (Figure 1). The present-day choice of geometry and material is based on several factors. Firstly, the need to withstand the high temperatures of the tip near the spark gap, sometimes up to 900 °C, yet maintaining a cool body temperature, not more than 150 °C, results in high thermal stresses. The high thermal conductivity and high-temperature properties of alumina ceramics compared with porcelains used earlier this century reduce thermal stress gradients and give improved reliability. Secondly, the higher mechanical strength of alumina means that the exposed area of insulator is more resistant to accidental impact and handling damage. Thirdly, the high hardness and strength makes possible the rolling process which curls the top edge of the metal housing to grip the insulator firmly into position. This reliable design was arrived at before any critical design procedures were developed, and is an example of simple design rules being used for a critical component. The shape of the insulator has certain features appropriate to its function and for reliability of manufacture. The exposed area, which is usually glazed for cleanliness, has sufficient diameter for robustness in handling. The central part of the body is designed to take the loads of the rolling process for gripping it in the housing. The electrode tip end is of small diameter to minimise radial thermal gradients. All edges are suitably rounded to avoid chipping. The manufacturing process devised to fabricate the insulator is one of fully automated dry-bag isostatic pressing onto a mandrel that is withdrawn to leave the hole for the central electrode. Final shaping of the outside dimensions is usually performed before firing by a dry form-grinding process.

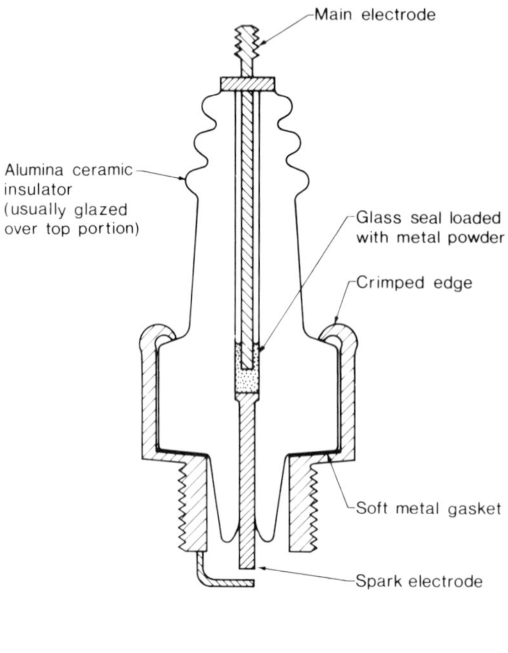

Figure 1 Schematic diagram of vehicle spark plug.

2.2 Geometrical rules of thumb

Discussions of simple factors affecting design are given by Creyke *et al.* (1982) and Morrell (1985), but the above example illustrates several of the rules-of-thumb. The shape is kept fairly compact. There are no long thin parts which would lead to a handling reliability problem. All edges are rounded to avoid chipping. The shape is appropriate for the most economical method of producing the central electrode hole. The process of joining allows typically as-fired ceramic tolerances of ± 2% on dimensions to be used because there are no surfaces which are required to be more accurately flat or dimensioned than this. The central electrode, for example is sealed into position using an electrically conducting glass-metal composite. The rolling process for the housing can adapt to the tolerance band, and works because of the high strength of the ceramic. There are thus no dimensional problems in joining the other parts to the insulator.

Removal of sharp edges, both external and internal, is a fundamental rule-of-thumb in avoiding stress concentrations in handling and use. Outside edges should be chamfered or rounded. The bottoms of internal corners or grooves should similarly be rounded. Another rule is that mechanical contacts which are localised, or that put edges in shear, such as keyways or pin drives, should be avoided. Holes that are too close to edges lead to loss of tolerances and to easy break-out.

2.3 Tolerances

For many applications the question of acceptability of as-fired tolerances arises. If such tolerances can be employed, then there are no relatively expensive post-firing machining operations, except perhaps surface cleaning by vibro-milling or "rumbling". While tolerances closer than the usual ± 2% can be achieved in certain circumstances, designs need generally to allow for such tolerances. To a large extent, the tolerances achieved depend on the geometry and the type of tooling used for the processing route adopted. For isostatically pressed shapes, the tolerances may be held more closely than on shapes pressed in a die where the green density may vary from point to point in the component, and the total die fill may vary from one pressing to the next. On the other hand, for long runs of small components, die pressing may have significant cost advantages over isostatic pressing. Long, thin or unsymmetrical shapes can lead to greater firing distortions than those already resulting from gravity induced slumping.

Tolerances can clearly be improved towards engineering-type norms by machining components. In view of the hardness of many engineering ceramics this is usually a slower process than for most metal alloys, and generally requires diamond tooling. It is therefore often more expensive than similar machining operations on metal components. Clearly there is a cost advantage in keeping the number and area of machined surfaces to a minimum consistent with the essential function of the component. Simple flat surfaces, or uniform cylindrical surfaces generally pose few problems. Examples include vehicle water-pump seals which are routinely ground and lapped on one or both surfaces to a controlled surface roughness, yet cost only a few tens of pence each. Pump components, such as for central heating pumps, may be cylindrically ground and bored out to tight tolerances, but geometries are usually uniformly cylindrical. On the other hand, there are considerable cost penalties for more complex geometries that require, for example, the drilling of very small holes, or the forming of complex cylindrical geometry. Because perfectly square internal corners are impossible to obtain by grinding, they should be undercut in a second operation when essential to the functioning of the component (Figure 2). Avoidance of them is desirable from the mechanical point of view, and they add to the machining cost.

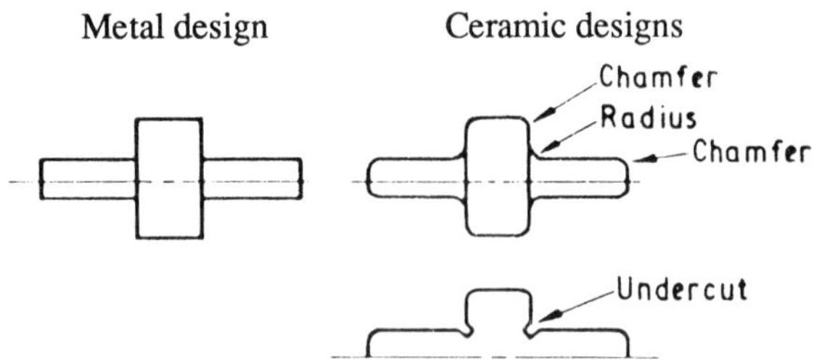

Figure 2 The metal design (left) is inappropriate for ceramics, and is modified (right) with and without undercut appropriate to function.

Although ceramics can in general be prepared with excellent surface quality as determined by roughness measurements, and can often retain this quality in service because of hardness properties, care has to be taken with surface specification. In particular, it has to be remembered that because grinding processes chip out fragments of material from the surface, the surface becomes damaged. In some cases, this can be advantageous. A surface compressive stress is introduced which enhances apparent fracture strength when tested in bending. This is desirable in certain types of transformation toughened ceramics. In others, the damage can reduce as-fired strength by the introduction of cracks which are not necessarily removed unless considerable subsequent lapping is undertaken. For electrical and electronic applications, the surface quality may be critical. Removal of trapped grinding fluids can require a high-temperature re-firing stage to avoid degradation of dielectric loss properties. Sharp angular fractured grains on the surface appear to give unreliable adhesion and wetting of thick film electronic circuitry. For mechanical engineering applications, the most important factor is the effect on fracture and fatigue properties. This is discussed further later in the paper. A balance has to be struck between technological requirements of surfaces, cost, tolerances and manufacturing method by negotiation with potential manufacturers.

2.4 Joining

A ceramic component usually has to be fixed to another component, often of dissimilar material. There is a very wide range of techniques by which this may be done, but clearly the overall factor dictating the viable options is the cost-effectiveness. Ceramic-to-metal joints using vacuum brazing are usually quite expensive because of the number of processes involved, but this method is usually the best if a vacuum-tight seal is required. Design of seals requires careful balancing of thermal expansion coefficients or, in mis-matched seals, the use of compliant deforming layers to minimise thermal stresses on cooling from the brazing temperature. Considerable technical advances have been made in the last decade with this technology, which is used for a very wide range of products, from small vacuum-tight electrical leadthroughs to large microwave windows. Perhaps the most complex joint in use is that between a silicon nitride turbocharger rotor and the steel shaft (Figure 3). The joint has to operate at temperatures up to 400 °C. This involves the use of two intermediate thermal expansion layers of rigid metal joined by a total of three layers of graded brazing, with an overlapping external sheath also brazed into position.

For applications at or near room temperature, adhesive bonding, especially with gap-filling cold-cure adhesives to take up allowances for ceramic-type tolerances, is often the simplest and most effective. Considerable care has to be taken with mechanical fixing, such as bolting (Figure 4), crimping, shrink-fitting (Figure 5) such that the clamping loads are well spread through compliant layers and that there are no excessive body forces exerted on the ceramic component which might lead to failure later on. Localised contact stresses are minimised by using a brazing system to fix a metal flange onto the component, and then the flange takes the local stress of bolting or welding.

Figure 3 Cutaway of Nissan ceramic turbocharger showing ceramic rotor joined to steel shaft through a multilayer brazed joint.

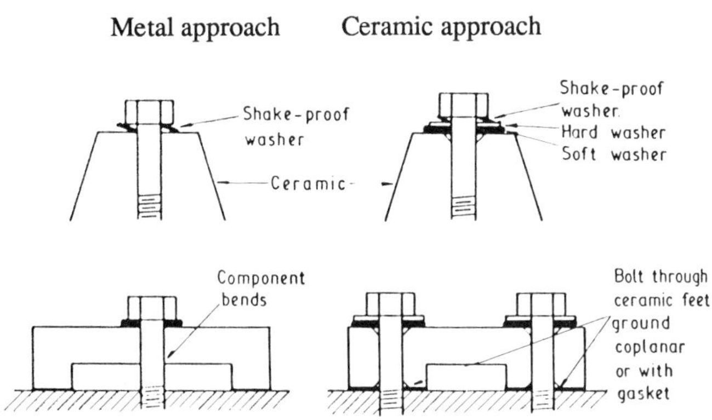

Figure 4 Techniques for bolting metal components (e.g. left) must be modified for ceramics (right).

2.5 Development of new applications

It is clear that in the development of a new application, there is a considerable number of factors that need to be considered. The direct substitution of a metal component by an identically-dimensioned ceramic one is usually neither a geometrical nor a cost-effective option. It is necessary to reconsider the entire design to make the most effective use of a ceramic material. There are a different set of limitations governing the availability of sizes and shapes compared with other materials. Methods of attachment of ceramic components may restrict flexibility. Costs of machining and joining need also be considered. Often, new designs or applications succeed only after a number of design iterations as problems are overcome. This is often best achieved by early discussions with possible ceramic component suppliers who have the necessary background knowledge about the likely success of their own products in new applications.

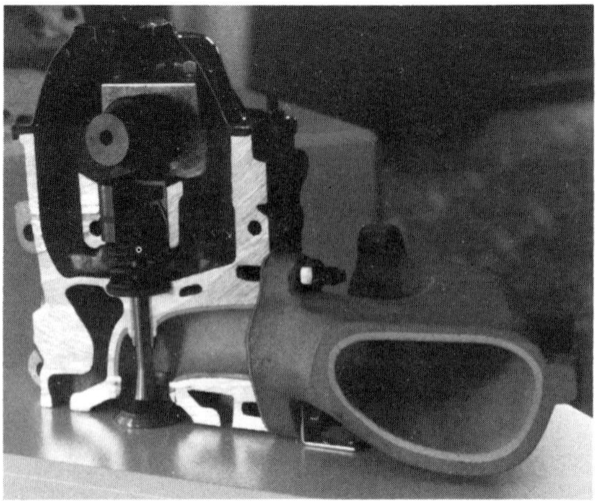

Figure 5 Cast in-situ aluminium titanate exhaust port thermal insulator cast into position in aluminium cylinder head and manifold. Design requirements are principally that the ceramic is not crushed on cooling after casting due the large thermal expansion mismatch.

3. CRITICAL DESIGN

The discussion so far has centred on the simple criteria for improving the reliability of component by attention to general shape factors. In situations where components are to be used with significant levels of tensile stress, the design may need to be optimised with regard to minimising the stresses, or going further and estimating the reliability of the component. In some cases, such as for a turbine component, the component performance required may restrict the possible geometry. Material choice is then determined by the way in which its properties contribute to the required reliability. Simple guidelines are then totally insufficient, and it is necessary to design from a numerical basis. This section reviews the factors influencing ceramic performance relevant to critical design.

3.1 Mechanical behaviour of ceramics

The fundamental problem with design, that of the unpredictability of ceramic components, arises from the nature of the microstructure and its interaction with surface modifying processes. It is almost impossible to eliminate completely all processing defects from ceramics, even with hot-pressing or hipping. There are always pores or surface flaws present, associated with which are stress concentrations which can lead to fracture. In conventional polycrystalline materials, strength is usually limited either by large internal defects or processing inhomogeneities, or by surface defects resulting from processing after firing, or from the progressive growth of pre-existing small cracks under the influence of the external environment. The strength σ_f recorded in any test is related to the defect size by:

$$\sigma_f = K_{Ic}/Y\,c^{1/2}$$

where K_{Ic} is the critical stress intensity facter at the tip of a sharp defect of half-length c, and Y is a geometrical factor related to flaw shape and orientation relative to the stress field. The size, orientation, and position of such fracture origins relative to the applied stress field determines whether fracture will occur. Thus in order to begin to assess the likelihood of fracture of a component, some assessment of the distribution of "worst defects" needs to be made as a starting point. Mechanical tests, such as the commonly used bend tests, are usually employed to obtain such data since NDE techniques are inadequate for this purpose. A sufficient number of test-pieces is tested to provide reasonable reliability to the statistical parameters chosen. Commonly, at least 30 tests need to be performed in order to estimate the spread of results to better than 10%. It is the spread, rather than the mean fracture strength, that is required in order that the small probabilities of failure can be estimated.

However, it is important to note that different types of test are sensitive to different types of defect. For example, a three- or four-point bend test places the highest stress at the surface and edges of the specimen. Strength data thus reflect the quality of surface and edge finishing if the density of internal defects is relatively low. If the equivalent volume of material is tested in uniform tension, there will be a higher probability that internal defects will be the "worst" ones, and the fracture strength recorded will probably be lower. In order to obtain useful design information, ideally both the internal and external defect distributions have to be characterised.

The types of test mentioned so far stress the material uniaxially, but components may be stressed multiaxially. The general effect of biaxial or triaxial stressing in tension is to increase the risk of fracture because flaws unfavourably orientated for fracture under a given uniaxial stress are now stressed in all directions. Stressing in tension with orthogonal compression may be advantageous in reducing the risk of fracture by the same token. The little information in the literature on this topic leads to the conclusion that there is no unique fracture criterion for ceramics because different types of defect (inclusions, pores, induced flaws, etc.) seem to behave in different ways in multiaxial stressing. The simple mechanical estimators reviewed below assume a unique criterion that is independent of the existence of any compressive effect. It is a moot point as to how far this can be taken.

In most materials, strength is time-dependent. Small flaws can grow in a way determined by the environment adjacent to their tips. If it is a surface flaw, the external environment can control the flaw growth. Under a given applied stress, the flaw growth will accelerate towards failure, giving an empirical relationship of the type:

$$v = A K_I^n$$

where v is the crack velocity, K_I is the stress intensity factor on the crack tip, n and A are constants. The value of n determines the stress sensitivity of the process; high values indicate little growth unless very close to catastrophic failure, while low values indicate significant sensitivity. This equation integrates to give the time to failure, t_f (Morrell (1985)):

$$t_f = 2/(Y^n A \sigma(n - 2)c^{(n-2)/2})$$

where the spread of values of c, the largest flaw size, results in a spread of times to failure. Values of c, A and n can be obtained from fast-fracture tests or from delayed failure tests under constant stress. A value of Y is normally estimated. Clearly this is a simplistic view. Other factors such as cyclic stressing effects, changes in defect population in service, e.g. at high temperatures, and plastic effects at elevated temperatures may need to be taken into account in the design of tests to acquire relevant data.

The final question that needs to be asked is whether the actual component being designed contains the same defect population as the test-pieces used to generate the statistical information. Since the majority of ceramics contain defects related to the exact processing and finishing methods, including the type of grinding undertaken, and even the machine used, there is a strong possibility that test bar data cannot be used with great confidence. Nevertheless, they are a valuable guide to the levels of performance that can be expected of good-quality and consistent materials.

3.2 Stress analysis

Clearly the first step in designing a component is to estimate the stress distribution the component will see while in assembly and/or service. Often this cannot be obtained analytically because of complexity of shape or temperature distribution, and recourse must be made to finite element analysis. For this, accurate data for elastic properties are required, and for thermal gradient situations, the thermal expansion and thermal conductivity data are also required. Advances in the ease of use of FEA computer packages means that complex situations can now be analysed fairly readily, the chief uncertainty remaining being the modelling of boundary conditions on the component. Assumptions need to be made about the exact distribution of surface stresses applied. Care needs to be taken over the modelling of these stresses, as simple assumptions of uniform pressure distribution may not always apply. The same is true for modelling of heat transfer to or from a component. For thermal transient situations, the process has to be repeated at closely-spaced time intervals in order to ascertain the nature of stress Some prior knowledge or some practical test is required in order to ensure that reasonable assumptions are being made. transients resulting.

The advantage of FEA is that the effects of changes in boundary conditions and component geometry on the stress distribution can be ascertained. Using such an iterative approach, the principal tensile stresses in the component can be minimised consistent with geometrical restrictions imposed by the application. This clearly maximises the likelihood of survival.

3.3 Estimating failure probabilities

Once the stress distribution has been obtained and the maximum principal tensile stresses have been minimised, the next step is to estimate whether over the design life of the component there is a significant probability of the component failing. The fracture properties of the ceramic material are required, in particular the statistical data from which failure probabilities can be estimated. The conventional approach is to apply empirically Weibull statistics as a tool for fitting the experimental fracture data to a convenient mathematical relationship (e.g. Bergman (1987)). The probability of failure, P_f, is given by:

$$P_f = 1 - \exp - ((\sigma_f - \sigma_u)/\sigma_o)^m$$

where σ_f is the fracture stress, σ_u is a lower stress level for failure, usually set to zero, σ_o is the so-called characteristic strength (where $P_f = 63\%$), and m is the Weibull modulus, a measure of the steepness of the distribution.

In order to minimise the risks of failure at low probabilities, the strength distribution should be narrow, i.e. with a high value of m. This factor is generally considered to be more important than obtaining a high value of characteristic or mean strength.

By modifying the above equation, the effects of volume or surface area (for the two basic types of failure, from volume or surface flaws) can be incorporated (Creyke *et al.* (1982)):

$$P_f = 1 - \exp\left\{ - \left(\frac{V}{V_o}\right)\left(\frac{1}{m}!\right)^m\left(\frac{\sigma}{\bar{\sigma}}\right)^m \right\}$$

where V is the volume under uniform stress σ, $\bar{\sigma}$ is the mean fracture stress, and V_o is a normalising term. To cope with this term and with the fact that stresses vary throughout a component, Stanley and coworkers (Stanley *et al.* (1976), Stanley (1987) devised the "unit strength" concept whereby the above equation can be integrated over the volume of the body:

$$P_f = 1 - \exp\left\{ - \left(\frac{1}{m}!\right)^m\left(\frac{\sigma_{nom}}{\bar{\sigma}_{ref}}\right)^m\left(\frac{V}{V_{ref}}\right) \Sigma(V) \right\}$$

where σ_{nom} is the nominal stress at some point in the component, $\bar{\sigma}_{ref}$ is the mean strength of the "unit" reference volume V_{ref}, and $\Sigma(V)$ is the stress volume integral defined by:

$$\Sigma(V) = \int_V \left(\frac{\sigma(V)}{\sigma_{nom}}\right)^m \frac{dV}{V}$$

The equivalent result can be obtained for stressed surface area for failure from surface flaws. For multiaxial stressing, an assumption that can be made is that the stress volume integral is summed for each of the three orthogonal axes in the component. For stressing in compression, the assumption can be made that for the vast majority of materials, failure is far more likely to result from tensile stressing than from compression stressing, since for strong ceramics, the strength ratio compression:tension is typically between 8 and 10. For weaker materials with pre-existing internal defects, this ratio could be rather less, down to about 3, and in such circumstances the compression failure characteristics may be required.

This basic equation, or others of equivalent type, can then be combined numerically with a stress analysis, and the probability of failure of the component can be estimated.

The above simple analysis predicts the failure probability at a given stress level in the component stressed under the same conditions as the data used for the estimation of m and $\bar{\sigma}_{ref}$. In order to extrapolate to longer times, the data need to be scaled according to the delayed failure mechanism described in 3.1 above. To account for changes in temperature, the effects of temperature have to be added.

This analysis applies to a totally non-ductile situation. If there is plastic flow or creep, as a result of high temperatures, for example, the situation is more complex to analyse. Stresses can relax, or the lifetime can be reduced by limited strain to failure. The plastic properties of the material must be compared with the stress levels being applied and an assessment made of the creep rupture life expectancy of the most highly stressed parts of the component. An example of this analysis applied to a ceramic turbocharger is given by Soma *et al.* (1987).

3.4 Uncertainties in prediction

As can be appreciated from the assumptions that have to be made in the above analysis, the uncertainties in prediction can be considerable. If calculations suggest that failure probabilities are five or six orders of magnitude less than the maximum acceptable, then there is a strong likelihood that the design will work without failures in practice, provided that all contributory factors have been foreseen and accounted for in the estimations. If the calculations are closer to the maximum acceptable probability, there is a possibility that the assumptions made are sufficiently in error that the actual risk could be significant. The principal risks are:

1. The long-term conditions of service are not adequately definable, and have an unquantified effect on the remaining strength of the component.
2. The distribution of defects in actual components made by the method most appropriate for the particular shape may be rather different to that of test bars used to obtain the basic failure data. Two particular aspects are (i) the presence of large defects or inclusions which are missed in testing small test-bars because of their infrequency, and (ii) the quality of machining may not be as controlled as for the surfaces of test bars.

The evidence becoming available in the technical literature suggests that these factors are not insignificant (see, e.g., Kawamoto *et al.* 1986, Kleer *et al.* 1986). There are instances where, for example, the Weibull modulus reduces as the size of the component increases. Once thought to be a material parameter, it must be recognised now that the Weibull modulus may be a function of how a material is tested as well as of the defect distribution. The relative significance and frequency of surface, near surface and bulk defects changes with stress field and size of component. Such factors are not usually examined in test bar type tests. Although fractography of bend test-pieces can assist with defect identification, the better test is one on the component itself. If this is of small dimensions then the cost of such an exercise is not prohibitive, but if the component is large, expensive and difficult to test effectively, there is a significant barrier to providing checks on the test-bar data.

3.5 Proof-testing

With such uncertainties, it is clear that guaranteeing the performance of any single component reliably is difficult. If a guarantee is required, the only possibility is to undertake a proof test on the component. Thus, unless the material properties are reliable and predictable, and the service conditions are well-defined, the estimation of probability of failure will act only as a guide. The stress analysis can serve to indicate areas in a design where there are stresses of sufficient magnitude to be potentially dangerous if they exceed, say, about one-quarter of the mean fracture strength of small test bars as a first approximation, and the design can be modified to reduce these stresses. This procedure coupled with proof-testing is at present the best option to guaranteeing that the component *as-made* is fit for purpose (Morrell 1989).

The principle of proof-testing is to stress any defects in the component at a level sufficiently greater than the service stress level, but only for a short time, so that if the component survives without fracture, it will have a minimum life at the lower service stress. The testing time should be kept short so that the strength level of survivors is not degraded by the stressing operation. The geometry of stressing should be the same as that the component sees in practice, and this needs to be modelled as closely as possible. The ratio of proof stress to service stress can be estimated for a given minimum survival period, $t_{f\,min}$, from the delayed failure relationship given above, from which it can be shown that:

$$t_{f\,min} \propto \frac{R^{n-2}}{\sigma_s^2}$$

where R is the proof to service stress ratio, σ_s is the service stress and n is the subcritical crack growth exponent. For materials which display little tendency for slow crack growth, a value of R close to unity is often sufficient. For materials with significant tendency to crack growth, the value of R could be as high as 4 for guaranteeing long life.

The important fact to note is that if proof-testing is used as the component guarantee, it should be ascertained that the material is capable of being proof-tested. There may be a need to check, firstly, that the material after proof-testing is as strong as the proof-test appears to indicate and that no unforeseen factor has caused weakening as a result of the test, and secondly, that the service conditions to be used do not induce a new or changing population of defects which overrides the original one guaranteed by the proof test. An example of this is the oxidation of silicon nitride. Proof-tested silicon nitride subsequently exposed to air at high temperatures has a strength distribution that shows some components much weaker than the proof-test level (e.g. Wiederhorn and Tighe (1978)). Oxidation has introduced a new population of defects at the surface. It can be argued that the proof-test has eliminated all the test-pieces with unacceptably large defects in them resulting from manufacture, and therefore still has value in quality assurance. However, under such conditions, there remains a service failure risk which cannot be guaranteed against. Such material behaviour needs to be quantified and allowed for in statistical design.

Other instances in which proof-testing may not be effective is in the stressing of joints to other components. These may suffer unacceptably by overstressing. Thus although proof-testing at the latest possible stage of assembly of the ceramic into a device is most desirable in order to take account of accumulated damage, it may not be effective at this stage. Finally, proof-testing involving thermal overload stress requiring higher than the normal service temperatures might also damage the component or joint.

Despite these possible drawbacks, proof-testing remains the only viable means of mechanical quality assurance of a given design. If the data are not available to define accurately a minimum time to failure, then setting a proof-test stress level at about 50-70% of the mean failure stress will usefully eliminate the weakest, typically 10%, of the population, It is this small proportion that causes the uncertainty in the numerical estimates of failure probability.

4. CONCLUSIONS

Simple design procedures based on a commonsense approach to eliminate risky features such as sharp edges or thin sections are extremely useful in helping to design reliable ceramic components. An appreciation of the limitations of manufacturing processes is desirable because the limitations of reliable processing impose marked restrictions on geometry. Amelioration, preferably elimination, of localised stresses caused by attachment methods to other components is essential. The basic rule is to design to minimise all tensile stresses as far as is practicable, transferring them to adjacent more-ductile components. Using such procedures allows reliable use of ceramics for the purposes for which they have desirable properties, such as hardness and resistance to wear, corrosion and high temperatures.

Where significant levels of tensile stress cannot be avoided, they should be reduced as far as possible by iterative stress analysis and geometry adjustment. Estimation of the reliability of the design can be undertaken using well-established routines based on Weibull statistics, but care has to be exercised that the basic mechanical property data employed are valid and realistic for the component in question, and that any degradation of those properties in service is understood. The accuracy of the result depends upon the care and attention paid to the materials testing, the quality of data obtained and the techniques analysis.

Without better understanding of these factors, at present the only method of providing some degree of reassurance that any individual component, once designed and fabricated, is reliable is to proof-test it. Proof-testing has a number of potential pitfalls arising from circumstances where the defect distribution is altered by testing or in service, or when connections to other components are unacceptably strained. Nevertheless, with a lack of viable NDE techniques which can not only find defects but also interpret their significance, proof-testing remains the most effective means of checking out a design and its manufacture before service.

References

Bergman B 1987 *Brit. Ceram. Proc.* **39** 175-86.

Creyke W E C, Sainsbury I E J and Morrell R 1982 *Design with Non-Ductile Materials*, Applied Science Publ., London.

Kawamoto H, Shimuzu M, Suzuki M and Miyazaki, H 1986 *Proc. Conf. Ceramic Materials and Components for Engines*, edited by Bunk W and Hausner H, Deut. Keram. Ges., Bad Honnef, F R Germany, pp 1035-42.

Kleer G, Richter H, Döll W and Heider W 1986 *ibid*, pp 979-86.

McLean A F 1986 *ibid*, pp 1023-34.

Morrell R 1985 *Handbook of Properties of Technical and Engineering Ceramics, Part 1, An introduction for the Engineer and Designer*, HMSO, London.

Morrell R 1989 *Mater. Sci. Eng.* **A109** 131-7.

Soma T, Ishida Y, Matsui M and Oda I 1987 *Advanced Ceramic Materials* **2**(4) 809-12.

Stanley P 1987 *Brit. Ceram. Proc.* **39**, 167-74.

Stanley P, Fessler H and Sivill A D 1976 *Proc. Inst. Mech. Engrs.* **190** (49/76), 585-95.

Weiderhorn S M, Tighe N J 1978 *J. Mater. Sci.* **13**, 1781-93.

Inst. Phys. Conf. Ser. No 111
Paper presented at Int. Conf. on New Materials and their Applications, Univ. of Warwick, 1990

A new machineable phosphate based glass ceramic

Tian He Wang and P F James

School of Materials, University of Sheffield, Elmfield, Northumberland Road, Sheffield, S10 2TZ

ABSTRACT: Crystal nucleation and growth was studied in the $CaO-P_2O_5-Al_2O_3-TiO_2-SiO_2$ system. Al_2O_3 in combination with TiO_2 and SiO_2 was particularly effective in promoting volume nucleation and hence glass ceramic formation. Fine crystals of $AlPO_4$ were first observed to precipitate. These acted as sites for the heterogeneous nucleation of needle-like crystals of the unidentified major phase. Tests indicated that the fine grained glass ceramics were biocompatible and had good machineability. Hardness, strength, thermal expansion and other properties of the materials are discussed.

1. INTRODUCTION

Glass ceramics are fine grained polycrystalline materials made by the controlled crystallization of glasses. Because of their hardness and type of fracture behaviour they are generally difficult to machine with standard metal working tools. Few examples of mechanically machineable glass ceramic compositions have been reported, the most notable exception being the silicate based material developed by Corning in which interlocking plate-like crystals of fluorophlogopite mica provide the major phase (Beall (1972), Grossman (1972)). In the present work a new machineable glass ceramic is described based on the $CaO-P_2O_5$ system. In addition to its machineability this material is of interest because of potential biomedical and dental applications. Previous studies have shown that it is difficult to promote volume nucleation in phosphate glasses, although surface nucleation is generally easier to produce. Abe (1979, 1983) and Abe and Saito (1977) prepared crystallized $CaO-P_2O_5$ glasses with an highly orientated microstructure but growth was initiated from the outer surfaces. Kokubo et al (1986) were able to prepare glass ceramics in the $MgO-CaO-SiO_2-P_2O_5$ system containing 16.3 wt% P_2O_5 from powdered glass compacts. These materials contained wollastonite and apatite crystal phases that appeared to nucleate within a layer near the surface of the glass particles. Evidence for the bulk precipitation of calcium apatite from boroaluminophosphate glasses was found by Osaka et al (1988). Vogel et al (1987) described the preparation of glass ceramics from $CaO-P_2O_5$ glasses containing up to 21 wt% Al_2O_3 and up to 25 wt% alkali oxide. The main crystal phases were apatite and aluminium phosphate ($AlPO_4$). Applications of glass ceramics containing CaO and P_2O_5 chiefly arise from their biocompatibility. Wilder et al (1982) have also produced $Na_2O-CaO-P_2O_5$ and $Na_2O-BaO-P_2O_5$ glass ceramics using platinum as a nucleating agent. These were suitable for sealing to high thermal expansion metals.

Recently James (1986) and Wang (1987) reported that additions of Al_2O_3 in combination with TiO_2 and SiO_2 were particularly effective in promoting bulk crystal nucleation and hence glass ceramic formation in the $CaO-P_2O_5$ system. The compositions studied had low or near zero alkali oxide contents but generally higher CaO and lower Al_2O_3 contents than used by Vogel et al (1987). Tests indicated that the fine grained glass ceramics were biocompatible and possessed good machineability. In this paper glass ceramic formation in the $CaO-P_2O_5-Al_2O_3-TiO_2-SiO_2$ system is described and the effects of microstructure on machineability and other properties discussed.

2. EXPERIMENTAL

Over fifty glasses were prepared in the $CaO-P_2O_5-Al_2O_3-TiO_2-SiO_2$ system. A selection of the compositions studied is given in Table 1. The batch starting materials were standard laboratory reagent (SLR) grade $CaH_4(PO_4)_2H_2O$, $CaCO_3$ and TiO_2; analytical reagent (AR) grade $Al(OH)_3$; and for high purity silica quartz powder. Melting temperatures were in the range 1100-1450°C depending on composition. The melts were generally very fluid and homogeneity could be obtained within a few hours without stirring. Initial difficulties in dissolving the refractory oxides TiO_2 and Al_2O_3 in the melt, particularly for the five component glasses, were overcome by the following procedure. The batch was first sintered at 1200-1250°C for 30 minutes in an alumina or platinum-rhodium crucible using an electric furnace. The temperature was then increased to 1400-1450°C and held for 1 hour to melt the glass. The melt was then cast into a glass block which was crushed and remelted at 1450°C for 1 hour. Finally the glass was cast into a preheated steel mould and transferred to a furnace set at the annealing temperature (typically 600°C), held for 1 hour and allowed to cool to room temperature. The resulting technique ensured that the glass block (typically 40 x 40 x 100 mm^3) was homogeneous, clear and bubble free. If the glass melt was stirred by a platinum blade during the final melting, the melting time could be reduced to 30 minutes. Most of the glasses were melted in alumina crucibles. Some were also melted in platinum-rhodium crucibles as only minor corrosion was observed even after using them for up to four times. The homogeneity of the glasses was also demonstrated by their uniformity of crystallisation behaviour which will be discussed below.

Differential Thermal Analysis (DTA) was used to determine the glass transformation temperature (Tg) (Table 1) and to study the crystallization behaviour of the glasses. A heating rate of 10°C/min was used.

All the glasses were heat treated and examined for evidence of volume (internal) crystal nucleation or surface crystallization. Recent studies have shown that maximum volume nucleation rates occur near, or slightly above, Tg (James (1982, 1985, 1989)). Heat treatments were therefore carried out at Tg and at successively higher temperatures for times usually up to 24 hours but in some cases for several days. Particular attention was given to the range Tg to Tg + 50°C where a first heat treatment was followed by a heat treatment at a higher temperature so that if crystals had been nucleated they could grow large enough to be observed by optical microscopy. At higher temperatures, where growth was more rapid, a single stage treatment was sufficient.

Frequently these procedures resulted only in the formation of a crystal layer at the glass surface, demonstrating that no volume nucleation had taken place. Instead, if the specimen showed a crystalline layer and also many isolated crystals inside it, this indicated volume nucleation had occurred. Hence the presence of such a surface layer was a good test that a sufficient growth treatment had been given. After heat treatment surface crystallization was removed by grinding and polishing prior to examination by transmission or reflection optical microscopy. For the latter the samples were lightly etched.

Glasses showing volume nucleation were examined by transmission electron microscopy (TEM) using ion beam thinned samples, by scanning electron microscopy (SEM) and by electron probe microanalysis (EPMA). Crystal phases were identified by powder x-ray diffraction (XRD). Percentage crystallinities of the glass ceramics were determined by quantitative XRD. Full details are given elsewhere (Wang 1987).

Tests of hardness, strength and machineability were carried out on the glass ceramics. A standard Leitz miniload hardness tester with a diamond indentor was used to obtain the Vickers hardness number. Bending strength was measured on rectangular beams (2 x 2 x 30 mm^3) using four point loading. The beam samples (10 to 12 for each glass ceramic tested) were cut, ground with SiC and polished with cerirouge powder prior to testing. Machineability was tested using an ordinary drilling machine with a 2.5 mm diameter tungsten carbide drill bit operating at 3800 rpm and with the drill under a constant 2.5 kg load. The time required to drill through 4 mm thick glass ceramic plates was measured. Twenty holes were drilled in each plate allowing the average time required to drill through 1 mm to be calculated.

TABLE 1

<u>Compositions of Phosphate Glasses Prepared (mole %)</u>

Code	CaO	P_2O_5	TiO_2	SiO_2	Al_2O_3	Other	Tg(°C)	SURFACE (S) OR VOLUME (V) CRYSTALLIZATION
CP1	50	50					496	S
CP2	60	40					594	S
CP3	66.7	33.3					-	Crystallized on cooling
CPF4	47.6	47.6				4.8 CaF_2	468	S, APS
CPZ5	46.5	46.5				7.0 ZrO_2	555	S
CPT6	47.6	47.6	4.8				565	S (CP)
CPT7	49.0	42.0	9.0				610	S
CPA8	48.0	48.0			4.0		538	S
CPA9	45.9	45.9			8.2		579	Low V ($AlPO_4$), S
CPA10	44.5	44.5			11.0		-	Crystallized on cooling
CPA11	48.0	45.0			7.0		582	Low V ($AlPO_4$), S
CPA12	42.0	48.0			10.0		-	Low V ($AlPO_4$), S
CPS13	53.0	41.0		6.0			562	S, APS
CPS14	50.0	40.0		10.0			574	S, APS

TABLE 1 CONTINUED

Code	CaO	P_2O_5	TiO_2	SiO_2	Al_2O_3	Other	Tg(°C)	SURFACE (S) OR VOLUME (V) CRYSTALLIZATION
CPTS15	45.0	45.0	5.0	5.0			578	S (CP)
CPTS16	48.0	42.0	5.0	5.0			584	S, (CP, C_2P)
CPTS17	52.0	40.0	4.0	4.0			586	S
CPTA18	42.5	42.5	12.8		2.3		607	S
CPTA19	45.7	45.7	5.7		2.9		578	S
CPTA20	43.3	43.3	10.3		3.1		614	S
CPTA21	47.4	41.2	8.3		3.1		609	S
CPTA22	47.3	40.5	8.7		3.5		616	S
CPTA23	49.7	39.4	7.3		3.6		609	S
CPTA24	45.6	41.4	8.4		4.7		617	Low V ($AlPO_4$), S
CPTA25	45.1	40.9	8.4		5.5		621	High V ($AlPO_4$)
CPTA26	46.6	42.2	5.5		5.7		614	V ($AlPO_4$)
CPTA27	42.9	42.9	8.2		6.0		613	V ($AlPO_4$)
CPTSA28	43.7	43.7	4.9	4.9	2.8		584	S
CPTSA29	42.7	42.7	5.2	5.2	4.2		589	S
CPTSA30	42.1	42.1	5.3	5.2	5.3		596	S
CPTSA31	41.8	41.8	5.0	5.0	6.5		609	V (Low), S (CP, C_2P)
CPTSA32	41.5	41.5	5.0	5.0	7.0		603	High V ($AlPO_4$, CP, C_2P), S (CP, C_2P)
CPTSA33	41.5	41.5	4.8	4.8	7.5		606	As CPTSA32
CPTSA34	41.0	41.0	5.0	5.0	8.0		608	High V ($AlPO_4$, CP, C_2P, some phase X) finer microstructure
CPTSA35	39.0	39.0	4.0	12.0	6.0		592	Low V ($AlPO_4$ + mainly CP), coarse microstructure
CPTSA36	41.0	41.0	4.5	4.5	7.0	2.0Na_2O	578	High V ($AlPO_4$), coarse microstructure
CPTSA37	40.5	39.3	5.6	7.8	6.8		608	High V ($AlPO_4$, phase X) fine microstructure
CPTSA38	44.0	39.5	5.0	5.0	6.5		609	As CPTSA37
CPTSA39	45.4	39.7	4.7	4.7	5.4		607	As CPTSA37
CPTSA40	46.0	39.0	5.5	5.5	4.0		614	Lower V ($AlPO_4$), coarser microstructure
CPTSA41	48.4	36.8	4.0	4.5	6.2		627	High V ($AlPO_4$, phase X), fine microstructure
CPTSA42	40.3	39.1	5.5	7.8	7.4		612	Similar to CPTSA37 and 38
CPTSA43	40.0	38.9	5.4	7.8	8.0		615	"
CPTSA44	39.7	38.7	5.4	7.7	8.6		622	"
CPTSA45	42.7	40.1	5.9	3.0	8.3		613	"

KEY: APS = Amorphous Phase Separation, CP = CaP_2O_6, C_2P = $Ca_2P_2O_7$

3. RESULTS AND DISCUSSION

3.1 Volume and Surface Nucleation (Table 1)

The $CaO.P_2O_5$ composition (CPI) with a liquidus of 982°C formed a glass easily, but heat treatments over a wide range of temperatures, including 30°C below Tg, followed by a higher temperature growth treatment only gave surface crystallization. CP2 formed a glass when cooled rapidly by pressing between steel plates but again showed no volume crystallization on heat treatment. CP3 did not form a glass even when the melt was pressed between plates.

Keeping the CaO/P_2O_5 ratio at 1 or greater than 1 TiO_2, SiO_2, CaF_2 and ZrO_2 were added as third components. None of the compositions (Table 1) exhibited volume crystal nucleation, although TEM revealed amorphous phase droplets in the glass containing CaF_2 (CPF4), and visible opalescence due to amorphous phase separation was detected in the glasses containing SiO_2 (CPS13, CPS14). The glasses containing both TiO_2 and SiO_2 (CPTS 15-17) likewise showed no volume crystal nucleation even after exhaustive heat treatments, but also showed no amorphous phase separation.

Addition of Al_2O_3 alone (CPA 8-12) did produce some low volume nucleation but only for larger additions and satisfactory glass ceramics could not be produced. For example, CPA9 heated at 650°C for 67 hrs showed a low density of volume nucleated crystals (about $10^2 mm^{-3}$). XRD indicated the presence of the $AIPO_4$ crystal phase. However after heating at higher temperatures such as 780°C, the major phases ($Ca_2P_2O_7$ and some CaP_2O_6) crystallized mainly from the sample surfaces rather than heterogeneously on the $AlPO_4$ crystals producing a thick crystal layer and a coarse internally crystallized microstructure.

In the $CaO\text{-}P_2O_5\text{-}TiO_2\text{-}Al_2O_3$ glasses (CPTA 18-27) volume nucleation of the $AlPO_4$ phase occurred for Al_2O_3 contents of 4.7 mole % and above. In CPTA 24 only low volume nucleation rates were observed. In CPTA 25 with a higher Al_2O_3 content more rapid nucleation of $AlPO_4$ occurred between 630 and 700°C (for example, 5 x 10^6 mm^{-3} crystals after 680°C for 50 min). Further heating at a higher temperature (such as 780°C for 12 min) gave heterogeneous nucleation of the major phases (CP and C_2P) producing a highly crystallized glass ceramic. However, large pieces of this glass could not be obtained due to devitrification during cooling, the poor glass forming ability being caused by the high Al_2O_3 and TiO_2 contents. This problem was improved by lowering the TiO_2 (CPTA 26) or reducing the CaO/P_2O_5 ratio to 1 (CPTA 27), but the heterogeneous nucleation of the major phases on the $AlPO_4$ was relatively slow. Only a second stage heat treatment at a lower temperature (below 760°C), where heterogeneous nucleation was probably more efficient, lead to a fine microstructure.

Comparison of the results for the CPA and CPTA glasses indicated that although TiO_2 did not act directly as a nucleating agent it did play an important indirect role in promoting heterogeneous nucleation on the $AlPO_4$ crystals, possibly by lowering the interfacial energy between the $AlPO_4$ and major phase.

For the five component glasses CPTSA 28 to 34 the CaO to P_2O_5 and TiO_2 to SiO_2 ratios were equal to one and the alumina content was gradually increased. The Tg increased with alumina content. Homogeneous glasses of CPTSA 28, 29 and 30 were produced after melting at 1400°C for 5 hours, whereas CPTSA 31-34 were remelted at 1450°C as described previously. No volume nucleation occurred when the alumina content was less than 6.5 mole %. For compositions with more than 6.5 mole % alumina a large number of fine (less than 1 μm) $AlPO_4$ crystals formed after the first stage treatment (typically at 630 to 700°C), their nucleation rate increasing rapidly (for a given temperature) with increase in Al_2O_3 content. For glass CPTSA 31 attempts to produce heterogeneous nucleation on the $AlPO_4$ crystals at higher temperature (eg 800°C) proved unsuccessful, although surface crystallization of CP and C_2P occurred readily. However, for CPTSA 34, which contained a higher Al_2O_3 content of 8 mole %, heterogeneous nucleation of other phases (CP, C_2P and an unidentified phase 'X') did occur after a second stage heat treatment, enabling a finer glass ceramic microstructure to be obtained.

For glasses CPTSA 35 and 36, the CaO/P_2O_5 ratio was kept at one. The SiO_2 content was increased for CPTSA 35 and Na_2O was added for CPTSA 36. Although $AlPO_4$ could be precipitated only coarse microstructures were produced after a second stage heat treatment.

Finally, for glasses CPTSA 37 to 45 the CaO/P_2O_5 ratio was increased. This resulted in increased nucleation and growth rates but the presence of SiO_2 enabled good glasses to be formed without devitrification. On heating these glasses above Tg at about 700°C a large number of $AlPO_4$ crystals were nucleated, with sizes less than 0.5 μm after 1 hour treatment. At this stage the sample was opaque and the crystallinity less than 10% by volume. On heating at a higher temperature (about 800°C) heterogeneous nucleation occurred rapidly on the $AlPO_4$ crystals. Unlike the previous compositions the predominant phase nucleated heterogeneously was the phase X referred to above, which had a fine needle-like morphology (Figure 1). For these glasses the best double stage heat treatments to produce glass ceramics were a first stage treatment in the range 670 to 720°C for 1 to 1.5 hrs followed by further treatment at 780 to 820°C for 20 to 40 mins. Among the highest nucleation rates of $AlPO_4$ and the major phases were observed for CPTSA 41 in this series.

Figure 2 shows a coarser microstructure glass in CPTSA 37 after heating at 840°C for 15 min. Only a low density of $AlPO_4$ crystals were nucleated at this temperature. These are visible at the centre of the larger crystals of the X phase. XRD of this sample gave the following 'd' spacings for phase X (in nm):

0.539, 0.403, 0.378, 0.325, 0.308, 0.269, 0.239
0.222, 0.206, 0.191, 0.179, 0.162, 0.154

These values were tentatively fitted to a tetragonal structure with unit cell parameters a_o and c_o of 0.537 and 0.610 nm respectively. Attempts to determine the composition of this phase X using EPMA were inconclusive.

Studies of the nucleation and growth kinetics of the $AlPO_4$ and phase X are given by Wang (1987).

This section may be briefly summarized as follows. Glass ceramic formation was best in the five component system with the CaO/P_2O_5 ratio

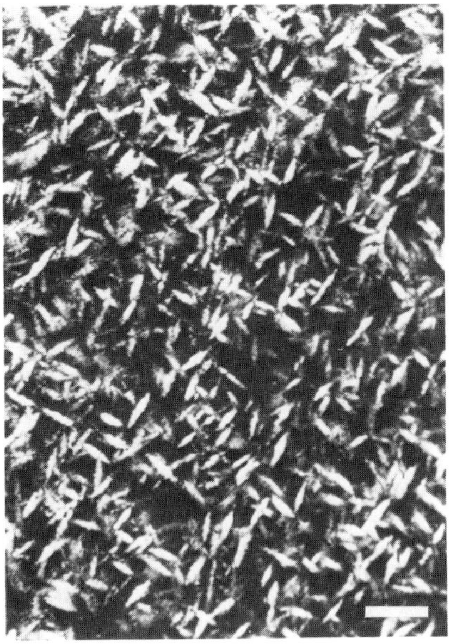

Fig 1. Glass CPTSA 42 heated at 700°C for 60 min and at 850°C for 12 min. Transmitted polarized light (marker 50 μm).

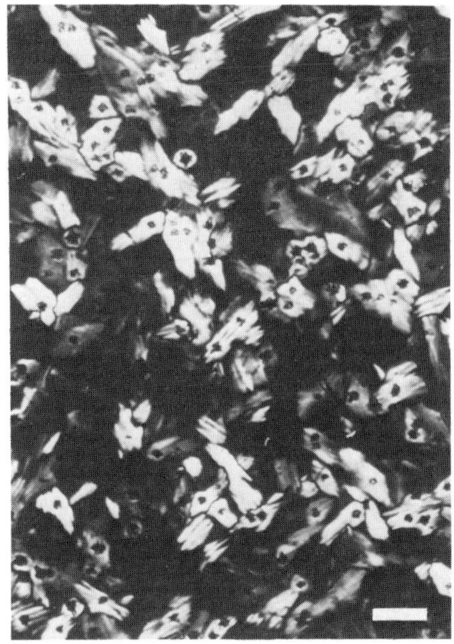

Fig 2. Glass CPTSA 46 heated at 840°C for 15 min. Transmitted polarized light (marker 50 μm).

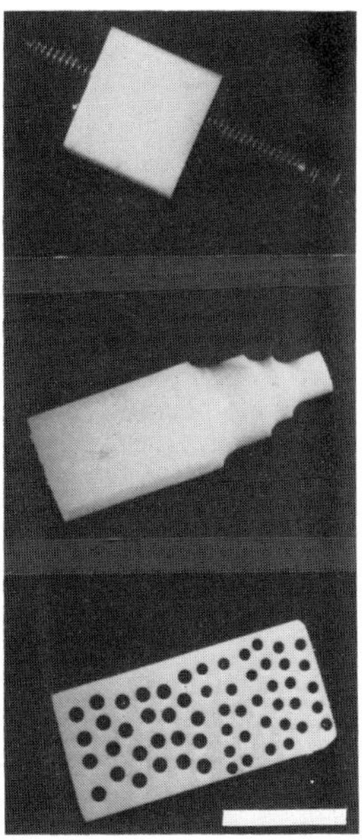

Fig 3. Machined samples of glass ceramic prepared from glass CPTSA 37 heated at 700°C for 60 min and 810°C for 25 min (marker 10 mm).

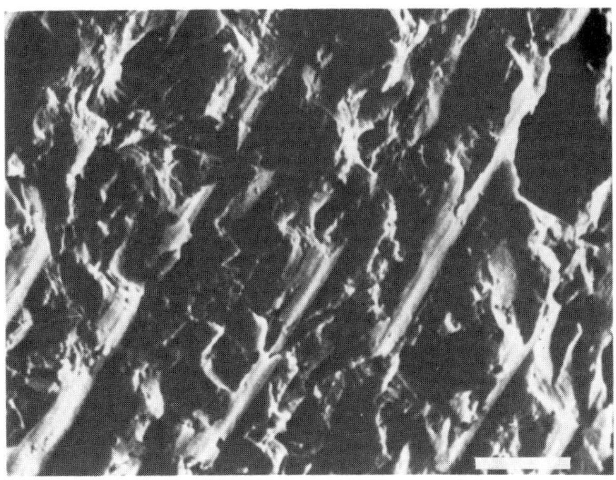

Fig 4. Surface of glass ceramic, prepared from glass CPTSA 42 heated at 720°C for 60 min and 830°C for 15 min, after cutting with a saw blade. Scanning electron micrograph (marker 20 μm).

greater than one. Sufficient Al_2O_3 was essential to enable the precipitation of $AlPO_4$ crystals. TiO_2 and SiO_2 in the glass tended to aid the nucleation of $AlPO_4$, and also to aid the heterogeneous nucleation of major phases, particularly the X-phase. SiO_2 also helped to increase glass forming ability.

3.2 Properties of the Glass Ceramics

Fine grain glass ceramics were prepared by heat treatment of glass CPTSA 42. Glass samples were held at a nucleation temperature between 670 and 735°C for 1 hr. The temperature was then raised at 10°C/min to the growth temperature (850°C) and held for 12 mins before cooling to room temperature. The results of some property measurements are given in Table 2, and include data for samples given the growth treatment only, which exhibited a coarser microstructure.

TABLE 2

Property	Nucleation Temperature (°C)					
	None	670	685	700	715	735
Degree of Crystallinity (Volume %)	65.9	73.2	76.3	74.7	70.0	67.1
Modulus of Rupture (MNm^{-2})	41.6	47.5	84.1	88.7	82.8	71.6
(Standard Deviation)	(6)	(4.9)	(9.2)	(10.3)	(12.1)	(10.2)
Vickers Hardness ($Kgmm^{-2}$)	390	427	478	492	482	460
(Standard Deviation)	(18)	(27)	(24)	(27)	(21)	13
Machineability time (s) to drill 1 mm	3.8	4.8	11.6	12.1	11.5	6.5
(Standard Deviation)	(0.4)	(0.3)	(0.4)	(0.7)	(0.6)	(0.4)
Thermal Expansion Coefficient ($°C^{-1}x10^6$) 20-185°C	13.3	14.09	14.21	12.57	13.27	13.40
185-650°C	9.3	8.64	9.12	9.20	9.60	9.37

Other properties: Young's modulus $41\text{-}45x10^3$ MNm^{-2}; Density 2650 kgm^{-3}; dielectric strength $>5x10^6$ Vm^{-1}; Volume electrical resistivity $7x10^{15}$ ohm m. For glass thermal expansion $8.57x10^{-6}$ (185-650°C) and liquidus temperature 1335 ± 5°C.

The materials produced were white and opaque (similar to ivory in appearance) and could be polished to a high degree. Both modulus of rupture and hardness had maximum values for nucleation treatments in the range 685-715°C. The highest crystallinities were also found approximately in this range, although the finest microstructure (grain size) was obtained for nucleation at 670°C.

Although the glass ceramics had only moderate strengths, their machineability was good and far superior to ordinary glass or ceramics. Thus samples could be easily drilled using conventional workshop equipment and tungsten carbide tipped drill bits cooled by water to produce fine holes typically several mm to 0.5 mm diameter. Samples could be turned and threaded (Fig 3). Also fine holes could be drilled and fine patterns engraved using a dental drill. The machineability time test (section 2 and Table 2) showed that materials nucleated between 685 and 715°C required longer to drill, indicating that the stronger materials were more difficult to machine. The good machineability of these glass ceramics is not yet clearly understood. It is likely that material can be removed as fine particles from the surface by a 'pulverisation' process, similar to that suggested for the Corning Macor material (Grossman 1972). The path of crack propagation through the glass ceramic is complex. One possibility is that fracture occurs preferentially through the needle-like crystal phase but the fracture path is diverted at crystal-crystal or crystal-glass interfaces so that small blocks are removed readily during machining. Alternatively fracture may occur preferentially through the glass-crystal interface or through the glass itself, again being deflected at interfaces, producing an irregular fracture path and leading to removal of small blocks of material. Comparison of machined surfaces of the glass ceramic with those of the untreated glass showed that the former were rougher on a fine scale indicating the more complex fracture path (Fig 4).

Some other properties of the glass ceramic are given in Table 2. The thermal expansion coefficient of the glass ceramic was greater than that of the parent glass. A distinct change in expansion was observed at about 185°C suggesting a reversible polymorphic phase transformation within the major (X) phase.

Another interesting property of these glass ceramics is biocompatibility, which has been indicated by some tissue culture experiments at Sheffield (see Wang 1987). This is not surprising in view of the high Ca and P contents, elements also present in bone. Biocompatibility and machineability suggest biomedical applications. In dentistry a possible use is as an implant material for augmentation of the alveolar ridge (Lamb 1986). Resorption of the ridge is often a severe problem for the wearers of dentures. Other possibilities are in orthopaedics for bone replacement or repair of prosthetic devices and other implants, and as coatings for orthopaedic appliances. In the dense solid form the glass ceramic could be expected to resist resorption or exhibit only slow resorption. However, a controlled microstructure of pores could be introduced into the glass during melting using gas bubbles. A highly porous glass ceramic could then be produced by heat treatment to provide a resorbable bioceramic.

The ability to machine these glass ceramics into intricate shapes and to produce highly polished surfaces suggests other possible applications in engineering, for example as glass ceramic to metal seals and as substrates in electronics, and even in the arts and crafts industries.

ACKNOWLEDGEMENTS

Thanks are due to Mr J W Smedley for help with the glass melting. T Wang wishes to thank the British Council for financial support.

REFERENCES

Abe Y and Saito H, 1977, J Japan Ceram Soc 85 45.

Abe Y, 1979 Nature 282 55.

Abe Y, 1983 US Patent 4417912.

Beall G H, 1972 in Advances in Nucleation and Crystallization in Glasses, Ed by Hench L L and Freiman S W, Amer Ceram Soc Inc, Columbus, Ohio, pp 251-61.

Grossman D G, 1972 J Amer Ceram Soc 55 446.

James P F, 1982 in Advances in Ceramics Vol 4, Ed by Simmons, J H, Uhlmann D R and Beall G H, Amer Ceram Soc Inc, Columbus, Ohio, pp 1-48.

James P F, 1985 J Non-Cryst Solids 73 517.

James P F, 1986 British Patent GB 2199028.

James P F, 1989 Glasses and Glass Ceramics, Ed Lewis M H, Chapman and Hall, London, pp 59-105.

Kokubo T, Ito S, Sakka S and Yamamuro T, 1986 J Mater Sci 21 536.

Lamb D, 1986 private communication.

Osaka A, Toho Y, Asada M and Miura Y, 1988 Materials Letters 6 347.

Vogel J, Holand W, Vogel W, 1987 US Patent 4698318.

Wang T, 1987, PhD thesis, University of Sheffield

Wilder Jr J A, Healey J T and Bunker B C, 1982, in Advances in Ceramics Vol 4, Ed by Simmons J H, Uhlmann D R and Beall G H, Amer Ceram Soc Inc, Columbus, Ohio, pp 313-333.

Inst. Phys. Conf. Ser. No 111
Paper presented at Int. Conf. on New Materials and their Applications, Univ. of Warwick, 1990

Cation environments and the mixed alkali effect in silicate glasses: An EXAFS investigation

G.N. Greaves
SERC Daresbury Laboratory, Warrington WA4 4AD, UK
C.R.A. Catlow and B. Vessal
Department of Chemistry, University of Keele, Keele, Staffs. ST5 5BG, UK
J. Charnock and C.M.B. Henderson, Department of Geology, University of Manchester, Manchester M13 9PL, UK
R. Zhu and S. Qiao, Institute of High Energy Physics, P.O. Box 918, Beijing, PRC
Y. Wang, Institute of Nuclear Research, P.O. Box 8204, Shanghai, PRC
S.J. Gurman, Department of Physics, University of Leicester, Leicester LE1 7RH, UK
S. Houde-Walter, Institute of Optics, University of Rochester, New York 14627, USA

ABSTRACT: EXAFS measurements have been used to obtain the environments of salient cations in a variety of binary alkali silicate glasses. Compared to SiO_2 a lengthening of the mean tetrahedral Si-O bond length is observed. K-O and Cs-O bond lengths and coordination numbers indicate these alkalis and their nearest neighbour oxygens occupy quasi-octahedral sites in single alkali glasses. Small differences are observed when both alkalis are present, indicating intimate mixing. This is predicted by the Modified Random Network (MRN) which is used to provide a unified picture for the Mixed Alkali Effect based on structural constraint.

1. INTRODUCTION

The power of Extended X-ray Absorption Fine Structure (EXAFS) for determining the local structure of multicomponent glasses is well-known (e.g. Greaves 1990). It has been particularly successful in differentiating metal environments - notably those of modifiers and intermediates - that are usually a minority component and thereby scarcely detectable by x-ray or neutron scattering. We have demonstrated, for instance, that Na and Ca which are ionically bonded in the glass structure exhibit well-defined near neighbour shells reminiscent of their environments in their crystalline counterparts (Greaves et al (1981), Binsted et al 1985). Moreover when more than one metal is present, for example Pb^{2+} and Fe^{3+} in phosphate glasses (Greaves et al 1988), we found that although there are compositional differences by and large each metal maintains its own specific local structure. In this study we have extended these measurements to the mixed alkali glass KCs Si_2O_5 and its single alkali derivatives $K_2Si_2O_5$ and $Cs_2Si_2O_5$. Preliminary analysis has been recently given (Greaves 1989). In addition to alkali environments in single and mixed alkali glasses we also consider the local structure of Si, exploring from the standpoint of both glass forming and glass modifying cations the possible structural constraints for ionic conduction and in particular the Mixed Alkali Effect.

Use will be made of the Modified Random Network (MRN) model for glass structure (Greaves 1985) introduced to reconcile the well-defined local structure of modifiers found from XAS experiments with that of glass forming units whose rigid geometry dominates diffraction measurements. Accordingly the bonding in the MRN is partly ionic and partly covalent. The ionically bonded modified regions link through non-bridging oxygens (NBOs) to the covalently bonded glass forming network. The latter is interconnected by bridging oxygens in the way originally conceived by Zachariasen (1932) in his Continuous Random Network (CRN). The MRN differs from Zachariasen's CRN picture in the following ways: (i) the coordination number, CN, of NBOs is in general greater than the 2-fold coordination of bridging oxygens, (ii) the higher CNs of modifying cations and NBOs means that these atoms will tend to cluster, (iii) ionic conduction takes place along modifier "channels" and is governed by percolation processes, (iv) viscosity and shear behaviour are related to the line density of NBOs each of which is a line of easy shear in the structure. In these respects the MRN provides a unified view of the salient properties of modified glasses, including, as we shall see, the Mixed Alkali Effect.

2. SILICON ENVIRONMENTS

We start by examining the local structures of Si obtained from K-edge EXAFS. Although broadly invariant amongst silicate glasses, Si environments exhibit small but significant differences, as a function of glass composition and in comparison with their crystalline polymorphs. Starting with the classic CRN glass, silica, the k^3 weighted free structure and Fourier transform are shown in Figure 1. Whilst XAS is well-known to be dominated by nearest neighbour correlations, Si-O in this case, more distant shells of atoms can be clearly seen, notably the first Si shell close to 3.2 Å and the second O shell at 3.9 Å. In particular the Si-Si peak, R_2, can be used in conjunction with the precise Si-O peak, R_1, to obtain values for the mean bridging oxygen bond angle θ and its variance. This was estimated to be ~ 160° ± 20° (Greaves et al 1981) on the assumption that the Si-Si feature was Gaussian. Multiple scattering effects (Gurman et al 1988) however were not included. When this correction is made there is a small reduction in θ of about 5° from the single scattering value. In addition some assumption must be made about the asymmetry or otherwise of the Si-Si correlations. In particular if these are symmetric, since $R_2 = 2R_1 \sin(\theta/2)$, the bond angle distribution $P(\theta)$ is necessarily asymmetric which may not be the case. Possible asymmetry in $P(R_2)$ has not been investigated and more detailed analysis is required before a reliable $P(\theta)$ distribution can be determined. EXAFS promises to be the preferred technique.

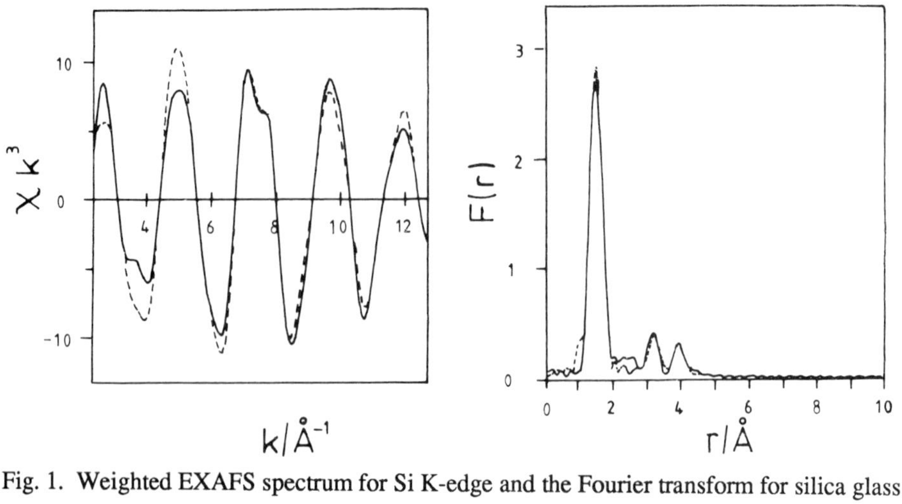

Fig. 1. Weighted EXAFS spectrum for Si K-edge and the Fourier transform for silica glass.

Si K-edge EXAFS spectra have also been measured in a variety of oxide glasses and crystals. In addition to the peaks at 3.2 and 3.9 Å characteristic of the silica network (Figure 1), the spectra also reveal Si-modifier correlations in the 3.4 - 3.8 Å range. These features, however, are more evident in the alkali spectra to be discussed later. Instead analysis here has been restricted to the position and width of the 4 nearest neighbour oxygens.

Such a single shell analysis however would be an oversimplification for many crystalline silicates. Unlike quartz, where all oxygens occupy bridging sites, silicates in general also possess non-bridging oxygens which increase in number as the network component diminishes. Moreover a particular SiO_4 configuration characterises the structure. However, as a result of the increasing steric hindrance, caused by juxtaposing corner sharing SiO_4 tetrahedra with more highly coordinated modifier sites in a crystalline array, these units become more distorted as the number of NBO's increase. Bridging oxygen bond lengths increase by up to 0.08 Å, Si-NBO contacts decrease in distance. On balance there is usually an overall dilation of the mean Si-O separation, as can be seen in Table 1.

From Si K-edge EXAFS we have detected similar increases in Si-O distances for glasses of the same compositions. The results are presented in Table 1 where they are contrasted with silica and quartz and their crystalline counterparts. There are significant increases in Si-O distances for $Li_2Si_2O_5$ and $K_2Si_2O_5$ glasses but not for $Na_2Si_2O_5$, mirroring equivalent behaviour in the binary crystalline silicates and pointing to the presence of similar SiO_4 units. MASNMR and Raman experiments provide more direct evidence for which configurations exist in the glass, indeed one of us has recently shown that their distribution can be predicted on the basis of a simple statistical thermodynamic model (Gurman 1990).

Table 1. Silicon environments in alkali silicates.

	EXAFS for Glasses		X-ray Diffraction for Crystals	
	R/Å	$2\sigma^2$/Å^2 total	R/Å	$2\sigma^2$/Å^2 static only
SiO_2	1.61(3)	0.004	1.61	0.001
$Li_2Si_2O_5$	1.62(4)	0.003	1.64	0.004
$Na_2Si_2O_5$	1.61(6)	0.004	1.61	0.008
$K_2Si_2O_5$	1.63(0)	0.002		
K_2SiO_3			1.63	0.004

Despite the qualitatively similar behaviour between crystal and glass there are significant differences in detail. In crystalline $Na_2Si_2O_5$ the increase in bridging oxygen distances almost exactly balances the decrease in NBO distances to give a mean Si-O distance the same as in quartz. The mean glass bond length behaves similarly. However, in the crystal this is achieved at the expense of a substantial static contribution to the bond length variance, σ^2, of 0.004 Å^2. The thermal component amounts to 0.001 Å^2 at room temperature (Greaves et al 1981) - only 25% of the static disorder present in crystalline $Na_2Si_2O_5$. The results in Table 1, however, demonstrate that the disorder around Si in $Na_2Si_2O_5$ glass is far less than this, viz. 0.002 Å^2; the disorder in Si-O distances measured in $Li_2Si_2O_5$ and $K_2Si_2O_5$ glasses is also less than in the crystals. We find this state of affairs almost universally true amongst modified oxide glasses and their crystalline equivalents. Figure 2 compares raw Si K-edge EXAFS for Diopside ($CaMgSi_2O_6$) and

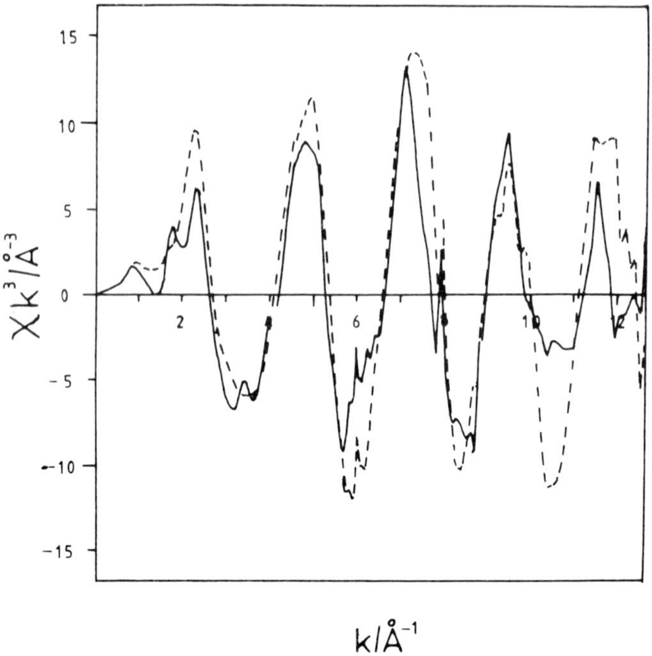

Fig. 2. Weighted EXAFS from the Si K-edge for crystalline Diopside (solid line) and Diopside glass (dashed line). The differences in fine structure amplitudes relate to the better local order exhibited by the glass.

Diopside glass. Where the underlying frequency is practically the same for both (R_1 = 1.62(8) Å for the crystal and 1.62(0) Å for the glass), the amplitudes for Diopside are all significantly less than for Diopside glass. The Si environment in the crystal is far more disordered: $\sigma^2_{crystal}$ = 0.004 Å^2, whilst σ^2_{glass} = 0.002 Å^2.

The explanation for these observations lies in the fact that, without the constraints of translational invariance, glasses can adopt more perfect SiO_4 configurations, the disorder being transferred to P(θ), the distribution of bridging oxygen bond angles and to a lesser extent in P(φ), the dihedral angle distribution. Accordingly the glass will simulate the sheet or chain morphology of the crystalline state in the short range, but lose this in the long range disorder - the polymeric network contorting as envisaged in the MRN model. This can have considerable "knock-on" effects for the alkali environments, which will now be discussed.

3. ALKALI ENVIRONMENTS

The K K-edge and Cs L_{III}-edge in the single alkali glasses $K_2Si_2O_5$ and $Cs_2Si_2O_5$ are shown in Figure 3 where they are compared with curved wave theory fits. We find that the mean alkali oxygen distances for the alkali silicate glasses are very similar to the corresponding crystalline carbonates. For $K_2Si_2O_5$, $R_{K\text{-}O}$ = 2.8(1) Å compared to 2.8(0) Å in K_2CO_3 whilst for $Cs_2Si_2O_5$, $R_{Cs\text{-}O}$ = 3.1(2) Å compared to 3.2(6) Å in $CsCO_3$. The coordination numbers are also well-matched lying between 5 and 7. Evidently both alkalis are approximately octahedrally coordinated in the silicate glasses as they are in alkali carbonates. Earlier results for $Na_2Si_2O_5$ glass (Greaves et al 1981) gave

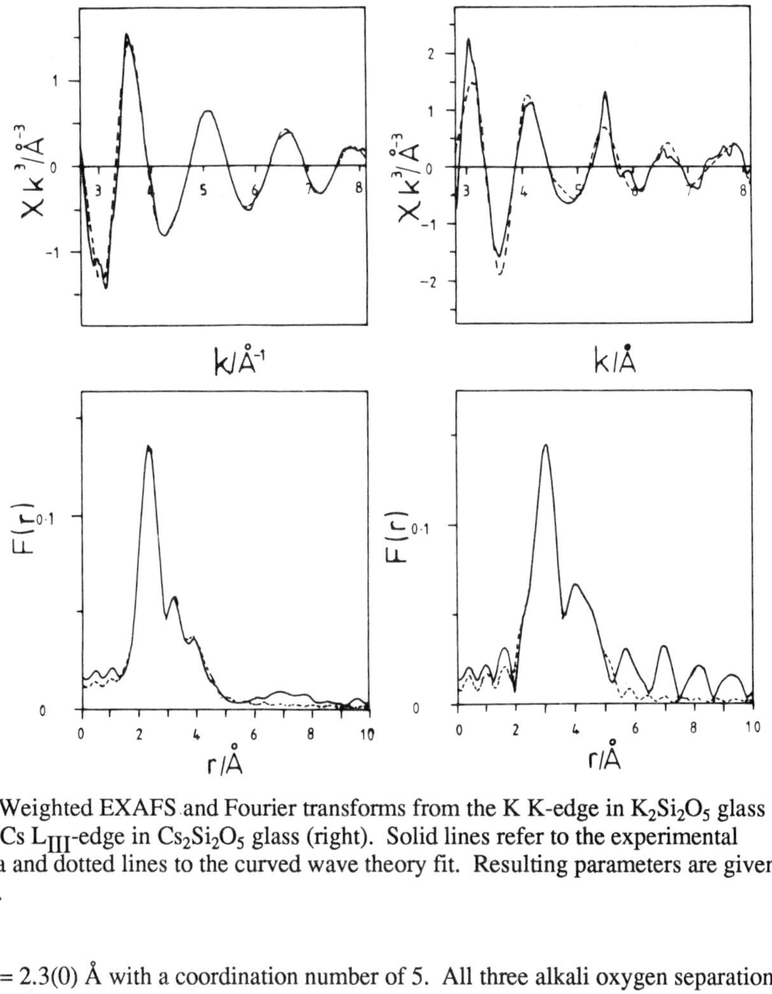

Fig. 3. Weighted EXAFS and Fourier transforms from the K K-edge in $K_2Si_2O_5$ glass (left) and the Cs L_{III}-edge in $Cs_2Si_2O_5$ glass (right). Solid lines refer to the experimental raw data and dotted lines to the curved wave theory fit. Resulting parameters are given in Table 2.

$R_{Na\text{-}O}$ = 2.3(0) Å with a coordination number of 5. All three alkali oxygen separations scale accurately with their respective ionic radii (Wells 1975).

Comparing the atomic distribution for Si in SiO_2 (Figure 1) with those for K and Cs in the alkali silicates (Figure 3), it is evident that (i) the oxygen distributions around the alkalis are much broader and (ii) the alkali-cation shells are more prominent. The analysed widths (FWHM) of the oxygen ligand distributions are 1.3 Å (K-O) and 0.9 Å (Cs-O) compared to 0.1 Å (Si-O). Although the strengths of the oxygen bonds for the heavier alkalis are weaker than for Si and will be associated with greater thermal disorder, both K and Cs centres suffer considerable static disorder. It is evident from the previous discussion, that some of this disorder must derive from the adjacent buckled network.

Turning now to the cation shells surrounding K and Cs, these occur around 3.6 - 3.8 Å and 4.3 - 4.5 Å and can be attributed with reasonable confidence to Si and alkali shells respectively. Taking these shell radii in conjunction with Si-O, K-O and Cs-O distances given earlier we can obtain values for the NBO bond angles, e.g. K-Ô-Si. These are much smaller than the bridging bond angles in the network and do not suffer from the non-linearity between R_2 and θ we have referred to above. NBO bond angles are listed in Table 2. Note in particular that these lie between 90° indicative of octahedral coordination and the tetrahedral angle of 109.6°. Evidently NBOs enjoy a coordination number of approximately 5 matching that of the modifying alkali and confirming bond consistency arguments. Note that NBO bond angles are greater in $K_2Si_2O_5$ than in $Cs_2Si_2O_5$ suggesting the coordination number increases with the size of the alkali, which is not unexpected.

Table 2. Alkali environments in alkali and mixed alkali silicate glasses. AÔSi values relate to the particular A-O bond length and the Si-O bond length of 1.63 Å measured in $K_2Si_2O_5$.

	$R_{A\text{-}O}$/Å	FWHM $\Delta R_{A\text{-}O}$/Å	A Ô A	AÔSi
$\underline{K}_2SiO_5$	2.81	1.3	104°	108°
$\underline{K}CsSi_2O_5$	2.71	1.0	94°	105°
$K\underline{Cs}Si_2O_5$	3.09	1.2	92°	100°
$\underline{Cs}Si_2O_5$	3.12	0.9	93°	102°

4. MIXED ALKALI ENVIRONMENTS

Turning finally to the mixed alkali glass $KCsSi_2O_5$, the K K-edge and Cs L_{III}-edge EXAFS and Fourier transformers are presented in Figure 4 along with theoretical fits. Values for the mean nearest neighbour distances $R_{A\text{-}O}$ and their widths $\Delta R_{A\text{-}O}$ together with the bond angles AÔA and AÔSi are listed in Table 2. Compared to the single alkali glasses the geometries of K and Cs are similar but there are clearly differences in detail. In particular disorder in the K environment, $\Delta R_{A\text{-}O}$, improves in the mixed alkali glass at the expense of the Cs environment. The K oxygen distribution in $K_2Si_2O_5$ glass we find is skewed to larger R but in KCs Si_2O_5 glass this narrows, becoming much more symmetrical, with an overall reduction in the mean oxygen distance. On the other hand for Cs in $Cs_2Si_2O_5$ glass almost the opposite is true: the oxygen distribution is skewed to smaller R becoming more symmetrical and broader in the mixed alkali glass. This is accompanied by a small decrease in Cs-O bond length. Further evidence that the two alkalis influence one another's atomic environment comes from the NBO bond angles that show some reduction in the mixed alkali glass, indicating a trend to higher coordination.

One would expect a random mixing of K and Cs on the basis of the Zachariasen model, but our observations indicate a closer association which requires the two alkalis to share a limited domain in the overall glass structure, just as the MRN model prescribes. Diffusion of alkalis of different size through the glass require particular sizes of interstitial site to be available in the locality of each cation. The transport mechanisms will mutually interfere if both alkalis share the same percolation pathways. In particular the mobility activation energy for a given alkali is expected to increase with the introduction of another alkali - hopping sites matched to the size of first alkali becoming more scarce as the second alkali is added to the structure. Statistically this activation energy should reach a maximum and the conductivity should fall to a minimum when the two alkalis have equal concentrations, as is observed (Day 1976). From consideration of the MRN we expect other dynamic properties to be similarly affected. So the viscosity will also increase if the ease of shear along NBO lines decreases. By contrast, though, static properties like the physical density which

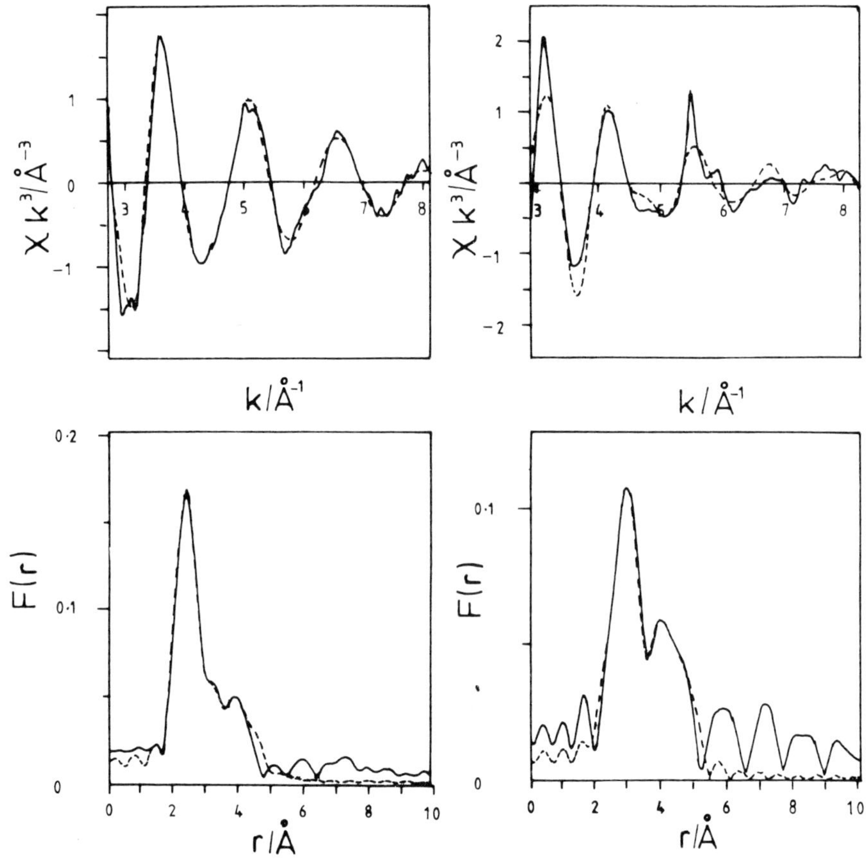

Fig. 4. Weighted EXAFS and Fourier transforms from the K K-edge (left) and the Cs L_{III}-edge (right) in K $CsSi_2O_5$ glass. Solid lines refer to experimental data and dotted lines to the theoretical fit. The structural parameters are listed in Table 2.

depend on mean bond lengths and CNs and which as we see are not strongly affected by mixing, are expected to follow Vegard's Law as the composition is changed in agreement with observation.

5. CONCLUSION

The EXAFS results we have presented taken in conjunction with the ideas introduced in the MRN model provide the first tangible evidence, at least in silicate glasses, that the Mixed Alkali Effect finds its origins in structural constraints.

ACKNOWLEDGEMENTS

SERC is thanked for financial assistance. The director and staff of the Synchrotron Radiation Source at Daresbury Laboratory are thanked for the provision of EXAFS and computing facilities.

REFERENCES

Binsted N, Greaves G N and Henderson C M B 1985 Contrib. to Mineral Petrol. **89** 103
Day D E 1976 J. Non-Cryst. Solids **21** 343
Greaves G N 1985 J. Non-Cryst. Solids **71** 203
Greaves G N 1990 Glass Science and Technology ed D R Uhlmann and N J Kreidl (London: Academic Press) p1
Greaves G N, Fontaine A, Lagarde P, Raoux D and Gurman S J 1981 Nature **293** 611
Greaves G N, Gurman S J, Gladden L F, Spence C A, Cox P, Sales B C, Boatner L A and Jenkins R N 1988 Phil. Mag. **58** 271
Gurman S J, Binsted N and Ross I 1986 J. Phys. C **19** 1845
Gurman S J 1990 J. Non-Cryst. Solids (in press)
Wells A F, Structural Inorganic Chemistry 1975 (Oxford: Clarendon Press) p259
Zachariasen W A 1932 J. Am. Chem. Soc. **54** 3841

Ceramic and glass ceramic superconducting thick films

L.A. Blunt, M.E. Yakinci, D. Holland, J.S. Abell*
T. Shields* F. Wellhofer.*

Department of Physics, University of Warwick U.K.
* School of Metallurgy, University of Birmingham U.K.

ABSTRACT: Fabrication of thick films of both YBCO and BISCCO superconductor on zirconia and alumina substrates is described. Their suitability for application is assessed in terms of electrical and physical properties. It was found that, for both materials, the choice of substrate played a critical role in fabrication due to chemical interaction. The best results for both materials were obtained when using a 3% yttria stablised zirconia substrate. Microstructural aspects of both materials are discussed.

1. INTRODUCTION

The relatively recent discovery of high temperature superconductors has led to a massive increase in research into these new materials, refinement of their properties and fabrication techniques. The $YBa_2Cu_3O_x$ and $Bi_2Sr_2Ca_2Cu_3O_x$ ceramic materials have received most attention with transition temperatures of 90K and 110K respectively being obtained for bulk specimens. The highest transition temperatures are found for the yttrium compound where the atomic ratio is $YBa_2Cu_3O_x$ The bismuth compound appears to consist of a homologous series $Bi_2(SrCa)_{n+1}Cu_nO_x$ 1,2] where the value of Tc increases with n. The n=3 phase has been prepared with a Tc of 110K.

To take advantage of these materials for use in electronic devices it is essential that they can be produced in wire form and as printed circuits. The printed circuit form of the superconductor is usually produced as a thin film (up to 2um thick) or thick film mounted on a number of substrates. The present study details an investigation into the properties of superconducting thick films produced by the screen printing method.

The formation process of a ceramic thick film resembles the bulk formation process but since the properties of the film are greatly dependant on the characteristics of the substrate, as in thin films, this thick film printing method is regarded as an intermediate process between those for a thin film and bulk material.

Table 1. fabrication conditions and electrical properties of YBCO and BISCCO tracks, temperature in °C.

Material	Substrate	Heat Treat. temp	time	Adhesion	thickness	Tc/K onset	zero
YBCO	3% YSZ	1025	6min	good	40um		90.6*
YBCO	3% YSZ	1025	6min	good	40um		89.5
YBCO	3% YSZ	975	6min	good	40um		88.0
YBCO	Alumina	1025	6min	good	40um		56.0
YBCO	Alumina	975	6min	good	40um		40.0
YBCO	8% YSZ	1050	6min	good	40um	95.9	91.6
YBCO	8% YSZ	1050	6min	good	40um	94.0	87.3
BISCCO	Alumina	All temperatures non-superconducting					
BISCCO	3% YSZ	855	2h	good	45um	92	72
BISCCO	3% YSZ	860	2h	good	45um	72	72
BISCCO	3% YSZ	865	2h	good	45um	96	75
BISCCO	3% YSZ	870	2h	good	45um	93	74
BISCCO	3% YSZ	875	2h	good	45um	93	76
BISCCO	3% YSZ	880	2h	good	45um	92	70
BISCCO	3% YSZ	890	2h	good	45um	87	45
BISCCO	3% YSZ**	855	2h	poor	45um	75	50
BIPSCCO	Alumina	All temperatures non-superconducting					
BIPSCCO	3% YSZ	845	2h	good	20um	95	56
BIPSCCO	3% YSZ	860	2h	good	20um	90	35
BIPSCCO	3% YSZ	875	2h	good	20um	85	25
BIPSCCO	3% YSZ	890	2h	good	20um	120	90 62***
BIPSCCO	3% YSZ	900	1h	good	20um	118	85 65***

* J_c = 750 A/cm^2 at 77K.
** polished 3% YSZ substrate.
*** Intermediate transition point.

In the present study the thick films were produced on various substrates using the screen printing technique. This technique is well established within the electronic device industry and hence an application of this standard technology to superconducting devices would clearly be an advantage. However certain problems have arisen with this approach mainly due to the interaction between the superconducting material and commercially available substrates resulting in properties to inferior to those of bulk material.

2. EXPERIMENTAL

The YBCO powders were prepared by solid state reaction of commercially available Y_2O_3 CuO and $BaCO_3$ at 950°C for 24 hours and slow cooling at 2°C/min. The powders were subsequently ball milled for 2 hours. The ink for screen printing was prepared by ultrasonic mixing of a constant weight ratio of powder with Blythe 63/2 organic binder. The screen printing was carried out through a stainless steel 325 mesh with a simple pattern of tracks of varying sizes, the track thickness was controlled by varying the number of passes. Controlled heating rates of 5°C/min, designed to burn off the organics, were followed by sintering at 975°C, 1025°C and 1050°C for 6 min in flowing O_2 followed by cooling at 2°C/min .

The BISCCO tracks were produced using the glass ceramic route. Initially Bi_2O_3, $SrCO_3$, $CaCO_3$, and CuO powders all of reagent grade were tumble mixed and melted in alumina crucibles for 1.5 hours. The resulting melt was then rapidly quenched between two graphite coated steel plates to produce a glass [3]. The resulting glass was then ground until able to pass through a 38um sieve. The glass powder was then screen printed in the method described above. The subsequent heat treatment of the tracks, carried out in order to promote devitrification consisted of a heating rate of 20°C/min to 840-900°C and a dwell of 2 hours followed by a cooling rate of 20°C/min to room temp. Electrical properties were determined by four point probe resistivity measurements. The current carrying capacities of the YBCO specimens was measured using a pulsed four point probe method [4].

3. RESULTS AND DISCUSSION

The main results and fabrication variations for both the YBCO and BISCCO specimens are summarised in Table 1. A comparison of the resistivity results for the YBCO specimens showed that the higher annealing temperatures consistently produced higher Tc zero values. This may be due to a higher degree of intergrain connectivity due to partial melting at the higher temperatures. The highest Tc0 result of 91.6K was obtained using a heat treatment temperature of 1050°C and a substrate of 8% YSZ, figure 1.

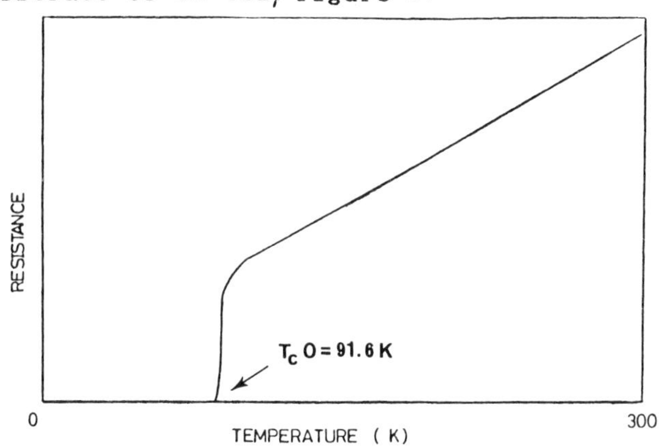

Fig. 1. Resistivity trace of YBCO track on 3% YSZ sintered at 1050°C.

X-ray analysis of the tracks showed preferential grain orientation along the c-axis direction and this was especially evident at the higher annealing temperatures Figure 2. Here it was noted that the c-axis direction was normal to the plane of the track surface. Metallographic cross-sections taken through the track and substrate and examination using an SEM showed that there was clearly an

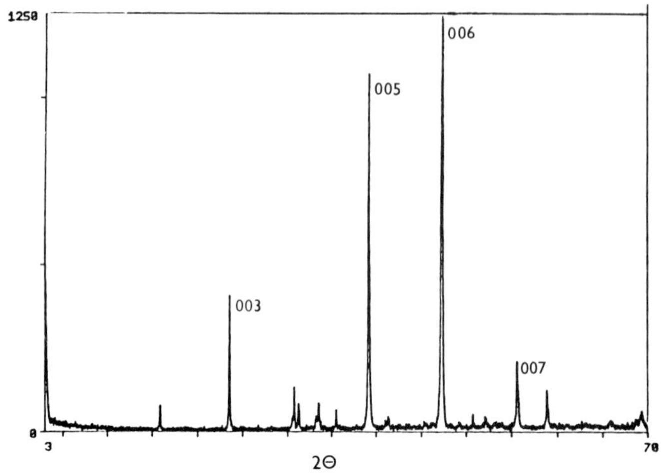

Fig. 2. XRD trace of YBCO track on 3% YSZ sintered at 1050°C, shows c-axis orientation.

interfacial layer between the substrate and superconducting material Figure 3. EDAX analysis together with XRD. data showed the layer to be $BaZrO_3$. The other phases visible were CuO phases. The significance of the elongated grain structure has been discussed earlier [5]. The YBCO tracks produced on alumina substrates had noticably inferior properties and showed evidence of substantial interfacial reactions probably resulting in $BaAlO_3$ type impurities throughout the track and these had a deleterious effect on the superconducting properties. SEM examination of the surface of the track is shown in figure 4a examination showed the structure to be granular in nature with the grains being interconnected by partial melting. At the highest annealing temperatures almost complete melting has occurred with the consequent reduction in inter-grain porosity.

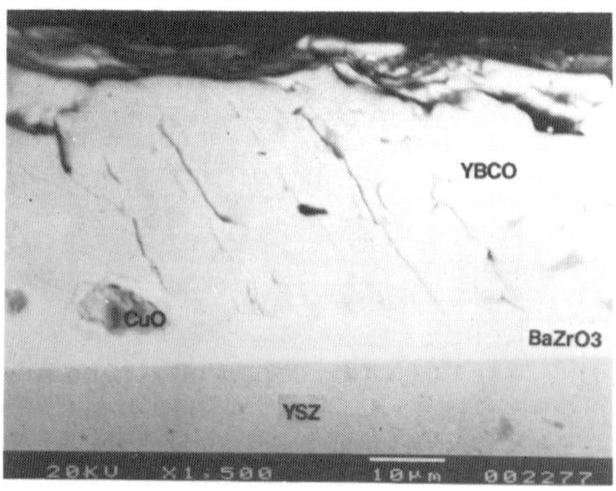

Fig. 3. SEM micrograph cross-section of YBCO track on 3% YSZ showing interlayer.

Resistivity results for the BISCCO thick films, Table 1 and figure 5 also showed the significance of the fabrication conditions. Here it was found that an optimum annealing temperature of 865°C for 2h produced the highest Tc onset temperatures. The highest annealing temperatures produced the poorest resistivity results with especially low temperatures for Tc zero. The specimens printed on the polished substrate showed very poor adhesion to the substrate indicatinng virtually no interaction, when the resistivity of this track was measured it gave a significantly lower Tc-onset and Tc zero than the sample printed on the as recieved substrate and annealed at the same temperature. All the specimens printed on the alumina substrate showed no superconducting properties.

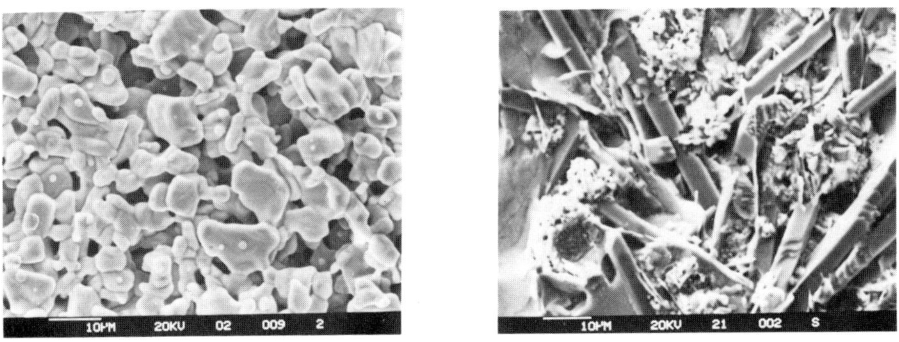

a) b)

Fig. 4. a) SEM micrograph of the surface of the YBCO track on 3% YSZ sintered at 975°C. b) SEM micrograph of the surface of the BISCCO track annealed at 865°C.

XRD. analysis indicated the presence of n=3 phases in all the specimens, Figure 6, and the specimens with the highest Tc values showed increasing amounts of the n=3 high Tc phase present within the specimen. No discernable grain orientation was detected within these specimens. Metallographic cross-sections of the specimens and substrates viewed using the SEM revealed a microstructure consisting of needle like grains of superconducting material, CuO phases and a clearly defined interlayer, figure 7. EDAX analysis showed that the needle phases were closely related in composition to the n=1,2 or 3 superconducting phases and that the interlayer consisted of $(Sr_{0.66}Ca_{0.33})Zr_3O_x$. As the anealing temperature increased it appeared that the thickness of the interlayer increased slightly therefore removing more material from the superconducting track and affecting its properties, however it should be noted that this was not confirmed due variation

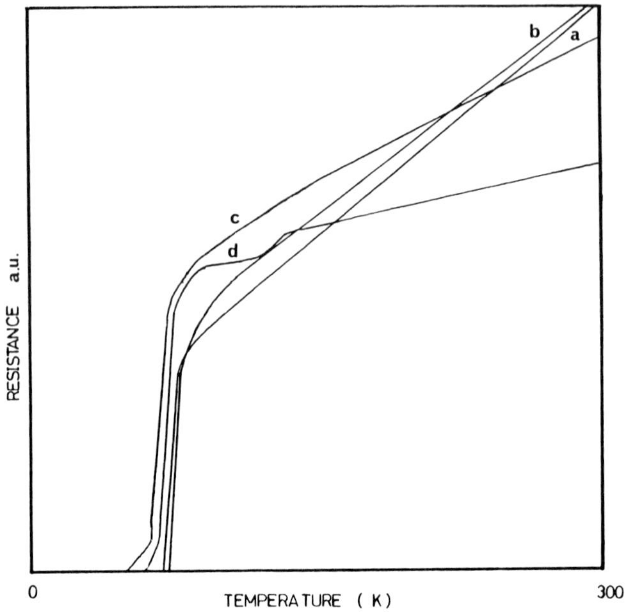

Fig. 5. Resistivity traces of BISCCO tracks, a) BISCCO track annealed at 855°C, b) BISCCO track annealed at 865°C, c) BIPSCCO track annealed at 845° C, d) BIPSCCO track annealed at 890°C.

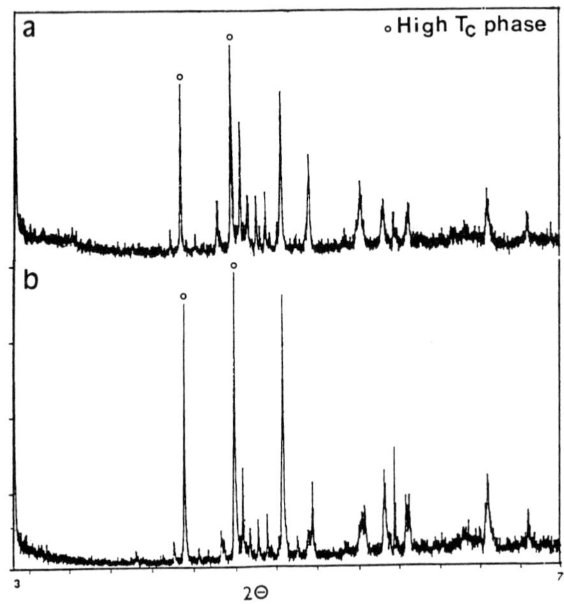

Fig. 6. XRD Traces of BISCCO tracks a) annealed at 855°C, b) annealed at 865°C

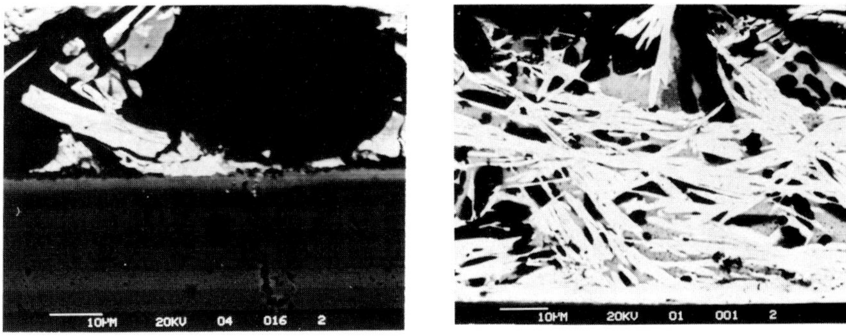

Fig. 7. SEM micrographs of cross section of BISCCO track on 3% YSZ, annealed at 865°C

and to inherent difficulties in measurment of this thin layer. Cross sections taken through the tracks produced on alumina substrates showed evidence of substantial interaction and interdiffusion resulting in large amounts of an alumina rich phase which appears to have destroyed the superconductive properties. It was thought that the Sr/Al reaction plays a significant role for the alumina substrates. Interdiffusion of Sr^{2+} and the Al^{3+} and evaporation of the Bi_2O_3 changes the composition of the films giving a Al-Cu rich structure which occurs throughout the track width thus destroying the superconducting properties. Surface examination of the tracks showed a tangled matrix of acicular superconducting grains inter-spaced with smaller granular grains of CuO and Sr/Al phases, figure 4b.

Other tracks were produced by substituting Pb for Bi at a ratio where the composition of the original glass was $Bi_{1.6}Pb_{0.4}Sr_2Ca_2Cu_3O_x$. These specimens were printed on 3% YSZ and following annealing they showed enhanced Tc onsets when annealed in the high temperature range, however they also showed a much lower Tc0 temperatures than the undoped specimens. The specimens annealed at 890°C and 900°C showed Tc onset temperatures characteristic of the n=3 phase also these specimens had intermediate transitions at 98K and 85K respectively. The specimen annealed at 900°C was anealed for the reduced time of 1h in order to reduce partial melting and consequent interaction with the substrate. Once again the specimens printed on alumina substrates showed no evidence of superconductivity and displayed clear indication of large amounts of substrate/track interaction.

4. CONCLUSIONS

Overall the screen printing method of producing superconductiing circuits has been shown to be feasible. However unlike bulk production of superconductors, where annealing temperature and atmosphere are critical, an additional factor is the interaction of the superconducting material and the substrate. Clearly a small amount of interaction is needed to provide the adhesion, however it is clear that this should be controlled in order to reduce the products of the interaction to a minimum and thereby decrease the effect on the superconducting properties.

REFERENCES

1. Rao C.N.R., Ganapathi L., Vijayaraghava R., Ranga Rao G., Murthy K., and Mahan Ram R.A., Physica C, (156),827,(1988).
2. Buckley R.G., Tallon J.L., Brown I.W.M. Presland M.R., Flower N.E., Gilbreed P.W. Bowden M., and Milestone N.B. Physica C (156),629,(1988).
3. Blunt L.A., Holland D.,Yakinci M.E. and Butler M. Proc. Int. Conf. Modern Aspects of Superconductivity. Edited by R. Suryranaryanan., Paris France. (1989).
4. Colclough M. to be published.
5. Abell J.S., Wellhofer F., Taylor K.N.R., Shields T.C., Holland D., Blunt L.A. and Yakinci M.E. Proc. 1st. Int. Conf. European Ceram. Soc. Maastrict Holland.(1989).

Table 1. Fabrication conditions and electrical properties of YBCO and BISCCO tracks

Material	Substrate	Heat Treat.°C temp	time	Adhesion	thickness	Tc/K onset	zero
YBCO	3% YSZ	1025	6min	good	40um		90.6*
YBCO	3% YSZ	1025	6min	good	40um		89.5
YBCO	3% YSZ	975	6min	good	40um		88.0
YBCO	Alumina	1025	6min	good	40um		56.0
YBCO	Alumina	975	6min	good	40um		40.0
YBCO	8% YSZ	1050	6min	good	40um	95.9	91.6
YBCO	8% YSZ	1050	6min	good	40um	94.0	87.3
BISCCO	Alumina	All temperatures non-superconducting					
BISCCO	3% YSZ	855	2h	good	45um	92	72
BISCCO	3% YSZ	860	2h	good	45um	72	72
BISCCO	3% YSZ	865	2h	good	45um	96	75
BISCCO	3% YSZ	870	2h	good	45um	93	74
BISCCO	3% YSZ	875	2h	good	45um	93	76
BISCCO	3% YSZ	880	2h	good	45um	92	70
BISCCO	3% YSZ	890	2h	good	45um	87	45
BISCCO	3% YSZ**	855	2h	poor	45um	75	50
BIPSCCO	Alumina	All temperatures non-superconducting					
BIPSCCO	3% YSZ	845	2h	good	20um	95	56
BIPSCCO	3% YSZ	860	2h	good	20um	90	35
BIPSCCO	3% YSZ	875	2h	good	20um	85	25
BIPSCCO	3% YSZ	890	2h	good	20um	120 90	62***
BIPSCCO	3% YSZ	900	1h	good	20um	118 85	65***

* Jc = 750 A/cm at 77K.
** polished 3% YSZ substrate.
*** Intermediate transition point.

Sialon U-phase: A new crystalline grain-boundary phase for β'-sialon ceramics

J. Grins and P.-O. Käll

Arrhenius Laboratory, Department of Inorganic Chemistry,
University of Stockholm, S-106 91 Stockholm, Sweden

and

K. Liddell, P. Korgul and D.P. Thompson

Wolfson Laboratory, Materials Division, Department of Mechanical, Materials and Manufacturing Engineering, University of Newcastle upon Tyne, U.K.

ABSTRACT: Sialon U-phase can be crystallized by low temperature (1000-1400°C) heat-treatment of grain-boundary glasses in β'-sialon compositions densified with yttria or rare earth oxide additions. U-phase is trigonal, Space Group P321, with typical unit cell dimensions a = 7.9Å, c = 4.9Å and a general composition of the type $Ln_3(Si,Al)_6(O,N)_{14}$ with the Al : Si ratio just greater than unity and between one and two nitrogen atoms per formula unit. The crystal structure has been determined from X-ray powder diffraction data using a Rietveld refinement program.

1. INTRODUCTION

The technology associated with the production of fully-dense nitrogen ceramics is now well established and the challenge facing researchers is to improve performance particularly as regards the retention of strength and toughness at high (>1300°C) temperatures. The fall-off in strength at 800-1000°C associated with grain boundary glass can be substantially avoided if the glass is thoroughly devitrified into crystalline oxynitride second phases, and the limit of refractoriness (and the associated fall-off in mechanical properties) is then determined by the eutectic temperature in the relevant region of the M-Si-Al-O-N system concerned. It is generally believed that compared with silicon nitride, an added advantage of a sialon starting composition (besides easier densification) is that on devitrification of the grain-boundary glass, there is a larger choice of possible crystalline grain-boundary phases. However, experience during the last 15 years has shown that in all current metal sialon systems, the number of 5-component crystalline oxynitrides is disappointingly small, and many of these are only stable at low ($\not>$1400°C) temperatures (Thompson, 1990). It is not surprising therefore that oxides, with their good oxidation resistance and high melting temperatures, often exhibit the best performance, and the β'-YAG($Y_3Al_5O_{12}$) combination, first developed some 12 years ago, still offers better strength retention at high (>1300°C) temperatures than other possible matrix-grain boundary phase combinations. However, recent research has shown (Thompson, 1990) that other crystalline phases do exist in metal sialon

systems, and the field is still wide open for exploration, certainly in more complex (e.g. 2 metal additive) sialon systems. From phase diagram considerations, it is expected that an increased number of components would lower eutectic temperatures, but this disadvantage can be offset if the crystalline phases produced have high melting temperatures. However, if the disparity in composition between matrix and grain-boundary phase is large, there is always a low eutectic between them, and for example, the β'-YAG eutectic of ~1300°C is not untypical, despite the high melting point (~1930°C) of YAG.
The present paper provides crystallographic, microstructural and some behavioural information on the sialon U-phase first reported 3 years ago (Spacie et al. 1988). This phase occurs in most Ln-sialon systems and also shows some stability in the yttrium sialon system.

2. EXPERIMENTAL

The initial exploration of sialon U-phase was carried out on samples prepared in the β'-YAG plane of the yttrium sialon system and in the β'-$CeAlO_3$ and β'-$NdAlO_3$ planes of the Ce- and Nd- sialon systems respectively (Spacie et al. 1988). Further preparative work was therefore continued along these lines, and for this purpose, mixtures of Si_3N_4*, Al_2O_3+, AlNø and either Y_2O_3¢, La_2O_3&, $CeO_2$$ or Nd_2O_3£ were mixed together in proportions corresponding to a wider range of compositions on the planes discussed above. Powders were ball-milled for 72 hours in isopropanol in rubber-lined mills using sialon grinding media and then uniaxially compacted into cylinders of mass ~3g. These were placed in BN-lined graphite crucibles and fired in a carbon-element furnace at 1700°C in nitrogen, and subsequently heat-treated in nitrogen at 1000-1400°C. Weight changes were noted and product phases identified by X-ray diffraction using Hägg-Guinier focusing cameras and $CuK\alpha_1$ radiation. In the light of these experiments, additional more oxygen-rich compositions were designed, melted and heat-treated to focus more precisely on the correct composition of U-phase. Subsequent microstructural studies were carried out using a JSM35 instrument equipped with EDX facilities. Crystal structure studies were carried out using powder X-ray intensity data collected from a Rigaku diffractometer equipped with a rotating anode (~10 kW, CuKα radiation). For structure refinement the Rietveld DBW program was used.

3. RESULTS AND DISCUSSION

3.1 Preparative studies

Table 1 lists the results of firing a range of compositions in the central region of the β'-$NdAlO_3$ plane of the Nd-Si-Al-O-N system. Clearly, U-phase is not a major product after initial firing, but appears after heat-treatment showing that it is formed by devitrification of glass. Heat-treatment runs were carried out at various temperatures in the range 1000-1400°C but Table 1 shows results for only 1100°C because this was found to be the optimum heat-treatment temperature for Nd U-phase. The results show very clearly that it is impossible to obtain pure U on this plane and that more oxygen-rich compositions are needed. Recently, Fernie et al. (1989) have described preparative studies on neodymium U-phase, concluding that the metal atoms

* Starck-Berlin, Grade LC10
\+ Alcoa, Grade A17
ø Starck-Berlin, Grade A
¢ Rare earth Products, 99.9%
& Koch-Light, 99.9%
$ Rare Earth Products, 99.9%
£ Rare Earth Products, 99.9%

Table 1. Preparative results for compositions fired on or close to the β'-$NdAlO_3$ plane

No.	Nd	Si	Al	O	N	T(°C)	t(h)	U	$NdAlO_3$	β'	15R	β-Al_2O_3	Others
1.	1.5	3.0	5.5	9.0	5.0	1700	1.7	–	w	–	w	w	
						1100	16	vs	vvw	vvw	vw	–	
2.	2.0	3.0	5.0	9.0	5.0	1600	1.0	–	vs	–	vvw	vw	
						1100	70	s	vs	–	vvw	vw	
3.	2.8	2.1	4.9	10.5	3.5	1710	1.0	–	vw	–	–	–	
						1100	16	s	s	vw	–	–	
4.	1.3	2.5	6.4	8.9	5.1	1710	1.0	–	vw	–	–	–	
						1100	16	vs	–	vvw	vw	vw	
5.	1.3	3.8	5.1	7.6	6.5	1720	1.0	–	vw	–	mw	vvw	12H(mw)
						1050	48	ms	–	w	vw	–	$NdSiO_2N$(w)
6.	1.3	1.3	7.6	10.2	3.8	1700	1.0	–	s	–	–	m	12H(m)
						1050	48	mw	s	–	–	m	$NdSiO_2N$(w)
7.	1.5	6.0	2.5	6.0	8.0	1700	1.2	–	–	w	–	–	
						1050	22	s	–	mw	–	–	$NdSiO_2N$(s)
8.	2.8	4.2	2.8	8.4	5.6	1700	0.5	–	–	vw	–	–	Mel (w)
						1050	22	m	–	–	–	–	$NdSiO_2N$(m) Mel(w)

s = strong; m = medium; w = weak; Mel = Nd-melilite($Nd_2Si_3O_3N_4$)

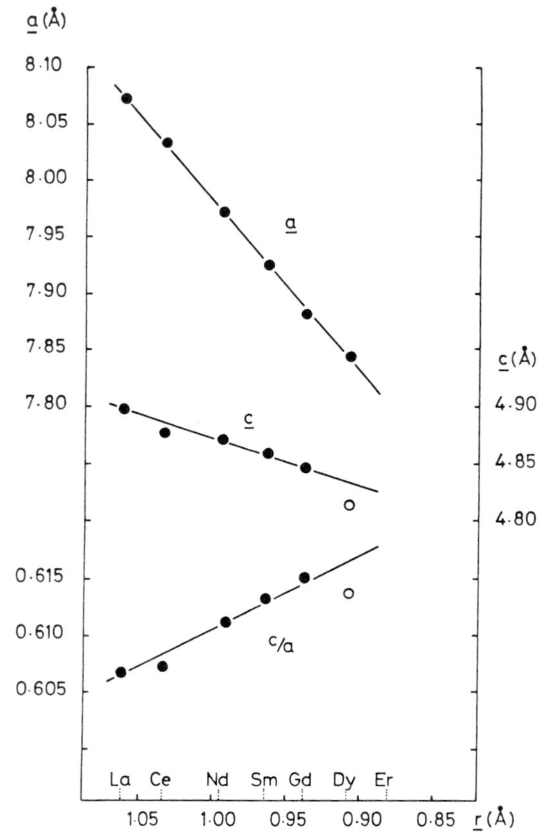

Figure 1. Unit cell dimensions of Ln-U-phases as a function of Ln ionic radius.

are in the ratio 1 : 1 : 1, and suggesting $Nd_3Si_3Al_3O_{12}N_2$ as the most likely composition. This is consistent with evidence obtained in the present study from EDX analysis and also from X-ray diffraction studies described below. Compositions of the type $Ln_3Si_{3-x}Al_{3+x}O_{12+x}N_{2-x}$ with x = 0,½,1 were made up from binary oxides and nitrides as described above and fired at 1600-1750°C. Glasses were readily formed, and at higher melting temperatures, nitrogen was lost, forming large amounts of $LnAlO_3$ in the products whereas at lower temperatures (1600-1650°C), essentially single-phase glasses were obtained after melting, with only small amounts of $LnAlO_3$ or the apatite phase $Ln_5(SiO_4)_3N$. These samples were heat-treated at 1000-1400°C and crystallized to give ~85% of U-phase with no other crystalline phases present apart from those in the as-melted glass. Clearly U-phase is compositionally close to the formula quoted above.

3.2 Unit cell dimensions

Although the original work (Spacie et al., 1988) described the indexing of the X-ray diffraction pattern in terms of a cubic unit cell, Fernie et al. (1989) showed from electron diffraction evidence that U-phase has a hexagonal unit cell, and this has been confirmed in the present work. The specimens used in the original study contained only small ($\not>$20%) amounts of U-phase and the X-ray diffraction lines were broadened because of the small crystallite sizes developed when U-phase is devitrified out of grain-boundary glasses on direct cooling from firing temperatures. When better quality samples were prepared, it became clear that some lines were broader than others and in some cases, resolution into doublets became more marked, confirming the non-cubic symmetry. Moreover, even despite the invoking of the body-centered absence rule, lines occurred at forbidden $h^2 + k^2 + l^2$ values (28, 60) which could not be accounted for by impurity phases. From careful measurements of diffraction lines, the same trigonal unit cell reported by Fernie et al. (1989) was independently deduced with lattice parameters ranging from **a** = 7.967Å, **c** = 4.869Å to **a** = 8.073Å, **c** = 4.894Å for the Nd- and La- U-phases respectively. The apparent pseudocubic nature of the powder X-ray diffraction pattern arises because of the almost exact equality of the **c**/**a** ratio with $\sqrt{3/8}$. Table 2 shows precise indexed X-ray diffraction data for La-, Ce- and Nd- U-phases. Similar preparative work was also carried out in other rare earth sialon systems and unit cell dimensions are listed in Table 3. The variation with ionic radius of the rare earth cation is shown in Figure 1 and follows the expected trend. In all these systems there is a small but significant variation in unit cell dimensions as a function of U-phase composition.

3.3 Microstructure/Composition

U-phase crystallizes from oxygen-rich rare earth sialon glasses of approximately equal Ln : Si : Al atomic proportions when heat-treated in the temperature range 1000-1400°C. Grain size increases with temperature of crystallization and Figure 2 shows microstructures for two cerium U-phases crystallized at 1150 and 1310°C respectively. The small dark regions are either pores or trace amounts of 15R sialon and the intermediate grey regions more prominent in the 1310°C micrograph are $CeAlO_3$. Fernie et al. (1989) discussed EDX analysis of Nd U-phase samples and concluded that contrary to their previous observations of a 4 : 3 : 3 Nd : Si : Al atomic ratio, a more precise estimate was 3 : 3 : 3. Several of the grains in Figures 2(a) and 2(b) were analyzed by EDX and gave a range of Ln : Si : Al ratios varying between 3 : 3 : 3 and 3 : 2.5 : 3.5. A consistent feature of the observations was always the slightly higher atomic proportions of aluminium

Table 2. Indexed X-ray diffraction data for La-, Ce- and Nd- U-phase

La				Ce				Nd			
Hexagonal: a = 8.073 Å, c = 4.897 Å.				Hexagonal: a = 8.033 Å, c = 4.878 Å				Hexagonal: a = 7.971 Å, c = 4.872 Å			
hkl	d_{calc}	d_{obs}	I_{obs}	hkl	d_{calc}	d_{obs}	I_{obs}	hkl	d_{calc}	d_{obs}	I_{obs}
100	6.991	6.990	w	100	6.957	6.963	w	100	6.903	6.896	w
001	4.897	4.894	m	001	4.878	4.881	m	001	4.872	4.864	s
110	4.036	4.037	m	110	4.017	4.019	m	110	3.986	3.983	s
101	4.011	4.012	vvw	101	3.994	3.996	vvw	101	3.980		
200	3.496	3.498	mw	200	3.478	3.477	mw	200	3.542	3.541	ms
111	3.115	3.114	vs	111	3.101	3.102	vs	111	3.085	3.086	vvs
201	2.845	2.845	vs	201	2.832	2.832	vs	201	2.816	2.816	vs
210	2.642	2.643	m	210	2.629	2.629	m	210	2.609	2.609	ms
002	2.448	2.448	w	002	2.439	2.439	w	002	2.436	2.435	m
300	2.330	2.330	w	300	2.319	2.319	m	300	2.301	2.299	ms
211	2.325	2.326	mw	211	2.315			211	2.300		
102	2.311	2.311	vw	102	2.301	2.302	vw	102	2.297		
301	2.104	2.104	vw	301	2.094	2.095	vw	301	2.081	2.078	ms
112	2.093	2.093	m	112	2.085	2.085	m	112	2.078		
220	2.018	2.018	w	220	2.008	2.008	w	220	1.993	1.990	ms
202	2.005	2.006	m	202	1.997	1.997	m	202	1.990		
310	1.939	1.939	ms	310	1.929	1.930	ms	310	1.915	1.914	s
221	1.866	1.867	vw	221	1.857	1.858	vw	221	1.844	1.846	m
311	1.803	1.803	ms	311	1.794	1.794	ms	311	1.782	1.781	vs
212	1.796	1.796	m	212	1.788	1.788	m	212	1.781		
400	1.748	–	–	400	1.739	–	–	400	1.726	–	–
302	1.688	1.687	mw	302	1.681	1.680	mw	302	1.673	1.672	mw
401	1.646	1.647	vvw	401	1.638	–	–	401	1.627	1.623	vw
003	1.632	1.632	vw	003	1.626	1.626	vw	003	1.624		
320	1.604	1.604	w	320	1.596	1.596	w	320	1.584	1.584	w
103	1.589	1.589	vvw	103	1.583	1.583	vvw	103	1.581		
222	1.557	1.558	vw	222	1.550	1.550	vw	222	1.542	1.543	w
.	.	.	.	.	.	.	.	.	.	.	.

Table 3. Unit cell dimensions of some rare earth sialon U-phases

Lanthanide	Unit cell dimensions (Å) a	c
La	8.073	4.897
Ce	8.033	4.878
Nd	7.971	4.872
Sm	7.925	4.859
Gd	7.882	4.847
Dy	7.842	4.812

compared with silicon.

3.4 Crystal structure

The X-ray pattern showed no systematic absences, consistent with a wide range of possible hexagonal and trigonal space groups. Literature search to identify compounds with similar unit cell dimensions yielded 2 groups of compounds with similar stoichiometries, namely rare earth sulphides of the type $Ln_6(Si,Al)_4S_{14}$ (De Saint-Giniez, Laruelle and Flahaut, 1968) and a series of gallogermanates of general formula $Ln_3Ga_5GeO_{14}$ or $Ca_3Ga_2Ge_4O_{14}$ (Belokeneva and Belov, 1981; Mill, Butashin, Khadzhabagyan, Belokeneva and Belov, 1982). The sulphide structure was rejected because the Ln content was much higher than expected either from preparative evidence or from EDX analysis, and also the calculated density for a possible U-phase composition of say $Nd_6Si_2Al_2O_{10}N_4$ (7.6 gcm^{-3}) is considerably higher than the observed values (4.2 - 4.6 gcm^{-3}). In contrast, the gallogermanate composition was consistent with preparative and EDX evidence, and even though the calculated density (5.1 gcm^{-3}) was larger than observed values, this could be explained by the not insignificant amounts of second phases (especially 15R, density 3.4 gcm^{-3}) in the samples measured.

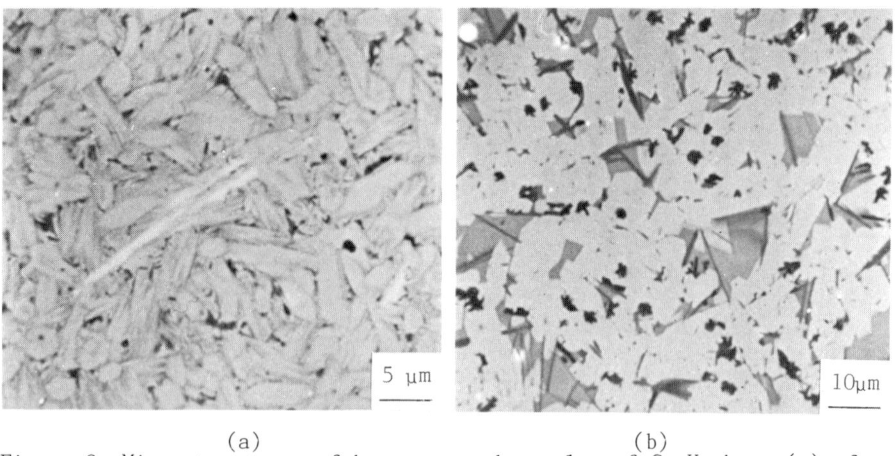

Figure 2. Microstructures of heat-treated samples of Ce U-phase (a) after 17 h at 1150°C in nitrogen and (b) after 5 h at 1310°C in nitrogen.

Table 4. Atomic coordinates and temperature factors for $Nd_3Si_3Al_3O_{12}N_2$. Random occupation of (Al,Si) and (O,N) sites has been assumed. Standard deviations are indicated in brackets after each coordinate.

Atom	N*	x/a	y/b	z/c	B (Å^2)
Nd	3(e)	0.4132(2)	0	0	1.98(4)
Al	1(a)	0	0	0	1.54(15)
(Al,Si)1	2(d)	1/3	2/3	0.5343(34)	1.54(15)
(Al,Si)2	3(f)	0.7603(10)	0	1/2	1.54(15)
(O,N)1	2(d)	1/3	2/3	0.2180(34)	1.51(24)
(O,N)2	6(g)	0.4410(29)	0.2984(14)	0.3252(20)	1.51(24)
(O,N)3	6(g)	0.2219(14)	0.0923(13)	0.7713(20)	1.51(24)

* Positions in Wyckoff notation (Space Group P321, No. 150)

X-ray intensity data were collected for both La- and Nd- U-phase. Despite the restrictions imposed by overlapping reflections and traces of impurity phases, good agreement was obtained between observed and calculated intensities for the gallogermanate model, and the structure refined to reasonable atomic parameters (Table 4). Projections of the structure down [001] and [120] are shown in Figure 3, but a detailed discussion is outside the scope of the present paper and is given in a separate publication (Käll et al. 1990). Clearly the crystal structure determination provides supportive evidence for the correctness of the composition deduced both from preparative evidence, EDX analysis and the previous evidence of Fernie et al. (1988).

3.5 Discussion

Sialon U-phase occurs as a crystalline phase in rare earth sialon systems of composition $Ln_3Si_{3-x}Al_{3+x}O_{12+x}N_{2-x}$ where x varies between 0 and just less

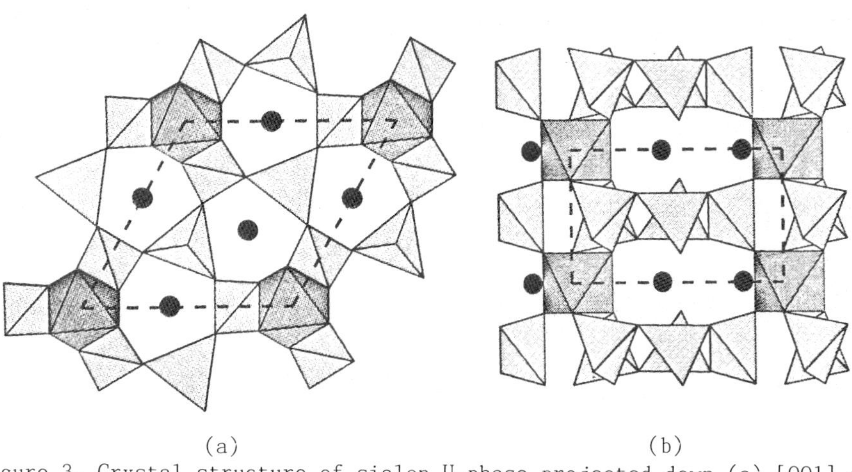

(a) (b)

Figure 3. Crystal structure of sialon U-phase projected down (a) [001] and (b) [120]. Rare earth cations are represented as closed circles.

than 0.5. Ift forms by devitrification of rare earth sialon glasses when heat-treated in the temperature range 1000-1400°C and appears to have a greater stability for lower Z rare earths than for higher ones, the larger cation size appearing to be the dominant factor. The upper limit of refractoriness of U-phase is system-dependent, and accurate estimates of this have not so far been obtained. Preliminary estimates suggest, however, that in the La- and Ce- systems, it is stable up to ~1400°C. Clearly the eutectic temperature for β'-sialon - U-phase ceramics (~1300°C) is no improvement on existing β'-sialon - YAG materials. Whereas the original paper (Spacie et al. 1988) suggested that U-phase was stable in the Y-Si-Al-O-N system, subsequent work has shown that it is difficult to prepare more than trace amounts; certainly low (1000-1100°C) temperatures are needed, and this is consistent with the observation in rare earth systems that smaller cation size has a destabilizing effect on the structure. Some range of Si→Al and O→N substitution is observed within the 3 : 6 : 14 Ln : (Si+Al) : (O+N) ratio, and the requirement of balanced valencies. Incorporation of more than two nitrogen atoms per cell (to achieve greater refractoriness) does not seem possible; if this were so, larger amounts of U-phase (with a greater variation in unit cell dimensions than has been observed) would have been seen in the preparative work.

4. CONCLUSIONS

The crystal structure of sialon U-phase is isomorphous with the rare earth gallogermanates of general formula $Ln_3Ga_5GeO_{14}$ and consists of individual AlO_6 octahedra and 5 $(Si,Al)(O,N)_4$ tetrahedra per unit cell linked together to form a three-dimensional network with the Ln atoms occupying the large interstices. U-phase is of interest from a crystal chemistry viewpoint, but is probably of restricted interest as a secondary crystalline phase for β'-sialon ceramics. These preparative studies show that in M-Si-Al-O-N and 6-component M_1-M_2-Si-Al-O-N systems, there may still be other unexplored crystalline oxynitrides which offer potential as grain-boundary phases in crystalline sialon ceramics.

Acknowledgements

Professors Thommy Ekström and Mats Nygren at the Arrhenius laboratory are thanked for their interest in and comments on this work. The work has been financially supported by the Swedish Board for Technical Development and by AB Sandvik Hard Materials in Stockholm.

REFERENCES

Belokeneva E L and Belov N V 1981 Sov. Phys. Dokl. 26(10) 931.
De Saint-Giniez D, Laruelle P and Flahaut J 1968 C. R. Acad. Sci. Paris 267C 1029.
Fernie J A, Leng-Ward G and Lewis M H 1989 Mat. Sci. Eng. Lett. 9 29.
Käll P-O, Grins J, Liddell K, Korgul P and Thompson D P 1990 "The Crystal Structure of Sialon U-phase" to be published.
Mill B V, Butashin A V, Khadzhabagyan G G, Belokeneva E L and Belov N V 1982 Sov. Phys. Dokl. 27(6) 434.
Perera D S, Korgul P and Thompson D P 1989 Proc. 1st ECerS meeting 'Euro-Ceramics ed G de With, R A Terpstra and R Metselaar (Elsevier Applied Science) pp 1.530 - 1.535.
Spacie C J, Liddell K and Thompson D P 1988 J. Mat. Sci. Lett. 7 95.
Thompson D P 1989 Brit. Ceram. Proc. 45 ed R W Davidge and D P Thompson (Institute of Cermaics) pp 1 - 15.

Inst. Phys. Conf. Ser. No 111
Paper presented at Int. Conf. on New Materials and their Applications, Univ. of Warwick, 1990

Optical characterisation of sol-gel glasses

C.McDonagh*, B.O'Kelly$, K.Devlin*, Z.R.Tang$ and J.McGilp$
*School of Physics, Dublin City University, Glasnevin, Dublin 9, Ireland.
$Department of Pure and Applied Physics, Dublin University, Trinity College, Dublin 2, Ireland.

ABSTRACT: The sol-gel technique has been used to make a range of partially densified glasses, doped with europium. It is well known that the optical properties of the europium ion depend on its immediate environment, in this case the silicate matrix. The europium fluorescence spectrum was monitored during the evolution of the partially densified glass from the initial sol. Changes in the $^5D_0 - ^7F_J$ fluorescence and decay times were interpreted in terms of the changing environment of the europium ion in the silica host. The effect of varying various experimental parameters during the gelation and densification process was investigated.

1. INTRODUCTION

The sol-gel process is a method of preparing glasses and ceramics at low temperatures by hydrolysis and polymerisation of organic precursors. The technique has been used extensively to make improved glasses and ceramic materials for applications as thin films, coatings, fibres, composites and improved bulk materials. There remains, however, much about the process that is not well understood. It is not clear how variables such as choice of precursors, pH of the starting solution and thermal history influence the final glass product. In this study, silica sol-gel glass is doped with the optically active europium ion and the optical properties of the ion are used to probe the structure of the glass at various stages of the process and under different preparation conditions.

Ion spectroscopy of solids has been widely used as a structural tool of both glasses and crystalline solids (Yen 1981, Reisfeld 1973). The free-ion energy levels are modified considerably by the field of the host material. Different hosts provide environments of different symmetry and ligand field strengths, and optical transitions will reflect these changes. In the sol-gel process, the environment of the dopant ion will change between the sol stage and that of the densified glass. In this paper results using trivalent europium as the dopant ion are reported.Europium is easy to incorporate in the trivalent state in glasses, and the f-f optical transitions produce fluorescence in the visible spectrum which is sensitive to the surrounding ligands. Both fluorescence and fluorescence decay times are reported in this work. A set of glasses produced by hydrolysis with water and with deuterium oxide are investigated. High field NMR was performed on a selection of the glasses.

2. EXPERIMENTS

Europium-doped silicate glasses were prepared with tetraethoxyorthosilane(TEOS),water at a preselected pH,ethanol and europium nitrate pentahydrate in the molar ratio 1:4.4:1:0.003. The mixture was stirred magnetically and cured at 72°C - 74°C. The cured gels were subsequently calcined at various temperatures up to 800°C. The deuterated samples were prepared using an identical procedure and were carefully sealed after the appropriate heat treatment. The europium concentration in all samples was 2% of the silicon concentration.

Fluorescence experiments were performed using the experimental arrangement illustrated in Figure 1.The excitation source was a fluorescent lamp coated with a phosphor with peak emission at 370nm. The detector was a Hamamatsu R928 photomultiplier tube. The arrangement for measuring fluorescent decay times is as shown in Figure 2. Light of wavelength 318nm from a frequency-doubled, pulsed, Nd-Yag-pumped dye laser is directed at the sample and the decays are measured using a Stanford

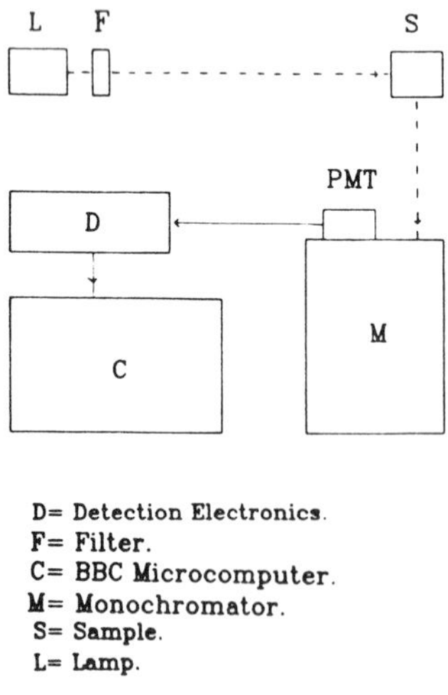

Figure 1. Experimental Arrangement for fluorescence.

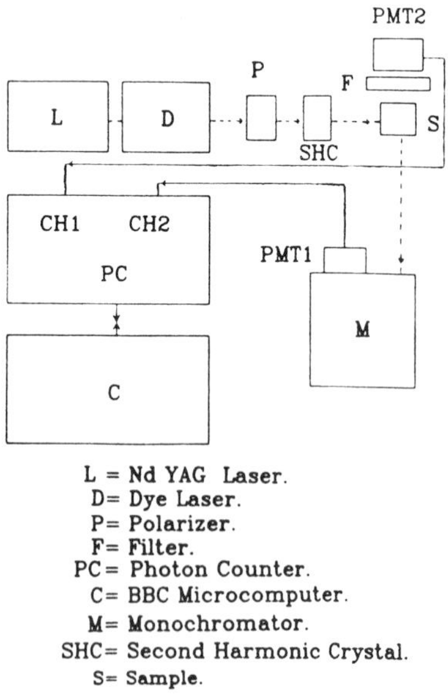

Figure 2. Experimental Arrangement for decay times.

Research SR400 photon counter. The pulse-to-pulse variation of the laser intensity in the lifetime data was corrected by recording the fluorescence of the sample at a fixed delay time in channel B of the photon counter and using this to normalise the signal in channel A. Data acquisition and handling were performed in both these experiments using a BBC Master 128 computer.

^{29}Si Magic Angle Spinning NMR was measured on a Bruker MSL 300 Spectrometer.

3. RESULTS

The fluorescence of Eu^{3+} in pure silica glass prepared by the sol-gel process is illustrated in Figure 3. Spectra a-d represent different stages of the glassification process. Fluorescence decay times corresponding to the above spectra are presented in Table 1. Figure 4 shows the fluorescence of thegeland partially densified glass for hydrolysis by deuterium oxide. Decay times for the D_2O series are also included in Table1. Figure 5 shows an example of the decay time data. The lifetime is calculated from the slope of the log-linear plot. ^{29}Si NMR data are summarised in Figure 6.

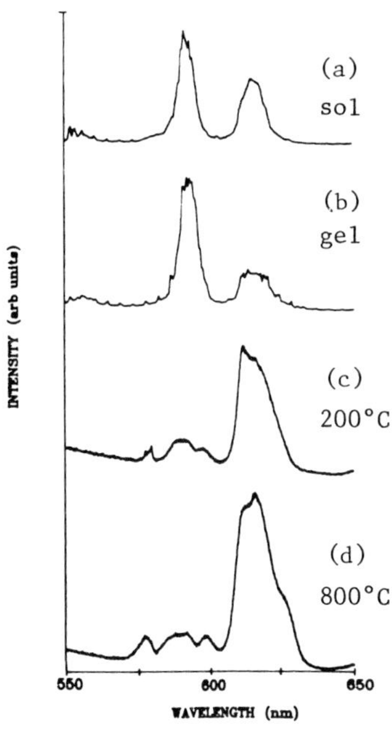

Figure 3. Fluorescence Spectra of H_2O silica samples.

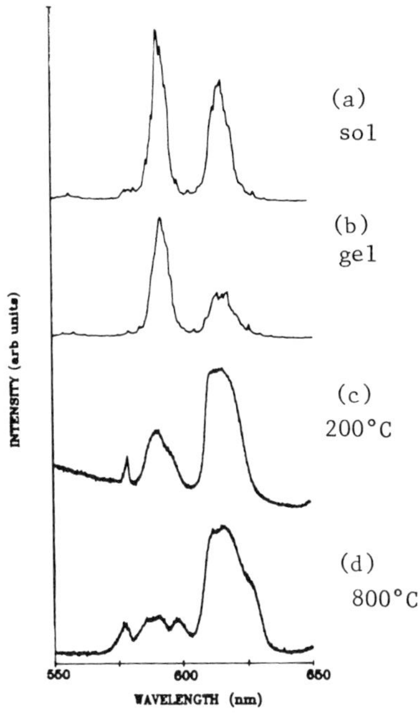

Figure 4. Fluorescence Spectra D_2O silica samples.

4. DISCUSSION

It is clear from the data in Figure 3 that the Eu^{3+} fluorescence transitions are sensitive to their environment. In spectrum (a) of the sol, two of the three relevant europium transitions $^5D_0 - ^7F_0$, $^5D_0 - ^7F_1$, $^5D_0 - ^7F_2$ are present. The shape and position of these two transitions and the general absence of structure on these lines,is consistent with the europium ion in the sol being surrounded by a symmetric solvation shell. As the glass densifies the J-splitting begins to appear,indicating that the degeneracy is being lifted by an asymmetric ligand field. The degeneracy appears to be fully lifted in the 800°C partially densified glass [spectrum (d)]The $^7F_1$ / $^7F_2$ ratio has been used as a measure of the symmetry of the europium surroundings: the higher the ratio the more symmetric the site (Gallagher 1964). In a glassy environment the lower energy $^7F_2$ transition should be considerably more intense than the other two as is indeed observed for spectra (c) and (d).The high energy singlet also appears in (c) and broadens out as the glass densifies (spectrum (d)) to reflect the multiplicity of europium sites which are present in the amorphous solid matrix. Spectrum (b), that of the gel,is anomalous in that the intensity ratio of the two lower energy transitions indicate an increase in the symmetry of the matrix at the gel stage, compared to the sol and the partially densified glass. This result is corroborated by the decay time measurement and is discussed further in the next section.

The fluorescent decay times of europium are even more sensitive to the various stages of the sol-gel process than the fluorescence spectra. A time of 0.132(2)ms,where estimated errors in the last figure are given in parenthesis, was recorded for the decay time of the ion in the silica sol (Table 1). This compares with a lifetime of 0.1ms for the ion in nitrate solution (Carnall 1979).Decays for the 200°C and the 800°C samples were 0.34(2)ms and 1.51(2)ms respectively. Europium in conventional silicate glass has a lifetime of approximately 2.3ms (Brecher et al 1980). This indicates that the microcopic environment of the europium ion changes considerably before it becomes fully densified. The gel sample has a shorter lifetime of 0.117(2)ms which is consistent with the anomalous intensity ratio in Figure 3. A shorter decay time indicates a higher non-radiative decay rate arising from an increase in hydroxyl groups in the immediate vicinity of the europium ion (Phillips 1969). This would produce a more symmetric environment compared with the sol which would explain the anomolous intensity ratio observed. This phenomenon needs further investigation.

The dynamics of the sol-gel-glass transition were investigated further by substituting D_2O for water in the sol.The O-D bond has smaller vibrational quanta than the O-H bond, making non-radiative decay less probable (Mackay 1975). This causes enhancement of the fluorescence intensity and an accompanying increase in the decay time. This isotope effect has been widely used to elucidate solution dynamics in photochemistry (Phillips 1969). Both the fluorescence enhancement and the increase in lifetime have been observed in this study. The D_2O-sol fluorescence intensity increases by a factor of ~5 while the sol lifetime increases from 0.132(2)ms to 0.760(2)ms. The difference in decay rates for H_2O and D_2O environments can be quantitatively related to the number of water molecules associated with the europium ion (Albin et al 1985). Using the expression

$$p = 1.05(1/\tau(H_2O)-1/\tau(D_2O))$$

the number of water molecules p is estimated as ~6 for the sol and ~7 for the gel. Since the coordination number for europium in solution is normally ~9, it would appear that the ion in the sol is surrounded by 6 water molecules and 3 organic groups, while at the gel stage, the number of waters increases to 7. The decrease in lifetime for the H_2O-sol is consistent with this as one extra OH-group in the solvation shell of the ion is sufficient to increase the non-radiative rate and hence decrease the decay time. The fact that a similar decrease in decay time (Table 1) occurs in the D_2O-gel suggests that an OH-group (not-OD) displaces one organic group in the europium shell. The D_2O gel contains less than 5% water making the probability that an OH would bond to the europium ion preferentially very small. There may be a kinetic isotope effect here which requires further investigation. Turning now to the glasses, the 200°C and the 800°C spectra in Figure 4 are similar to those of the undeuterated samples in Figure 3 but the longer lifetime of the 200°C deuterated sample (Table 1) suggests that some OD groups remain in the matrix at this temperature. The 800°C glass has a lifetime which is similar to that of the undeuterated sample, implying that no OD groups exist in the vicinity of the europium ion at this temperature. On the other hand, because the lifetime is still considerably shorter than that of fully densified glass (Table 1) it is clear that non-radiative paths still exist, and it is suggested that hydroxyl groups from the ambient may reappear in the glassy matrix.

Finally, Magic Angle Spinning ^{29}Si NMR (Figure 6) demonstrates that the 200°C glass contains silicon which is coordinated to two, three and four other silicon atoms, via the oxygens, while the 800°C glass contains four-fold coordinated silicon only. These data corroborate the decay time results which suggest that the environment of the europium ion at 200°C is still quite different from a normal glassy environment. The NMR for the 800°C sample indicates that the silicon is fully coordinated but the lifetime data suggest that the environment is still not that of a normal glass.

Table 1 (decay times in ms)

	H_2O	D_2O
Sol	0.132(2)	0.760(2)
Gel	0.117(2)	0.610(2)
200°C	0.34(2)	0.47(2)
800°C	1.51(2)	1.55(2)

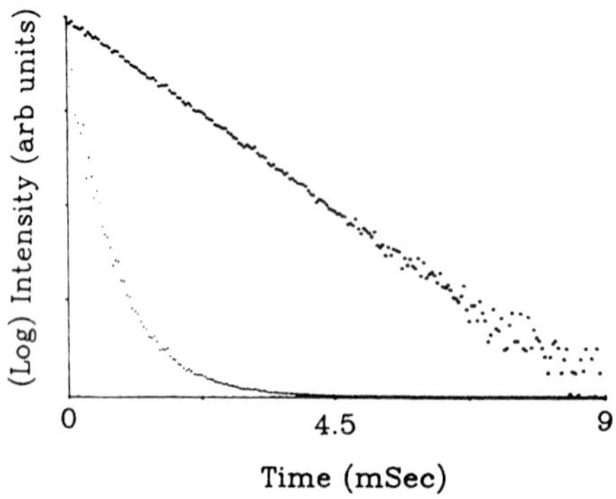

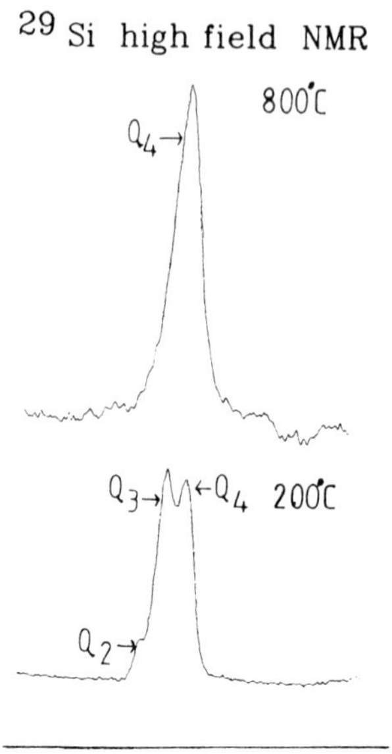

Figure 6 NMR data. ⟶

Figure 5. Decay spectrum

5. CONCLUSION

In this work, optical studies of dopant ions in sol-gels have been shown to have great potential for investigating the sol-gel process. The information obtained using optical techniques complements and enhances that obtained by other methods such as NMR. From this preliminary study it is clear that fluorescence decay times are particularly sensitive to the changing structure of the sol-gel materials.

6. ACKNOWLEDGEMENT

The authors wish to acknowledge Eolas (The Irish Science and Technology Agency) who funded this project.

7. REFERENCES

Albin M. 1985, Inorganic Chemistry, Vol.24, No. 26, p4591

Brecher C.and Riseberg L.A., 1980, J.Non – Cryst.Sol., vol 40 p 469

Carnall W.T., 1979, Handbook on the Physics and Chemistry of Rare-Earths, p171, Ed. Geschneider K.A. North Holland.

Gallagher P.K., 1964, J.Chem.Phys, vol.41, no.10

Mackay K. and Dove M.F., 1975, Comprehensive Inorganic Chemistry, Vol.1 p77 Pergamon.

Phillips D., 1969, Photochemistry Vol.1, p1, ed. The Chemical Society.

Reisfeld R. 1973, Structure and Bonding, Vol.13, p53, Springer Verlag.

Yen W.M. 1981, Laser Spectroscopy of Solids, Springer Verlag.

Inst. Phys. Conf. Ser. No 111
Paper presented at Int. Conf. on New Materials and their Applications, Univ. of Warwick, 1990

Crystallization studies of multicomponent AlF_3 glasses

A B Seddon and A V Cardoso

School of Materials, University of Sheffield, Elmfield, Northumberland Road, Sheffield, S10 2TZ, UK

Glasses based on $AlF_3[BaF_2,CaF_2,MgF_2,SrF_2]YF_3$ have been prepared with additions of LiF up to 13 mol%. The kinetics of devitrification upon reheating the glasses above the glass transformation region have been studied by means of non-isothermal Differential Thermal Analysis and Differential Scanning Calorimetry. The data have been treated by the Augis Bennett method and the results from powdered and particulate samples are compared with those obtained by measuring crystal growth rates via optical microscopy.

1. INTRODUCTION

During the last decade the development of infrared (IR) transmitting glasses based on zirconium/barium fluorides for potentially low loss optical fibre telecommunications has given impetus to the research of related glasses. Many new vitreous halide systems have been identified (Comyns 1989) but equally some older formulations have been revived: one example is multicomponent aluminium trifluoride glasses first patented by Sun in 1949. The high level of aluminium, which has a low relative atomic mass, limits the IR transparency of these glasses and hence their overall intrinsic optical loss will tend to be greater than that of the zirconium/barium fluoride glasses mentioned above, thus short length fibreoptic or bulk glass applications are envisaged.

Aluminium fluoride based glasses tend to devitrify both during cooling from the melt and on subsequent reheating eg for fabrication processes. We have previously reported on minimising the formation of crystallites during casting from the melt of glasses based on $AlF_3[BaF_2,CaF_2,MgF_2,SrF_2]YF_3$ by tailoring both glass composition and procedure (Seddon and Cardoso 1989). For fabrication into optical fibres it is useful to have a measure of the crystal growth rate in the fibre drawing temperature range ie at a glass melt viscosity of about 10^4 Pa s. Kinetic data of crystal growth may be generated by means of isothermal and non-isothermal thermoanalysis of small glass samples where the enthalpy evolved during devitrification is detected and a variety of algorithms may be used to compute the value of the effective activation energy for crystal growth. Alternatively glass specimens may be isothermally heat treated and quenched, and crystal growth observed and measured directly via microscopy.

In this paper we compare kinetic data obtained using both non-isothermal thermoanalysis and direct microscopy to examine the change in glass stability on reheating above Tg (the glass transformation range) as alkali fluoride is added to a base multicomponent AlF_3 glass.

2. EXPERIMENTAL

Glass melting, casting and annealing have been described previously (Seddon and Cardoso 1989). Batches are listed in Table 1 and were designed to retain the molar ratio of constituents in the base AlF_3 glass (batch 1) as alkali metal fluoride is added. BDH "Fluortran" or "Crystran" chemicals were used without further purification. IR spectra of polished glass specimens of 2mm pathlength were collected by means of a Perkin Elmer 680 Spectrophotometer.

TABLE 1 Batch compositions (mol%) of prepared glasses.

Batch	AlF_3	[BaF_2	CaF_2	MgF_2	SrF_2]	YF_3	[LiF	NaF]
1	37.0	12.0	15.0	12.0	9.0	15.0	-	-
2	35.2	11.4	14.3	11.4	8.6	14.3	4.8	-
3	33.7	10.9	13.6	10.9	8.2	13.6	9.1	-
4	33.7	10.9	13.6	10.9	8.2	13.6	-	9.1
5	32.2	10.4	13.1	10.4	7.8	13.1	13.0	-

Glasses were subjected to Differential Thermal Analysis and Differential Scanning Calorimetry using a Stanton Redcroft 671 Analyser and Perkin Elmer DSC7 respectively. Glass specimens were either ground to ≤150 mesh or fragmented into pieces with dimensions between 0.5-3.0 mm. For DTA 0.265g of glass was analysed against calcined alumina in platinum crucibles under flowing nitrogen at various isochronal heating rates in the range 2-20°C min^{-1}. For DSC 10mg of glass was analysed against an empty aluminium pan in dynamic argon at isochronal heating rates between 15-40°C min^{-1}; for each heating rate the baseline flatness was optimised, and temperature and energy calibrations were performed. In some cases measurements were repeated on glasses made from different batches of nominally the same composition.

To obtain crystal growth data isothermal heat treatments under nitrogen were carried out on cuboid pieces of glass which had been cut from the bulk and polished. The temperature during the heat treatment was controlled to better than ±1°C. Samples were subsequently air-quenched and one surface was polished away to reveal the crystal fronts which had grown into the bulk glass from specimen surfaces perpendicular to that removed (see Figure 3). The mean of 20 measurements of the distance from the specimen surface to the crystal front within the specimen interior was taken.

3. RESULTS

Figure 1 shows the similarity of IR spectra of the multicomponent AlF_3 base glass alone and with additions of lithium or sodium fluoride.

3.1 Non-isothermal thermoanalysis

We have collated the non-isothermal DTA or DSC data for several isochronal heating rates (α) for each glass composition and two levels of subdivision. There are a variety of methods available to mathematically treat the data in order to obtain an effective crystallisation activation energy (Ec). These have been critically reviewed by Yinnon and Uhlmann (1982) who advocate the treatment of Augis and Bennett (1978). This consists of an extension to the Kissinger method to encompass solid state reactions obeying an Avrami law (equation 1) for any value of n, the dimensionless exponent:

$$x = 1-\exp\{-(kt)^n\} \qquad \ldots\ldots(1)$$

where x is the amount of material transformed in time t, and k is the effective rate constant defined by an Arrhenius type relationship:

$$k = v\exp\frac{-Ec}{RT} \qquad \ldots\ldots(2)$$

(v is the frequency factor; R is the gas constant; T is absolute temperature).

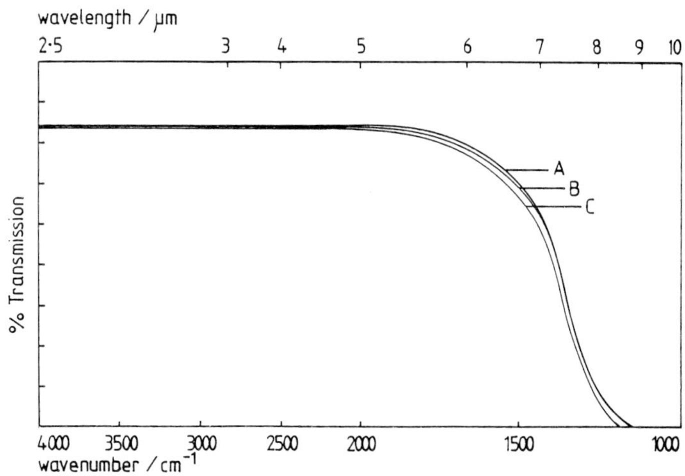

Figure 1. IR spectra of multicomponent AlF_3 glasses containing: A) 0 mol% LiF; B) 9.1 mol% NaF and C) 9.1 mol% LiF. Optical pathlength was 2mm.

Augis and Bennett (1978) showed that for Tp and some lower temperature To:

$$v\exp\left(\frac{-Ec}{RTp}\right) \approx \left(\frac{Tp - To}{\alpha}\right)^{-1} \qquad \ldots\ldots(3)$$

Following equation 3, we plotted $\ln(\alpha/(Tp - To))$ versus $(1/Tp)$ (see example in Figure 2) and Ec values were calculated from the linearly regressed gradients. These are given in Figure 5. To is taken as a temperature lower than Tp at which reaction rates are negligible: we have assumed To = 273K.

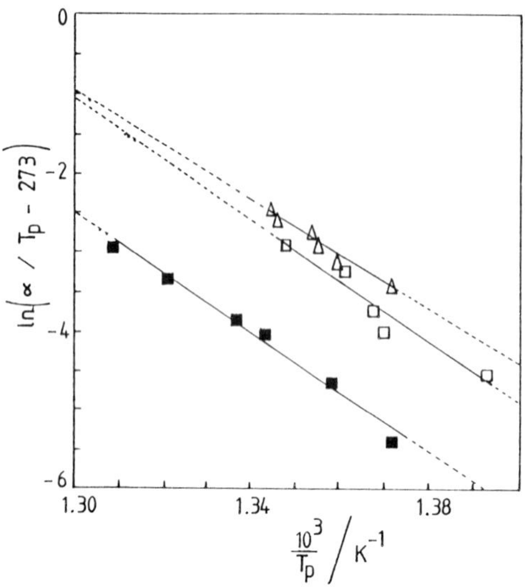

Figure 2. Augis Bennett plot of

$$\ln\left(\frac{\alpha}{Tp-To}\right) \text{ vs } \frac{1}{Tp}$$

(symbols defined in text) for the multicomponent AlF_3 glass containing 13.0 mol% LiF (batch 5). Δ DSC/powder; □ DTA/powder; ■ DTA/fragments.

3.2 Crystal growth measurements of the base AlF_3 glass (batch 1)

Crystal growth was measured from the surface of glass samples (see Figure 3) which had been isothermally heated for various times at different temperatures. An Arrhenius plot of the crystal growth rates is given in Figure 4 and Ec calculated from the gradient is included in Figure 5.

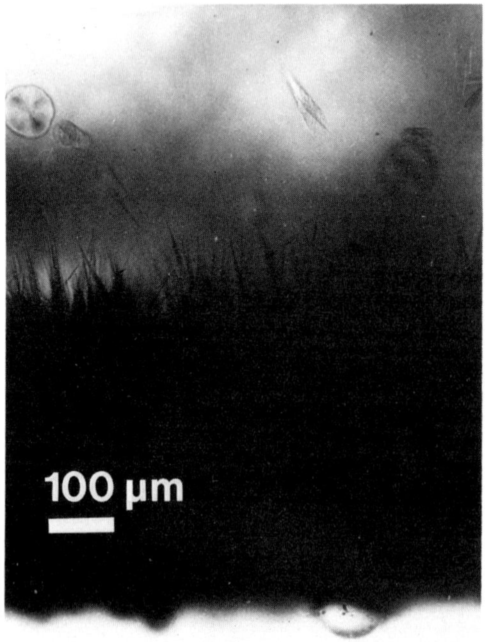

Figure 3. Optical micrograph of crystals in the base AlF_3 glass (batch 1) after heating at 508°C/2h and quenching.

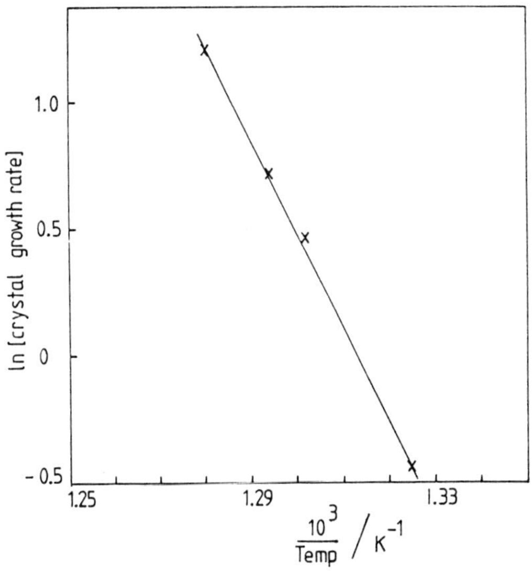

Figure 4. Arrhenius plot of ln (crystal growth rate) vs 1/temperature for the base AlF_3 glass (batch 1) from optical microscopy.

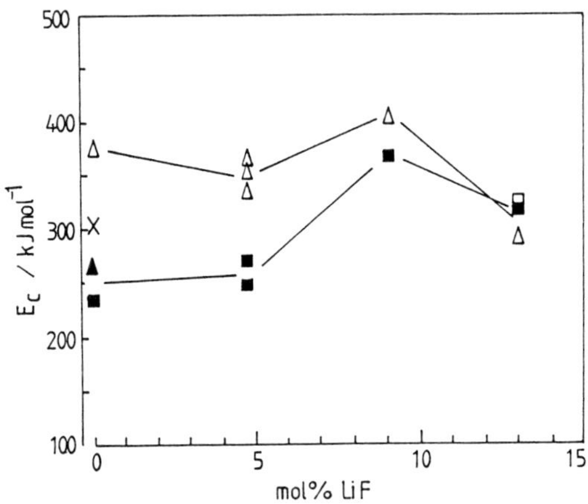

Figure 5. Effective activation energy of crystal growth (Ec) vs LiF concentration for multicomponent AlF_3 glasses determined by: Δ DSC/powder; ▲ DSC/fragments; □ DTA/powder; ■ DTA/fragments; X from optical microscopy.

During the microscopy studies we also observed some volume crystallisation involving at least two different crystal morphologies and these are under investigation. In addition we were able to observe that "as-cast" glass surfaces, including ones which had been in contact with the mould surface, failed to crystallise from the surface.

From XRD of the base AlF_3 composition after heat treatment at 500°C / 2 hours the following phases were identified: β CaF_2,AlF_3; $CaF_2,4YF_3$ and $2BaF_2,MgF_2,2AlF_3$.

4. DISCUSSION

As may be clearly seen from the typical DSC curves of the base glass given in Figure 6, there is a shift to lower temperatures of the onset temperature and peak temperature (Tp/K) of the first crystallisation exotherm above Tg as the level of subdivision of the glass sample is increased. The temperature shift indicates that surface nucleated crystallisation is important and therefore that heterogeneous nucleation is dominating. The observed absence of crystallisation on heat treating "as cast" surfaces of the base AlF_3 glass suggests that these surfaces are intrinsically resistant to heterogeneous nucleation during the temperature/times considered here. Fracturing, grinding and polishing of the glass surfaces may lead to mechanically induced crystal nuclei and also to nuclei resulting from chemical contamination such as oxyhalides. The levels of oxygen in halide glass systems have recently been shown to be critical with regard to glass stability (Mitachi and Tick 1988).

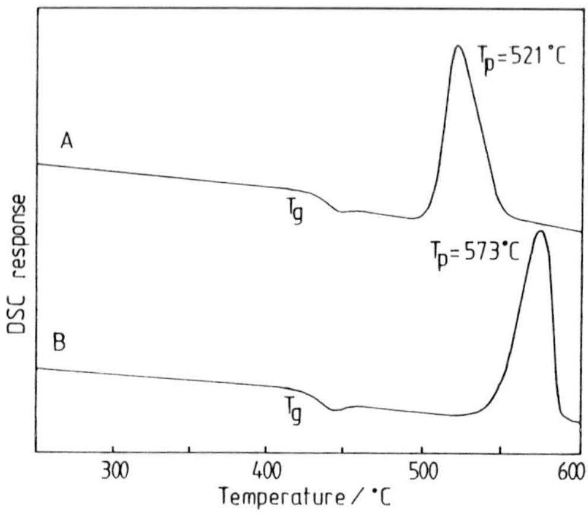

Figure 6. DSC curves of the base AlF_3 glass (batch 1) heated at 20°C min^{-1} for: A) sample ground to ≤ 150 mesh and B) fragments of glass of dimensions 0.5-3mm.

From Figure 5, repeat measurements on different melts of the same composition and subdivision using the same thermoanalytical technique gave Ec values within 10%, cross comparing between DSC and DTA values were within 15%. For the base AlF_3 composition (batch 1) DTA and DSC for fragments of glass gave Ec = 233,268 kJ mol^{-1} respectively. However DSC analysis of the powdered base glass gave a far higher value of Ec = 376 kJ mol^{-1}. A possible explanation for this is that the temperature ranges over which Ec data were collected differ depending on the level of subdivision of the sample: for powdered base glass analysed by DSC, Tp ranges from 518-531°C; whereas for specimen lumps analysed by DSC or DTA, Tp lies between 548-580°C and 523-574°C respectively. The same trend occurs for the glass containing 4.8 mol% LiF.

Implicit in the Augis Bennett method are the assumptions that the nucleation rate is negligible over the temperature range under consideration (site saturation) and that the crystal growth rate has an Arrhenian temperature dependence. The latter arises because the temperature dependence of crystal growth is assumed to be dominated by a melt viscosity which is Arrhenian over a short temperature interval. The viscosity of multicomponent AlF_3 glasses is non-Arrhenian (Lin et al 1988) and this may explain the large change in Ec when the temperature range of data collection is altered. Zirconium/barium fluoride containing glasses containing LiF exhibit a similar trend of increasing activation energy as temperature is decreased (Seddon et al 1990). Ec from crystal growth data for 482-508°C might therefore have been expected to have a higher value than that actually obtained which was 305 kJ mol^{-1}.

For the glass containing 13 mol% LiF, Ec values from fractured and powdered specimens begin to converge as do the temperature ranges of data collection (see Figure 2). These glasses partially devitrify as they are cast from the melt and observation of heat treated specimens via optical microscopy shows that volume crystallisation appears to be favoured.

As LiF is added to the base glass, Ec tends to increase up to 9.1 mol% LiF and thereafter slightly decreases. However Ec for the 13 mol% LiF glass is either lower than the zero LiF mol% glass or higher depending on whether specimens were powdered or not. It is concluded that the level of subdivision of these glasses influences the trends of Ec derived from non-isothermal techniques.

ACKNOWLEDGEMENT

We are grateful to CAPES, Brazil for funding this work.

REFERENCES

Augis J A and Bennett J E, 1978, J Therm Anal 13, 283.

Comyns AE (ed), 1989, Fluoride Glasses, Critical Reports on Applied Chemistry, Vol 27, (Chichester: John Wiley and Sons).

Lin F, Zhange H, Hu H and Gan Fuxi, 1988, Mat Sci Forum, 32-33, 595.

Mitachi S and Tick P A, 1988, Mat Sci Forum, 32-33, 197.

Seddon A B and Cardoso A V, 1989, Proc 6th Int Symp on Halide Glasses, Clausthal, Germany.

Seddon A B, Williams D L and Clare A G, Phys Chem Glasses (in press).

Sun K N, 1949, US Patent 2466509.

Yinnon H and Uhlmann D R, 1983, J Non-Cryst Solids, 54, 253.

Inst. Phys. Conf. Ser. No 111
Paper presented at Int. Conf. on New Materials and their Applications, Univ. of Warwick, 1990

Oxychloride glasses for infra-red transmission

M R Sahar, M M Ahmed and D Holland
Physics Department, University Of Warwick.

ABSTRACT : Heavy metal oxychloride glasses can be prepared with good IR transmission out to 8.8 microns. Scattering losses will prohibit their use for long distance transmission, compared to fluorozirconates, but they have some attractions in terms of improved thermal stability. Tc-Tg gaps of up to 149°C can be achieved and compositions containing less than ~30 at % Cl appear to be stable in air. The thermal and chemical properties of ternary glasses based on the Sb_2O_3-$PbCl_2$-$ZnCl_2$ system are discussed in relation to composition.

1.0 INTRODUCTION

To be a good candidate for IR transmission glass, glass must show good transparency at longer wavelengths. In order to shift the IR absorption edge to longer wavelengths it is necessary to have glass formers which contain heavy elements (high reduced mass), interacting by weak force constants (low field strength and bond strength). Oxide glasses show very strong absorption bands in the IR region mainly caused by fundamental M-O vibrations while most halides show extended IR transmission but are severely attacked in the presence of water. The limiting factor for chalcogenide glasses is the inability to reduce losses to the level required for telecommunication. Oxyhalide glasses may provide a compromise between oxide properties and the halide glass behaviour. However, lack of information on their thermal and chemical properties has restricted interest in these materials. In the present work, ternary glasses based on Sb_2O_3-$PbCl_2$-$ZnCl_2$ have been prepared and their properties are reported.

2.0 EXPERIMENTAL

2.1 Glass Preparation

The appropriate, well - mixed glass constituents (from 99% Sb_2O_3, 99% $PbCl_2$, 99% $ZnCl_2$) weighing about 30 gm in total were put into an alumina crucible with a lid to reduce the losses of volatile components. Then the mixture was melted in air at temperatures ranging from 1000°C - 1200°C depending on composition (the temperature is higher when the Sb_2O_3 content is larger). After about 5 minutes, the crucible was taken out and held for about 1 - 2 minutes and well agitated until it reached the required viscosity. The liquid was then poured into moulds of different shapes to suit various experimental purposes, before being annealed at 250°C for 1 hour and then allowed to cool down to room temperature. The full procedure has been reported elsewhere by Ahmed and Holland (1987). Approximately the same procedure was used for all glass compositions.

2.2 Thermal Analysis

The thermal characteristics of all glasses were determined by DSC (Differential Scanning Calorimetry). Experiments on powder samples were carried out using a Setaram HT 1000 Calorimeter with a heating rate of 0.4 °C/min over a temperature range from 100°C - 500°C.

2.3 Crystallisation studies

The crystallisation studies were carried out by heat treating the glass samples at various crystallisation temperatures Tc for 24 hours in an electric furnace. The crystallised product was then characterised by XRD (X-ray diffraction) as well as EDX (Energy Dispersive X - ray Analysis) for phase identification.

2.4 Crystal Growth rate

A sample of each composition was heat treated at four different temperatures around Tc for four different times ranging from 4 - 20 minutes. The crystal growth was calculated using an averaging method described by Underwood (1970), applied to an electron micrograph.

2.5 Viscosity

The viscosity measurements were carried out using an indentation method at four different temperatures from slightly above Tg to slightly below Tc. The indenter displacement was measured using an LVDT (Linear Voltage Displacement Transducer) and the voltage signal sent to a chart recorder. The viscosity was calculated using the following equation given by Douglas et al (1965);

$$\text{viscosity} = \frac{9Pt}{32\ (2R)^{0.5}\ L^{1.5}}$$

where P is the applied force, t is the time after the indenter is applied to the sample, R is the indenter radius and L is the penetration depth.

2.6.0 Durability

2.6.1 Normal Atmosphere

Samples of about 4 gm in weight were left exposed to normal atmospheric conditions at room temperature, 25°C. After 7 days, the samples were washed with xylene , leaving only the cleaned glass surface. Samples were then dried at 80°C for 2 hours before being re-weighed. Weight loss per unit area was then calculated.

2.6.2 Corrosion Rate

Seven samples with a diameter ~11 mm and length ~8 mm, were prepared and their weights were measured. Each sample was then immersed in 100 ml of distilled water with pH = 5.2 at 20 °C. Sample 1 was taken out in the first day and after being washed with xylene to remove the very thin white surface layer, the weight loss was measured. The corrosion rate was calculated using the equation,

$$\text{Corrosion Rate} = W / (S.T) \quad (g\ cm^{-2}\ d^{-1})$$

where W is the weight loss, S is the total surface area and T is soak time (days). Sample 2 was taken out in the second day and so on. Similar experiments were applied to other glasses.

2.6.3 Water Permeation In Glass

A cylindrical glass sample with one polished flat surface was encapsulated in a transparent plastic tube so that only the polished surface was exposed. This was then connected to a flow of deionised water, pH = 5.2 at 20 °C. The position of the " diffusion front " was observed using a vernier travelling microscope. The reading was taken every 24 hours up to 7 days.

3.0 RESULTS AND DISCUSSION

3.1 Thermal And Crystallisation Characteristics

Values of Tg and Tc, of most glass compositions, are given in table 1. In some samples, more than one crystallisation temperature Tc was observed. The values of Tc − Tg are in the range from 53°C in S_8 to 149°C in S_3, showing good stability for the glass system. The values of Tg (Transformation temperature) were in the range from 190°C up to 318°C. These values decrease however when the nominal concentration of $ZnCl_2$ is increased. This is perhaps due to the ability of chlorine to disrupt the glass network.

The nominal and EDX analysed compositions of the glasses can be shown by two ternary diagrams. given in figure 1. Because of differential losses of volatile components during melting, the analysed components of the glass differ from the nominal.

Table 1. Glass compositions and thermal parameters.

Samp. No.	Glass composition (mol %)	Tc (	Tg ± 1)	Tc - Tg °C
A_1	0.8 Sb_2O_3 - 0.2 $ZnCl_2$	381	281	100
A_2	0.6 Sb_2O_3 - 0.4 $ZnCl_2$	411	279	132
S_1	0.7 Sb_2O_3 - 0.3 $PbCl_2$	459	318	141
		485		
S_2	0.5 Sb_2O_3 - 0.5 $PbCl_2$	315	265	50
		388		
S_3	0.6 Sb_2O_3 - 0.3 $PbCl_2$ - 0.1 $ZnCl_2$	455	306	149
S_4	0.6 Sb_2O_3 - 0.2 $PbCl_2$ - 0.2 $ZnCl_2$	393	281	111
		425		
S_5	0.5 Sb_2O_3 - 0.3 $PbCl_2$ - 0.2 $ZnCl_2$	343	283	60
S_6	0.4 Sb_2O_3 - 0.3 $PbCl_2$ - 0.3 $ZnCl_2$	332	269	63
S_7	0.3 Sb_2O_3 - 0.3 $PbCl_2$ - 0.4 $ZnCl_2$	281	235	46
		330		
S_8	0.2 Sb_2O_3 - 0.3 $PbCl_2$ - 0.5 $ZnCl_2$	243	190	53
		290		
		390		
		445		
A_3	0.4 Sb_2O_3 - 0.6 $ZnCl_2$	315	253	62

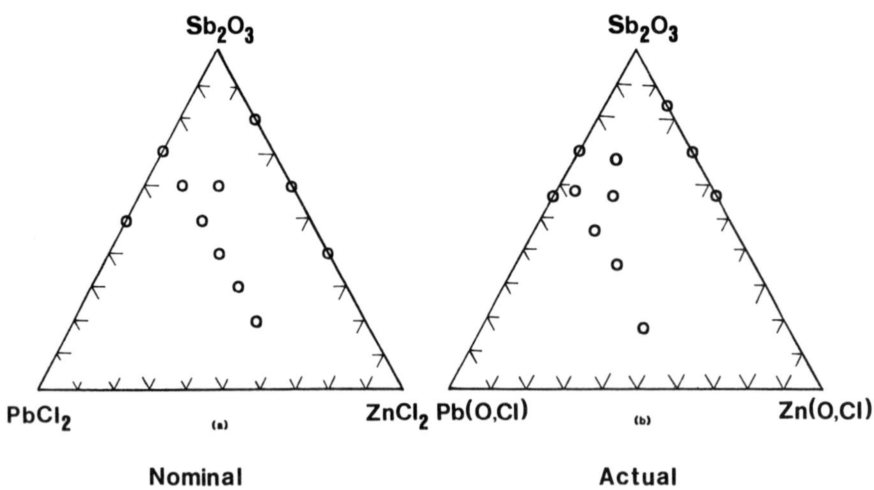

Figure 1. Nominal and analysed compositions of glasses studied.

The crystallisation studies by XRD, give only a family of phases, thus further investigation using EDX is required. Results of these investigations (Table 2) show two phases, Sb_2O_3 and lead antimony oxychloride with a variation in oxygen-chlorine ratio. In S_8, $ZnCl_2$ was also observed. Confirmation of these identifications has been discussed previously by Sahar et al (1990).

Table 2. Phase analysis from XRD and EDX.

Sam. No.	Tc (± 1) °C	Phase Occurrence XRD Analysis major	minor	EDX Analysis major	minor
A_1	381	Sb_2O_3	-	Sb_2O_3	-
A_2	411	"	-	"	-
S_1	459	$PbSbO_2Cl$	Sb_2O_3	$PbSb_2O_{3.5}Cl$	Sb_2O_3
	485	"	"	"	"
S_2	315	"	"	"	"
	388	"	"	"	"
S_3	455	"	"	"	"
S_4	393	"	"	"	"
	425	"	"	"	"
S_5	343	"	"	$PbSb_2O_3Cl_2$	"
S_6	332	"	"	"	"
S_7	281	"	"	$PbSb_2O_{2.5}Cl_3$	"
	330	"	"	"	"
S_8	243	$PbCl_2$	β - $ZnCl_2$	$PbCl_2$	β - $ZnCl_2$
	290	"	"	"	"
	390	"	"	"	"
	445	"	"	"	"
A_3	315	Sb_2O_3	.	Sb_2O_3	"

3.2 Crystal Growth Rate

The crystal growth rates were calculated by measuring particle size before impingement occurs, divided by time of exposure. From the initial slopes of the growth rate urves, the maximum growth rate can be obtained. These values can then be plotted against the analysed chlorine content of each sample. This relation is presented in figure 2. From that figure, a linear relationship can be observed in spite of the fact that the phase produced is changing in stoichiometry. These results show that maximum growth rate is chloride dependent.

The maximum growth rate occurs at a certain temperature, T_{max} and in figure 3 it can be seen that T_{max} decreases as chlorine content increases.

Both results indicate that it is chloride concentration which affects crystallisation.

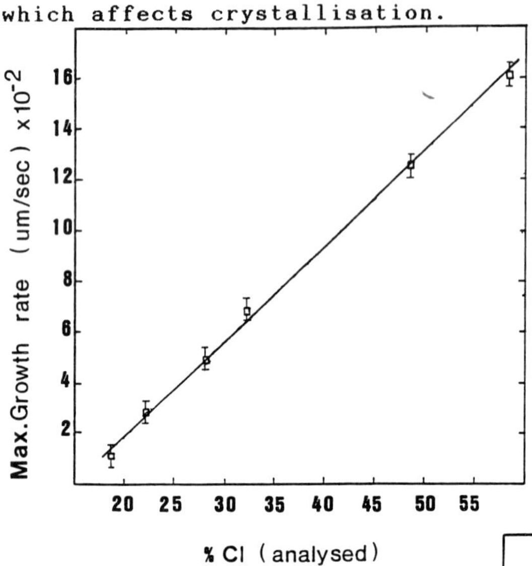

Figure 2. Maximum crystal growth rate as a function of %Cl.

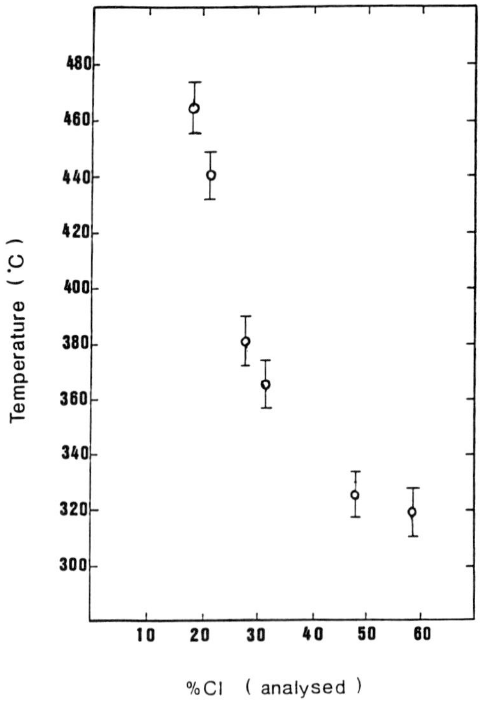

Figure 3. Temperature of maximum crystal growth as a function of %Cl.

3.3 Viscosity

The values of log viscosity at a fixed temperature (in this case 322°C) are plotted against chlorine content in figure 4. It is clear from this that, as chlorine increases, the viscosity decreases indicating changes which can be related to glass structure. As the chlorine content increases, the viscosity reduces and consequently the crystal growth rate increases. This has long been recognised (Hillig and Turnbull 1956).

3.4 Durability

Most glasses containing less than ~ 30 at % Cl (analysed) appear to be stable in air. This can be seen from the results given in table 3.

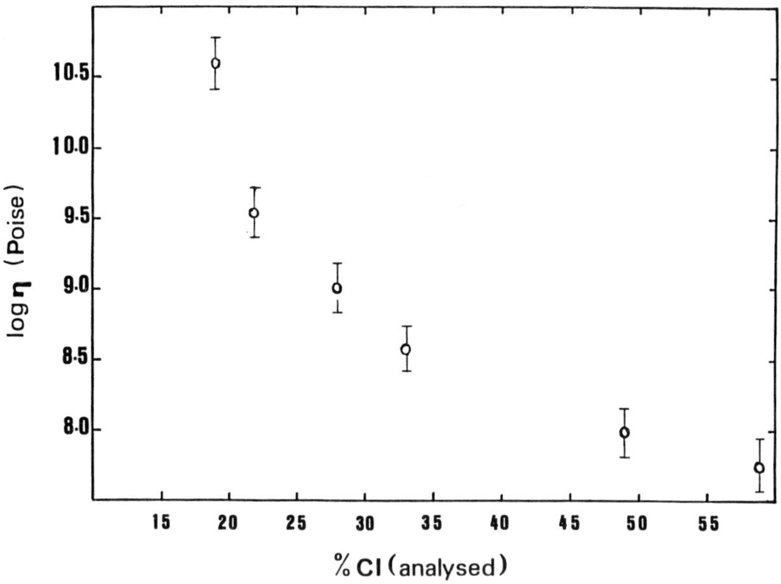

Figure 4. Viscosity at 322°C as a function of %Cl.

The relation between corrosion rate after 4 days in distilled water against analysed chlorine content is presented in figure 5. This result indicates that although some glasses resist attack under atmospheric conditions, most of them were reactive in the presence of water and the degree of reactivity increases with analysed chlorine content. Addition of more $ZnCl_2$ at the expense of Sb_2O_3 increases the corrosion rates as expected from the higher solubility of $ZnCl_2$, studied in detail by Gorre et al (1983). The corrosion rates of the glasses (after 4 days soak time) are in the range of $(2 - 10) \times 10^{-4}$ $gcm^{-2}d^{-1}$. In comparison, these values are an order of magnitude lower than those for the fluorozirconate glasses studied by Simmons and Simmons (1986).

Samp. No	Weight loss (gm cm-2)
S_1	$< 10^{-6}$
S_2	$< 10^{-6}$
S_3	$< 10^{-6}$
S_4	$< 10^{-6}$
S_5	$< 10^{-6}$
S_6	1.5×10^{-5}
S_7	2.3×10^{-5}

Table 3 : Solubility of glasses in normal atmosphere

(exposed for 7 days at 25 °C)

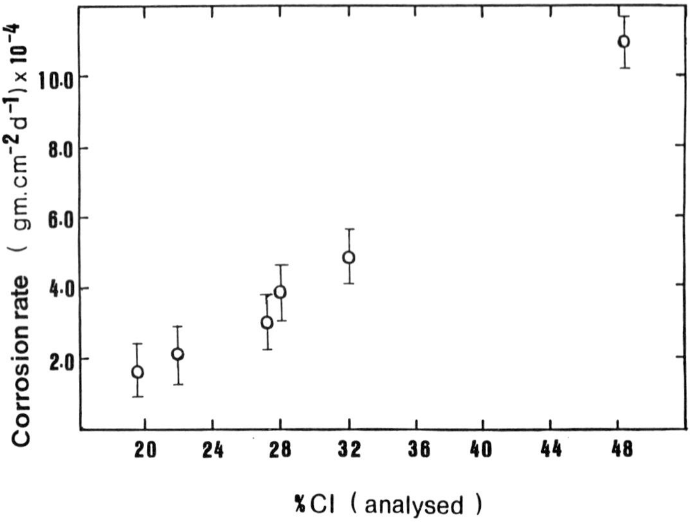

Figure 5. Corrosion rate in distilled water as a function of %Cl.

The relation between water penetration depth (after 4 days) and chlorine content is shown in figure 6. The result indicates that the water permeation is chlorine controlled. If the diffusion coefficients are calculated using the method described by Tomozawa et al (1984),and plotted against chlorine content, (figure 7) it can be seen that, it

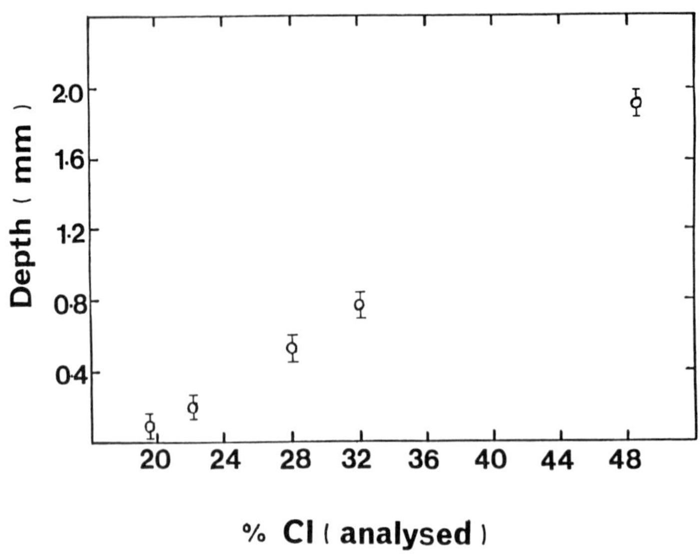

Figure 6. Penetration by deionised water after 4 days as a function of %Cl.

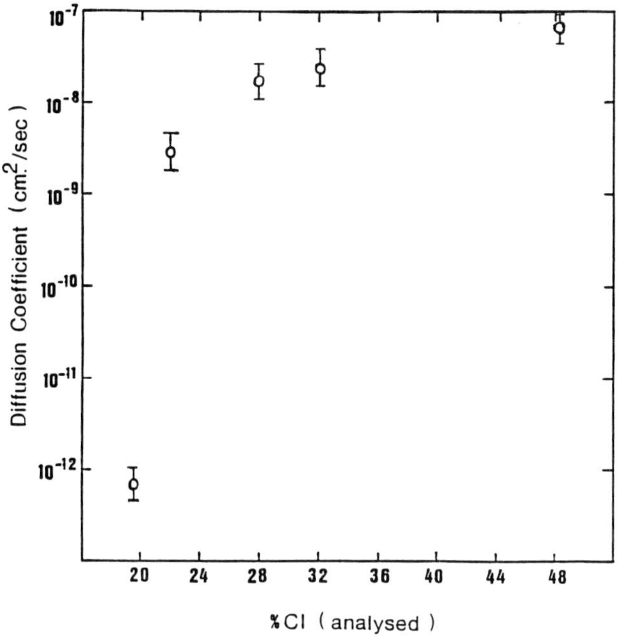

Figure 7. Diffusion coefficients in glasses as a function of %Cl.

is chlorine that controls the glass durability. It was suspected that the corrosion process is initiated by the reaction of glass with water in the glass matrix and as a consequence releases Cl^-, H^+, Pb^{2+} and Zn^{2+} to the solution by a diffusion process. Qualitative analysis for the presence of cations in the solution using spot tests and flame tests described by Feigl (1956) and Vogel (1961) respectively, can be seen in table 4. Quantitative analysis of corrosion products and a postulated mechanism of corrosion will be presented elsewhere.

Sample	Cations		
	Pb^{2+}	*Sb^{3+}	Zn^{2+}
S_1	Y	N	-
S_3	Y	N	-
S_5	Y	N	-
S_6	Y	N	-
S_7	Y	N	Y
S_8	Y	N	Y

Table 4: Qualitative analysis for the presence of cations in the solution using Spot Tests and Flame Tests (*)

Y - Detected N - Undetected

The precipitated crystal deposit which was removed from the corroded glass was identified by XRD. The phases for S_1 to S_6 were Sb_2O_3 and " $PbSbO_2Cl$ ", while in S_7 and S_8, the phases were found to be " $PbSbO_2Cl$ " and surprisingly a new phase $Zn_5(OH)_8Cl_2$. Since EDX results show that the crystallised glass contains a compound with varying oxygen - chlorine ratio and infra-red shows the presence of hydroxyl groups in the corrosion layer, it is suggested that the deposits consist of $PbSb_2Sb_2O_{4-n}(OH)_nCl_n$ where n has a value which depends on the chlorine content. Further, EDX analysis and infra-red spectroscopy show that in the higher chlorine content glasses, S_7 and S_8, the corrosion phase should be of the form $Zn_5(OH)_{8-n}Cl_{2+n}$. This result is consistent with the XRD investigation on the hydrated layer.

4.0 CONCLUSIONS

Most of thermal and chemical properties of oxychloride glasses were chlorine concentration dependent. The decrease both in thermal and chemical stability reflect the disruption of glass network.

References

Ahmed M M and Holland D 1987 Glass Tech. vol 28, no 3 141

Douglas R W, Armstrong W L, Edward J P and Hall D 1965 Glass Tech. vol 6 no 2 52 Fiegl F 1956 Spot Test In Organic Analysis (London: Elsevier)pp 473-507

Gorre L E and Pastor R C 1983 Mater. Res. Bull. vol 18 1391

Hillig W B and Turnbull D 1956 J. Chem. Phys. 24 4

Sahar M R, Ahmed M M and Holland D 1990 Phys. Chem. Glasses (to be published).

Simmons C J and Simmons J H 1986 J. Amer. Ceram. Soc. 69(9) 661

Tomozawa M,Ito S and Molenelli J 1984 J. Non-Cryst. 64 269

Underwood E E 1970 Quantitative Stereology (Addison Wesley Pub. Co.) pp 103-105

Vogel A I 1962 Macro And Semimacro Qualitative Inorganic Analysis 4th ed.(Longmans)pp 145-150

Inst. Phys. Conf. Ser. No 111
Paper presented at Int. Conf. on New Materials and their Applications, Univ. of Warwick, 1990

Glass-ceramic coatings for extended performance

D Holland, F Hong, E Logan and S Sutherland
Physics Department, Warwick University, Coventry CV4 7AL

ABSTRACT: Many materials have admirable bulk properties for a given application but their surfaces are incompatible and require some form of barrier coating. Glass-ceramic coatings are easily produced - generally by the application of a glass which is then converted to a glass-ceramic of superior properties by a heat treatment proces. By careful adjustment of glass composition and thermal history, the phase content and microstructure of the glass-ceramic can be tailored to give particular values of desired physical properties. The design of such coatings and the requirements for good adhesive strength are reviewed with reference to electrical and mechanical applications.

1. INTRODUCTION

There are many materials which have bulk properties which are particularly attractive for certain applications but which cannot be used in practice because their surfaces are not compatible. This can be because of adverse chemical or mechanical reactions between the surface and the environment during use, or because of physical incompatibility between the surface and adjacent material. For instance, mild steel will corrode under normal atmospheric conditions, titanium or silicon carbide will oxidise if used at elevated temperatures in air and metals cannot be used as substrates to support microelectronic devices unless some means is found of isolating the conducting elements of the device from the substrate ground plane. These problems have traditionally been solved by the application of coatings to seal the surface completely, or locally, from its surroundings. For example, the enamelling of metals is a centuries old process which has progressed from decorative to protective use.

1.1 Glass-ceramics

Glass-ceramics are but one possible candidate for use as protective coatings. They are not always the optimum selection but they do have the advantage that their properties are

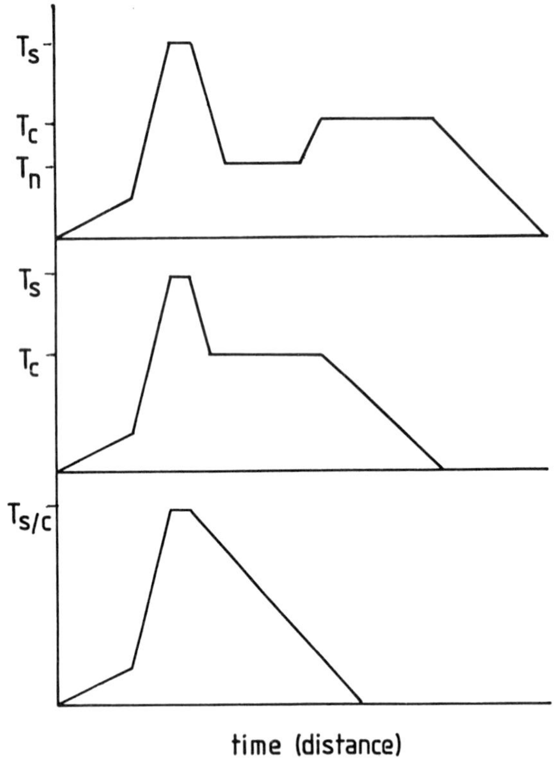

Figure 1. Thermal cycles for glass-ceramic coating.

T_s - sealing temperature

T_n - nucleation temperature

T_c - crystallisation temperature

Temperature v. time for batch furnacing or v. distance for belt furnacing.

capable of fine tuning over a wide range in order to achieve compatability with the substrate material in question and the application in view (McMillan 1979).

Glass-ceramics are formed by the controlled crystallisation of a glass to give a polycrystalline material. Fig. 1 illustrates the general ways in which this can be brought about. Choice of composition of the parent glass and control of the thermal history used to produce crystallisation, allow a range of phase compositions and microstructures to be generated. The properties P of such polyphase materials are often the summation of the values for the individual phases P_i modified by their volume fraction V_i.

$$P = \sum^{i} P_i V_i$$

As a result of this, properties can be tailored to particular desired values.

1.2. Thermal expansion matching.

One particular property, of great importance to all coating applications, is the thermal coefficient of expansion (TCE). If the material to be coated is to see significant thermal cycling, either during fabrication or during actual use, then the thermal expansion of the substrate and the coating must be closely matched to avoid the generation of stress. This stress is proportional to the difference in TCE values of the materials as given approximately by equation 1.

$$\sigma = \frac{E(a_2 - a_1)(T - T_o)}{(1 - v)} \quad ......(1)$$

where
a_2 and a_1 are the TCE values of the materials
T and T_o are the upper and lower temperatures to which the materials are exposed
E and v are the Young's modulus and Poisson ratio. For successful coating, E and v should be similar for the two materials.

Such stress levels can lead to cracking, either in the bulk of the coating, or at the interface between the coating and the substrate. Even if the adhesive bond strength between the coating and the substrate is high, the stress can lead to bowing of thin section joints, the degree of bow being given by (Timoshenko 1925, Glascock and Webster 1984):-

$$\frac{1}{p} = 6(a_2 - a_1)(T - T_o)G \quad ......(2)$$

where
p is the radius of curvature of the bow

G is a geometric factor $= \dfrac{(1 + m)^2}{h[3(1 + m)^2 + (1 + mn)(m^2 + \frac{1}{mn})]}$

where
m is the ratio of the thicknesses of the materials d_2/d_1
n is the ratio of the elastic moduli and Poisson's ratios

$$n = \frac{E_2(1 - v_1)}{E_1(1 - v_2)}$$

$h = d_1 + d_2$

This is a serious problem in substrates for microelectronics where further screen printed features need to be applied to the coating surface (conductors, resistors, capacitors etc). The screen printing technique, particularly for fine line conductors, requires a high degree of flatness.

1.3. Adhesion.

It is essential that the coating should adhere to the substrate during the lifetime of the component. Two major

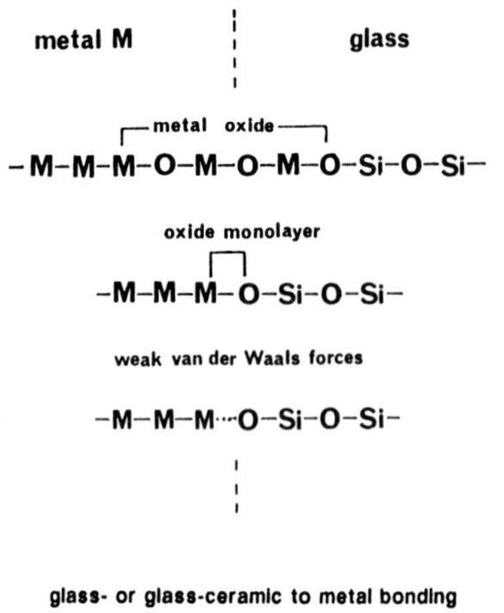

Figure 2. Possible transitionary bonding between a metal substrate and a silicate coating. If there is no continuity of bonding, then only weak van der Waals forces or mechanical keying provide some adhesion.

factors here are the TCE match described above and the strength of the bond between the coating and the substrate. The contributions to the adhesive bond strength are complex but it is generally agreed that chemical bonding is more important than simple mechanical keying, although this can make an additional contribution. Chemical bonding implies a continuity in the bonding from bulk substrate through to the bulk coating. It is reasonable to expect continuity between an oxide substrate and a silicate glass-ceramic coating but less so if the substrate is a metal, nitride or carbide. The fact that these substrates can be coated successfully with glass-ceramic relies on the formation of a transition layer between the two materials with some attributes of each. In the case of metals, this interlayer is usually the metal oxide although metal silicides have also been observed and multiple layers can also be produced, for example metal---metal oxide--(metal/metal') silicate---silicate coating. Metal nitrides and carbides form an interlayer again of metal oxide to give SiC---SiO_2---coating. Fig 2 illustrates the possible sequence of bonds formed. The necessary thickness of the transitional layer depends on several factors. A monomolecular layer is sufficient to link say metal and silicate structures but its integrity depends on the means of formation of the layer and on its stability with respect to reaction with the coating. A thick layer requires less accurate control during fabrication but then itself becomes a significant structural component and the durability of the coating may then depend on the adhesion of the oxide to the substrate or on the mechanical strength of the layer. However, even thick layers can be destroyed if interaction with the coating is extensive.

An essential part of adhesion is the presence of a region, adjacent to the oxide/coating interface, which is saturated with metal oxide (Pask 1987). Saturation prevents further dissolution of the oxide interlayer and hence loss of adhesion. This saturation is only maintained if diffusion from this region is low. Therefore the optimum interlayers are formed from metal ions of low solubility and diffusivity in typical silicate glasses. Even if interlayer dissolution occurs, there is still the possibility of continuous generation of the oxide if a suitable redox couple exists between the substrate and the coating. For a metal substrate, this takes the form:-

$$M'_{substrate} + M''O_{glass} = M'O_{interface} + M''_{precipitate} \quad ...(3)$$

Thus the interlayer can be preformed before coating of the substrate, or may form during the coating process. The former is necessary if a significant thickness of interlayer is required or if specific conditions are required to produce the desired composition of the interlayer.

The surface to be coated is first cleaned, then possibly abraded and preoxidised if necessary. The coating itself is produced by preparing the glass of the required composition. Normal glass production techniques are employed and the combination of oxides and carbonates are mixed together and melted (usually in a platinum crucible for purity) at a temperature sufficiently high that a fluid melt is produced. The melt is cast into deionised water to form a frit which is remelted and refritted and finally milled to a fine (<10 microns) powder. This powder can be applied to the substrate surface by spraying, dipping, electrophoresis or screen printing. In all these techniques the powder is used as a suspension in an organic vehicle which then must be removed by oxidation before the coating is fired at a temperature at which it will flow and wet the surface of the substrate/interlayer. Devitrification of the glass, to form the glass-ceramic, can then be achieved either by a separate heat treatment cycle, or in some cases it may be possible to achieve flow, wetting of the surface and crystallisation during the firing or on cooling from the firing temperature. Schematic firing schedules are illustrated in Fig 1.

2. APPLICATIONS

Examples of the use of glass-ceramic coatings are most often found in the electronics industry, in high temperature engineering applications and as protection from chemical attack. The following case histories serve to illustrate some of the advantages of glass-ceramic coatings and some of the problems associated with their development.

2.1 Electronics

Heat removal from high power VLSI chips is essential to maintain device characteristics and to prevent damage. The most efficient way of removing heat is by conduction through a high thermal conductivity substrate in contact with the chip. The most readily available materials for such work are metals but, being also good electrical conductors, they need to be isolated from the conductor tracks which are carrying signals

Choice of metal substrate.

Si - TCE = $3x10^{-6}$ °C^{-1}

Metal / alloy	TCE x10^{-6} °C^{-1}	Thermal conductivity $Wcm^{-1}K^{-1}$
Cu	16.6	3.98
Al	25	2.37
W	4.5	1.78
Mo	5	1.4
Invar (Fe/Ni)	0.9	0.16
Cu / Invar / Cu sandwich	3 - 4	

Table 1. Properties of some metals used in substrate manufacture for VLSI packaging.

to and from the device. This means that the surface of the substrate (other than the chip site) must be coated with a material of high electrical resistivity, low permittivity (for high frequency work) and low dielectric loss (to prevent signal deterioration). In addition to these requirements, there is a need to match the TCE of the metal substrate, and therefore the coating, to the TCE of silicon (~30 10^{-7} °C^{-1}). The major restriction here is the availability of metals or alloys with a TCE close to this value and which also have good thermal conductivity. Table 1 shows the properties of some metals which have been considered and even used for microelectronic substrates. In terms of heat removal the most attractive choice would be copper but the TCE is much too high for the pure metal. This can be reduced by incorporating it into a sandwich with Invar, of very low thermal expansion, or into a composite material with tungsten. The Cu/W material is difficult to coat because of the varying surfaces involved and in addition is metallurgically difficult to handle. The copper/Invar/copper sandwich, with layer thicknesses chosen to give the desired overall TCE, gives high thermal conduction in the plane of the substrate and heat can be removed by the use of edge-connector heat sinks.

In the case of the pure metals, the choice lies with tungsten and molybdenum. Tungsten has the closer TCE match but molybdenum is an easier metal to fabricate. Both copper/Invar/copper and molybdenum have recently been investigated as metal cored substrates for microelectronics (Logan et al 1990). The glass-ceramics developed as coatings were different in each case. A lithium aluminosilicate glass-ceramic was developed to TCE match the copper/Invar/copper and a magnesium aluminosilicate for the molybdenum substrate. In both cases the coating was applied by screen printing successive layers until a fired thickness of ~100 microns was achieved. Firing was carried out under a

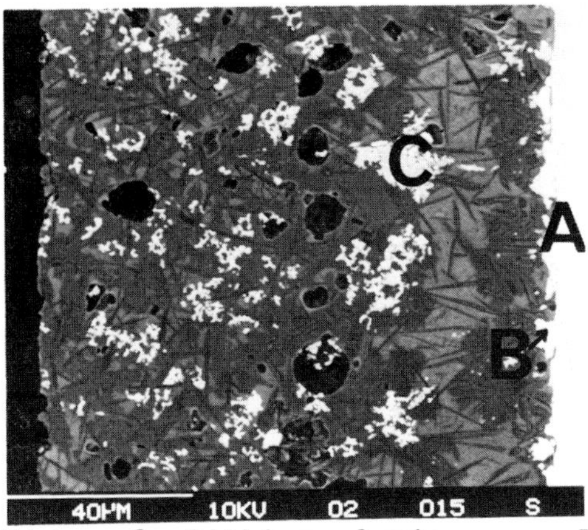

Figure 3. Lithium alumino-silicate coating on copper/invar/copper (A) showing interfacial oxide (B), dissolved and reprecipitated oxide (C).

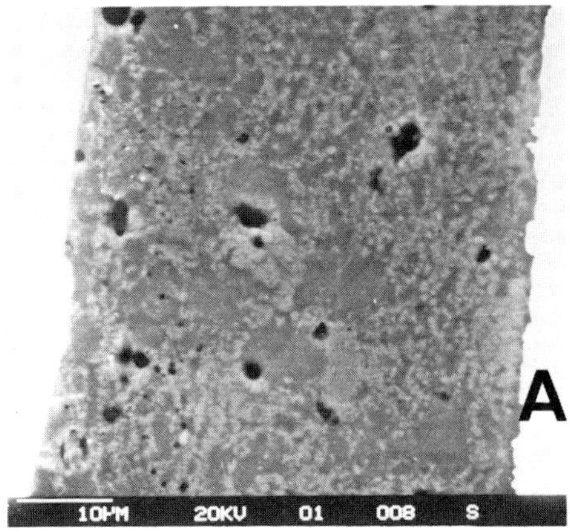

Figure 4. Magnesium alumino-silicate coating on molybdenum (A) showing no discernible diffusion

nitrogen atmosphere to prevent excessive oxidation of the metal at high temperatures but a small amount of oxygen had to be introduced at low temperatures (<600°C) to enable oxidative removal of the organic vehicle.

Figs 3 and 4 show cross-sections through the coated metals. Both coatings are smooth and dense but in the case of the interface between the lithium aluminosilicate and copper there has been extensive diffusion of copper ions into the glass-ceramic and these have precipitated out as copper oxide crystals on cooling. This interaction leads to instability of the transitional layer and also presents the problem of increased permittivity and reduced electrical resistance - particularly if furnace atmospheres lead to further reduction of the copper oxide to metallic copper. In the case of the magnesium aluminosilicate coating on molybdenum, there is no visible interfacial layer at the magnifications employed and therefore it must be significantly thinner than 1 micron. This is important, for molybdenum oxides are mechanically weak and a thick layer would lead to loss of coating by crack propagation through the oxide layer. There is no interdiffusion detectable by EDX analysis (Energy Dispersive X-ray analysis) and we might therefore expect the layer to be stable. In fact there is some diffusion,giving an oxide saturated interface. The saturation concentration for high valency molybdenum ions is small and is below the detection limit of EDX. There is however a much more sensitive detector of molybdenum diffusion, and that is measurement of the relative permittivity of the coating as a function of frequency for different thicknesses. Such a measurement (Fig 5) shows that there is a rise in the imaginary part of the permittivity at low frequencies when the coating thickness falls below 60 microns. This indicates the presence of ionic impurities (i.e. Mo^{4+}) in a significant volume of the coating (Thorp et al. 1990).

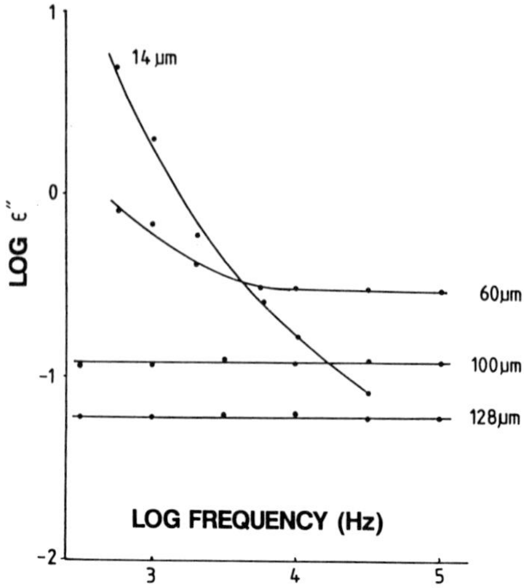

Figure 5. Imaginary part of the permittivity of a glass ceramic on molybdenum as a function of frequency for different thicknesses. Low frequency losses become significant for thin layers as a result of the ionic contamination arising from the dissolution of oxides of molybdenum.

3.2. High temperature materials

The exploitation of the refractory, high strength materials required in many areas of engineering is restricted by their susceptibility to oxidation under normal atmospheric conditions at high temperatures. Tianium alloys, superalloys, silicon carbide and many of the SiC fibre reinforced materials could operate at higher temperatures than at present feasible if a suitable protective coating could be found. Such coatings must themselves be refractory and yet be capable of application under conditions which will not degrade the substrate. Glass-ceramics are frequently much more refractory than the parent glasses from which they are formed. For example, Fig 6 shows the thermal expansion curves of a nimonic alloy, the lithium silicate glass-ceramic coating designed to TCE match the metal and the parent glass from which the coating was made. The expansion curve for the glass turns over just below 500°C, where softening begins. The softening of the glass-ceramic does not commence until >850°C.

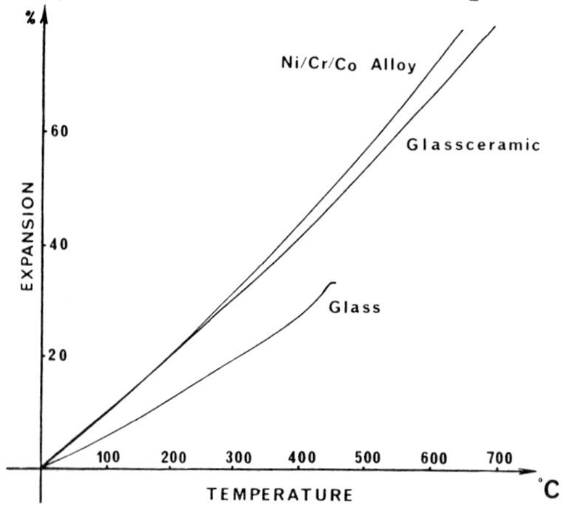

Figure 6. Expansion curves for nimonic, the matched glass-ceramic and the parent glass.

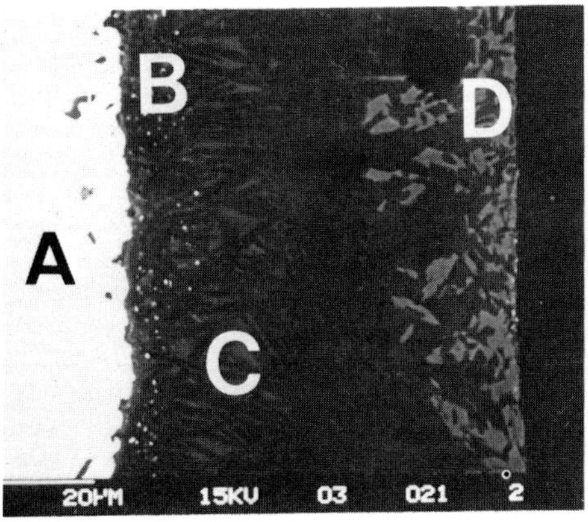

Figure 7. Cross-section of coating on nimonic (A) showing Cr_2P precipitates (B), coarse microstructure (C) and $LiCr(SiO_3)_2$ (D).

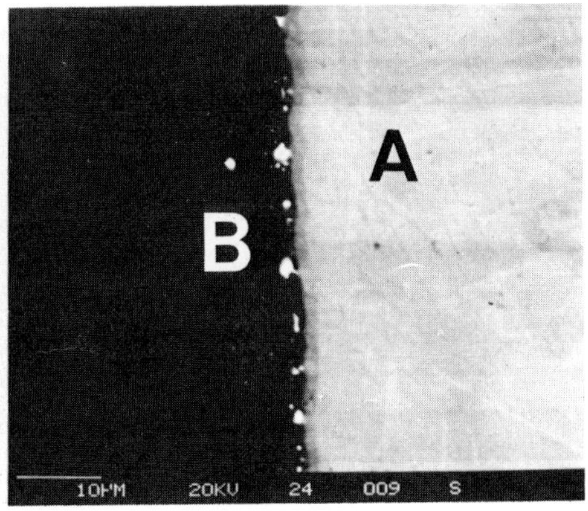

Figure 8. Cross-section of coating on titanium (A) showing Co precipitates (B).

This system also illustrates the problems which can be presented by the action of a metal/glass redox couple. If the nimonic is preoxidised to give a Cr_2O_3 surface layer, then the metal can be coated with glass-ceramic with no difficulty. If however the glass is applied to a bare metal surface then the following reaction takes place:

$$Cr_{substrate} + ZnO_{glass} \rightarrow CrO_{interface} + Zn_{alloy} \quad ....(4)$$

Now Cr^{2+} has a higher solubility and diffusivity in silicate glasses than Cr^{3+} and the CrO interfacial layer rapidly disappears thus destroying the transitional bonding arrangement. There is an additional reaction which creates further problems. This involves a redox couple between Cr in the alloy and the phosphate component in the glass. This can be nominally expressed as :-

$$9Cr + "P_2O_5" \rightarrow 5CrO + 2Cr_2P$$

The chromium phosphide produced can be seen as bright contrast precipitates in Fig 7. P_2O_5 is an essential part of the coating composition as this provides nuclei, in the form of lithium orthophosphate crystallites, upon which the major phases will grow. Removal of P_2O_5 as Cr_2P near to the interface reduces nucleation and the corresponding crystal microstructure is very coarse in this region. The results of a further redox reaction can be seen at the surface of the coating in Fig. 7. Cr^{2+}, which has diffused to the surface region is oxidised to Cr^{3+}:-

$$2Cr^{2+} + Zn^{2+} \rightarrow Zn + 2Cr^{3+}$$

The Cr^{3+} precipitates as $LiCr(SiO_3)_2$ and Zn is lost to the atmosphere by evaporation at the firing temperature. The precipitate has a very different thermal exapnsion from the coating and this results in disruption of the surface. Thus the consequences of incorrect choice of substrate pretreatment are complex and severe.

Titanium is an important engineering material since it retains a high strength to weight ratio at high temperatures. However, the solution of oxygen into the surface of the metal at high temperatures has an adverse effect on the fatigue behaviour and shortens life. An oxygen barrier is required but titanium ia a difficult metal to coat because it is a strong reducing agent and will react with silicates to give Ti_5Si_3 and oxygen gas which foams the fluid glass. In addition, the oxide TiO_2 is mechanically weak and poorly adherent. Therefore neither bare metal nor a thick preformed oxide layer are chemically or mechanically acceptable. An alternative approach (Hong 1989) is to produce "in-situ" oxidation of the interface. This is a technique employed by the enamelling industry where an "adherence" oxide (often NiO or CoO) is added to the glass melt and is reduced by the substrate metal e.g:-

$$Ti + 2CoO \text{ ------> } TiO_2 + 2Co$$

The metal produced forms precipitates near the interface (Fig 8) and may also alloy with the substrate. The competing siicide reaction is largely suppressed and by minimising the coating temperature, a successful glass-ceramic oxygen barrier can be applied.

CONCLUSIONS

Individual substrates require the development of coatings tailored to their particular requirements. Glass-ceramics can provide a readily accessible solution.

REFERENCES

Glascock H H and Webster H F, 1984, IEEE CHMT-6, 460.
Hong F and Holland D, 1989, Surface and Coatings Technology 39/40, 19.
Logan E A, Holland D and Partridge G, 1990, submitted to Glass Technology.
McMillan P W, 1979, "Glass Ceramics", Academic Press, New York.
Pask J A, 1987, Ceram Bull. 66, 1587.
Thorp J S, Akhtaruzzaman M, Logan E A and Holland D, 1990, J. Europ. Ceram. Soc. (to be published).
Timoshenko S, 1925, J.O.S.A. & R.S.I. 11, 232.

Inst. Phys. Conf. Ser. No 111
Paper presented at Int. Conf. on New Materials and their Applications, Univ. of Warwick, 1990

Bonding of lithium zinc silicate glass-ceramics to nickel-based superalloys

B L Metcalfe and I W Donald

MOD(PE), Atomic Weapons Establishment, Aldermaston, RG7 4PR, UK

ABSTRACT: Lithium zinc silicate glasses containing a high proportion of zinc are of interest in the manufacture of hermetic seals to nickel-based superalloys. The influence of glass composition and process parameters on the microstructural and thermal properties of the glass-ceramic have been studied by a number of techniques including DTA, XRD, SEM and dilatometry. The presence of cristobalite, quartz and γ and β lithium zinc silicate has been confirmed by XRD. Thermal expansion has been shown to be markedly dependent on the thermal history of the glass-ceramics.

1. INTRODUCTION

For many years oxide glasses have been used in the electronics industry for the production of hermetic seals on such devices as vacuum tubes. Their ability to bond well to metal and their high electrical resistivity coupled to the relative ease of manufacture ensured their continued use. In the early 1960's McMillan and co-workers (1963, 1966a, 1966b) demonstrated that glass-ceramics could be used for the same purpose. Glass-ceramics had the same advantages as glasses but offered the additional advantages of higher mechanical strength, superior resistance to static fatigue and the ability to tailor the thermal expansion coefficient to match that of the metal being bonded to. One system identified by McMillan as being particularly suitable for bonding to moderately high thermal expansion metals was based on lithium zinc silicate containing a high proportion of zinc oxide. Recently it has been reported by Metcalfe and Donald (1989), Donald and Metcalfe (1989) and Donald et al (1989) that this system can be used for bonding to nickel-based superalloys.This work also studied a variety of glasses based on a single lithium zinc silicate composition but to which various transition metal oxides had been added to improve both the microstructure and bonding properties of the glass-ceramic. From this work a modified lithium zinc silicate composition containing 2 mole % of selected transition metal oxides was chosen for more detailed investigation.

2. EXPERIMENTAL PROCEDURES

2.1 Preparation of glasses and glass-ceramics

The starting material for all the work was a 200kg batch of glass frit (LZS) manufactured by Ceramic Developments (Midlands) Ltd, (CDM). The composition of this glass is given in Table 1 and for comparison the analysis of the lithium zinc silicate glass reported in previous work by these authors is included. 500gm batches of glass modified to contain 2 mole % of either copper, nickel or molybdenum were prepared by melting the appropriate amounts of frit and metal oxide in a Pt-5%Rh crucible at 1450°C for 1-2 hours, casting into demineralised water and remelting in order to obtain a homogenous product. After remelting the glass was poured into either water to produce quenched frit for thermal analysis studies or into pre-heated graphite moulds to produce rod which was immediately annealed at 450°C for 1 hour and then cooled at ≈0.5°/ minute to room temperature.

Table 1-Composition of glasses (mole %)

	Li_2O	Na_2O	ZnO	B_2O_3	P_2O_5	SiO_2
Current	17.84	5.25	17.73	4.31	1.23	53.64
Previous	18.74	4.81	18.67	3.51	0.92	53.35

Conversion of the glass into glass-ceramic was carried out using a three stage heating schedule consisting of a simulated sealing schedule, nucleation and crystallization; a typical schedule is shown in Figure 1. Different nucleation and crystallization temperatures were investigated but the time the samples were held at these temperatures remained at 1 hour throughout the investigation. Samples for thermal expansion measurements were in the form of square bars ≈5mm x 40mm long.

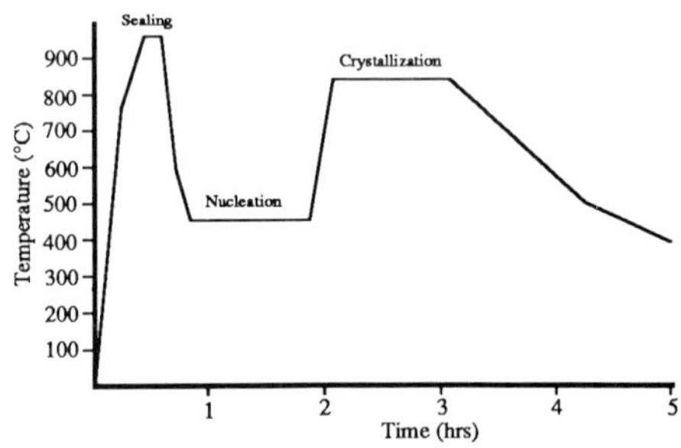

Figure 1. Typical heat-treatment schedule

2.2 Thermal characterization of glasses and glass-ceramics

The thermal properties of as-quenched glass were determined by both traditional DTA methods using a Stanton Redcroft Differential Thermal Analyser and by Dynamic Mechanical Thermal Analysis (DTMA). Athough established as a useful technique in the polymer field, DMTA has not been widely used in the glass field where its one important advantage over DTA, its greater sensitivity to second order reactions, makes the determination of T_g very much easier. As the response is dependent on the mechanical properties of the sample it is possible to determine the T_g of the small quantity of residual intergranular amorphous phase in the glass-ceramic. The characterization included an assessment of nucleating efficiency using the method of Thakur and Thiagarajan (1966), determining the optimum nucleating temperature by the method of Marotta et al (1981) and determining apparent activation energies of crystallization using the modified Kissinger method of Matusita et al (1975 and 1980). Measurements of the thermal expansion characteristics were made in air over the range ambient to 650°C using a CDM dilatometer. Thermal expansion coefficients were determined over three temperature ranges, 20-250°C, 20-460°C and 20-650°C.

2.3 Microstructural characterization of glass-ceramics

Pieces cut off the ends of the thermal expansion bars were mounted in bakelite resin, polished and examined using both optical and scanning electron microscopy techniques. Optical examination of the specimens proved difficult because of their low reflectivity. A thin (≈50Å) coating of gold/palladium alloy sputtered onto the surface of the specimens increased the reflectivity sufficiently to allow photographs to be taken of the microstructures although some detail was lost. Transverse sections of the cylinder seals were also prepared in a similar manner although these were not sputter coated as optical micrographs were not required. More detailed examinations were carried out using a Cambridge S360 SEM at 20kV and imaged using back-scattered electrons. This technique highlighted variations in the elemental composition of the many phases present. X-ray spectra were obtained using the Energy Dispersive Spectrometer (EDS) fitted to the SEM. With this it was possible to obtain qualitative analysis of specific microstructural features and the elemental variations across the bonding interface using line scans. Quantitative analysis of selected phases was performed by microprobe analysis. X-ray diffractometry was carried out at ambient temperatures to identify the many phases present. High temperature XRD was also used to investigate the features observed on the dilatometer traces.

2.4 Sealing

Sealing experiments were carried out using 7.5mm diameter glass rod and either Nimonic 90 or Inconel 625 cylinders with an 8mm diameter hole. Both metal and glass parts were thoroughly cleaned prior to assembly. Sealing was carried out under argon in a vacuum furnace modified to operate at a pressure of ≈35kPa above ambient. All experiments were carried out using a sealing temperature of 950°C for 5 or 10 minutes and a crystallization temperature of either 850 or 875°C. After sealing all samples were leak-tested using a mass-spectrometer and helium gas.

3. RESULTS AND DISCUSSION

3.1 DTA/DMTA

Typical DTA traces of as-quenched and nucleated LZS glass are shown in Figure 2. These show two crystallization peaks which also appeared on the DTA traces of the LZS glass containing the oxide additives. Nucleating the glass alters the position of both crystallization peaks although the effect is more profound on the first peak. These differences are also supported by the loss tangent data obtained from DMTA (Figure 3) although this was obtained from glass which had been annealed not quenched.

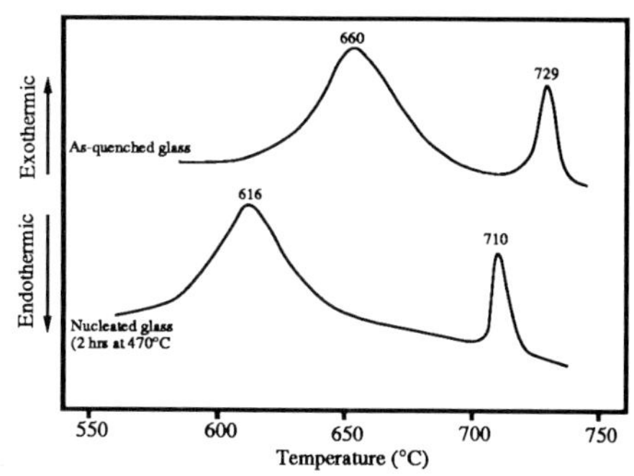

Figure 2. Typical DTA traces

Optimum nucleating temperatures were found by plotting the difference in peak temperature between as-quenched and nucleated samples, ΔT, against nucleating temperature (Figure 4). The plot shows a maximum ΔT value and therefore greatest nucleating efficiency, at ≈465°C for LZS. Above 520°C there appears to be an increase in nucleating efficiency but this is thought to be due to a combination of nucleation followed by some crystal growth.

Activation energies have been calculated for both crystallization peaks for all the glasses and these are given in Table 2.

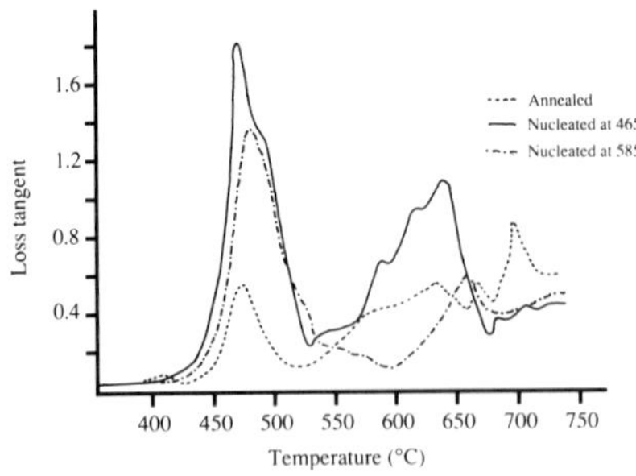

Figure 3. DMTA traces

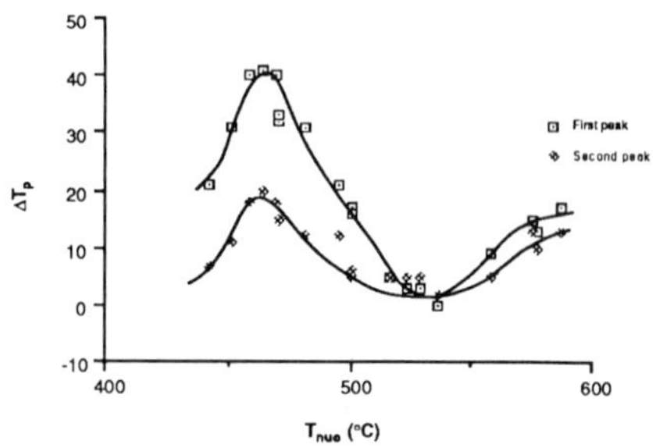

Figure 4. Optimum nucleating temperature

Table 2-Activation energies for crystallization ($kJmole^{-1}$)

Glass	First peak	Second peak
LZS	316	389
LZS+ $2^{m}/_{o}$ Cu	341	362
LZS+ $2^{m}/_{o}$ Ni	306	342
LZS+ $2^{m}/_{o}$ Mo	317	220

Adding copper and nickel oxide to the base glass changes slightly the activation energies calculated for the first peak but there is no difference in the energy of the corresponding molybdenum peak. With the second peak however a much greater effect is apparent. Although the activation energies of both the copper and nickel second peaks have increased, the values are lower than the corresponding LZS peak especially in the case of nickel. Molybdenum is anomalous in that it is the only additive to decrease the activation energy of the second peak and this shows a dramatic drop when compared to the other glasses.

3.2 Dilatometry

A typical expansion curve for the base glass-ceramic nucleated at 465°C and crystallized at 850°C is shown in Figure 5. The trace shows four distinct regions which have been identified as follows. Region A is the α to β cristobalite inversion. Between A and B is a region of moderate expansion ($\approx 11\times10^{-6}{}^{\circ}C^{-1}$) which extends to just above 500°C. At this point there is a second phase change which from the high temperature XRD results was originally thought to be a combination of the α- β quartz inversion and the γ_0 - β_{11} lithium zinc silicate conversion. However DMTA results suggested that this was not the whole interpretation. From the results of a multifrequency experiment, the activation energy for the transition was determined using an Arrhenius plot of ln frequency

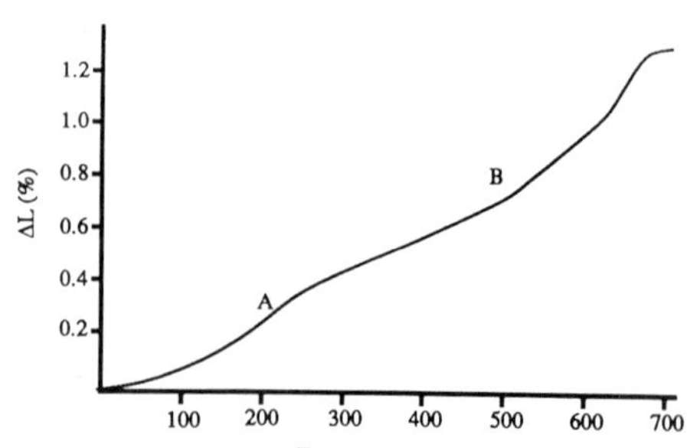

Figure 5. Typical thermal expansion trace for LZS glass-ceramic

against reciprocal peak temperature. The value of 542 kJmole^{-1} obtained is far too high for the displacive α - β quartz inversion and it has been proposed that this corresponds to the glass transition of residual amorphous phase. If this is correct then it would appear that there are several reactions taking place simultaneously around 530°C. Above this temperature is a high ($\approx$25 x10^{-6}°C^{-1}) expansion region which becomes very high immediately prior to softening. These regions were observed in all of the base glass-ceramic samples regardless of heat-treatment schedule.

The thermal expansion coefficients, over the range 20-460°C, for LZS glass-ceramic nucleated at 465°C and 585°C are plotted in Figure 6. The curves show a minimum at 875°C which corresponds to the required expansion coefficient. One criterion for the production of stable hermetic seals is that the expansion coefficient of the glass-ceramic should match or be slightly lower than that of the metal. As the expansion coefficient of Nimonic 90 is $\approx$14 x 10^{-6}°C over the temperature range 20-460°C this would allow processing to be carried out at 865±15°C.

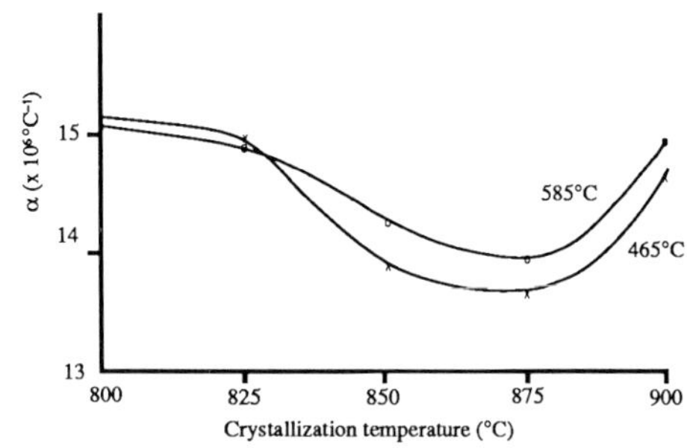

Figure 6. Expansion coefficient as a function of crystallization temperature

3.3 MICROSTUCTURE

All the glasses were X-ray amorphous both as-quenched and after nucleation for 1 hour at 465°C. Following a simulated sealing schedule the glass-ceramic showed the presence of the three silica polymorphs cristobalite, quartz and tridymite plus β_1 and γ_0 Li_2ZnSiO_4, the latter two probably being present as a solid solution. High temperature XRD revealed that the γ_0 Li_2ZnSiO_4 converts to β_{11} Li_2ZnSiO_4 above 530°C. The patterns obtained were complex and many of the peaks did not correspond exactly to those reported by West and Glasser from their phase studies on either the equilibrium (1970a) or non-equilibrium (1970b) ternary systems making interpretation difficult. The data showed an increase in the tridymite : cristobalite ratio as the crystallization temperature increased from 800 to 875°C followed by a decrease. This correlates well with the observed thermal expansion data.

Optical examination of LZS nucleated at 465°C and crystallized at 800, 850 and 900°C showed the microstructure to consist of a fine crystalline structure with some larger acicular grains (Figure 7) but the low reflectivity of the samples prevented a more detailed examination at higher magnifications. Sputtering with gold increased the reflectivity but obscured the fine structure. Optical examination of the glass-ceramics containing nickel and copper oxides nucleated at 465°C and crystallized at 850°C showed that the addition of these oxides had not changed the microstructure significantly when compared with LZS (Figure 8) although the large acicular crystals in the glass containing copper oxide are much more obvious. SEM micrographs of specimens heat-treated under identical conditions (Figure 9) confirmed the similarity between the LZS, nickel and copper glass-ceramics. In these the microstructure consists of a mixture of equi-axed grains with a mean grain size $\approx$1μm, residual glassy phase and some very large acicular grains. Addition of molybdenum trioxide completely changes the microstructure, perhaps not surprising when the activation energy of the second crystallization

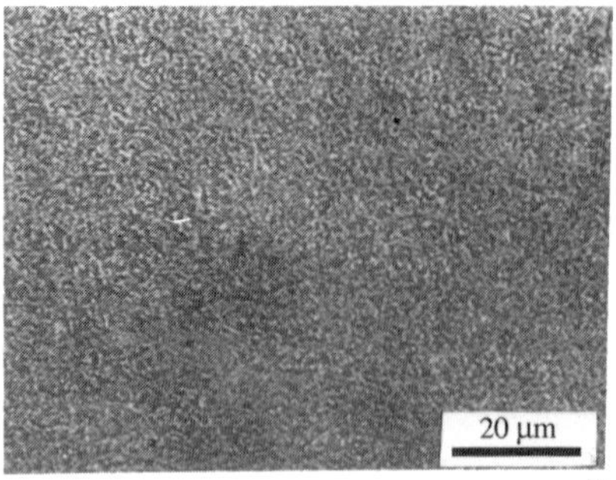

a

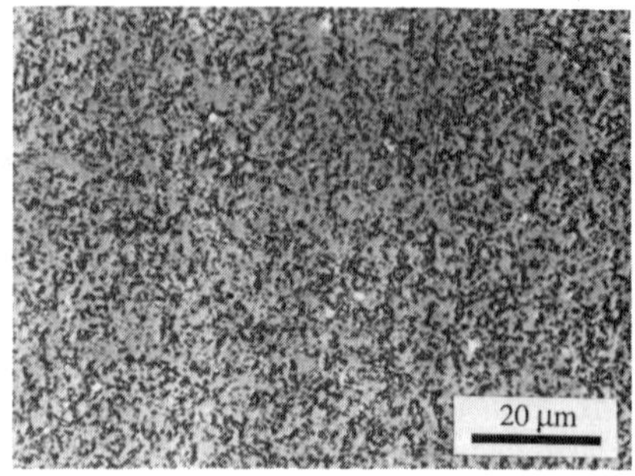

b

Figure 7. Optical micrographs of LZS glass-ceramics crystallized at (a) 800°C, (b) 850°C and (c) 900°C

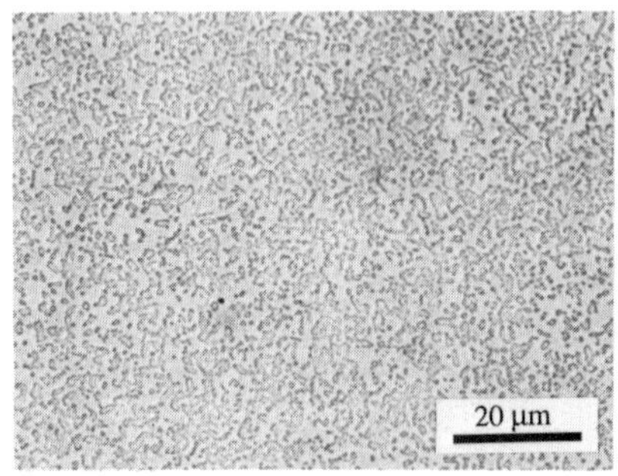

c

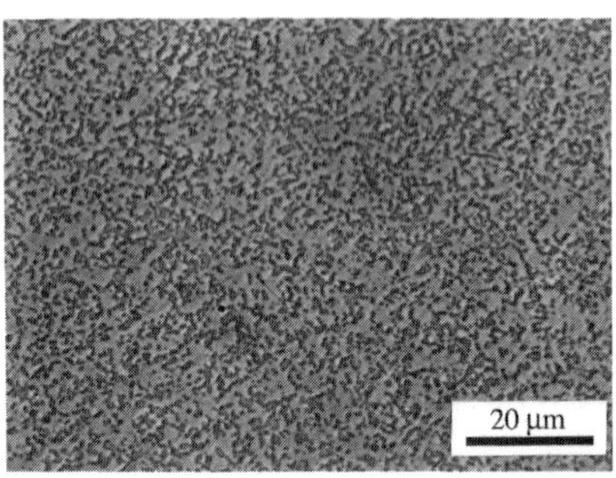

a

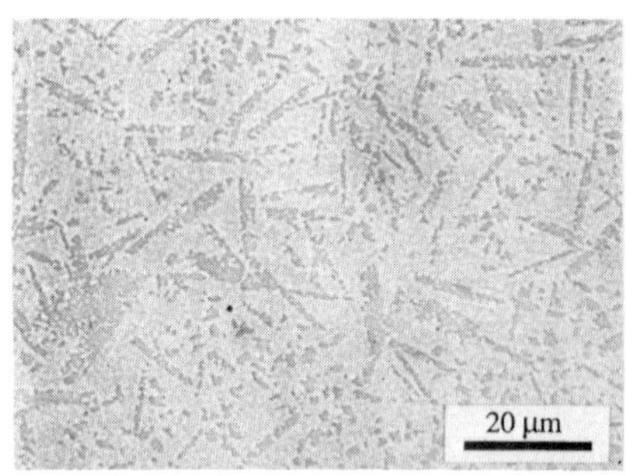

b

Figure 8. Optical micrographs of LZS glass-ceramics crystallized at 850°C and containing 2mole % of (a) NiO, (b) CuO and (c) MoO_3

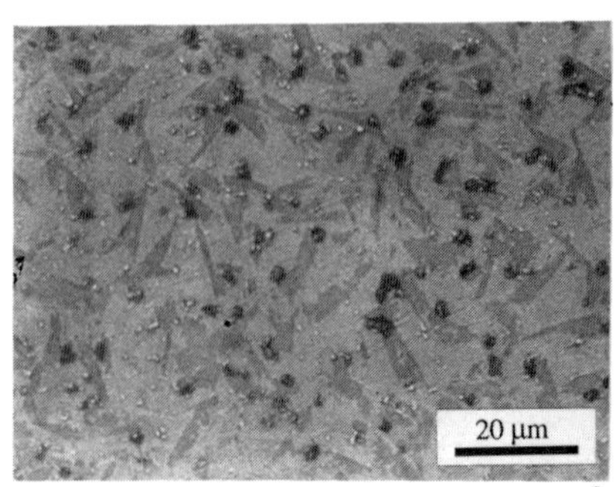

c

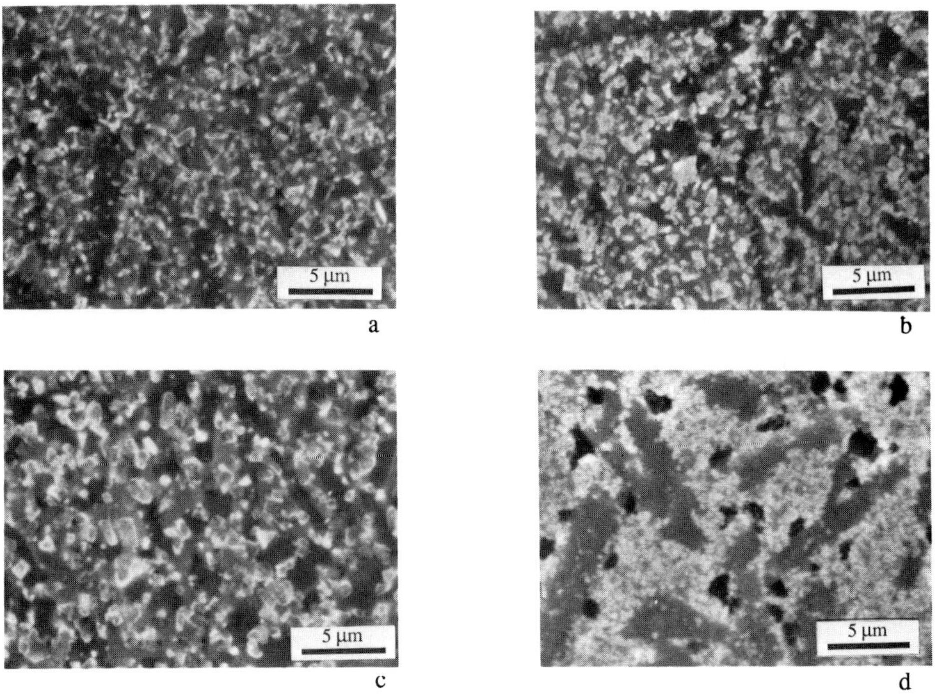

Figure 9. SEM micrographs of (a) LZS, (b) CuO, (c) NiO and (d) MoO_3 glass-ceramics crystallized at 850°C for 1 hour

peak is so different. The large acicular grains are still present but the equi-axed grains are now sub-micron. Also visible are a number of voids adjacent to the acicular grains. X-ray energy dispersive analysis of the lining of the voids has shown these to be the remnants of a zinc rich phase. Comparison of the three LZS specimens showed no increase in the grain size of the equi-axed grains between an 800 and 850°C crystallization temperature but by 900°C there was a better defined acicular structure and the small equi-axed grains had grown to ≈ 2µm (Figure 10).

3.4 SEALS

Line scans across the Nimonic 90/LZS glass-ceramic interface showed that the sub-micron thick granular layer observed on the surface of the metal (Figure 11) was enriched in chromium and titanium eg particle A but depleted in nickel and cobalt when compared to the bulk metal. Fluctuating concentrations of zinc were detected in the Nimonic 90 which suggests that it has diffused from the glass-ceramic into the metal along grain boundaries. Three phases with differing contrasts were observed, labelled B, C and D in Figure 11, which had differing zinc / silicon ratios the palest having the highest ratio. This phase was also found to contain an appreciable quantity of phosphorus. As it was not possible to detect lithium using either the

SEM or microprobe it was assumed that these phases were lithium zinc silicates as there was no XRD evidence of a zinc silicate phase.

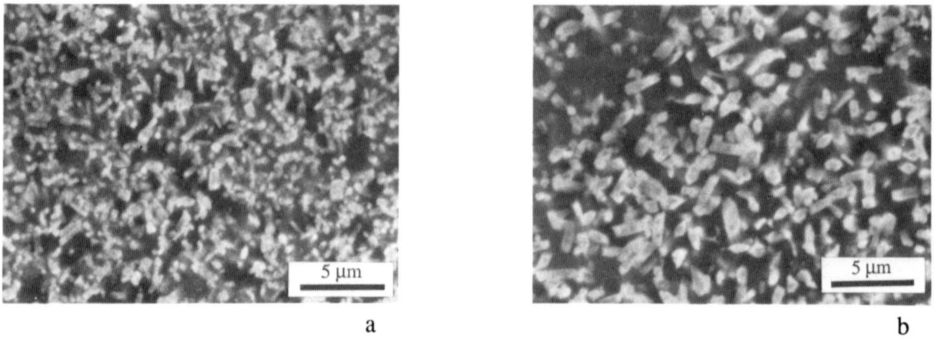

Figure 10. SEM micrographs of LZS glass-ceramics crystallized at (a) 800°C and (b) 900°C for 1 hour.

Figure 11. LZS / Nimonic 90 interface

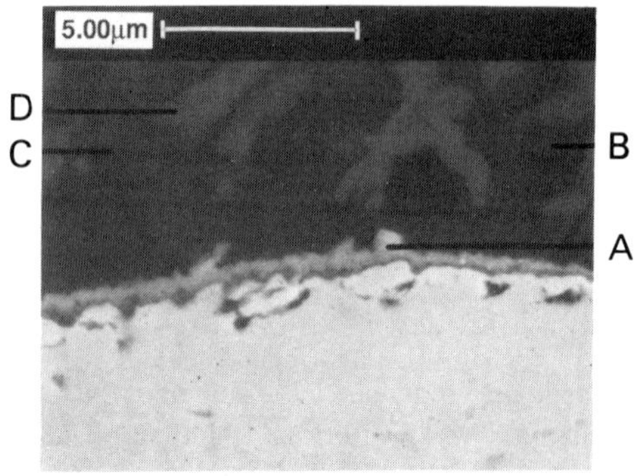

The interface between LZS and Inconel 625 (Figure 12) although coarser, was similar to that described above in that there was a granular layer present on the metal which had a different composition to the bulk alloy. As previously this layer had an increased chromium : nickel ratio compared to the alloy and zinc had diffused into the substrate metal. Particle A contains chromium with some silicon, nickel, zinc and molybdenum also present.

SEM examination of the interfaces in the copper oxide / Nimonic 90 system (Figure 13) revealed a much thicker interface than in either of the previous systems. Severe corrosion of the Nimonic 90 (A, Figure 13) had occured to a depth of approximately 4µm although the composition of this region (B, Figure 13) is very similar to the unaffected substrate. Along the edge of the glass-ceramic is a sub-micron thick layer (C) rich in chromium and silicon with smaller amounts of titanium, nickel, cobalt and zinc.

In none of these systems could the diffusion of chromium into the glass-ceramic be demonstrated. This has serious implications on the stress gradients across the interfaces as these are related to the concentration gradients. Diffusion of some of the major components of the alloys would minimise these stress gradients.

Figure 12. LZS / Inconel 625 interface

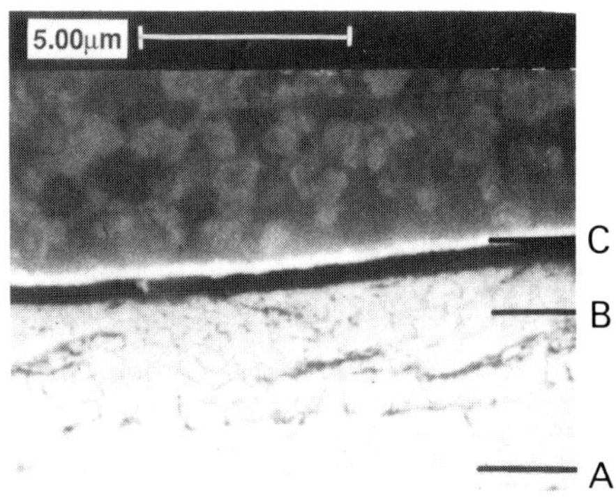

Figure 13. LZS + CuO / Nimonic 90 interface

4 CONCLUSIONS

1 These studies have demonstrated the possibility of using lithium zinc silicate glass-ceramics for bonding to Nimonic 90 alloy and the temperature range over which hermetic seals can be produced has been established.
2 Addition of 2 mole% copper and nickel oxides does not enhance bonding to either Nimonic 90 or Inconel 625 alloys.
3 Addition of 2 mole% copper or nickel oxide does not alter the grain structure when compared to the parent lithium zinc silicate glass-ceramic.
4 Addition of 2 mole% molybdenum trioxide produces an anomalous decrease in the activation energy of the second crystallization peak.
5 LZS glass-ceramic containing 2 mole% molybdenum trioxide has a finer grain structure

than that found in either the parent glass-ceramic or those containing nickel or copper oxides.

6 Chromium and titanium levels are enhanced along the interface

7 Some evidence exists for the diffusion of zinc from the glass into the Nimonic 90 and Inconel 625 alloys.

8 The phases present in the glass-ceramic are very dependent on its thermal history and this in turn influences the thermal expansion coefficients.

9 No significant diffusion of chromium from the metals into the glass-ceramics had occured.

Acknowledgements

The authors are grateful to Messrs. M. Clay and T. Carter, AWE, and Dr S Tarling, Birkbeck College, for the X-ray diffraction studies, Messrs. C. Ruckman and G. Lawrence, AWE, for scanning electron microscopy and microprobe analysis and Dr R Hill, Thames Polytechnic, for the Dynamic Mechanical Thermal Analysis studies.

References

Donald I W, Metcalfe B L, Wood D J and Copley J R 1989 *J Mater Sci* **24** 3892

Donald I W and Metcalfe B L "*Preparation and properties of lithium zinc silicate glass-ceramics*" Paper presented at the Symposium on Glass in Electronics, Society of Glass Technology, Blackpool 11-12 May 1989

Marotta A, Buri A and Branda F 1981 *J Mater.Sci.***16** 341

Matusita K, Sakka S and Matsui Y 1975 *J. Mater.Sci.***10** 961

Matusita K and Sakka S *J. Non-Cryst. Solids* **38 & 39** 741

McMillan P W and Hodgson B P 1963 *Engineering* **196** 366

McMillan P W, Hodgson B P and Partridge G 1966 *Glass Technol.* **7** 121

McMillan P W, Partridge G, Hodgson B P and Heap H R 1966 *Glass Technol.* **7** 128

Metcalfe B L and Donald I W "*Effect of composition on the sealing behaviour of lithium zinc silicate glass-ceramics*" Paper presented at the Symposium on Glass in Electronics, Society of Glass Technology, Blackpool 11-12 May 1989

Thakur R L and Thiagarajan S 1966 *Glass ans Ceramic Bull.* **13** 33

West AR and Glasser F P 1970 *J. Mater. Sci.* **5** 557

West AR and Glasser F P 1970 *ibid* **5** 676

Paper presented at Int. Conf. on New Materials and their Applications, Univ. of Warwick, 1990

Sol-gel derived highly porous SnO_2 for semiconductor gas sensors

Martin Forster, Jakob Eberle, Sigfrid Strässler and Gustav Pfister
Cerberus AG, Research Dept. CH-8708 Männedorf, Switzerland

Small chips of highly porous (>50 % v/v) semiconducting SnO_2:Pd are produced by a sol-gel process combined with photochemical Pd-deposition. Heated to 300-400 °C in air the chips exhibit an electrical conductance which is nearly linear with ethanol-concentration and shows only a low humidity-dependency. High uniformity of the material yields gas sensors with a narrow distribution of sensitivity.

1. MATERIAL AND GASSENSOR PREPARATION

Commercial semiconducting gas sensors for the detection of combustible gases consisting of a sintered SnO_2-layer show high variation in sensitivity for the individual sensors. In order to enhance the uniformity of the SnO_2 we developed a sol-gel process with which large pieces of porous (> 50% v/v) SnO_2 are obtained according to scheme I):

$$SnCl_4 \xrightarrow[H_2O]{+\,OH^-} SnO_2\text{-colloid} \xrightarrow[+\,PdCl_2]{+\,h\nu} SnO_2\text{:Pd-colloid} \xrightarrow[\text{drying, sintering}]{\text{dialysis, gelation}}$$

$$\underset{0.4 \times 0.3 \times 0.15\ mm^3}{SnO_2\text{:Pd-chips}} \xleftarrow{\text{cutting}} SnO_2\text{:Pd-wafer} \xleftarrow[\text{electroding}]{\text{lapping}} SnO_2\text{:Pd-xerogel} \qquad \text{I)}$$

Due to the photochemical reaction the Pd is formed as nanometersized particles which are uniformly distributed and colour the SnO_2 completely black - even in as low a concentration as 2% (w/w). The SnO_2:Pd-chips are sintered onto a planar heater of Pt deposited on an aluminia substrate 1.6x1.2x0.4 mm³. The Pt-heater serves both as heater and as thermometer thereby ensuring an exact temperature control of the SnO_2:Pd-chip. Fig. 1 shows such SnO_2:Pd-chips and Fig. 2 a complete gassensor. The power-consumption of the sensor is 450 mW at 380 °C .

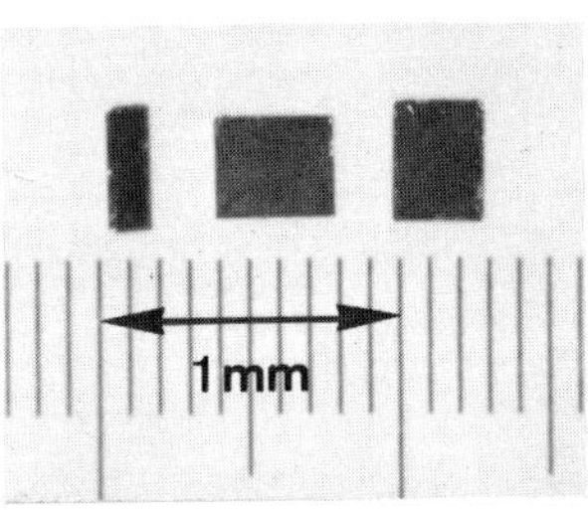

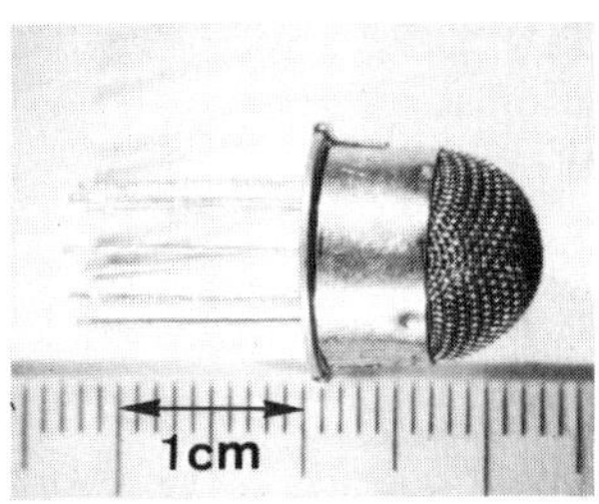

Fig.2: Complete Cerberus gassensor

2. GASSENSOR PROPERTIES

During one year the electrical conductances of 48 gassensors have been measured in synthetic air with controlled humidity as a function of the concentration of various gases. With appropriate electronics and using the Pt-heater the temperature of the SnO_2:Pd-chips was stabilized to 380 ± 2°C. Fig. 3 shows the conductance of such a sensor in relation to the concentration of ethanol with relative humidities of 5%, 25% and 95% at 22 °C. Fig.4 shows similar data for hydrogen and methane.

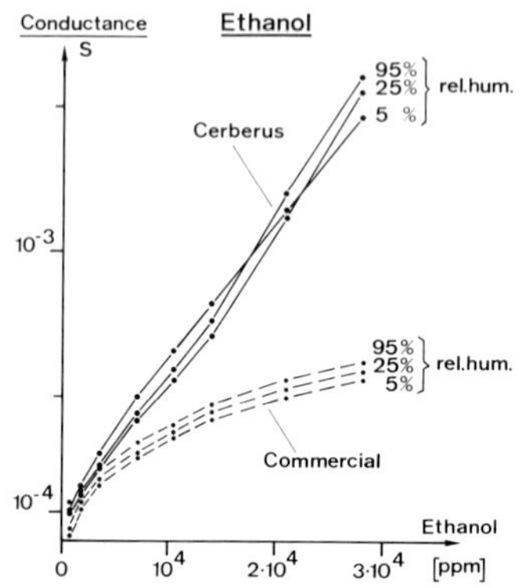

Fig.3: Conductance vs. concentration of ethanol

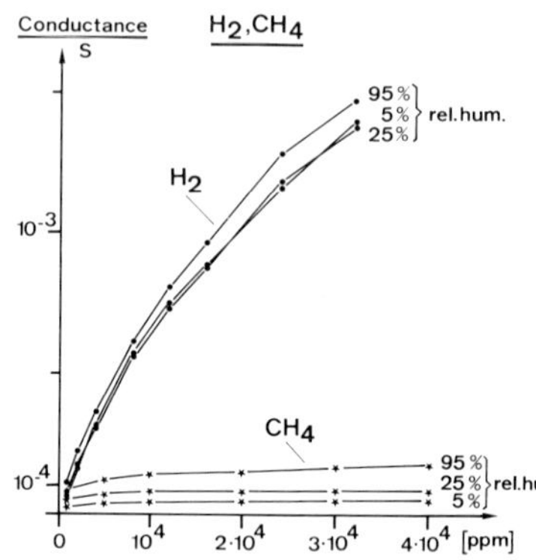

Fig.4: Conductance vs. concentration of hydrogen and methane

Obviously our sensors are very well suited for the detection of ethanol and hydrogen but not for methane. For ethanol the response of our sensors is nearly linear with concentration and shows almost no humidity dependency, by contrast with commercial semiconducting gassensors.
Fig. 5 shows the conductances of 46 gassensors in relation to the concentration of hydrogen. The signals have been normalized to 16000 ppm hydrogen. Fig. 5 clearly indicates a sensitivity enhancement after 7 months of operation. Furthermore the similar behaviour of these sensors demonstrates the high degree of uniformity achieved by using this sol-gel derived highly porous SnO_2:Pd as a sensing material. This fact is also confirmed by Fig. 6 indicating a narrow distribution of sensor resistances.

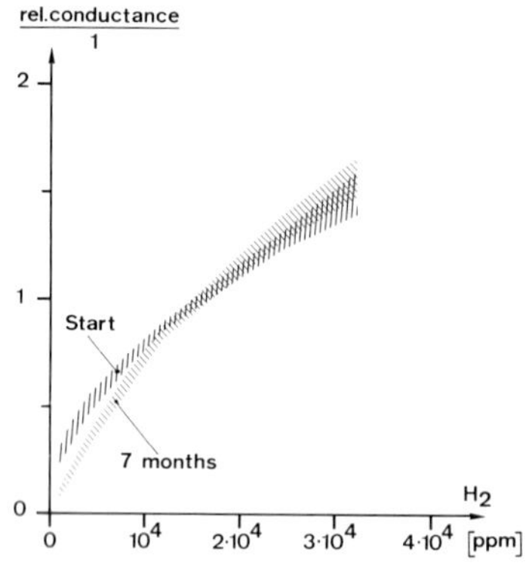

Fig.5: Relative conductances of 46 sensors at start and after 7 months

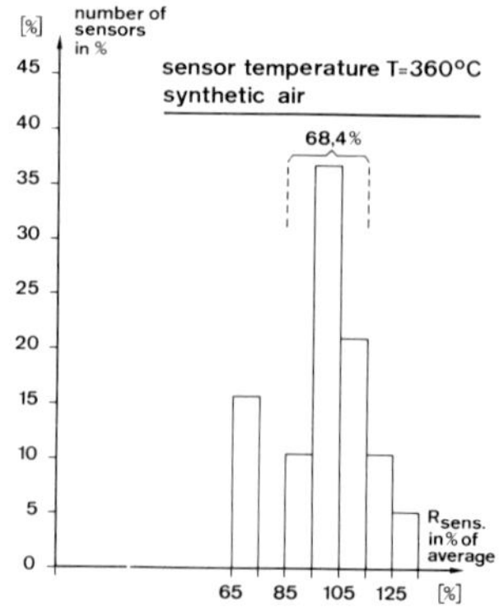

Fig.6: Distribution of sensor resistances

Inst. Phys. Conf. Ser. No 111
Paper presented at Int. Conf. on New Materials and their Applications, Univ. of Warwick, 1990

Near edge structure of dopants in amorphous silicon

G N Greaves and P. Stephenson
SERC Daresbury Laboratory, Daresbury, Warrington WA4 4AD, UK
S. Kalbitzer
Max-Plank-Institut für Kernphysik, D-6900 HNeidelberg 1, FRG
S. Pizzini and K.J. Roberts
Department of Pure and Applied Chemistry, University of Strathclyde, Glasgow G1 1XL, Scotland

ABSTRACT: The near edge structure of As and Ga K-edges in amorphous Si are considered in conjunction with that of substitutional As in crystalline Si. Multiple scattering calculations indicate that the predominant configuration for As is 3-fold (r=2.37 Å) and for Ga 5-fold (r=2.51 Å). The minority of tetrahedral sites that are electronically active are also discussed.

In contrast to crystalline Si, the doping of amorphous Si with Group III and Group V impurities is inefficient. In his model for doping Street (1982) proposed that impurities in amorphous Si acquire tetrahedral coordination through charge exchange with dangling bond Si atoms, Group IIIs becoming negatively charged and Group Vs positively charged. The quasi-equilibrium between the charged pairs and their neutral counterparts is presumed to be that prevailing during deposition (Street (1985)).

We have recently performed EXAFS measurements on amorphous Si ion implanted with Ga and As with the aim of detecting the structure of electronically active centres. By careful comparative analysis using substitutional As in crystalline Si as a standard we have determined the concentration of such centres at the level of a few percent for impurity loadings of 0.1 to 2%. Ga-Si and As-Si distances were found to be foreshortened in amorphous Si compared to substitutional sites in crystalline Si for which a near neighbour distance of 2.42 Å was deduced. A preliminary description of this work has recently been given (Greaves et al (1990)).

An outcome of these experiments has been the determination of the geometry of the electronically inactive Ga and As sites in amorphous Si. Mott (1967) proposed that the coordination number, CN, of impurities in amorphous semiconductors should be given by the 8-N rule where N is the valency. The basis for this being that each atom adopt the bonding configuration that minimises the local energy. For a covalently bonded site this is achieved when bonding states are filled and antibonding states are empty. This will leave the Fermi level in the centre of the energy gap and it clearly precludes substitutional configurations. In particular our measurements show that for most As centres in amorphous Si CN <4 with an As-Si bond length of 2.37 Å. On the other hand for Ga centres CN >4 and the Ga-Si bond length is 2.51 Å. Taken literally, the 8-N site would predict CNs of 3 and 5 for As and Ga respectively.

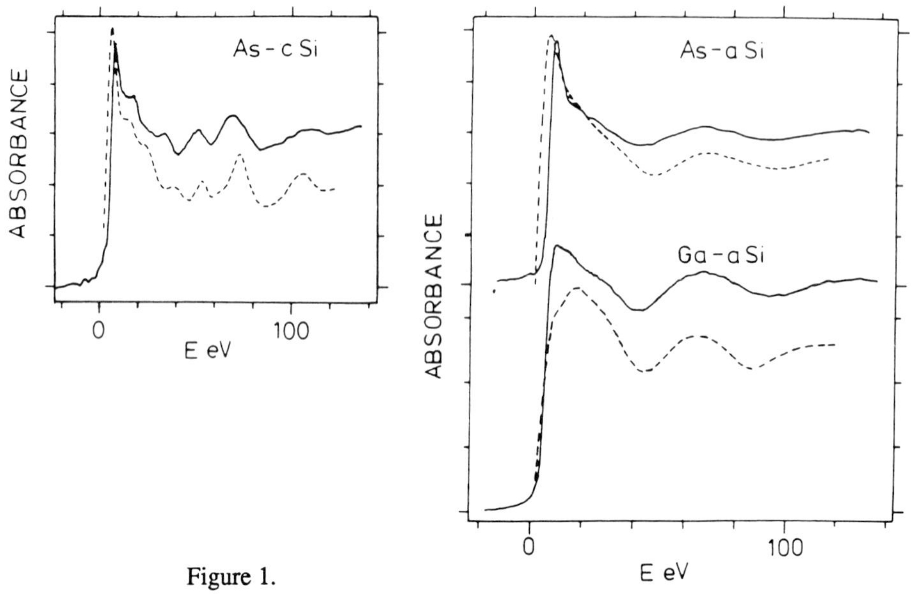

Figure 1.

In this note we consider the near edge structure of As and Ga impurities in amorphous Si and compare them with that of substitutional As in crystalline Si. Multiple scattering calculations have been performed using the theory of Durham et al (1982) and these are compared with experiment in Figure 1. The calculations employ the environments for As and Ga obtained from EXAFS analysis (Greaves et al (1990)). In particular As centres are taken as 3-fold coordinated and Ga centres as 5-fold coordinated. The good agreement between theory and experiment obtained for substitutional As in crystalline Si enabled detailed geometries for As and Ga sites to be explored. This exercise revealed that Si-As-Si bond angles were significantly less than 120°, being most probably close to 97° the value found in elemental As. Near edge structures for 4- 5- and 6-fold coordinated Ga sites were calculated, increasing the bond length from 2.42 to 2.56 Å. Of these CN=5 emerged as the most likely configuration. Finally the strength of the peak at threshold was investigated. This is strongly affected by the matrix elements for 1s-4p transitions which are in turn governed by the extent to which the final state wave function overlaps the atomic core. This increases with N and is clearly greater for As than for Ga as Figure 1 demonstrates.

The fraction of electronically inactive sites remains roughly constant for different impurity loadings. As a result the concentration of active sites varies linearly with the concentration of the impurity. Similar behaviour is observed from measurements of dangling bond densities (Street (1985)). Indeed we find the absolute concentrations of active sites match those of dangling bonds to within a factor ×5 or better, confirming approximate pairing in accordance with the doping model of Street (1982).

REFERENCES

Durham P J, Pendry J B and Hodges C H 1982 *Comp. Phys. Commun.* **25** 193.
Greaves G N, Kalbitzer S, Müller G, Pizzini S and Roberts K J 1990 *Il Nuovo Cimento* (in press)
Mott N F 1967 *Adv. Phys.* **16** 49
Street R A 1982 *Phys. Rev. Lett.* **49** 1187.
Street R A 1985 *J. Non-Cryst. Solids* **77** & **78** 1.

Diamond thin films and their applications

J. Mort
Webster Research Center, Xerox Corporation, Webster, N Y 14580, U. S. A.

ABSTRACT: Polycrystalline diamond thin films can now be fabricated at low pressures and temperatures by the thermal or plasma-assisted dissociation of hydrocarbon-hydrogen gas mixtures. This has generated considerable worldwide interest because of the remarkable properties of diamond in terms of hardness, thermal conductivity, semiconductivity, electronic transport, optical transparency and thermal and chemical stability. This paper reviews these issues with particular emphasis on the status of progress towards realizing some of the potential applications of these films.

1. INTRODUCTION

Until recently successful fabrication of synthetic diamond (the tetrahedrally-bonded form of carbon) required combinations of enormous pressure (100,000 atmospheres) and high temperature (2000C). Diamond single crystals have unique combinations of properties, [hardness, highest room temperature thermal conductivity, high melting point, high refractive index and large optical bandgap coupled with large electronic carrier mobilities] (Bachmann and Messier 1989); all of which derive from the properties its tetrahedral building-blocks. To date, however, their potential exploitation has been thwarted by the inability of the conventional synthesis to yield crystals of sufficient size or quality. The situation changed in the 1970s when Russian scientists reported the synthesis of diamond by vapor deposition from dissociated hydrocarbon gases. This involves essentially a low pressure and temperature process and, as importantly, enables the production of large-area diamond thin films. Today, this is a worldwide activity with both scientific and technological contexts (Bachmann and Messier 1989).

2. FABRICATION OF DIAMOND THIN FILMS

Although incomplete, a detailed explanation of the new growth process hinges on the fact that carbon can bond in two forms, diamond or graphite. Diamond involves tetrahedra while graphite results from a planar trigonal carbon bond; weak inter-planar Van der Waals type forces causing the familiar properties of graphite. The growth process involves a competition favoring the formation of tetrahedral bonding with the suppression of graphitic bonds by their etching or rearrangement. The fabrication process involves dissociation of hydrocarbon (typically methane)/hydrogen gas mixtures, sometimes with the addition of oxygen, by a variety of means such as a hot-filament, flame or microwave excitation (Bachmann and Messier 1989). In all cases, the predominant gas, by a ratio of 10:1, is hydrogen and it is the etching action of atomic hydrogen which favors diamond growth. Growth of the diamond film requires the provision of nucleation sites either by a physical abrasion of the substrate surface, commonly silicon wafers. Substrate temperatures for growth can range from 400-800C although, depending on this and other parameters such as gas composition, film morphology including graphitic content can vary substantially.

3. PROPERTIES OF DIAMOND THIN FILMS

The principal means of characterizing the films are scanning electron microscopy (SEM) and Raman spectroscopy. The former gives observation of the surface morphology of the films which are composed of individual diamonds several microns in diameter. Surface roughness and defect density within individual crystallites result in such light scattering that current films are dull gray. In the Raman spectra the 1332 cm^{-1}, characteristic of crystalline diamond and broader peaks (e.g., 1540 cm^{-1}) associated with graphitic and disordered bonds can be detected. Ratios of these peak amplitudes are taken as indicative of film quality and used to optimize fabrication parameters. Films grown with no lines detectable other than the 1332 cm^{-1} have been shown to have a thermal conductivity equivalent to that of a single crystal, which exceeds that of silver or copper at room temperature (Bachmann and Messier 1989). Room temperature electrical conductivities less than about 10^{-11}, much larger than observed in type 2A natural diamonds, are difficult to achieve and in films for which the Raman spectrum shows only the diamond peak, studies reveal that the electrical conductivity is still defect-controlled by graphite-like bonding (Mort et al 1989). Several studies have unequivocally demonstrated that the conductivity of the films can be controlled by boron doping by introducing varying amounts of diborane into the precursor gas mixture (Mort et al 1989).

4. APPLICATIONS OF DIAMOND THIN FILMS

Given the properties of diamond, the possible applications are to some degree self-evident and many suggestions have already appeared in the technical and patent literature (Bachmann and Messier 1989). Their practical realization hinge on a number of advances including patterned deposition, reduced substrate temperatures, improvements in deposition rate, film quality, adhesion and ultimately heteroepitiaxial growth. Some progress has already been made in patterning and reduced substrate temperatures (Bachmann and Messier 1989) but much remains in the area of film quality and heteroepitaxy, particularly if diamond thin films are to be useful as active electronic materials. For applications such as electronic packaging, films of less electrical quality can still exhibit sufficient thermal conductivity to make them efficient heat sinks for high density semiconductor chips or LED image bars. Recent studies have demonstrated that diamond thin films can be efficiently sensitized to visible light by combining them with plasma-deposited amorphous silicon films (Mort and Okumura 1990) and suggestions have appeared in the Japanese patent literature that such hetero-structures function as useful xerographic photoreceptors.

REFERENCES

Bachmann, P. K. and Messier, R., 1989, *Emerging Technology of Diamond Thin Films*, Chemical Engineering News, May 15, p 24-39

Mort, J., et al., 1989, *Boron Doping of Diamond Thin Films*, Appl. Phys. Letters, 55, 1121

Mort, J. and Okumura, K., 1990, *Photosensitization of Diamond Thin Films*, Appl. Phys. Letters, (to be published May 9 issue)

Paper presented at Int. Conf. on New Materials and their Applications, Univ. of Warwick, 1990

New composite tool materials based on PM high speed steels

A F Sanin, V A Domoratsky, V P Pedan, L L Scheglova, J J Dunkley*

State University, Gagarin Ave 72, Dniepropetrovsk-10, GSP 320625, The Ukraine, USSR.
* Davy McKee, Sheffield S9 4EX, UK

1. INTRODUCTION

The possibilities of improving the performance of high speed steels (HSS) by increasing the alloy content of W, Co, V, and C are now exhausted. The wider application of cemented carbides is limited by their poor mechanical properties. Improving HSS using hard phases such as TiC introduced by powder mixing does allow some increase in hardness but workability and mechanical properties are reduced.

One way of solving this problem may be a refinement of second phase dispersion and the elimination of coagulation by the use of prealloyed steel powders.

2. EXPERIMENTAL PROCEDURE

The following methods were compared:-
1 The water atomisation of HSS grades such as M2S and M15, but with Ti and carbon added to the melt before atomisation.
2 Atomisation with water having chemicals dissolved in it which are decomposed at high temperatures or which subsequently either form dispersed phases by reaction with alloying elements, or give solid solution strengthening.
3 Water atomisation of M2 HSS with an elevated carbon content and coating the powder with a thin film of titanium.

2.1 MELT ALLOYING WITH TITANIUM

The main problem of this method is a sharp rise in the liquidus temperature and of melt viscosity with even small additions of titanium. The rising viscosity leads to an increase in the atomised particle size which adversely affects sintering behaviour.

To eliminate these phenomena low melting point metals with a strong effect on the properties of the molten steel were introduced at up to 0.5% concentrations. Titanium levels were up to 3.2%.

2.2 ADDITIONS TO THE ATOMISING WATER

Inorganic compounds and solutions of metal salts capable of decomposing below the temperature of molten steel were used as additions to the atomising water. Additions were at levels of 0.5-2%.

2.3 POWDERS WITH COATED PARTICLES

Annealed HSS powder was mixed with Ti in a reactor and treated by ductile metal salt vapours. The titanium film thickness on the particles was controlled by varying the size of the Ti particles and the time of treatment.

In methods 1 & 3 the carbon content was raised by 0.25% for each percent Ti addition. Preforms and parts were produced using cold pressing, sintering and hot extrusion.

3 RESULTS AND DISCUSSION

Powders with particles sizes above 90 microns results when a Ti alloyed melt is water atomised at 1620-1650 degC. Introduction of low melting metals to the melt reduces the median particle size to 70 microns. The microstructure of the vacuum sintered preforms with 10-15% porosity shows a refinement of the carbide dispersion when using low melting metals (fig 1a, 1b). Note the absence of any depletion of alloying elements from the matrix such as is seen in the sintering of mixtures of HSS and TiC powders.

Heat treatment was carried out after hot extrusion. Hardness is raised compared to M2S without Ti. For example, with 0.86% Ti hardness was 66-67 HRc, and the treatment of the melt with low melting point metal also increases hardness a little.

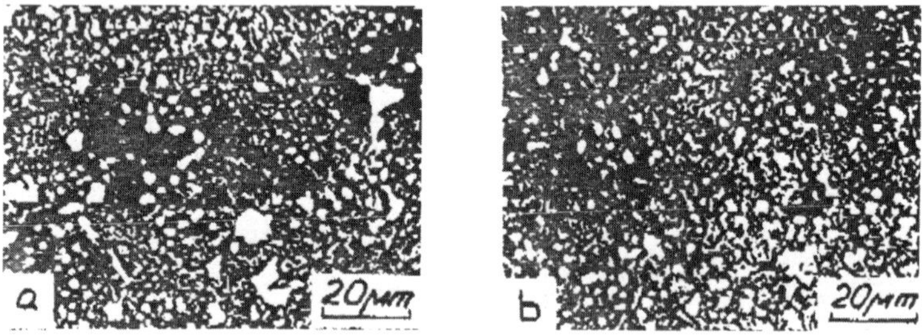

Fig 1 Structures of PMHSS with titanium additions by method 1.

a) Without low melting metal additions.
b) With low melting metal additions.

Producing powder by the second method particle size in unaffected and oxygen content slightly reduced. The structure after sintering and hot extrusion is fine but the fineness and phase distribution depend on the concentrations of additions in the water. This is clearly shown in fig 2, the structures of steels quenched from 1180 degC and tempered. Fig 2a is with 0.05% boric acid in the water, and fig 2b with 0.12%. In the latter case a boride phase is formed in a banded structure and the size of the martensite needles is increased.

Hardness is 66.5-67.5 HRc (at 0.05% Boric acid, 66.5 HRc), but further increases in the boric acid level doesn't increase hardness significantly, but leads to a sharp drop in toughness and strength.

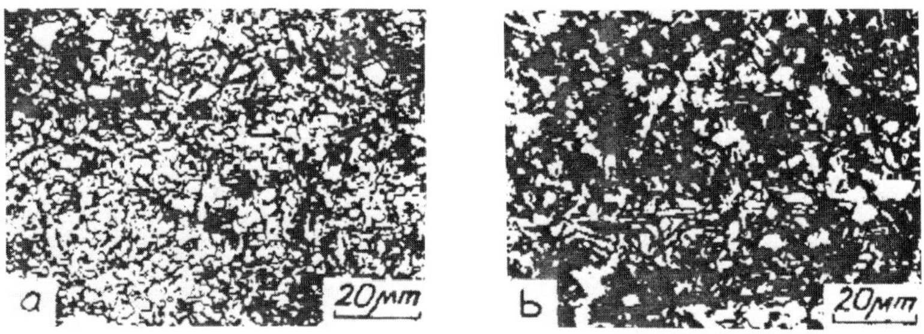

Fig 2. Microstructures of extruded PMHSS by method 2
a) 0.05% boric acid in the atomisation water
b) 0.12% boric acid in the atomisation water

Using the third method it is found that local eutectic melting and liquid phase formation occurs during sintering of the preforms leading to the formation of intermetallic compounds. Thus sintering at various temperatures can cause either shrinkage or growth of the preforms. The sintering temperature to control the liquid phase content and layer thickness must be controlled very precisely. The minimum porosity after vacuum sintering was 20% but was eliminated after hot extrusion with a reduction ratio of more than 12:1.

A discontinuous network of intermetallics formed in sintering at the grain boundaries and initial particle boundaries is destroyed by extrusion. After heat treatment maximum hardness is found at a titanium level of 6.2% - 69 HRc, but mechanical properties are inadequate with strength less than 2,000 MPa.

Inst. Phys. Conf. Ser. No 111
Paper presented at Int. Conf. on New Materials and their Applications, Univ. of Warwick, 1990

The use of glass coatings in HIPing of cold pressed HSS preforms

M Y Verkhhovtsev, A F Sanin, V K Nikolayev, V A Domoratsky, I I Shpirka.

State University, Gagarin Ave 72, Dniepropetrovsk-10, GSP 320625, The Ukraine, USSR.

ABSTRACT: This paper discusses the application of glass coatings to cold isostatically pressed preforms prior to HIPing. The glass properties needed are discussed and different methods of use are tested. The resulting structures of the HIPed product are reviewed.

1. INTRODUCTION

The large scale production of high strength components by HIP is being held back by the need to make complex steel or glass containers to transmit the pressure to the powder. These problems get worse for small components of complex shapes.

1.1 BASIC CONCEPT AND MATERIAL SELECTION

In view of the above problems, it is attractive to produce the complex shape by cold pressing and to subsequently coat the surface of the preform. The coating will transmit the HIP pressure while at the same time sealing the surface against gas leakage into the component. Different glasses or mixtures of glasses can be used as coatings. The main requirements for such materials for this application are:-

1 The softening temperature must be below the HIP temperature.
2 Gas must not pass through the coating into the component while HIPing.
3 The coating must permit gaseous products of reduction to escape at the temperatures at which they are evolved.
4 The expansion coefficient should match the component.
5 The glass must wet the material being HIPed.
6 The coating must be simple to apply.

2. METHOD

Powders of M2S, M15, and T15 high speed steels (HSS) were used. They were made by water atomisation to obtain an irregular powder suitable for cold pressing after annealing. The preforms were cold isostatically pressed to the shape required in a flexible elastomeric tool which was reused many times. The preforms so produced were treated in three different ways:-

1 Vacuum sintering; glass coating; HIPing.
2 The preform was placed in a crucible which was filled with glass powder and the assembly vacuum sintered and HIPed
3 A glass powder slip was applied to the surface followed by vacuum sintering and HIPing.

Glass mixtures based on Ba, Al, Si and P oxides with particle sizes finer than 50-100 microns were used. Vacuum sintering of the HSS preforms was carried out at temperatures that leave about 15% porosity so that no coarsening of the microstructure is observed, but oxides are carbothermically removed. The temperatures used for HIP were in the range 1100-1200 degC, and pressures in the range 80-120 MPa.

3. RESULTS AND DISCUSSION

Sintering using methods 2 and 3 was accompanied by considerable grain growth and by the emergence of a carbide network with eutectic appearance. This was apparently due to the influence of the glass melt on diffusion in the structure and the creation of eutectic by reaction at the glass-metal interface. The most significant influence on the eutectic network morphology was the content of P_2O_5 in the glass; higher levels of P_2O_5 lead to a greater thickness of the eutectic network.

The structure of the sintered steel made in the first method is satisfactory with fine grain and carbide sizes and no eutectic network. The porosity is open and interconnected.

Steel density increases after HIP. However, parts made using methods 2 & 3 are not of uniform density. At about mid-radius an area of high porosity is found, its thickness decreasing as HIP pressure rises. However, if the pressure rises above 100 MPa, radial cracks are found to start from this area. This defect is connected with incomplete removal of the gaseous products of reduction from the porosity during sintering due to blockage of the surface porosity by molten glass, and with the relaxation of stresses created by gas trapped in the component.

Using the first method these defects were not observed as the reduction products were completely removed during the vacuum sintering stage. In this case the steel structure is fully dense throughout the component.

Inst. Phys. Conf. Ser. No 111
Paper presented at Int. Conf. on New Materials and their Applications, Univ. of Warwick, 1990

Frequency response of resonant beam sensors

Rebwar M A Fatah
British Telecom Research Laboratories, Martlesham Heath, Ipswich IP5 7RE, UK.

Abstract: The equation of mechanical resonance can be applied to micromechanical resonators and relationships between the resonant frequency and the material properties can be established for the main engineering materials.

1) Introduction:

A resonant structure can be used as an accurate and a stable sensor. Several methods have been reported to sustain a microresonator at resonance but the most attractive method has used laser beams guided via optical fibre. If resonance can be sustained then external environmental parameters such as temperature and pressure can affect the mechanical state of the beam and thus change the frequencies at which resonance occurs. The shift in resonant frequency with some parameter of interest can be calibrated, and the device can be used as a sensor.

2) Theory and discussions:

The resonant frequency, f_i, of flexural vibrations of a homogeneous prismatic beam is given by:

$$f_i = \frac{\lambda_i^2}{2\pi L^2}\left(\frac{EI}{m}\right)^{\frac{1}{2}} = \frac{\lambda_i^2}{4\sqrt{3}\pi}\left(\frac{t}{L^2}\right)\left(\frac{E}{\rho}\right)^{\frac{1}{2}} \qquad (1)$$

where λ_i is a dimensionless parameter which is a function of the boundary conditions applied to the beam; L, t and ρ are the length, the thickness the density of the beam; m is mass per unit length. The resistance of a beam to bending depends upon its flexural rigidity EI, where I is the second moment of area and E is Young's modulus. According to (1), the resonant frequencies of a beam are determined by three factors. Firstly the end-conditions, secondly the geometry of the device and lastly the material properties. The resonant frequency does not depend upon the width of the beam. Assuming clamped end-conditions, the dependence of the resonant frequency upon the length and the thickness is shown in Fig 1.

Fig 2 shows the dependence of the resonant frequency upon the material properties, namely (E/ρ)value, for clamped end-conditions. Three points may be deduced from this graph:

(i) All the resonator materials obey a simple equation which can be given by:

$$f_i = \text{constant}\left(\frac{E}{\rho}\right)^{\frac{1}{2}} \qquad (2)$$

The constant is given in (1) and can be found from the slope of the line.

(ii) Unsuitable materials for mechanical resonators such as lead, wood and rubber do not lie on the line.

(iii) Keeping the geometry constant, it is possible to control the resonant frequency by changing the material, for example silicon beams have higher resonant frequencies than silicon dioxide. Aluminium coated beams have higher resonant frequency than expected due to the formation of aluminium oxide (Al_2O_3), at room temperature, the latter has a higher (E/ρ)value.

The devices studied here were coated with a thin metal layer. The metal coating changes the resonant frequency of the beam, and the equivalent value for EI must be used. Fig 3 shows the effect of common metal coatings on a silicon resonator beam.

In conclusion, we have shown that for a constant geometry there is a relationship between the resonant frequency and the (E/ρ) value for the main engineering materials. The theoretical effect of the metal coatings on silicon beams is also shown.

3) Acknowledgements:
The author wishes to thank K H Cameron, A L Matthews and P Cochrane from BTRL for all their comments, discussion and support. He is also grateful to the computing help of N W Pyke.

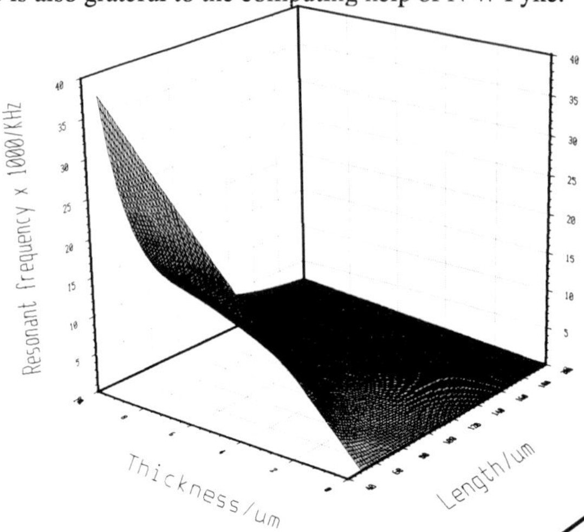

Fig 1:
The dependence of the resonant frequency upon the length and the thickness in a three dimensional plot.

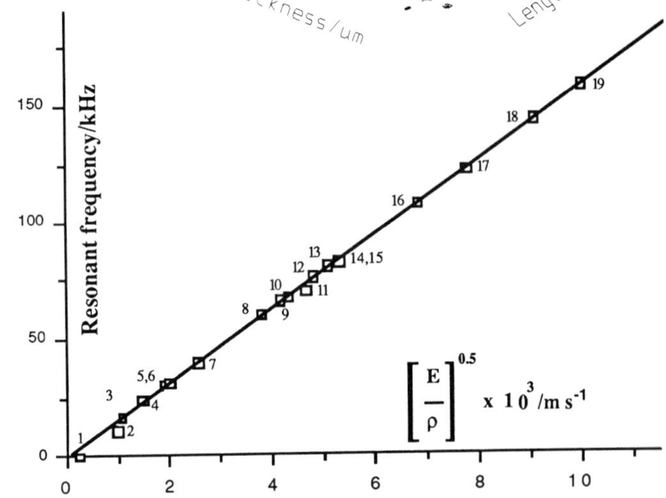

Fig 2: The relationship between material properties and resonant frequencies for the materials given below. It is assumed that the beam is made of silicon with dimensions of 180μm long and 0.5μm thick, and clamped end-conditions is considered. The materials are 1: rubber, 2: Pb, 3: PVF_2 , 4: Kapton, 5: Cr, 6:Au, 7: Tin, 8: Cu, 9: Invar, 10: Tungsten, 11: Ash wood along the grain, 12: GaAs, 13: Al, 14:SiO_2, 15: Stainless steel, 16:Si_3N_4, 17: Mica, 18: Si and 19: Al_3O_4.

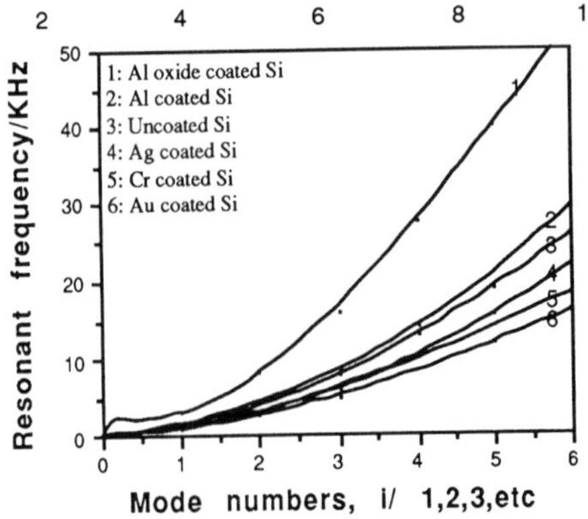

Fig 3:
The dependence of a silicon beam on different metal coatings. The beam assumed to have dimensions of 180μm long and 0.5μm thick, and coated with 0.2μm of metal.

Inst. Phys. Conf. Ser. No 111
Paper presented at Int. Conf. on New Materials and their Applications, Univ. of Warwick, 1990

Strength degradation of ground and etched Si surfaces

D.G. Chetwynd and S.T. Smith

Centre for Microengineering and Metrology, University of Warwick, Coventry CV4 7AL.

ABSTRACT: The design of a three point bending test machine for assessing the strength of small specimens is outined. Preliminary results are reported of the strength of sawn and ground silicon as a function of the surface depth removed by etching.

1. INTRODUCTION

There is currently much interest in the use of brittle materials within high performance mechanical systems. Ceramics provide unusual combinations of strength, stiffness, density and high temperature tolerance. Recent studies of the degradation of strength from ideal values of components produced by grinding brittle materials reveal that while isolated cracks on the surface can reduce the strength by a factor of ten (Puttick and Hosseini 1981, Lawn et al. 1981), ceramics ground under certain conditions have breaking strengths approaching the ideal (Cook et al. 1981, Marshall et al. 1983). This is probably because a surface compressive layer overlays a multiply fractured structure, evidence being that polishing glass ceramics can reduce their strength. This work seeks similar behaviour in single crystal silicon both because it is a 'model' material for machining studies and to establish practical guides to the manufacture of silicon elastic mechanisms.

Preliminary studies have been performed on specimens diamond machined by both sawing and high speed milling/grinding. The saw used is a Meyer and Burger external slicing machine using a 400 mm diameter blade. The milling uses miniature diamond burrs of up to 10 mm diameter mounted in high speed, air-driven spindles. These are held in the chuck of a CNC vertical milling machine and used as plunge and side cutters. The diamond wheels are dressed by machining a sacrificial workpiece to remove sharp asperities that are known to induce large scale cracking.

2. BREAKING STRENGTH TEST RIG

The fracture strength of the silicon specimens is measured using 3-point bending, a method chosen because it provides a localised maximum stress so that failures due to a large surface flaw are less likely. As shown in figure 1, a specimen is mounted horizontally on two hardened steel supports, a roller and a ball both 6 mm in diameter, spaced 35 mm apart. Two backstops provide accurate transverse alignment and complete a kinematic location which leaves free only axial translation. The specimen is loaded centrally between the supports by a 1 mm diameter roller which is mounted through a pivot and a load cell onto a micrometer driven by a D.C. motor and gearbox. Two LVDT displacement gauges contact the load roller to either side of the specimen, to monitor its deflection and rotation about a longitudinal axis as load is applied.

3. TESTING

An initial batch of 120 specimens of dimension 1.2 x 6 x 40 mm were cut from a 100 mm diameter silicon boule using the slicing saw. The crystal was oriented to better than 0.5° so that the tension surface during bending was in the (111) plane. Half of this batch was etched in a hydrofluoric and nitric acid solution for five minutes to remove a depth of about 25-30 μm from all the surfaces. One face of this set was then ground with a 10 mm diameter cutter rotating at 65000 RPM with a feedrate of 100 mm/s and a cut of 0.05 mm. The final pass was always in climb or downcut mode.Sub-sets of the sawn and ground specimens were then given short, timed etches to remove portions of the damaged surface layer. A small spot of varnish at the end of the ground ones protected that part from etching to give a step that directly measures the etch depth. All specimens were then loaded to breaking on the test rig. After fracture actual dimensions were measured and the maximum stress at failure calculated from linear beam theory.

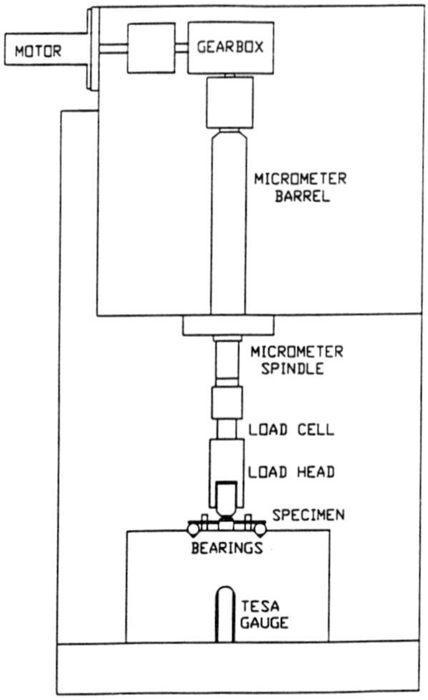

Fig. 1. Schematic of test rig.

4. RESULTS

For the sawn specimens, strengths scatter between 80 and 170 MPa, compared to an ideal expectation of about 190 MPa. Unetched specimens show the widest scatter, but also some strong samples. Etching for longer periods tends to reduce scatter and so increase typical strengths somewhat. Figure 2 shows strength against etch depth for the ground specimens. A more consistent and, on average, stronger surface is produced. Etching up to about 2 μm tends to slightly weaken the typical specimen but strength is regained as greater depths are removed.

Electron micrographs of the fracture surfaces of sawn specimens show three distinct regions. The surface and subsurface to a depth of about 20 μm consists of a mosaic fracture topography. Then there is a sharp transition region of low damage followed by the bulk material which has striations caused by overlapping cracks. The first few micrometres at the surface may also have a structure distinct from the initial mosaic topography. The layered structure was not seen on ground specimens and so is absent or extremely thin.

5. DISCUSSION

The rougher surfaces produced by the slower surface speeds of the saw can be at least as strong as those after etching unless a large depth is removed. This seems to confirm that a high density of cracks can be beneficial because their interaction prevents the degree of weakening expected from an isolated crack of similar size. All cracks seem to penetrate to about the same depth, so the dislocation free zone expected ahead of them (Hirsch et al. 1989) may produce a distinct layer.

High speed grinding is a more readily characterisable process providing a higher average strength with less variation between specimens. Surfaces etched down a few micrometres tend to be weaker than the as-ground ones. This suggests that a thin surface compressive layer is constraining initial crack opening at moderate stresses. This behaviour is consistent with x-ray studies of crystal damage (Smith 1990).

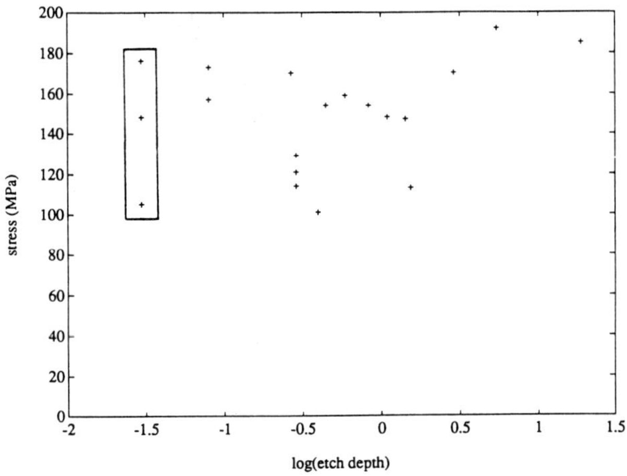

Fig. 2. Strength v log(depth) refered to 1 μm; unetched specimens boxed.

The surface appearance and the high strengths obtained suggest that the high speed grinding is cutting in a manner similar to the 'ductile grinding' mode observed in brittle materials on very stiff machines taking light cuts. As these conditions do not prevail here, it is believed that beneficial dynamic effects are occurring. These may have considerable practical significance and further investigations are proceeding.

6. ACKNOWLEDGMENT

The authors are very grateful for the considerable help provided by Mr. John Chattock in commissioning and operating the test rig.

7. REFERENCES

Cook R.F., Lawn B.R., Dabs T.P. and Chantikul P. (1981) Comm. Am. Ceram. Soc. **C64** 121-2.

Hirsch P.B., Roberts S.G. and Samuels J. (1989) Proc. Roy. Soc. Lond. **A421** 25-53.

Lawn B.R., Marshall D.B. and Chantikul P. (1981) J.Matl.Sci. **16** 1769-75.

Marshall D.B., Evans A.G., Khuri Yakub B.T., Tien B.T. and Kino G.S. (1983) Proc. Roy. Soc. Lond. **A385** 461-75.

Puttick K.E. and Hosseini M.M. (1981) J.Phys.D:Appl.Phys. **13** 875-80.

Smith S.T. (1990) J.Phys.D:Appl.Phys. **23** in press.

Inst. Phys. Conf. Ser. No 111
Paper presented at Int. Conf. on New Materials and their Applications, Univ. of Warwick, 1990

Design, construction and test of a fast light source for transient photoconductivity experiments

S M VAEZI-NEJAD+, D B Shiers+ and C Juhasz*

+ Electronic Research Group, School of Engineering, Thames Polytechnic, Wellington Street, London, SE18 6PF, UK.

* Solid State Group, Department of Electrical Engineering, Imperial College of Science, Technology and Medicine, Exhibition Road, London, SW7 2BT, UK.

ABSTRACT: The purpose of this paper is to describe the construction and performance of a simple and inexpensive light source based on a capacitor discharge gap which can produce light pulses of about a hundred nano-seconds duration. The device is particularly suitable for the Time of Flight (TOF) type of transient photoconductivity experiments.

1. Introduction

There is considerable interest in charge transport in amorphous semiconductors using the Time of Flight (TOF) type of transient photoconductivity experiments. In the previous 'New Materials and their Applications Conference' at Warwick University, two of the authors presented TOF experimental data in amorphous selenium based photoreceptors using a modified commercial xenon flash as the light source for the photogeneration of charge carriers [1]. However, minimum achievable Full-Width-Half-Height (FWHH) for such a light source is ~ 0.5μs which causes two fundamental limitations. Firstly, the material under test must be of low mobility so that for a given thickness (typically a few tens of micrometers), the transit time is greater than 1μs [2,3]. Secondly as the transit time is inversely proportional to the applied voltage across the sample, it might not always be possible to study charge transport over a reasonable rang of applied fields (10^4-10^6 Vcm^{-1}). In this paper, we describe a simple and inexpensive light source based on a capacitor discharge gap which produces light pulses much shorter than that of an xenon flash.

2. The Light Source

Capacitor discharge spark gaps have been used as a high speed light source for ultra-high speed photography. It has been reported that intense short light pulses can be produced reliably [4]. The apparatus we used to construct a capacitor discharge spark gap type of light source were; a high voltage dc power supply (~5kV), a high voltage RC network, a spark gap cavity and a trigger circuit to initiate firing of the gap.

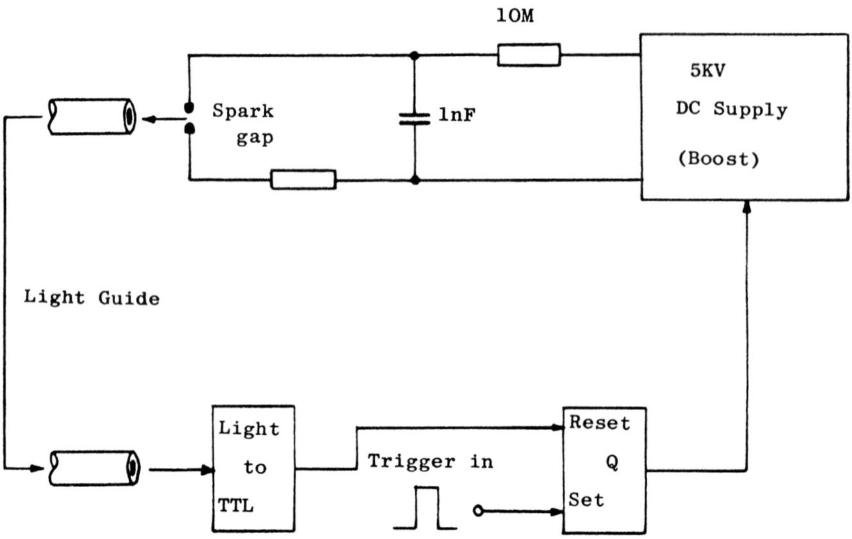

Fig 1. A schematic diagram of the fast light source

The circuit shown in Fig 1 was used to charge the capacitor to within 20% of the firing point when the light pulse was required. As soon as the spark ignites, a signal from the light source is sent back to the control. The high voltage (H.V) boost is obtained by the logic control of a relay in the primary circuit of the H.V transformer. The relay is arranged to short one arm of a potential divider across the mains supply. The transformer is connected across the other arm of the potential divider and it receives full mains voltage when the relay is closed. The precise timing of the capacitor discharge in the spark gap is not essential in our application but it can be done by a third electrode connected to a high voltage switch.

As shown in Fig 1, the logic circuit forms the basis of the trigger. The condition of the bistable (set/boost or reset/-20%) is determined by the trigger signal and the feedback signal respectively. The trigger signal can external TTL or manually via a debounced switch. The feedback or reset signal is derived from the light pulse received via the light. An Op-amp in the current to voltage conversion mode is used to convert the light pulse into a TTL compatible signal.

The spark gap cavity itself consisted of electrodes which were machined from 6mm brass studding to facilitate easy adjustment of the gap length. The electrodes were screwed horizontally into a tube through the insulating rings at 180 degrees to each other on the circumference. The electrode tips were both machined hemispherically with a 2mm radius. A 50mm tube was used to allow insertion of the standard optical component. A spherical mirror was placed at a distance of one radius from the spark gap to collect the back light and focus it on the source, thereby adding to the forward light. With the tube sealed by the back reflecting spherical mirror at one end and the lens at the other, the spark gap is protected from the outside interference (dust, objects etc).

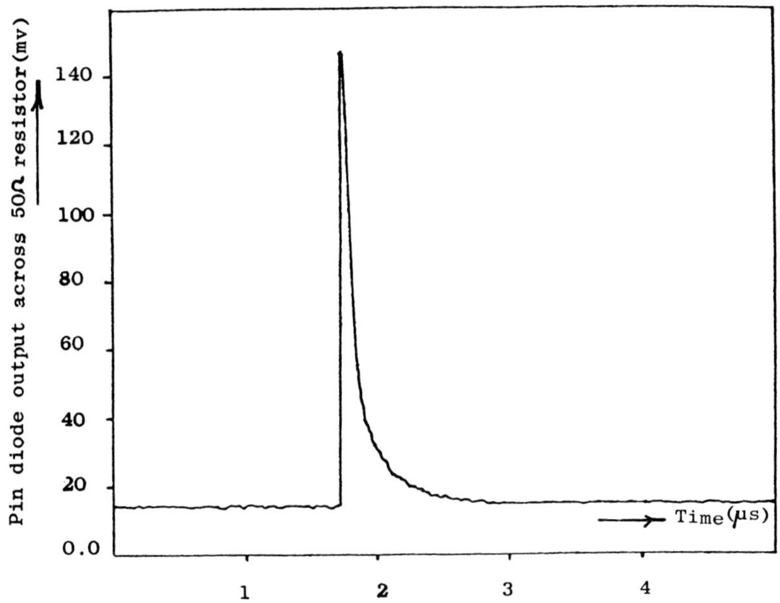

Performance of the light source was examined using a high speed PIN diode, an Op-amp and a short length of optical fibre. Under a single shot operation, short light pulses with FWHH of 100ns were detected. A typical output is shown in Fig 2. When the system switched to permanently boost, it free ran at about 10Hz. The light can be focussed onto a 1cm visible spot with a 5cm focal length bi-convex lens.

3. Conclusion

Construction, design and performance of a reliable and economic light source for transient photoconductivity experiments were described. It was shown that short light pulses of the order of 100ns can be easily obtained, The authors strongly believe that shorter light pulses of order of a few tens of nano-seconds can be achieved by further improvements.

4. References

[1] VAEZI-NEJAD S.M. and JUHASZ C, New Materials and their Applications, 1987. Proceedings of the Institute of Physics Conference held at the University of Warwick, 22-25 September 1987, p.283.

[2] VAEZI-NEJAD S.M. 1987, Int. J Electronics, Vol. 62, No. 3, p. 361.

[3] VAEZI-NEJAD S.M. and JUHASZ C, 1987, Semicond. Sci. Technol. Vol. 2, p. 809.

[4] FISHER H. 1963, 6th Int. Symp. on Ionisation Phenomena in Gases, July 1963. Paris.

Paper presented at Int. Conf. on New Materials and their Applications, Univ. of Warwick, 1990

Spray pyrolysis of $Bi_2Sr_2CaCu_2O_{8+\delta}$

T. O'Brien, W. Blau
Dept. of Pure and Applied Physics,Dublin University, Dublin 2, Ireland

Abstract. We investigated spray pyrolysis as a method of production of superconducting films of the BiSrCaCu family , and identified the process variables which govern deposition. The properties of the sprayed films are discussed and possible improvements to the process are suggested.

Introduction.

In the excitement following the discovery of the high T_c superconductors, researchers spent considerable effort in attempts to prepare films of the new materials. Among the many possible techniques (laser ablation/deposition, sputtering, C.V.D. etc.) spray pyrolysis has several compelling advantages for the researcher. First and foremost of these is cost. This is not the whole story of course. Sprayed films are unlikley to compare favourably with the best films produced using laser ablation or by M.B.E., but the initial investment is correspondingly smaller. Pyrolysis is a suitable method for the production of dense polycrystalline, oriented superconducting films approximately 1µm thick. Polycrystalline superconducting films will be useful in many applications, particulary in the field of passive microwave devices. The role of grain boundarys or other defects in many of the fast photoconductive signals reported is unclear. It is possible that well controlled polycrystalline films will be neccessary for photodetector applications, from microwave to visible wavelengths.

The technique of spray pyrolysis is very straightforward[1,2].The following diagram (Figure 1) illustrates the general experimental set up. First, a solution of salts of the elements required is prepared. In this case the elements are Bismuth, Strontium, Calcium, Copper and sometimes Lead. This solution is atomised by passage through a nozzle with a high pressure gas. The spray is directed onto a substrate, which is heated to a predetermined temperature. The droplets of liquid evaporate in the hot region near the substrate, depositing grains of salts or oxides. Any inhomogeneity at this stage will ultimately be seen in the quality of the superconducting film.
The process variables must be carefully controlled to produce single phase films of the $Bi_2Sr_2Ca_{n-1}Cu_nO_{2n+2}$ familywith good mechanical and electrical properties.

Experimental

We will discuss the results under two headings. Firstly our observations about the process itself, and secondly the properties of the superconducting films. Precise values of process variables such as substrate temperature or gas pressure, depend on specific details of the spraying set up, so in addition to the values we found, I will indicate which factors in the process affect them.

Pyrolysis followed by high temperatue firing, produces polycrystalline films, so the choice of substrate depends on the intended application of the film, as well as on the probability of film substrate reactions at the heat treatment stage. Single crystal or polycrystalline MgO is suitably inert at the temperatures typicaly employed, and its optical properties suited our requirements. Films on $SrTiO_3$ have very similar properties to those on MgO, though the actual deposition conditions depend on the substrate's thermal conductivity and thickness.

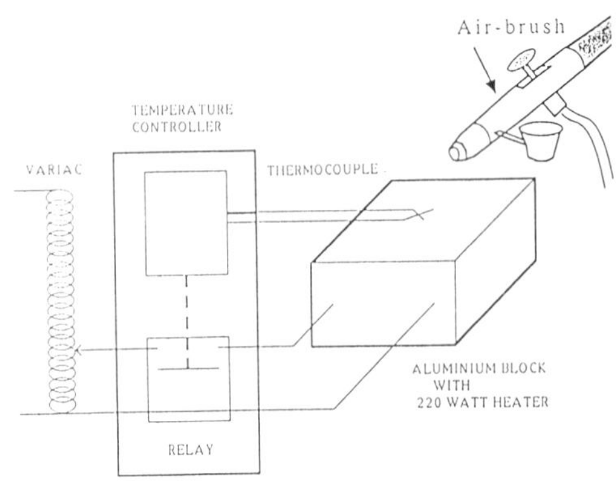

Figure 1. The experimental apparatus

The spray nozzle is maintained 5cm from the substrates. During spraying the nozzle is moved continuously.

We sprayed films from nitrate, carbonate, and acetate solutions. Of these both the nitrates and the acetates produced good films, though the nitrate solution produced noxious gasses at the spraying stage. Because of this we used acetate solutions for most of our experiments. Both nitrates and acetates require acidic solutions to remain fully dissolved, and 0.05M total metal ion concentration was the strongest solution we found practical to use. Solutions with ion stoichiometries (Bi:Sr:Ca:Cu) between 4:4:2:4 and 4:3:3:4 produced the best results.

The substrate temperature was regulated by a proportional+integral term type temperature controller, which switched a 200W resistive heater in the aluminium block (see figure 1). This arrangement doesn't give an accurate measurement of the substrate surface temperature, since this will depend on the cooling load. However this approach allows much greater repeatability than a surface temperature measurement, and avoids the control problems that the rapid surface variations produce.

Evaporation of spray droplets causes very rapid cooling of the substrate. The cooling rate depends directely on the spray gas pressure. As a result we found that substrate temperature and spray gas pressure could not be adjusted independently while maintaining film quality. For a given gas pressure there is a minimum temperature below which droplets of liquid can form on the substrate. There is also a maximum temperature above which very little material is deposited. Figure 2 indicates the region in which homogenous films can be deposited with our apparatus.

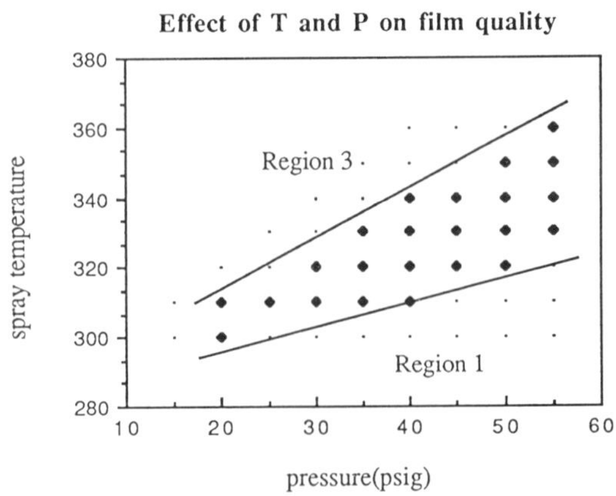

Fig. 2 Effect of substrate temperature and gas pressure on film quality.

At low temperatures the rate of evaporation at the substrate is less than the rate of arrival of spray droplets (which is fixed by gas pressure and solvent addition rate). In this region, while spraying continues, droplet size will increase, leading to inhomogeneity of the film (region 1 of Fig.2). Increasing the substrate temperature increases the rate of evaporation, thus decreasing the droplet size and improving uniformity of the film, until a temperature is reached at which the average droplet is likely to evaporate rather than grow. This temperature corresponds to the lower line in Fig. 2. In addition, we find that above a certain temperature the rate of deposition is extremely low (corresponding to the upper line in Fig. 2) . This suggests that, either the evaporation products of droplets of less than a certain size do not remain on the substrate, or that above some temperature the spray droplets themselves do not reach the surface, and are carried away in the gas stream. The second explanation accounts for the fact that we find a range of temperatures in which films of similar character can be deposited, rather than a gradation of properties up to the temperature at which deposition effectivley stops. With this interpretation, the region of successful deposition on Figure 2. corresponds to the region in which single droplet evaporation builds up the film.

As spray gas pressure is increased, or as the rate of liquid addition is increased, the range of temperatures at which good films can be deposited moves up in temperature. This is probably largely due to the higher cooling rate , requiring a higher temperature at the back of the substrate, in order to maintain the same deposition conditions.
The above considerations apply to deposition of the first layers on the substrate. It is the quality of these layers which is crucial to the quality of the final film. We usually raise the temperature of the substrate by 20-30°C for further deposition [3].

The solution concentration we used was the result of a compromise. If the concentration is too low, long spraying times are required, which makes temperature control more difficult. High concentrations on the other hand lead to very rapid and non-uniform deposition. We found total metal ion concentrations of 0.02M for acetates, and 0.04M for nitrates to be satisfactory.Once the concentration is fixed, the rate of addition of solution to the spray gas, determines the deposition rate. It also largley determines the cooling load on the substrate. The lowest possible addition rate consistent with a uniform spray pattern is best for ease of control of the process.

After spraying, the film consists of an even layer of oxide grains loosely adhered to the surface. Films sprayed in the 'single droplet' region, have grains from 0.5-1µm. If the film is uniform and homogeneous, a simple firing in a preheated furnace at 850°C for 1 hour followed by a rapid cool down to 700°C (50°C/min) forms the superconducting 2212 phase. Any inhomogeneity however in the as sprayed film can lead to segregation of the constituents in the final film,

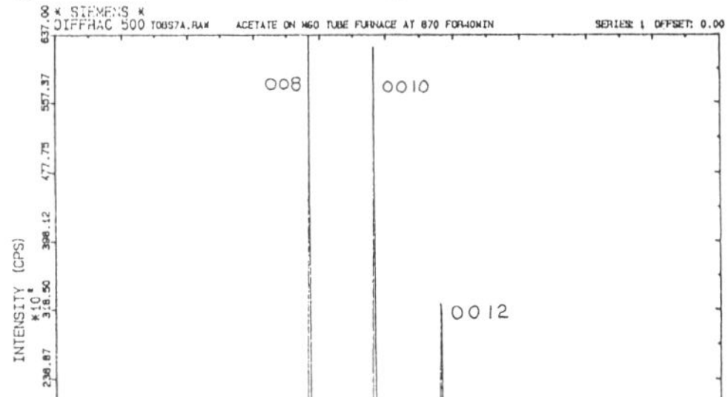

Fig. 3 X.R.D. of a film sprayed from acetate precursers onto MgO. Only the 00l reflections are evident.

producing a mixture of phases and unwanted intermetallic oxides .

The Films

The films produced were assessed using optical and scanning electron microscopy, X ray diffraction and energy disspersive X ray analysis, and by measurement of their electrical characteristics.Table 1 summarises the conditions which produced the best films on each substrate. The films produced on all substrates were polycrystalline, with a preferential orientation, c axis normal to the substrate. See Figure 3. The crystallite size of the purer films was generally 2-5µm.(Figure 5.)Most films produced contain more than one member of the BiSrCaCuO family [4], the n=2 phase (the 85K superconducter) predominates, often with some n=1 material present. The resistance curve of a sample which appeared virtually pure 2212 by X.R.D. is shown in Figure 4. The behaviour is metallic above the transition temperature, and the transition itself is quite sharp. The drop from 90% to 10% of the pre-transition resistance occurs within just 2K. This curve is typical of the better films. Table 1 summarises the results for various substrates.

Films containing a small amount of the n=1 phase (2201), show a non metallic, often distinctly semiconducting type of resistance behaviour above transition. In addition the transition for these impure films may be very broad, and may sometimes be incomplete.

With the purer films, we have measured current densities of $5x10^3 A/cm^2$ at 77K. This however may not represent the critical current of these films , as no systematic measurements of I_c have been made, and the problem of high contact resistance is severe for this measurement.

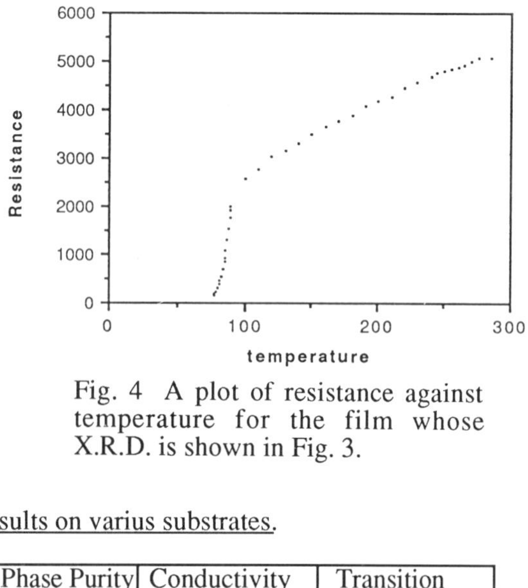

Fig. 4 A plot of resistance against temperature for the film whose X.R.D. is shown in Fig. 3.

Table 1 Summary of results on varius substrates.

Substrate material	Spray Temp @50psig.	Firing temperature	Phase Purity	Conductivity above Tc	Transition
MgO	340	850	single phase	metallic	Tco > 80K
Periclase	330	850	single phase	metallic	Tco > 80K
YSZ	400	860	single phase	metallic	Tco > 80K
SrTiO3	360	850	2212+2201	metallic/semi	Tco > 77K
Al2O3	350	830	multi phase	semiconductor	No Transition
Saphire	350	830	multi phase	semiconductor	No Transition

Conclusions:

Spray pyrolysis is a viable method for the production of granular superconducting films of BiSrCaCuO on inert substrates. The superconducting layer is formed from a melt reaction of the deposited layer of mixed oxides. The only parameters which affect the film homogeneity, are spray droplet size, and solution concentration, as long as deposition takes place in the 'single droplet ' region of Figure 2 . Achieving reliable chemical purity of the films is the major difficulty. A reduction in spray droplet size, or an ability to spray more dilute solutions successfuly could improve this situation.

Having said that, the films' properties compare well with those of films produced by other thick film processing techniques, such as screen printing[5].

Fig. 5 This is an SEM of an edge scratched into a 2212 film. The flat crystallites of superconducter can be clearly seen.

1 A.Gupta, G.Koren, E.A.Geiss, N.R.Moore,E.J.M.O'Sullivan and E.I.Cooper, A.P.L. 52, 163 (88)

2 Dale F Vaslow, Gunter H. Diekmann, David Dawson Eli, A.P.L. 53, 324

3 X.Z.Wang, D. Baurle, J.F. Lawlor, E. Davitt and J.M.D. Coey

4 M.Tarrascon,W.R.McKinnon,P.Barboux,H.W.Hwang,R.G.Bagley,L.H.Greene, G.W.Hull,Y.LePage,N.Stoffel,M.Giroud, Phys. Rev. B38, 8885

5 Kazutomo Hoshino Hidefusa Takahara and Masao Fukutomi J.J.Appl.P Vol 27,No.7, ppl1297,1299

Inst. Phys. Conf. Ser. No 111
Paper presented at Int. Conf. on New Materials and their Applications, Univ. of Warwick, 1990

Phlogopite micas as thermal actuators

L. Cartz, F. G. Karioris and X. Yang
Marquette University, Milwaukee, Wisconsin, USA

ABSTRACT: Selected phlogopite micas display anomalous reversible thermal expansion, as much as 300% , perpendicular to the cleavage planes, on heating up to $900°C$. The vaporization of traces of occluded H_2O accounts for an abrupt increase ($\approx$ 100%) in volume at the boiling point of water,and above this onset temperature, the thermal expansion is quantitatively similar to that of an ideal gas. The onset temperature increases with applied pressure approximately as the vapour pressure of H_2O. The formation at temperature of lenticular bubbles on dislocation networks can be observed by transmission electron microscopy.

The thermal expansion behaviour of selected phlogopite micas has been studied using a Perkin-Elmer Series 7 thermomechanical analyzer. Typical data of expansion versus temperature after 14 reproducible heating-cooling cycles are show in Fig. 1. The thermal behaviour observed for the phlogopite mica is anomalous in several respects when compared to most other solids. In the region of the curve OA, heating from room temperature to about $110°C$, the expansion is like that of other solids with a temperature coefficient of $\approx 10^{-5}\ °C^{-1}$.

In the region AB, $110°C - 130°C$, the expansion is abrupt and very large; the expansion coefficient in this region is $\approx 10^{-2}\ °C^{-1}$ which is $10^2 - 10^3$ times greater than that of typical solids. The region BC shows a linear relationship of expansion with temperature, but the expansion coefficient in this region is about 10 times that of typical solids. The cooling portion of the cycle CDE does not retrace the heating curve; this hysteresis effect is another anomalous aspect of the behaviour. Finally, if the maximum value of the temperature is increased, a new cyclic pattern is followed. Thermal behaviour of phlogopite micas very similar to that shown in Fig. 1. is reported by Hidnert and Dickson [1] where, in some cases, the cooling curve lies below the heating curve.

The anomalous expansion of the phlogopite mica can be explained in terms of the water occluded between the crystal structure layers. Natural phlogopite micas are known to contain varying amounts of water, up to 4 weight per cent[2]. There is a very rapid expansion of $\approx$ 100% from $100°C$ *to* $200°C$, and another 200% linear increase of the expansion from $200°C$ *to* $600°C$. Since the expansion is perpendicular to the silicate layers, the ordinate in Fig. 1. represents the volume expansion as well.

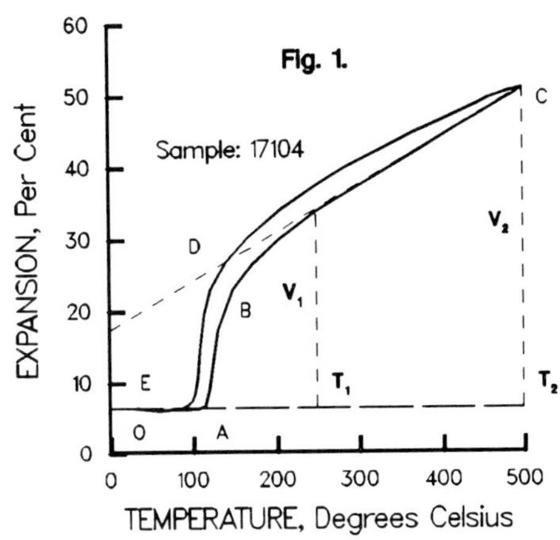

The water content needed to account for the initial $X\%$ of the expansion, segment AB in Fig.1., can be estimated from the change in volume $\Delta V = (X/100)V_M$ where V_M is the volume of mica. Let m represent the mass of water in the test sample and ρ the density of mica. Assuming that the ideal gas law holds for water in mica, each gram of water will produce $\approx 1723\ cm^3$ of water vapour at $373\ K$ and at one atmosphere pressure, it follows that the mass fraction of water content needed to produce an abrupt $X\%$ increase in volume is

$$m/M = (X/100) \times (1/1723) \times (1/\rho).$$

This calculation suggests that as little as 0.02% by weight content of water in the mica can account for a 100% expansion in volume at AB.

The effect of increasing pressure on the thermal expansion of the mica specimen is given in Table 1. where the onset of the maximum expansion AB moves to higher temperatures at higher pressures. The boiling points of water with pressure, taken from standard tables, are compared in Table 1. with the temperatures of maximum expansion at different applied pressures, when good agreement is obtained.

Table 1.

Pressure psia	Boiling Celsius	Onset Celsius
15	100.6	103
22	111.5	116
29	120.2	125
42	132.3	130

The behaviour of the *expanded mica* is quantitatively similar to that of an ideal gas in the region BC of Fig. 1. For example, the ratio of volumes $(V_2/V_1) =$ 1.6(3) where the ratio of absolute temperatures $T_2/T_1) = 1.4780$), in reasonably good agreement with Charles' law for a gas at constant pressure. The slope of the section BC varies with applied pressure approximately as one would expect.

Thermogravimetric analysis up to $1100°C$ shows that the mica is stable with no loss in weight below $900°C$. Transmission electron microscopy studies have been carried out at high temperatures [3,4] when observations have been made of bubble formation on dislocation networks and of the growth of lenticular flaws.

Different phlogopite micas exhibit different thermal expansion, varying from below 1% up to 300% at $600°C$. These variations depend on the water content of the natural mineral. Studies are in progress concerning the quantity, nature and environment of the occluded H_2O. Thermal actuator devices are envisaged using synthetic micas.

This work has been supported by the U.S. Army Research Office and by Johnson Controls, Inc. Milwaukee, WI, USA.

REFERENCES

1. Hidnert,P. and Dickson, G., *J. Nat. Bur. Stan.*, **35**, 309-353 (1945).
2. Beer, W. A., Howie, R. A. and Zussman, J., *Rock-Forming Minerals*, **3**, (Longmans 1963), pp. 46-47.
3. Cartz,L. and Trooper,B., *J. Appl. Phys.*, **36**, 2783-2787 (1965).
4. Cartz, L. and Trooper, B., *Proceedings of the Third European Conference on Electron Microscopy*, Czechoslovak Academy of Sciences, Prague 1964, p. 335.

Inst. Phys. Conf. Ser. No 111
Paper presented at Int. Conf. on New Materials and their Applications, Univ. of Warwick, 1990

Some features of the fracture behaviour of structural ceramics

A S Shatalin
Research and Industrial Firm "Tekhnologia", Obninsk, USSR

ABSTRACT

Some specific features of the fracture behaviour of structural ceramics manufactured on the basis of silicon nitride and silicon carbide, zirconia and silicon oxide will be discussed in comparison with metal - a classical structural material.

Effects of chemical and phase compositions of the materials, of their structure, defects and test temperature on deformation and fracture of ceramics under loading conditions are studied. A qualitative agreement of the fracture character with the lattice properties is observed. It is supposed that the grain boundary sliding with the turning of the fragments formed by the cracks and grain boundaries can change stress concentrations in local areas and partially compensate the absence of plasticity in ceramics.

KEY WORDS

Structural ceramics, composition, defects, load, fracture, deformation, lattice properties, brittleness, plasticity.

INTRODUCTION

Information about specific features of the fracture behaviour of structural ceramics is important for the development of mathematical models, improvement of the degree of certainty of the load bearing capacity, service life and reliability evaluations of ceramic components, and for the choice of research trends for the development of ceramic materials. Some properties of the tested materials are given in Table 1. Fig. 1 shows a typical structure of materials manufactured on the basis of silicon carbide with the exception of the reaction-bonded silicon carbide.

TABLE 1.

Property	SiO_2	OTM-355 $ZrO_2+6\%Y_2O_3$	OTM-904 PC Si_3N_4	OTM-907 densified PCSi_3N_4-Y_2O_3-MgO	OTM-912 SiC-AlN-Y_2O_3	OTM-906 HPSi_3N_4+9% Y_2O_3	OTM-914 HPSi_3N_4+ 18%Y_2O_3
Density, g/cm^3	1.96-2.05	5.68	2.5-2.7	3-3.3	3.15-3.30	3.34	3.45
Porosity, %	9-11	0	10-16	0-5	0-1	0	0
σ, MPa, 20 °C	40	520	210	500	360	660	830
1500 °C	-	160*	220	400**	360	390	620
σ 1200 and 100, MPa	-	100**	-	-	-	300	450
σ 1300 and 100, MPa	-	-	150	-	-	300	450
σ 1400 and 100, MPa	-	-	150	-	-	250	400
E st., MPa 10^{-4}	4.0±1.5	17	14.7±0.5	21.6±0.9	27.4±2.6	24.2±0.6	25.6
E d., MPa 10^{-4}	-	-	16.4±0.5	24.8±0.8	31.9±1.7	28.8±0.6	29.4±1
	-	0.29	0.22	0.25	0.21	0.27	-
Kic, MPa $m^{-1/2}$, 20 °C	-	8.25	2.0±0.5	5.5±0.4	5.4±1.3	7.5±1.5	7.9±1.5
1200 °C	-	2.4	2.5±0.5**	-	4.1±0.5	6.0±1.0	6.6±0.5
1400 °C	-	-	3.5±0.5	-	4.1±1.2	8.5±2.5+	7.7±2.2
m	-	-	5.9	-	5.8	8.4	8.6
γ_{ef}, j/mole, 20 °C	1.2±0.3	268	19±3	-	-	36±4	-
1200 °C	-	-	34±5	-		48±6	-
, k^{-1} 10^6, 20-1500 °C	0.9	10	3.7±0.2	3.2±0.1***	4.8	3±0.2	4.4±0.5
Oxidizability, 50 h, 1300 °C, %	-	-	1.0	0.07±0.03	0.16	0.06	0.1±0.06

Note: *at 1300, **-1000, ***-20-90+, -1500 °C

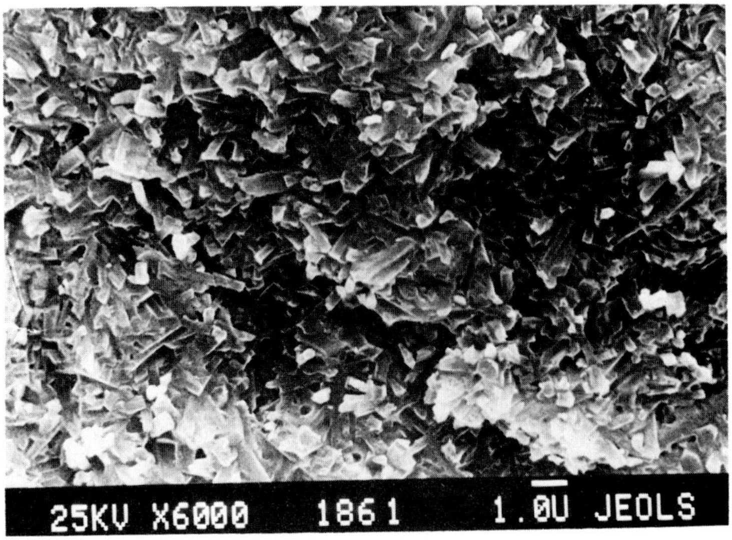

Fig. 1. Typical structure of materials made from silicon carbide and silicon nitride with elongated grain structure.

EFFECTS OF LATTICE PROPERTIES

The rate of dislocations density growth and the mobility of dislocations are limited by the following: bonding forces between atoms in lattices with covalent and ionic bonds are stronger in comparison with metallic bonds, for example, energy of interaction between atoms in chrome diboride (6.05 ev/at.) is more than two times higher than in pure chrome (2.16 ev/at.) [1]; stress fields are more complex; dislocation sliding systems are smaller in number; densities of dislocations in newly grown crystals are by several orders of magnitude smaller; the bonds in covalent crystals are directed [2] - all these factors are responsible for low plasticity, brittleness of fracture, large scatter in strength data, more significant effect of the defects and local stresses on the fracture of ceramic materials in comparison with metals.

EXPERIMENTAL RESULTS AND DISCUSSION

Experimental studies of the materials and components in the stages of design, testing and usage allowed us to establish the following features of ceramics fracture process. All the materials experienced brittle fracture when tested at temperatures up to 1000-1200°C and load rates in bending of 0.2 mm/min and more. The linear dependence of deformation with load illustrates this

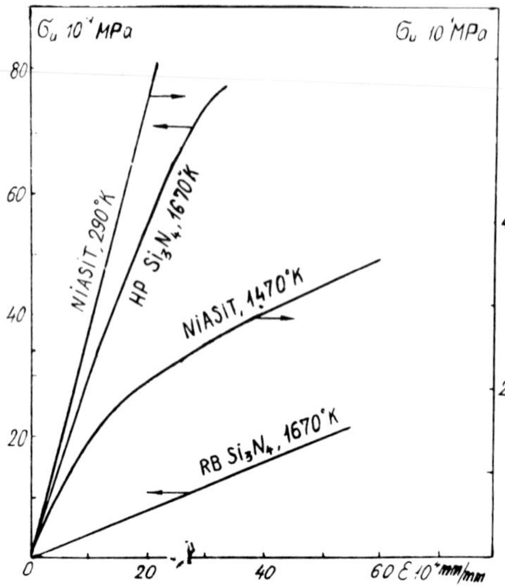

(Fig.2).

Fig. 2 Deformation curves for ceramic materials, load rate ≥ 0.2 mm/min.

Oxide materials experience significant deformation at ≥ 1200°C. Deformation of the samples made from ZrO_2-6wt%Y_2O_3 with the grain size 0.1-0.3 μm at the load rate 0.08 mm/min at temperatures ranging from 1200 to 1300°C is so significant that they flow almost without any increase of the applied load.(Fig.3).
Pure reaction-bonded silicon nitride exhibits brittle fracture both during short duration and long duration tests. Residual deformation did not exceed 0.1% after 150 hours at 1500°C and a stress value of 75% of σ_b. After testing for 100 hours at 1550°C and under stress of 0.5 σ_b residual deformation remained the same.

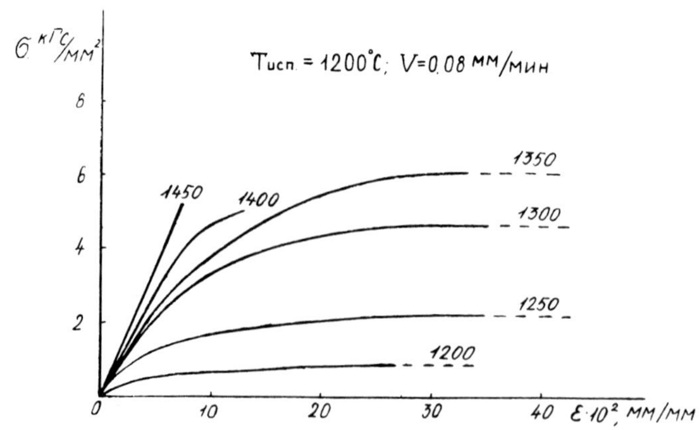

Fig.3 Deformation curves for the samples made from ZrO_2-6%Y_2O_3 fired at different temperatures and tested in bending.

Hot-pressed silicon nitride with the addition of 9 and 16wt% Y_2O_3 begins to show significant deformation at 1400°C during short time bending tests, but during long time tests deformation occurs already at 1200°C (Table 2).

Table 2

Material	T, °C	σ_b in tensile surface, MPa	τ,h	deformation of tensile surface,%	σ_b^{20} after test, MPa
1	2	3	4	5	6
$HPSi_3N_4$+8% Y_2O_3	1200	200	300	0.037	525
		250	300	0.047	730
		300	300	0.055	625
	1300	200	300	0.21	660
		250	300	0.24	650
		300	300	0.41	685
	1400	200	115	0.51	1015
		250	115	0.50	555
	1200	350	300	0.09	710
		400	300	0.11	810
		450	300	0.09	840
	1300	350	300	0.31	840
		400	300	0.40	760
		450	300	0.44	790
	1400	200	120	0.49	630
		250	120	0.76	580
		300	120	0.88	650
		400	156	1.19	520

Residual deformation increases with an increase of temperature and yttria content. After testing most of the samples had strength not lower than the strength they had before testing. Retention and even an increase of the samples strength above the average residual strength value in comparison with the starting strength value may be attributed to the processes of oxidation, recrystallization, phase content changes and, perhaps, diffusion of individual phases in the near surface layer with the thickness up to 100 μm (Table 3).

Table 3.

Phase composition	d, HKL, Å	Starting sample		1400 °C, 156 h, σ =350 MPa					
				Lateral surface		Tensile surface		Compression surface	
		CSA, μm	J, imp./sec	CSA, μm	J, imp./sec	CSA, μm	J, imp./sec	CSA, μm	J, imp./sec
$10Y_2O_3$-$9SiO_2$ Si_3N_4	3.39	2	226	15	434	3.5	226	15	2055
$Y_2Si_2O_7$	5.32	0.06	2.477	0.05	3049	0.4	72.17	0.05	2477
β-Si_3N_4	2.66	0.5	3833	15	5556	15	13824	15	1800
	2.49	0.5	3471	15	775	15	2843	15	1800
α - crys-tobalite		-	-	0.05	646	0.06	1045	0.03	836

Changes of the area of coherent scattering of the phases (CSA) and intensity of the peaks in X-ray photographs depend on the state of stresses as can be seen from table 3. Additionally, cracks with a depth up to 200 μm were found on the tensile surface, and there were healed cracks in the middle of the samples (Fig.4). Other structural changes were not observed.

Fig.4. Microcrack tace in the OTM-914 material after testing at 1350°C, x 9000.

Stoppage of large crack growth, healing of microcracks, retainment of the starting level of strength give evidence to the importance of the effects these processes have on fracture and, perhaps, on local stress redistribution during long term tests. Steady-state creep rate in the 16% Y_2O_3 material can be observed after load exposure for not less than 60 hours (Fig.5).
Calculated creep rates accounted to 0.31; 1.37; 5.45 x $10^{-5}h^{-1}$, under the stress of 350 MPa and at 1200, 1300, 1400°C correspondingly. Creep activation energy Q = 70-92 kcal/mol, exponent n ≈ 6.

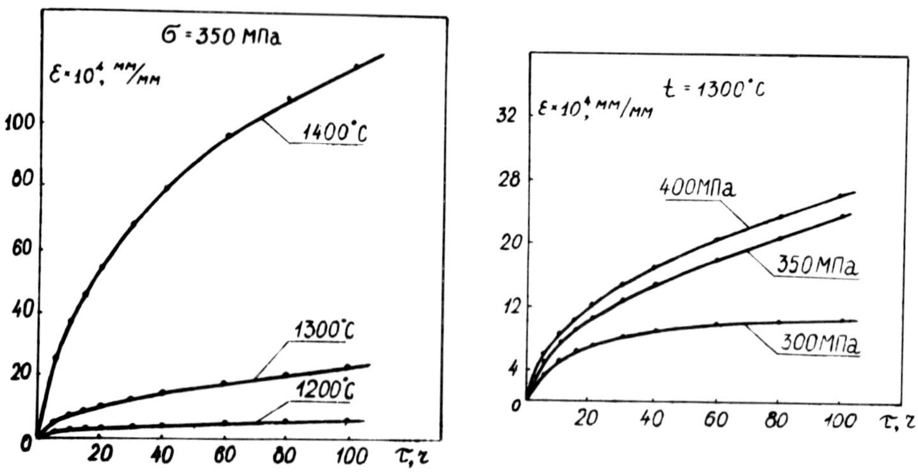

Fig.5. Deformation curves for the OTM-914 material at different temperatures (a), and stresses (b).

As compared with metals, fracture of ceramics is to a greater extent concerned with the presence of defects, crack formation under load, growth and healing of cracks. Internal defects are usually areas of large grain and pore aggregations, single large grains or pores with elongated crystals of the main phase (Fig.6), or inclusions of some other phases.

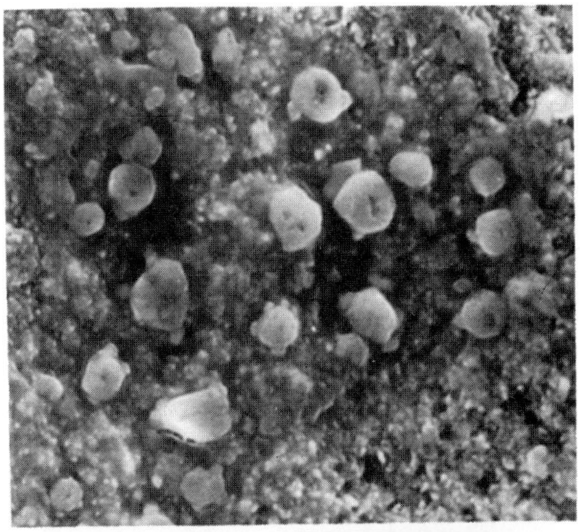

Fig.6 Typical structure defects examples: (a) large grains and pores aggregations; x 1000, (b) a large pore.

Sizes of the defects depend on the grain size, structure homogeneity and reproducibility of production procedures and techniques. In the OTM-914 material with the grain size in cross-section ~1 μm the sizes of the defects responsible for the fracture are in the range of 5 - 30 μm. As a rule one mirror and one fracture focus are observed when the samples are fractured (Fig.7a), in agreement with the fracture concepts developed by Griffiths and Weibull's weakest unit model. When structurally complex parts, for example, nozzle vanes, are fractured under the influence of cyclic thermal and mechanical loads and complex stressed states several fracture foci can be formed at various times (Fig.7b).

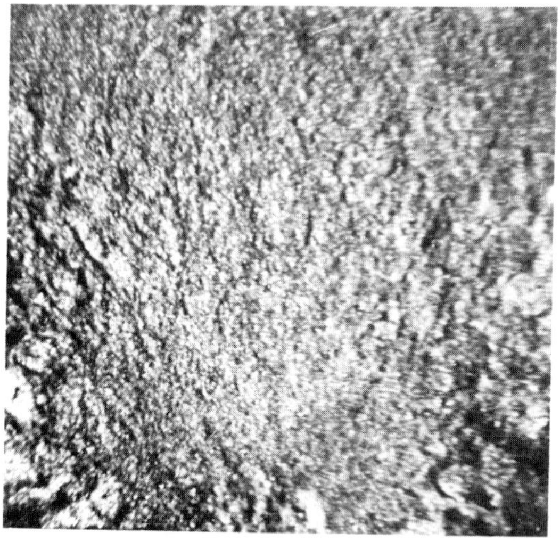

Fig.7 A sample with one fracture focus, x 60; (a), several foci of fracture in nozzle vane (b).

Ceramic materials fracture primarily along grain boundaries by microcrack formation and junction of microcracks (Fig.8). Observed deformations are caused by grain boundary sliding, dislocation movements along grain surface (Fig. 9), and, perhaps, by turning of the fragments formed by the cracks and grain boundaries.

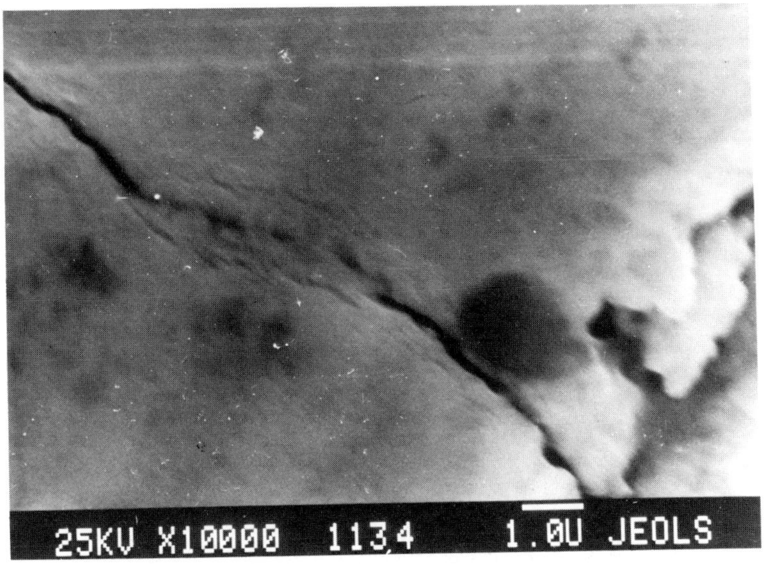

Fig. 8. Formation of a crack from microcracks.

Fig. 9. Dislocation structure of the surface of ZrO_2-6%Y_2O_3 grains after deformation at 1200°C, x 22000.

Similar results were obtained when investigating creep in α-SiC[3]. Grain deformation was not observed even when large deformations of the samples made from ZrO_2 took place. Large deformations are accompanied by the formation of many cracks (Fig. 10a), parallel displacement and stoppage of separate cracks (Fig. 10b).

As a rule fracture of ceramic components starts from the surface in areas of the largest local stresses and corrosion effects of the environment.

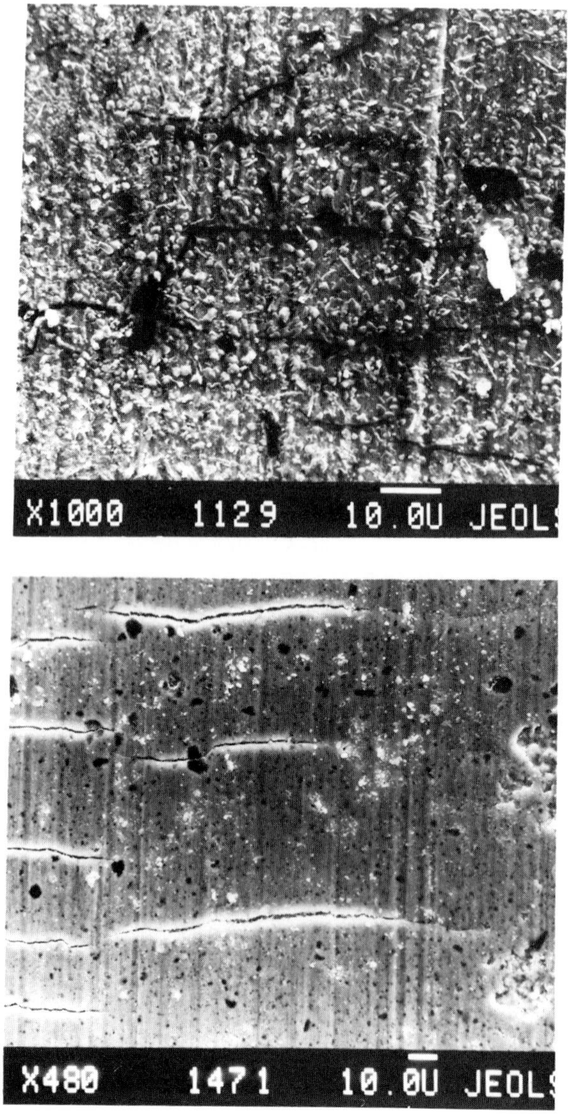

Fig. 10. Crack network on the tensile surface (a) and on the lateral surface (b) of the ZrO_2-6%Y_2O_3 sample after deformation at 1200°C.

CONCLUSION

1. Ceramic materials fracture practically without any deformation at temperatures below 1200°C. Fracture process and probability of fracture can be described in terms of Griffiths' energy theory and Weilbull's weakest unit model.

2. Oxygen-containing materials show their deformation ability at 1200°C. Pure silicon nitride exhibits no deformation under loading at temperatures up to 1500°C.

3. Deformation at elevated temperatures is due to grain boundary sliding and, perhaps, turning of fragments formed by cracks and grain boundaries.

4. Various physico-chemical processes, formation of a great number of cracks, their growth, stoppage and healing, grain boundary sliding and redistribution of local stresses hinder the development of mathematical models, reduce the degree of certainty of evaluations of service life and reliability of ceramic parts at elevated temperatures, under long term loading and complex stressed state.

5. Specific features of fracture of various ceramic materials show qualitative agreement with lattice properties and they are defined by the composition and structure of the material.

REFERENCES

1. Andreeva, T.V., Goryachev, Y.M., Kovenskaya, B.A., "Electronic structure and physical properties of refractory compounds", in Inorganic materials, 1979, v. 15, N4-583-589 P.
2. Shuvalov L.A., Urusovskaya A.A., Zheludev, I.S., et al., "Modern crystallography", 1981, v.4, P.495.
3. Lane, J.E., Carter, C.H., Jr. and Davis, Kinetic and Mechanisms of High-Temperature Creep in Silicon Carbide: 111, Sintered - Silicon Carbide, J.Am.Ceram.Soc., 1988, v.71, N4,281-295 pp.

Inst. Phys. Conf. Ser. No 111
Paper presented at Int. Conf. on New Materials and their Applications, Univ. of Warwick, 1990

The fabrication of carbon–carbon composites using a liquid crystal matrix precursor

Authors: A J Hosty*, B Rand[+] and F R Jones*

* School of Materials, University of Sheffield
[+] School of Materials, University of Leeds

ABSTRACT: Carbon-carbon composites can be fabricated by the infiltration of an organic precursor into a fibre preform followed by pyrolysis to generate the carbon matrix. When the infiltration liquid is isotropic pitch, a transient liquid crystal phase is generated within the fibre preform during pyrolysis. An alternative approach is to produce suitable liquid crystal precursors separately, and then to infiltrate these directly into the preform. Such liquid crystal precursors can have higher carbon yields than the original pitch and pressure carbonisation can be avoided. Also a wider variety of microstructural features can be developed in the matrix. The problems arising in the fabrication processes are discussed and illustrated with examples from a current study.

1. INTRODUCTION

Carbon-carbon composites are a class of materials designed to make effective use of the exceptional properties of the graphite crystal. These are, high specific in plane stiffness, high thermal and electrical conductivity parallel to the graphite layer plane and a high coefficient of thermal expansion in the 'c' direction. The latter, in synthetic graphites, can be partially accommodated by expansion into lamellar cracks generated within the material during fabrication. The closure of some of these cracks at high temperatures also results in an increase in strength at elevated temperatures. Hence, these composite materials may have good toughness and exceptional thermal shock resistance and may be manufactured with a high thermal conductivity and low electrical resistivity in the fibre direction, combined with thermal insulating properties and high electrical resistance in the other direction (McAllister 1983). These factors together with the extreme refactoriness of graphite (T sublimation > 3300°C) make carbon-carbon composites attractive materials for high temperature applications in inert atmospheres. Table 1 lists some typical properties of carbon-carbon composites and Figure 1 compares their temperature stability with other ceramic composites, showing that they are the only fibre reinforced composites that retain their strength at temperatures above 2000°C.

Thus, carbon-carbon composites have found application in solid fuel rocket nozzles, the nose cone and leading edges of the wings of the space shuttle, as well as for the brake discs of aircraft such as Concorde. Increasing demand for these materials is expected from the next generation of hypersonic aircraft, eg HOPE (Hasegawa 1990) and for Biomedical applications (Huettner and Claes 1989).

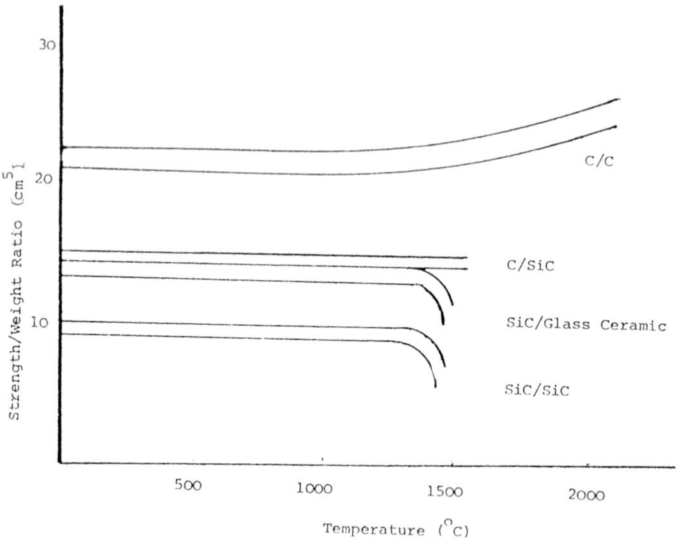

FIGURE 1: Temperature stability of ceramic composites.

Table 1. Range of properties for carbon-carbon composites.

Bulk density	1.3 - 2.0 gcm^{-3}
Porosity	5 - 35 %
Young's modulus	40-300GPa
Flexural strength	< 250 MPa
Fatigue	75-80% retention of strength after 10^7 flexural cycles
Interlaminar shear strength (2D)	8 - 20 MPa
Transverse strength (2D)	,,
Work of fracture	up to about 60 kJm^{-2}
Thermal expansion coefficient (K^{-1})	-2 - 10 10^{-6}
Thermal conductivity (W/mK)	4 - 140
Thermal shock resistance	extremely high.

For these reasons intensive research has been made into their fabrication. The two major aspects of interest are to establish a better understanding of microstructure-property relationships, enabling a more precise control of thermomechanical properties, and improvement in processing procedures leading to a reduction in the severe processing costs. The use of liquid crystalline precursors to form a graphitic matrix is a significant development (Huettner 1989 and Huttinger 1989) and is the subject of section three of this article, while the next section serves as a review of current fabrication technology.

2. FABRICATION ROUTES FOR CARBON-CARBON COMPOSITES

Synthetic carbons and graphites are derived by the pyrolysis of organic precursors, in an inert atmosphere. The structure and morphology of the resultant carbons are determined by the composition of the precursor compound and particularly by whether the pyrolysis takes place in the gaseous, liquid or solid state (Rand and McEnaney 1985).

Solid state pyrolysis (of crosslinked polymeric precursors) results in isotropic, hard, glassy carbons which do not normally graphitise on thermal treatment. On the other hand chemical vapour deposition and liquid phase pyrolysis of hydrocarbons results in highly graphitizable carbons that may show large regions of preferred orientation.

In general the manufacture of carbon-carbon composites is based on the routes shown in Figure 2. The chemical vapour infiltration route has been extensively used, especially in the manufacture of aircraft brake discs. Both isothermal and thermal and pressure gradient techniques are used (Buckley 1988). In all cases, the process is slow and costly, and interuptions may be necessary from time to time for machining away an outer surface coating which can seal the surface and limit densification. Constrictions within the porous matrix to be infiltrated can also be blocked creating internal closed porosity (Fitzer 1987). The slowness arises from the need to operate at low temperatures when the reaction is controlled by the chemical deposition processes (zone I in Figure 3), rather than the mass transport effects that become progressively more significant as the temperature rises, through zones II and III in the classical Wicke (1955) Arrhenius plot for gas-solid reactions involving a porous solid. There is, therefore, much interest in liquid infiltration routes utilising pitches and/or resins. In this route carbon fibres are combined with either a phenolic resin, coal tar or petroleum pitch, by using such techniques as wet winding, and vacuum impregnation. Resin impregnated composites are then crosslinked prior to carbonisation, which then takes place in the solid state and thus, as explained above, this normally results in a non-graphitizing precursor. However, it has been observed that the interaction of the resin with the fibres leads, on carbonisation, to regions of lamellar structure around the fibres, (Hishiyama 1974 et al).

Pitches are thermoplastic. Thus, composites with such matrices are not normally cured, but are carbonised directly.

For both pitch and resin precursors, the volumetric yield of carbon is limited by the volatiles released during pyrolysis and the higher density of the carbon product relative to that of the organic precursor, causing internal matrix shrinkage and consequent internal porosity.

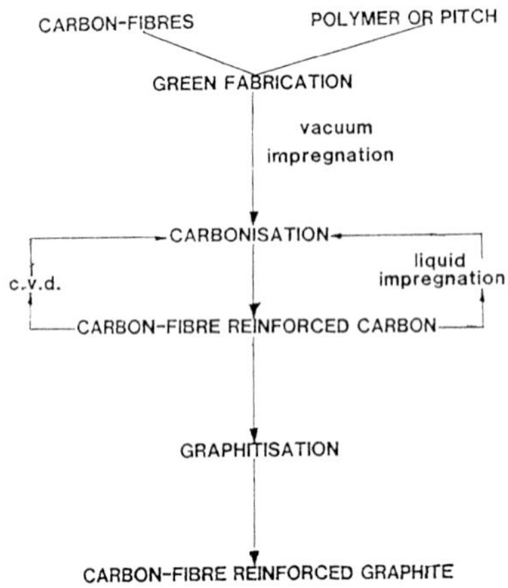

FIGURE 2: Fabrication routes for carbon-carbon composites.

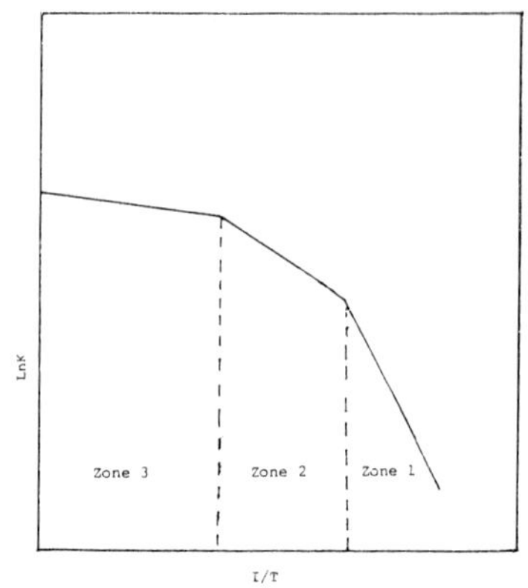

FIGURE 3: HEDDON & WICKE Arrhenius plot for gas-solid reactions involving a porous solid.

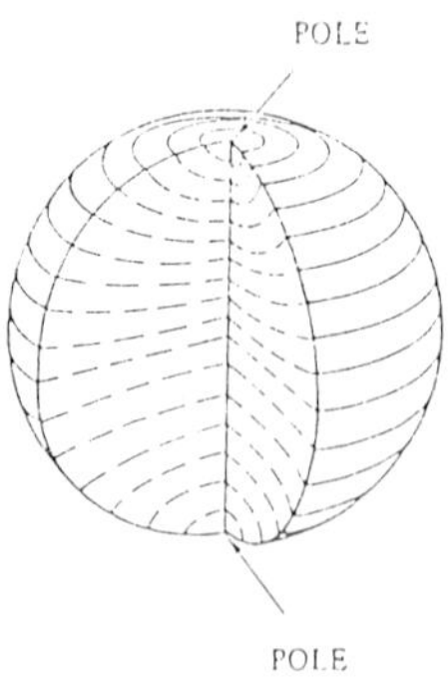

FIGURE 4: Schematic representation of the carbonaceous mesophase.

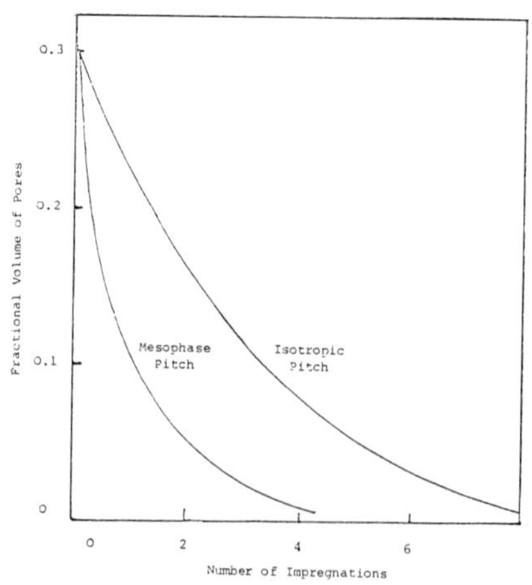

FIGURE 5: Plot of residual porosity versus number of impregnations for mesophase and isotropic pitches.

The internal cracks and pores so developed may be filled by successive impregnation and carbonisation cycles. As can be seen in Figure 2 it is not necessary that the same precursor is used for each infiltration stage and thus a wide range of combinations are possible leading to a range of carbon composites with complex matrix microstructures, particularly evident in the matrix-rich regions. Many cycles are normally required to fully densify the carbon composite.

The volumetric yield of carbon from a pitch precursor is lower than that of a resin precursor of similar mass carbon yield because of the higher density of the carbon phase. Thus, greater pososity results. However, in the pyrolysis of pitch materials, a substantial fraction of the mass loss is due to the evaporation of low molecular weight aromatic species. If these are retained in the composite during pyrolysis a higher mass and volumetric yield results. Hence, there has been a tendency to use isostatic pressure during pyrolysis to reduce the number of impregnation cycles (McAllister 1983). This process is expensive and limits the orientation of fibres and shapes of artefact that can be produced.

The graphitisability of carbon derived from pitches arises because pitch passes through a fused state in the early stages of carbonisation and a discotic nematic liquid crystal phase develops from the initially isotropic melt (Brooks and Taylor 1968). This carbonaceous mesophase comprises polynuclear aromatic molecules stacked in approximately parallel arrays, see Figure 4, and is, in effect, the embryonic graphite. Pitch precursors can be modified by heat treatment, by solvent extraction, by hydrogenation and by combinations of these procedures to produce materials with higher carbon yields. Mesophase pitch may be used as an impregnant and since this has a higher carbon yield the number of densification cycles required to reach a specific density is reduced, see Figure 5. Mesophase pitches can have higher densities and carbon yields than isotropic pitches or polymeric resins.

Some mesophase impregnants can also lead to an increase in Young's modulus of the composite of up to twice the theoretical value, due to the alignment of the mesophase with the carbon fibres (Huettner and Fitzer 1981), essentially encasing the fibres in a graphitic sheath. The microstructure which a mesophase develops on carbonisation is determined by its molecular composition. Mesophases in which the domains (regions of preferred orientation) are small tend not to show the same extent of preferred orientation around fibres. This domain texture can be varied by the incorporation of oxygen containing species into the mesophase, so altering its rheology and reactivity.

Clearly liquid crystal impregnants are attractive precursors for carbon-carbon composites. Requiring fewer impregnations and permitting the control of anisotropic physical and mechanical properties.

The use of liquid crystalline mesophase precursors for the carbon matrix requires a sound understanding of the pitch-mesophase-carbon transformation. The change in volume fraction and properties of this phase as a function of heat-treatment is described by Whitehouse and Rand (1982), and Rand (1987), in terms of a transformation diagram, eg Figure 6. One description of this transformation is as a lyotropic transformation, in which the isotropic pitch is considered to be a mixture of mesogens (mesophase forming material) and non-mesogens. Riggs and Diefendorf (1979, 1980a, 1980b) propose that the development of mesophase begins when a critical concentration of mesogens are present in the pitch. The

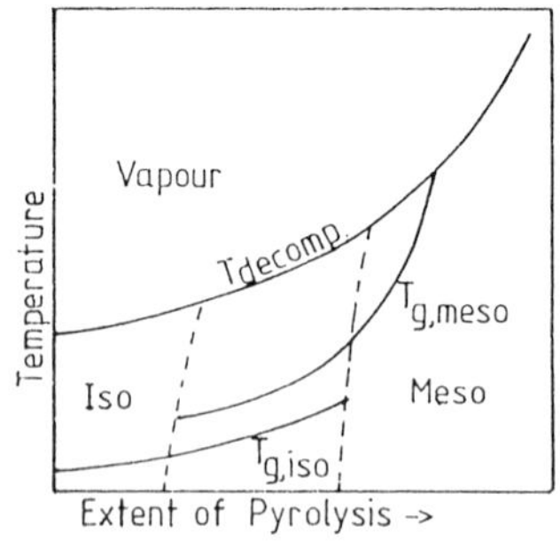

FIGURE 6: Schematic transformation diagram for the pitch → mesophase → carbon transformation.

TABLE 2: CARBON YEILD OF PITCHES + FRACTIONS

PITCH	CARBON YIELD w/o
A240	48
Aerocarb70	68
A240 Ti's	69
Aerocarb Ti's	74

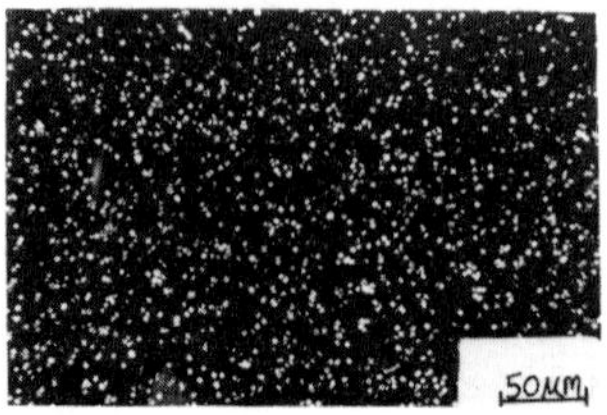

FIGURE 7: Aerocarb 70 heat treated to 400°C.

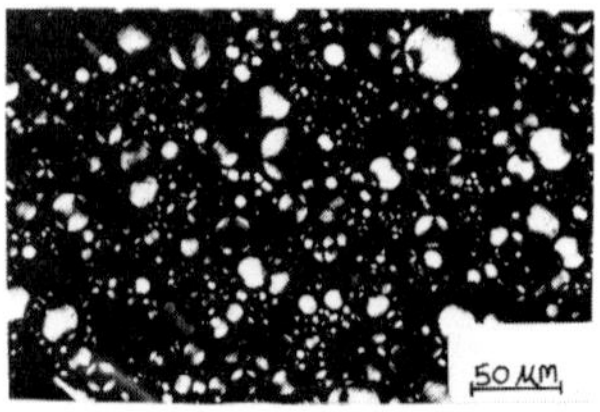

FIGURE 8: Aerocarb 70 toluene insolubles heat treated to 350°C.

concentration of mesogens can be raised in one of two main ways. By thermal treatment or by solvent extraction followed by heat treatment. Both these techniques and varieties based on these themes have been used to fabricate carbon fibre precursors. (Diefendorf and Riggs 1980, Singer 1979). However, solvent extraction allows the formation of an "instant" mesophase relative to the long heat treatments of around 40 hours for the straight heat treatment route. The use of solvent extraction followed by heat treatment to produce a matrix precursor for carbon-carbon composites is described below.

3. PRODUCTION OF CARBON-CARBON COMPOSITES FROM A LIQUID CRYSTAL PRECURSOR

2.1 Development of Liquid Crystal Based Carbon-Carbon Composites

Two commercially available petroleum pitches were investigated as sources of the liquid crystal precursor to be used in the fabrication of the carbon-carbon composites. Solvent extraction with toluene was used to remove the light aromatic materials and so increase carbon yield. The toluene insoluble materials (TI's) were then pyrolysed in a thermobalance to determine the increase in carbon yields, see Table 2. The mesophase development was then followed by polarised light microscopy on polished samples that had been heat treated to 350 and 400°C. Figures 7 and 8 show that solvent extraction results in the promotion of mesophase development. The TI fraction from Aerocarb has the highest carbon yield, ie 74w/o and was therefore selected as the matrix precursor.

2.2 Fabrication of Composites

The carbon composites were fabricated using vacuum impregnation of the toluene insoluble material into a 2-D plain weave carbon fibre cloth made from high strength, sized, PAN based carbon fibres in an autoclave followed by carbonisation in nitrogen. Using this technique good wetting and penetration of the carbon fibres is achieved. However, after impregnation at 350°C the matrix is still thermoplastic and carbonisation alone at this stage can lead to flow and bloating of the matrix.

Oxidation of pitch in air can cross-link the mesophase molecules to prevent flow and matrix bloating as in the production of carbon fibres. White (1988) has shown that mesophase matrices within uncarbonised composites can also be cross-linked. However, the maximum temperature at which this process can occur is limited by the glass transition temperature (Tg) of the matrix. Tg was determined by Dynamic Mechanical Thermal Analysis (DMTA) (Hosty et al 1989). The Tg of Aerocarb TI's was found to be 215°C and this temperature was used to carry out oxidation for periods of 24, 48, 120 and 168 hours. The oxidised samples were then carbonised at 1°C min^{-1} to 1000°C. Figure 9 shows that after 5 days oxidation a stabilisation depth of only 17 μm has been achieved. From these results, oxidation would appear to be an unsatisfactory method for stabilising these materials to any great depth. Indeed it has been suggested (White 1988) that this approach is only really effective if there is an open crack network in the matrix to facilitate the ingress of oxygen. The development of such a crack network when it does not already exist seems an undesirable operation.

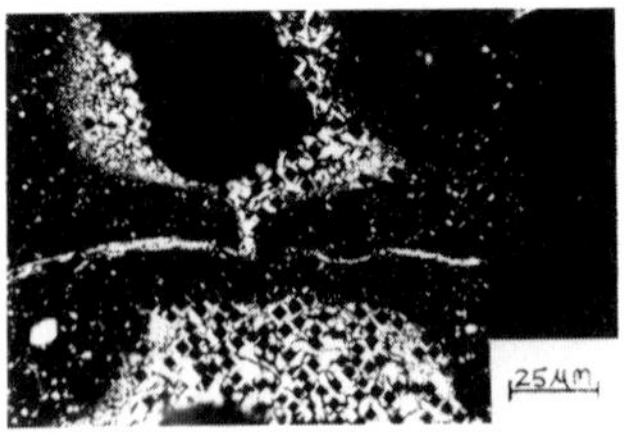

FIGURE 9: Partial stabilisation of pitch by air.

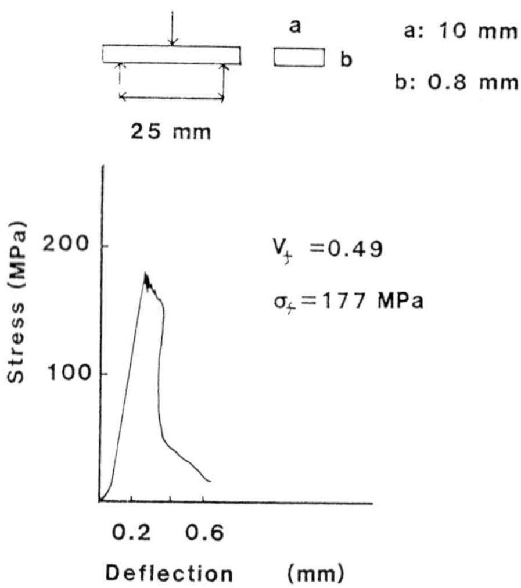

FIGURE 10 Fracture of 2-D carbon-carbon composites

As an alternative to oxidation as a way of preventing bloating a mild uniaxial pressure of 500 Pa was applied to the composite between room temperature and 600°C. The composite was then carbonised to 1000°C in nitrogen. Carbon fibre cloth has been densified through 3 impregnation steps using the above precursor each time.

The resultant composite had 49v/o carbon fibres and 19v/o porosity. Specimens of the composite measuring 50x10x0.8mm were tested in three point flexure with a test span of 25 mm, in accordance with CRAG test methods (Royal Aircraft Establishment technical report 1985). A typical load displacement curve for this material is shown in Figure 10. A flexural strength of 177 MPa and a mean flexural modulus of 58 GPa were calculated from the mean values of three test specimens.

Comparison of these results with those in Table 1 shows that the measured values of strength, modulus and porosity are within the range published. These are preliminary results as the matrix microstructure is unknown and in particular the presence and extent of matrix microcracking has yet to be determined.

Future work will concentrate on the fabrication of unidirectional composites, to study microstructure-property relationships, and to measure mechanical properties as a function of fibre type and orientation of the matrix and in particular to determine the effects, if any, of the fibre type on properties and specifically on the ultimate strength.

CONCLUSIONS

Carbonaceous mesophases are attractive matrix precursors for the matrix of carbon-carbon composites.

Solvent extraction of pitch with toluene increases mesogen concentration, facilitating the development of mesophase, and this increases the carbon yield. This toluene insoluble material has been used as a matrix precursor for carbon-carbon composites. A mild uniaxial pressure during fabrication is effective in preventing bloating of the matrix during carbonisation. This enables composites to be produced with high strength and modulus after successive impregnation and carbonisation steps.

ACKNOWLEDGEMENT

The authors acknowledge with thanks the receipt of a grant in aid of research from the Ministry of Defence. The views expressed, however, are those of the authors.

REFERENCES

Brooks J D and Taylor G H, 1968, Chem Phys Carbon 4 p243.

Buckley J D, 1988, Ceramic Bulletin 67 2.

Diefendorf R J and Riggs D M, 1980, US Patent 4,208,267.

Edie D D and Dunham M G, 1989, Carbon 27 5 pp 647-655.

Fitzer E, 1987, Carbon 25, 2.

Hasegawa K, et al, Materially speaking, 6 2.

Hishiyama Y, et al, 1974, Carbon 12, pp 249-258.

Hosty A J, Rand B and Jones F R, 1989, Extended Abstracts of 19th Biennial Conf on Carbon, Penn State (Amer Carbon Soc), pp 106-107.

Huettner W, 1989, "Potential of Carbon-Carbon Composites as Structural Materials", in Carbon fibres, filaments and composites, Eds Figueiredo J L, Bernardo C A, Baker R T K and Huttinger K J, NATO ASI Series E177 (London:Kluwer), pp 275-300.

Huettner W and Claes L E, 1989, "Carbon Based Materials in Medical Applications", in Carbon fibres, filaments and composites, Eds Figuiredo J L, Bernardo C A, Baker R T K, and Huttinger K J, NATO ASI Series E177 (London:Kluwer), pp 337-366.

Huettner W and Fitzer E, 1981, J Phys D Appl Phys, 14, pp 347-71.

Huttinger K J, 1989, "Theoretical and Practical Aspects of Liquid Phase Pyrolysis as Basis of the Carbon Matrix of CFRC" in Carbon Fibres, Filaments and Composites, Eds Figuiredo J L, Bernardo C A, Baker R T K and Huttinger K J, NATO ASI Series E177 (London:Kluwer) pp 301-326.

McAllister L E and Lochman W L, 1983, "Multidirectional Carbon-Carbon Composites", in The Fabrication of Composites 4, Eds Kelly A and Mileiko S T, (Amsterdam:Elsevier).

Rand B, 1987, Fuel, 66, pp 1491-1503.

Rand B and McEnaney B, 1985, Br Ceram Trans J, 84, pp 157-198.

Riggs D M and Diefendorf R J, 1979, Extended Abstracts of 14th Biennial Conf on Carbon, Penn State Univ, (Amer Carbon Soc), pp 147-148.

Riggs D M and Diefendorf R J, 1980a, Carbon '80 preprints of 3rd Int Carbon Conf, Baden-Baden (Deut Keram Ges), pp 326-329.

Riggs D M, Diefendorf R J, 1980b, Carbon '80 preprints of 3rd Int Carbon Conf, Boden-Boden (Deut Keram Ges), pp 330-333.

Royal Aircraft Establishment Technical Report, 1985, TR 85099.

Singer L S, 1979, "High Modulus Carbon Fibres from Mesophase Pitch", US Patent 4,005,183.

White J L and Schaeffer P M, 1988, Carbon '88, Proceedings of 7th Int Carbon Conf, Newcastle Upon Tyne, (Soc Chem Ind), pp 455-457.

Wicke E, 1955, Proc 5th Symposium on Combustion, Rheinhold, New York, p 245.

Whitehouse S and Rand B, (1982), Carbon '82, Extended abstracts of 6th Int Conf on Industrial Carbon and Graphite (Soc Chem Ind), London, pp 183-185.

Author Index

Subject Index